Urban Human Mobility

This comprehensive handbook covers human mobility within urban contexts, integrating academic theories with pragmatic insights and offering a detailed analysis of the diverse facets of human mobility and its substantial impact on the urban landscape, economy, and societal structures. It explains key fundamental concepts, methods, and models, presenting an in-depth exploration of predictive analytics, clustering patterns, advanced trajectory embedding techniques, artificial intelligence, machine learning, geographic information systems (GIS), Internet of Things (IoT), and smart city innovations. The authors include many case studies and examples of urban mobility in practice, making the content relatable and practical for educators, students, researchers, and practitioners.

Features

- Provides a multidisciplinary and holistic understanding of urban mobility with systematic introductions and discussions of theory, methods, technologies, tools, and applications.
- Covers a wide range of real-world case studies of urban mobility in practice globally that include data, programming code, and tools.
- Discusses cutting-edge technologies involved in mobility data analytics.
- Addresses practical challenges in data collection and the ethical implications of mobility research, which are crucial for professionals in the field.
- Offers future directions of human mobility research under the big data and artificial intelligence (AI) revolution.

Urban Human Mobility: Practices, Analytics, and Strategies for Smart Cities is for professionals, academics, and upper-level undergraduate and graduate students in the fields of urban planning/design, GIScience, data mining, and social sciences.

Urban Human Mobility

Practices, Analytics, and Strategies for Smart Cities

Edited by
Xiao Huang, Xinyue Ye,
Kathleen Stewart, and Subasish Das

CRC Press
Taylor & Francis Group
Boca Raton London New York

CRC Press is an imprint of the
Taylor & Francis Group, an **informa** business

Designed cover image: iStock Photo

First edition published 2025
by CRC Press
2385 NW Executive Center Drive, Suite 320, Boca Raton FL 33431

and by CRC Press
4 Park Square, Milton Park, Abingdon, Oxon, OX14 4RN

CRC Press is an imprint of Taylor & Francis Group, LLC

Library of Congress Cataloging-in-Publication Data
Names: Huang, Xiao (Professor of environmental sciences) editor
Title: Urban human mobility : practices, analytics, and strategies for smart cities /
edited by Xiao Huang, Xinyue Ye, Kathleen Stewart, and Subasish Das.
Description: First edition. | Boca Raton, FL : CRC Press, 2025. | Includes
bibliographical references and index.
Identifiers: LCCN 2024062304 (print) | LCCN 2024062305 (ebook) | ISBN 9781032821627 hardback |
ISBN 9781032821641 paperback | ISBN 9781003503262 ebook
Classification: LCC HE305 .U68355 2025 (print) | LCC HE305 (ebook) |
DDC 388.4--dc23/eng/20250327
LC record available at https://lccn.loc.gov/2024062304
LC ebook record available at https://lccn.loc.gov/2024062305

ISBN: 978-1-032-82162-7 (hbk)
ISBN: 978-1-032-82164-1 (pbk)
ISBN: 978-1-003-50326-2 (ebk)

DOI: 10.1201/9781003503262

Typeset in Times
by KnowledgeWorks Global Ltd.

Access the Support Material: www.routledge.com/9781032821627

Contents

SECTION III Technologies and Techniques Shaping Urban Human Mobility Studies

SECTION IV Advanced Topics and Future Directions

Preface

In an era defined by rapid urbanization, technological breakthroughs, and societal shifts, understanding human mobility has become more critical than ever. The movement of people—across streets, cities, and nations—shapes the very fabric of our urban lives, influencing economies, environments, governance, and societal well-being. As cities evolve into smart, interconnected hubs, urban human mobility presents both profound opportunities and formidable challenges that demand innovative solutions grounded in rigorous research, multidisciplinary collaboration, and emerging technologies.

Urban Human Mobility: Practices, Analytics, and Strategies for Smart Cities brings together a comprehensive collection of knowledge from leading experts in the fields of urban studies, data science, and smart city research, capturing the state-of-the-art frameworks, tools, and practices that address the complexities of human mobility in modern cities.

A DYNAMIC STRUCTURE FOR A COMPLEX PHENOMENON

The book is organized into distinct yet interconnected themes: foundational concepts, contemporary issues, and technological advancements in urban mobility studies. It begins with an introduction to urban human mobility, offering readers a conceptual foundation to navigate the intricate dynamics of urban movement. subsequent chapters delve into key pillars such as equity and accessibility, social and cultural dimensions, and the role of natural disasters—shedding light on mobility through diverse lenses.

Recognizing the role of data-driven advancements, the book explores cutting-edge topics such as generative artificial intelligence (GenAI)-powered multi-agent paradigms, trajectory privacy protection, and the use of deep learning for human trajectory uncertainty. Bridging theory with practice, case studies on ride-hailing demand analysis and public transit resilience provide tangible examples of mobility analytics in action.

This book does not shy away from addressing the pressing challenges of our time: the influence of COVID-19 on human daily mobility, climate change, and the need to rethink urban transportation systems for resilience. It culminates with forward-looking perspectives on sustainability, social segregation insights, and emerging digital twin technologies, offering readers a roadmap for the future of urban mobility.

A BRIDGE ACROSS DISCIPLINES

At its core, this book reflects the interdisciplinary nature of urban mobility studies. It weaves together methodologies from geography, transportation engineering, computer science, economics, and public policy, offering a holistic framework to analyze, predict, and optimize human mobility patterns. This integrative approach ensures its relevance not only for researchers and practitioners but also for urban planners, policymakers, and students eager to shape the future of smart, inclusive, and sustainable cities. Urban mobility is more than a technical problem; it is a reflection of how societies organize themselves, distribute opportunities, and prioritize equity. By capturing the interplay between human behavior, technology, and policy, this book aims to inspire solutions that are not only efficient but also ethical and inclusive.

As you turn these pages, you will gain insights into the ever-evolving world of urban human mobility—its challenges, triumphs, and untapped possibilities. Let this book be a guide and inspiration to those committed to creating cities that move seamlessly, sustainably, and equitably for all.

Xiao Huang, Xinyue Ye, Kathleen Stewart, and Subasish Das
Editors

About the Editors

Xiao Huang

Dr. Xiao Huang is an assistant professor in the Department of Environmental Sciences at Emory University. His research expertise encompasses human-environment interaction, computational social sciences, urban informatics, disaster mapping and mitigation, geospatial artificial intelligence (GeoAI), and disaster remote sensing. Dr. Huang has contributed extensively to his field, authoring over 200 peer-reviewed journal articles and over 20 book chapters and playing a pivotal role in editing five books. He is among the World's Top 2% Scientists by Stanford/Elsevier's rankings. In his professional capacity, he serves as an associate editor for *Computational Urban Science* and is a member of the editorial board for several prestigious journals. Dr. Huang's research has garnered significant attention and received coverage in renowned media outlets such as *Nature News*, NASA, NBC, and Fox. His work has attracted substantial funding from NSF, NASA, the Bill and Melinda Gates Foundation, the National Academies, and the Centers for Medicare and Medicaid Services.

Xinyue Ye

Dr. Xinyue Ye is the Harold Adams Endowed Professor in Urban Informatics and Stellar Faculty Provost Target Hire at Texas A&M University (TAMU). His research integrates computational social science, urban data science, and GeoAI to address issues ranging from infrastructure resilience and climate change to social justice and community perceptions, underscoring the dynamic interplay between technology, policy, and human behavior in shaping sustainable and livable cities. Dr. Ye was named one of the top 10 young scientists by the World Geospatial Developers Conference in 2021. His work has been funded by the National Academies, National Science Foundation, National Institutes of Health, National Institute of Justice, National Aeronautics and Space Administration, National Oceanic and Atmospheric Administration, Department of Commerce, Department of Defense, Department of Energy, Department of Health and Human Services, Department of Housing and Urban Development, Department of Transportation, Department of the Treasury, Microsoft, Baptist Health Foundation, Robert Wood Johnson Foundation, and the Canada Social Sciences and Humanities Research Council.

Kathleen Stewart

Kathleen Stewart is a professor in the Department of Geographical Sciences and director of the Center for Geospatial Information Science. She works in the area of geographic information science with a particular focus on geospatial dynamics. She investigates movement and mobility for a number of different application domains, for example, health and transportation where movement patterns or behaviors and spatial access are key topics. Her research is supported in part by grants from the National Institutes of Health, the National Science Foundation, the Federal Highway Administration, and Maryland Department of Natural Resources, among other organizations. Dr. Stewart serves as a member of the National Geospatial Advisory Committee (NGAC), a federal advisory committee sponsored by the Department of the Interior. She served on the Mapping Science Committee of the National Academies of Sciences, Engineering and Medicine (2016-2023). She is a fellow of the University Consortium of Geographic Information Science (UCGIS) and a member of the editorial boards of *The International Journal of Geographical Information Science* (IJGIS), *Geographical Analysis*, *Journal of Spatial Information Science* (JOSIS), and *International Journal of Geo-Information and Geomatics*, among others.

Subasish Das

Dr. Subasish Das is an assistant professor of the Civil Engineering Program in the Ingram School of Engineering at Texas State University. He is also involved with Texas A&M Transportation Institute (TTI) as a part-time associate research scientist. He has more than 13 years of experience related to roadway safety, traffic operation, and connected and automated vehicle (CAV) technologies. He is a systems engineer by training, with hands-on experience in Six Sigma and Lean Engineering. His major areas of expertise include database management, statistical analysis, and machine learning with emphasis on safety and transportation operations, spatial analysis with modern web GIS tools, interactive data visualization, and deep learning tools for CAV technologies. Dr. Das is a prolific author. He has published more than 160 technical reports and journal articles. He is the author of the book *Artificial Intelligence in Highway Safety*, which was published by CRC Press in 2022. The AASHTO Research Advisory Committee (RAC) awarded one of his research reports as 2014 AASHTO Sweet Sixteen High Value Research Project. Dr. Das is an active member of ITE and ASCE. He is an Eno Fellow. He served as vice president of membership of the Young Professionals in Transportation Houston chapter. He is currently a member of three TRB Committees: Information and Knowledge Management (AJE45), Artificial Intelligence and Advanced Computing Applications (AED50), and Impairment in Transportation (ACS50).

Contributors

Mahdi Aghaabbasi
Urban Digital Twin Lab, IST, School of Modeling, Simulation and Training
University of Central Florida
Orlando, Florida, USA

Bokolo Anthony Jr.
Department of Applied Data Science
Institute for Energy Technology
Department of Computer Science and Communication
Østfold University College
Halden, Norway

Xuegang (Jeff) Ban
Department of Civil and Environmental Engineering
University of Washington
Seattle, Washington, USA

Jiannan Cai
Institute of Space and Earth Information Science
The Chinese University of Hong Kong
Hong Kong, China

Ruizhi Chen
State Key Laboratory of Information Engineering in Surveying
Wuhan University
Wuhan, China

Xinghua Cheng
Department of Natural Resources and the Environment
University of Connecticut
Storrs, Connecticut, USA

Yunhe Cui
Department of Geography, Sustainability, Community and Urban Studies
University of Connecticut
Storrs, Connecticut, USA

Subasish Das
Ingram School of Engineering
Texas State University
San Marcos, Texas, USA

Ehsan Dorostkar
Department of Urban Planning
University of Guilan
Rasht, Iran

Chao Fan
School of Civil and Environmental Engineering and Earth Sciences
Clemson University
Clemson, South Carolina, USA

Song Gao
Department of Geography
University of Wisconsin-Madison
Madison, Wisconsin, USA

Konstadinos G. Goulias
Department of Geography
University of California Santa Barbara
Santa Barbara, California, USA

Henyu Gu
School of Geography and Ocean Science
Nanjing University
Nanjing, China

Qianwen Guo
Department of Civil and Environmental Engineering
Florida State University
Tallahassee, Florida, USA

Zijin Guo
School of Geography and Information Engineering
China University of Geosciences
Wuhan, China

Samiul Hasan
Department of Civil, Environmental, and Construction Engineering
University of Central Florida
Orlando, Florida, USA

Yiyi He
School of City and Regional Planning
Georgia Institute of Technology
Atlanta, Georgia, USA

Tao Hu
Department of Geography
Oklahoma State University
Stillwater, Oklahoma, USA

Qunying Huang
Department of Geography
University of Wisconsin
Madison, Wisconsin, USA

Xiao Huang
Department of Environmental Sciences
Emory University
Atlanta, Georgia, USA

Mei-Po Kwan
Department of Geography and Resource Management
Institute of Space and Earth Information Science
The Chinese University of Hong Kong
Hong Kong, China

Wan Li
Oak Ridge National Laboratory
Oak Ridge, Tennessee, USA

Xiang Li
School of Geosciences
University of South Florida
Tampa, Florida, USA

Linbo Liu
Center for Geographic Analysis
Harvard University
Cambridge, Massachusetts, USA

Luyu Liu
Department of Geosciences
Auburn University
Auburn, Alabama, USA

Zhewei Liu
Department of Land Surveying and Geo-Informatics
The Hong Kong Polytechnic University
Hong Kong, China

Zhongfu Ma
Department of Geography, Environment and Society
University of Minnesota
Twin Cities, Minnesota, USA

Grant McKenzie
Department of Geography
McGill University
Montreal, Quebec, Canada

Harvey J. Miller
Department of Geography
The Ohio State University
Columbus, Ohio, USA

Mahsa Najarsadeghi
Department of Urban Planning, Science and Research Branch
Islamic Azad University
Tehran, Iran

Malene Freudendal-Pedersen
Department of Sustainability and Planning
Aalborg University
Aalborg, Denmark

Raj Ponnaluri
CONSOR Engineers
Tallahassee, Florida, USA

Yi Qiang
School of Geosciences
University of South Florida
Tampa, Florida, USA

Jinmeng Rao
Department of Geography
University of Wisconsin-Madison
Madison, Wisconsin, USA

Madan B. Regmi
Transport Division
United Nations ESCAP
Bangkok, Thailand

Sudipta Roy
Department of Civil, Environmental, and Construction Engineering
University of Central Florida
Orlando, Florida, USA

Soheil Sabri
Urban Digital Twin Lab, IST, School of Modeling, Simulation, and Training
University of Central Florida
Orlando, Florida, USA

Yaxiong Shao
Center for Governmental Studies
Northern Illinois University
DeKalb, Illinois, USA

Hui Shi
Department of Geography
University of California Santa Barbara
Santa Barbara, California, USA

Kathleen Stewart
Department of Geographical Sciences
University of Maryland
College Park, Maryland, USA

Mengqi Sun
Department of Land Surveying and Geo-Informatics
The Hong Kong Polytechnic University
Hong Kong, China

Priyanka Verma
Department of Geography
McGill University
Montreal, Quebec, Canada

Changzhen Wang
Department of Geography
University of Alabama
Tuscaloosa, Alabama, USA

Pingping Wang
Department of Geography and Environmental Studies
Texas State University
San Marcos, Texas, USA

Sicheng Wang
Department of Geography
University of South Carolina
Columbia, South Carolina, USA

B. van Wee
Department of Engineering Systems and Services
Delft University of Technology
Delft, Netherlands

Meiliu Wu
School of Geographical and Earth Sciences
University of Glasgow
Glasgow, United Kingdom

Yuyang Wu
Department of Geography
University of California Santa Barbara
Santa Barbara, California, USA

Guanhao Xu
Oak Ridge National Laboratory
Oak Ridge, Tennessee, USA

Haowen Xu
Oak Ridge National Laboratory
Oak Ridge, Tennessee, USA

Junkang Xu
School of Civil and Environmental Engineering and Earth Sciences
Clemson University
Clemson, South Carolina

Yang Xu
Department of Land Surveying and Geo-Informatics
The Hong Kong Polytechnic University
Hong Kong, China

Yao Yao
School of Geography and Information Engineering
China University of Geosciences
Wuhan, China

Xinyue Ye
Department of Landscape Architecture & Urban Planning
Texas A&M University
College Station, Texas, USA

Kai Yin
Zachry Department of Civil and Environmental Engineering
Texas A&M University
College Station, Texas, USA

Yue Yu
Department of Land Surveying and Geo-Informatics
The Hong Kong Polytechnic University
Hong Kong, China

Jinghui Yuan
Oak Ridge National Laboratory
Oak Ridge, Tennessee, USA

Yihong Yuan
Department of Geography and Environmental Studies
Texas State University
San Marcos, Texas, USA

Jiechao Zhang
Department of Civil, Environmental, and Construction Engineering
University of Central Florida
Orlando, Florida, USA

Anye Zhou
Oak Ridge National Laboratory
Oak Ridge, Tennessee, USA

Di Zhu
Department of Geography, Environment, and Society
University of Minnesota
Twin Cities, Minnesota, USA

Section I

Foundations of Human Mobility and Urban Studies

1 Introduction to Urban Human Mobility

Xiao Huang, Xinyue Ye, Kathleen Stewart, and Subasish Das

1.1 OVERVIEW OF URBAN HUMAN MOBILITY

Urban human mobility refers to the dynamic patterns of movement of individuals within urban environments, encompassing various modes of transportation such as walking, cycling, driving, and public transit. It is influenced by a complex interplay of social, economic, cultural, and technological factors. Understanding urban mobility is crucial for urban planning and policymaking, as it directly affects traffic congestion, environmental sustainability, economic vitality, and quality of life. The concept extends beyond mere physical movement to include the accessibility and connectivity of different parts of a city, the efficiency of transport systems, and the overall urban experience. In modern cities, mobility patterns are increasingly shaped by technological advancements, such as mobile apps and real-time data analytics, which enhance the management and efficiency of transportation networks. Additionally, urban mobility studies consider the impact of infrastructure design, land use planning, and social equity, aiming to create sustainable, inclusive, and resilient urban environments that cater to the diverse needs of all residents. As cities continue to grow and evolve, optimizing human mobility remains a fundamental challenge and opportunity for urban development. By integrating innovative technologies, equitable planning, and sustainable practices, cities can enhance mobility, improve quality of life, and foster economic growth while reducing environmental impacts.

1.2 ASPECTS OF HUMAN MOBILITY IN MODERN CITIES

The study of human mobility in modern cities involves analyzing the various dimensions and factors that influence how people move within urban environments. This section introduces the multifaceted aspects of urban mobility, emphasizing the social, cultural, psychological, behavioral, economic, and energy-related considerations that shape urban systems and mobility patterns. Understanding these aspects is crucial for developing policies and strategies that promote sustainable, equitable, and efficient urban mobility.

1.2.1 Social and Cultural Dimensions of Urban Mobility

Urban mobility is deeply intertwined with the social and cultural fabric of cities, reflecting the values, lifestyles, and interactions of urban dwellers while shaping the community's social dynamics (Furnham 2010). The modes of movement within a city—be it public transport, private vehicles, bicycles, or walking—are influenced by and, in turn, influence social norms and cultural practices. For instance, in cities like Amsterdam and Copenhagen, bicycles serve as more than just a mode of transport; they symbolize cultural identity, reflecting values of environmental sustainability and healthy living (Gössling 2013). Conversely, in cities where car ownership signifies status, such as Los Angeles or Dubai, mobility patterns highlight social stratification and a cultural emphasis on personal freedom and convenience.

Public spaces like squares, parks, and streets are crucial for fostering social interactions, acting as meeting points for people from diverse backgrounds, and facilitating cultural exchange and

DOI: 10.1201/9781003503262-2

community bonding. The design and accessibility of these spaces can enhance or hinder social cohesion. Pedestrian-friendly areas encourage walking and socializing, contributing to a vibrant street life essential for urban vitality. Cultural events and festivals significantly impact urban mobility by increasing pedestrian traffic and demand for public transport. Efficient management of these surges is crucial for maintaining accessibility and the flow of movement. Additionally, the inclusivity of transport systems—considering the needs of all societal segments, including the elderly, disabled, and economically disadvantaged—reflects a city's commitment to social equity and cultural sensitivity (Wang et al. 2021).

1.2.2 Psychological and Behavioral Aspects of Urban Mobility

The psychological and behavioral aspects of urban mobility encompass the cognitive processes, emotional responses, and social behaviors that influence how individuals navigate urban environments, which are crucial for understanding travel choices, satisfaction with transportation systems, and overall well-being. Cognitive mapping, a mental representation of the urban environment, plays a fundamental role in how people move through cities. Individuals rely on these mental maps to navigate, make decisions about routes, and estimate travel times. Factors such as familiarity with the environment, previous experiences, and perceived safety significantly influence these cognitive processes. For example, well-designed signage and wayfinding systems can enhance navigability and reduce anxiety, especially for newcomers or tourists (Alkharoubi 2020).

Emotional responses to the urban environment also affect mobility behaviors. Feelings of safety, comfort, and pleasure can encourage walking, cycling, or using public transport, while fear, discomfort, or stress may deter individuals from certain modes of transport or areas. For instance, well-lit streets, clean and comfortable transit facilities, and aesthetically pleasing surroundings can create positive emotional experiences that promote the use of sustainable transport modes. Behavioral economics offers insights into how individuals make travel decisions. Concepts such as bounded rationality, where decision-making is limited by cognitive constraints, and choice architecture, which shapes how options are presented, are essential for designing effective mobility policies (Marsden et al. 2012).

Social behaviors, including the influence of peers and social norms, also play a significant role in urban mobility. People often mimic the travel behaviors of their social groups, which can be leveraged to promote sustainable transport practices. For instance, if cycling becomes a socially accepted and admired mode of transport within a community, it is likely to encourage more people to cycle. Understanding these psychological and behavioral aspects is essential for creating user-centered transport systems that not only meet the functional needs of urban dwellers but also enhance their overall travel experience and well-being.

1.2.3 Economics of Urban Mobility

The economics of urban mobility encompasses the analysis of financial and economic factors that influence transport systems and travel behaviors in cities. This includes examining the costs associated with various modes of transport, such as fuel, fares, vehicle maintenance, parking fees, and indirect costs like time spent traveling and environmental impacts. Understanding these costs is crucial for policymakers to design pricing strategies that encourage efficient and sustainable travel behaviors.

The economic impacts of mobility extend to broader urban development and economic growth. Efficient transport systems are vital for the functioning of urban economies, facilitating the movement of people and goods, and enabling access to jobs, education, and services. Conversely, poor transport infrastructure can hinder economic development by increasing travel times, reducing accessibility, and limiting opportunities for economic activities. Investment in transport infrastructure often has a multiplier effect, stimulating economic growth through job creation, increased productivity, and enhanced property values (Lakshmanan 2011). Funding mechanisms for urban transport systems vary widely and can include public funding, private investment, and public-private partnerships. Sustainable funding models are essential for maintaining and expanding transport infrastructure. Innovative financing approaches, such as value capture financing—where the increase in property values due to transport improvements is used to fund the projects—can provide additional resources for urban mobility projects.

Economic analysis also involves assessing the equity of transport systems. Ensuring affordable and accessible transport for all socioeconomic groups is crucial for social equity and economic inclusion. Transport policies need to address disparities in access and affordability to ensure that all urban residents can benefit from economic opportunities and services. The economics of urban mobility is a complex and dynamic field that requires a holistic approach, considering both the financial aspects and the broader economic impacts on urban development and social equity. By integrating economic analysis into transport planning, cities can create more efficient, sustainable, and inclusive mobility systems.

1.2.4 Energy and Urban Human Mobility

Energy consumption is a critical aspect of urban mobility, influencing both the environmental impact of transport systems and the sustainability of cities. The transport sector is a major consumer of energy, primarily derived from fossil fuels, which significantly contributes to greenhouse gas emissions and air pollution (Colvile et al. 2001). Understanding the relationship between energy use and urban mobility is essential for developing strategies to reduce energy consumption and promote sustainable transport practices.

Different modes of transport exhibit varying energy efficiencies. Private cars, especially those powered by internal combustion engines, are among the least energy-efficient, consuming significant amounts of fuel per passenger-kilometer traveled (Kenworthy and Svensson 2022). In contrast, public transport systems such as buses and trains are generally more energy-efficient, particularly when operating at high occupancy levels. Nonmotorized modes of transport, such as walking and cycling, are the most energy-efficient, requiring no external energy sources. The shift toward electric vehicles (EVs) is a significant trend in reducing the energy consumption of urban mobility, offering the potential to reduce dependence on fossil fuels and lower emissions, especially when powered by renewable energy sources.

Urban planning and transport policies play a crucial role in influencing energy consumption. Compact city designs with mixed land use and high-density development can reduce travel distances and promote the use of energy-efficient transport modes. Policies encouraging public transport use, cycling, and walking can significantly lower the energy footprint of urban mobility. Technological innovations also contribute to energy efficiency in urban mobility; intelligent transport systems (ITS) optimize traffic flow, reduce congestion, and lower energy consumption through real-time data and adaptive traffic management. Advances in vehicle technology, such as lightweight materials and aerodynamic designs, further enhance energy efficiency. Reducing energy consumption and emissions from transport contributes to climate change mitigation, improved air quality, and public health, supporting cities in achieving their environmental and sustainability targets.

1.2.5 Equity and Accessibility in Urban Human Mobility

Equity and accessibility are fundamental principles in urban mobility, ensuring that all urban residents, regardless of their socioeconomic status, have fair access to transport systems and the opportunities they provide. Addressing disparities in transport access is crucial for promoting social inclusion, economic opportunities, and overall quality of life. Equitable transport systems connect underserved areas with essential services, including job opportunities, education, and healthcare, thereby reducing social and economic disparities.

Equity in urban mobility involves providing affordable, reliable, and safe transport options for all residents. This requires designing transport systems that cater to the needs of diverse population groups, including low-income individuals, people with disabilities, the elderly, and other marginalized communities. Accessible transport systems are designed with universal design principles, ensuring usability by everyone, regardless of their physical abilities. The ease with which people can reach desired destinations using various modes of transport, including physical access to infrastructure, availability of information, affordability, and suitability of services, defines accessibility in this context.

Public transport plays a crucial role in promoting equity and accessibility in urban mobility (Zhang and Zhao 2021). Affordable and efficient public transport services can provide mobility options for those who do not own private vehicles. Policies such as subsidized fares, targeted services for underserved areas, and inclusive design features can enhance the accessibility and equity of public transport. Additionally, investments in infrastructure for active transport modes, such as safe and accessible pedestrian and cycling paths, promote low-cost and healthy mobility options. Urban planning that prioritizes mixed land use and high-density development can reduce travel distances, making walking and cycling more feasible and attractive.

1.3 PILLARS OF MODERN URBAN MOBILITY STUDIES

The study of modern urban mobility is underpinned by several critical pillars that collectively offer a comprehensive understanding of how citizens move within cities and how these movements can be optimized for efficiency, sustainability, and inclusivity. These pillars encompass technological advancements, evolving data sources, sophisticated analytical methods, and multidisciplinary engagement, each uniquely contributing to the advancement of urban mobility research and practice. Exploring these foundational elements allows us to better grasp the complexities and opportunities inherent in contemporary urban mobility studies.

1.3.1 Technological Advancements

The impact of information and communication technologies (ICT) on urban mobility is profound, revolutionizing transportation management through enhanced data collection, real-time analysis, and operational efficiencies (Giannopoulos 2004). Sensor networks, mobile applications, and advanced data processing have made urban transportation systems more dynamic and responsive. ICT integration streamlines traffic management, reduces congestion, and improves user experiences by providing timely information and personalized travel suggestions. Mobile devices with advanced global positioning system (GPS) capabilities generate rich data on individual movement patterns, invaluable for analyzing travel behaviors and congestion zones. The Internet of Things (IoT) has further transformed urban mobility by connecting devices and systems within smart cities, optimizing traffic signals, alleviating congestion, and enhancing public transportation schedules. Autonomous vehicles (AVs), equipped with advanced sensors and algorithms, promise to significantly reshape urban mobility by reducing accidents and emissions and improving accessibility. Smart city infrastructures, leveraging extensive sensor networks and mobile applications, provide actionable insights for traffic management, emergency responses, and urban planning.

1.3.2 Data Source Transitions

Traditional data sources for urban mobility studies, such as household travel surveys and manual traffic counts, are foundational but limited in scope and frequency. Modern data sources, including social media platforms, mobile applications, and IoT devices, enable continuous, real-time data collection, offering a nuanced view of urban dynamics (Huang et al. 2024a; Huang et al., 2024b). Social media platforms provide immediate insights into public sentiment and mobility trends through geotagged posts. Mobile applications like Strava and Citymapper generate detailed datasets on walking, cycling, and public transit usage. Wearable devices and smart card transaction records are increasingly important for understanding physical activity patterns and transit network usage. Ride-sharing apps contribute data on travel routes and user preferences, while GPS data from mobile devices enhance spatial dynamics analysis. Despite challenges such as data privacy and computational resource requirements, integrating big data into urban mobility studies presents substantial opportunities for real-time analysis, improved decision-making, and policy development.

1.3.3 Analytical Methods

Analytical methods in urban mobility studies have significantly advanced with the advent of big data and sophisticated computational techniques. Data-driven models now leverage diverse datasets to analyze and predict urban mobility patterns, enabling dynamic adjustments to transportation systems for optimized traffic flow and public transit schedules. These approaches facilitate evidence-based decision-making and policy development. By fusing data from various sources, such as public transit systems and ride-sharing services, researchers can obtain a comprehensive view of urban mobility, optimizing transportation networks and enhancing efficiency. For example, integrating ride-sharing data with public transit schedules helps identify service gaps and areas requiring additional transit options. Additionally, advanced intelligent algorithms process and analyze vast amounts of data to generate precise predictions and adaptive urban mobility solutions, handling high-dimensional data and uncovering complex patterns. Furthermore, models incorporating spatio-temporal dynamics provide deeper insights into urban dynamics, with advanced machine learning techniques and geospatial data enabling dynamic visualization of changes over time. This enhances predictive accuracy and supports effective policymaking and urban development strategies.

1.3.4 Multidisciplinary Engagement

In recent years, the study of urban mobility has grown increasingly complex, necessitating a multidisciplinary approach to comprehensively address the diverse factors influencing movement within urban environments. This section underscores the shift toward multidisciplinary engagement in urban mobility studies, particularly in the era of big data. Fields such as urban planning, engineering, data science, public policy, sociology, environmental science, transportation economics, geography, psychology, public health, computer science, and civil engineering are integral to advancing our understanding and management of urban mobility. Urban planners strategically design transportation networks aligned with overarching city development goals, while engineers focus on constructing and maintaining infrastructure using advanced materials and technologies. Data scientists and information technology (IT) specialists leverage big data analytics and machine learning to decode complex transportation datasets, devising predictive models to enhance public transit systems. Public policy experts develop regulatory frameworks that promote sustainable transportation and ensure equitable access, necessitating continuous dialogue between policymakers, researchers, and the public. Sociologists and behavioral scientists investigate human factors that influence transportation behaviors, which is crucial for designing systems catering to diverse urban populations. Environmental scientists assess the impacts of transportation on air quality and greenhouse gas emissions, ensuring mobility strategies align with broader environmental goals. Economists analyze the economic aspects of urban mobility, optimizing resource allocation and evaluating the economic viability of projects. Geographers contribute spatial analysis techniques to understand resource distribution and travel patterns, while psychologists study cognitive and emotional factors influencing travel behaviors. Public health experts examine the health impacts of transportation systems, supporting the development of systems that promote mental and physical well-being. Together, these disciplines provide a comprehensive approach to understanding and improving urban mobility, emphasizing the importance of interdisciplinary collaboration in addressing the multifaceted challenges of urban transportation.

1.4 KEY CHALLENGES IN URBAN HUMAN MOBILITY

1.4.1 Multimodality: Addressing the Disparity of Different Mobility Data Sources

Multimodal transportation systems, which integrate various modes of travel such as walking, cycling, public transit, and private vehicles, present unique challenges in urban mobility. One significant issue is the disparity in data quality and availability across different mobility sources (Huang et al. 2021). While some modes, like public transit, are often well-documented with comprehensive data on routes, schedules, and ridership, others, such as walking and cycling, may lack detailed data. This inconsistency hampers the ability to create a cohesive and efficient multimodal transport system. To address this, cities must invest in comprehensive data collection and integration strategies that encompass all modes of transport. This includes deploying advanced sensors, mobile applications, and IoT devices to gather real-time data, as well as collaborating with private-sector mobility providers to access relevant data. Harmonizing disparate data sources allows urban planners to gain a holistic view of mobility patterns, enabling more effective planning and optimization of transport networks.

1.4.2 Digital Divide: Who Speaks for the Underrepresented?

The proliferation of big data in human mobility research and the implementation of advanced analytics offer significant benefits, yet they also highlight the digital divide. Advanced mobility data collection and analysis require access to technology and digital literacy, which not all urban residents possess. This creates a scenario where data-driven policies and innovations may primarily benefit those with digital access, potentially exacerbating existing inequalities. Marginalized communities, including low-income populations and the elderly, may find themselves underrepresented in mobility data, leading to policies that do not address their needs (Huang 2024). To mitigate this divide, it is crucial to develop inclusive digital strategies that ensure broader access to technology and digital literacy programs. Policymakers must also actively engage with underrepresented communities to understand their mobility challenges and ensure their voices are included in the planning process. By addressing the digital divide, cities can ensure that advancements in urban mobility benefit all residents, promoting social equity and inclusion.

1.4.3 The Rising Urban AI: Who Controls Model Biases?

Artificial intelligence (AI) is increasingly being integrated into urban mobility systems, offering capabilities such as predictive analytics, AVs, and smart traffic management. However, the rise of AI brings concerns about model biases and their impact on urban mobility. AI models are often trained on large datasets that may reflect existing biases, which can result in unfair outcomes. For instance, biased data can lead to AI systems that prioritize certain areas over others for transport services or that do not adequately consider the needs of marginalized communities. Addressing these biases requires a multifaceted approach. It is essential to ensure diversity in the datasets used to train AI models and to implement rigorous testing to identify and mitigate biases. Additionally, involving a diverse range of stakeholders in the development and deployment of AI systems can help ensure that multiple perspectives are considered. Transparent governance frameworks are also needed to oversee AI applications in urban mobility, ensuring they are used ethically and equitably.

1.4.4 Data Governance: Who Protects Our Mobility Privacy?

With the increasing reliance on data to drive urban mobility solutions, privacy concerns have become a significant challenge. Collecting detailed data on individuals' movement patterns raises questions about how this data is stored, used, and protected. Ensuring the privacy of individuals while leveraging mobility data requires robust data governance frameworks. These frameworks should include clear policies on data anonymization, secure storage, and restricted access to sensitive information. Additionally, there must be transparency in how data is collected and used, with mechanisms for individuals to understand and control their data. Public trust is crucial for the successful implementation of data-driven mobility solutions, and this trust can be fostered through stringent data privacy measures and open communication with the public about data practices. Prioritizing privacy and implementing strong data governance facilitate harnessing the benefits of mobility data while respecting individuals' rights and fostering a sense of security and trust among residents.

1.5 CONCLUSION

As we conclude this introductory chapter on urban human mobility, we have explored the complex web of factors that influence how people move within cities. From the social and cultural dimensions that shape collective behaviors to the psychological and economic considerations driving individual choices, urban mobility emerges as a multifaceted phenomenon. It reflects societal values, technological advancements, and the very design of our urban environments. Understanding urban human mobility demands a comprehensive approach, integrating diverse disciplines and leveraging advanced analytical methods. The interplay between technological innovations and evolving data sources offers unprecedented opportunities to optimize transportation systems, enhance accessibility, and promote sustainable practices. However, these advancements also pose challenges, such as data privacy concerns, digital divides, and potential biases in AI systems. Addressing these issues is crucial for creating equitable and inclusive urban environments.

As we progress through this book, each chapter will delve deeper into specific aspects of urban human mobility, providing detailed analyses, case studies, and forward-thinking strategies. From examining the latest technologies reshaping our cities to exploring the economic, social, and environmental impacts of mobility, we aim to equip you with a holistic understanding of this dynamic field.

REFERENCES

Alkharoubi, Amer Mousa. "Wayfinding for pedestrians in the crowded areas of Al-Hajj: How can wayfinding system designs increase the efficiency of wayfinding and navigation performances for pedestrian pilgrims during the Islamic pilgrimage (Al-Hajj)." PhD diss., University of Minnesota, 2020.

Colvile, R. N., Emma Jane Hutchinson, J. S. Mindell, and R. F. Warren. "The transport sector as a source of air pollution." *Atmospheric Environment* 35, no. 9 (2001): 1537–1565.

Furnham, Adrian. "Human Mobility in a Global Era." In: S. Carr (ed) The Psychology of Global Mobility, pp. 23–45. New York: Springer, 2010.

Giannopoulos, Georgios A. "The application of information and communication technologies in transport." *European Journal of Operational Research* 152, no. 2 (2004): 302–320.

Gössling, Stefan. "Urban transport transitions: Copenhagen, city of cyclists." *Journal of Transport Geography* 33 (2013): 196–206.

Huang, Xiao. "Computational urban science needs to go beyond computational." *Computational Urban Science* 4, no. 1 (2024a): 1–5.

Huang, Xiao, Zhenlong Li, Yuqin Jiang, Xinyue Ye, Chengbin Deng, Jiajia Zhang, and Xiaoming Li. "The characteristics of multi-source mobility datasets and how they reveal the luxury nature of social distancing in the US during the COVID-19 pandemic." *International Journal of Digital Earth* 14, no. 4 (2021): 424–442.

Huang, Xiao, Siqin Wang, Di Yang, Tao Hu, Meixu Chen, Mengxi Zhang, and Guiming Zhang et al. "Crowdsourcing geospatial data for earth and human observations: A review." *Journal of Remote Sensing* 4 (2024b): 0105.

Kenworthy, Jeffrey R., and Helena Svensson. "Exploring the energy saving potential in private, public and non-motorized transport for ten Swedish cities." *Sustainability* 14, no. 2 (2022): 954.

Lakshmanan, Tiruvarur Ramaiyer. "The broader economic consequences of transport infrastructure investments." *Journal of Transport Geography* 19, no. 1 (2011): 1–12.

Marsden, Greg, Karen Trapenberg Frick, Anthony D. May, and Elizabeth Deakin. "Bounded rationality in policy learning amongst cities: Lessons from the transport sector." *Environment and Planning A* 44, no. 4 (2012): 905–920.

Wang, Chihuangji Herbert, Edward Steinfeld, Jordana L. Maisel, and Bumjoon Kang. "Is your Smart city inclusive? Evaluating proposals from the US Department of Transportation's Smart City Challenge." *Sustainable Cities and Society* 74 (2021): 103148.

Zhang, Mengzhu, and Pengjun Zhao. "Literature review on urban transport equity in transitional China: From empirical studies to universal knowledge." *Journal of Transport Geography* 96 (2021): 103177.

2 The Pillars of Modern Urban Human Mobility Studies

Xiao Huang

2.1 INTRODUCTION

In the era of rapid urbanization and technological advancement, the study of urban mobility has become increasingly significant. Urban human mobility, which encompasses the movement of people within urban areas, is a critical component of urban planning and development. The efficiency and sustainability of urban transportation systems directly impact economic vitality, environmental sustainability, and the quality of life for urban residents. As cities continue to grow and evolve, addressing the challenges associated with urban mobility requires a multifaceted approach that leverages technological advancements, innovative data sources, and multidisciplinary engagement. This chapter explores the fundamental pillars of modern urban mobility studies, focusing on technological advancements, data source transitions, analytical methods, and the importance of multidisciplinary collaboration (Figure 2.1). By examining these pillars, we aim to provide a comprehensive understanding of the current state and future directions of urban mobility research, emphasizing the critical role of integrated and adaptive approaches in creating smarter, more sustainable urban environments. The subsequent sections delve into the transformative impact of information and

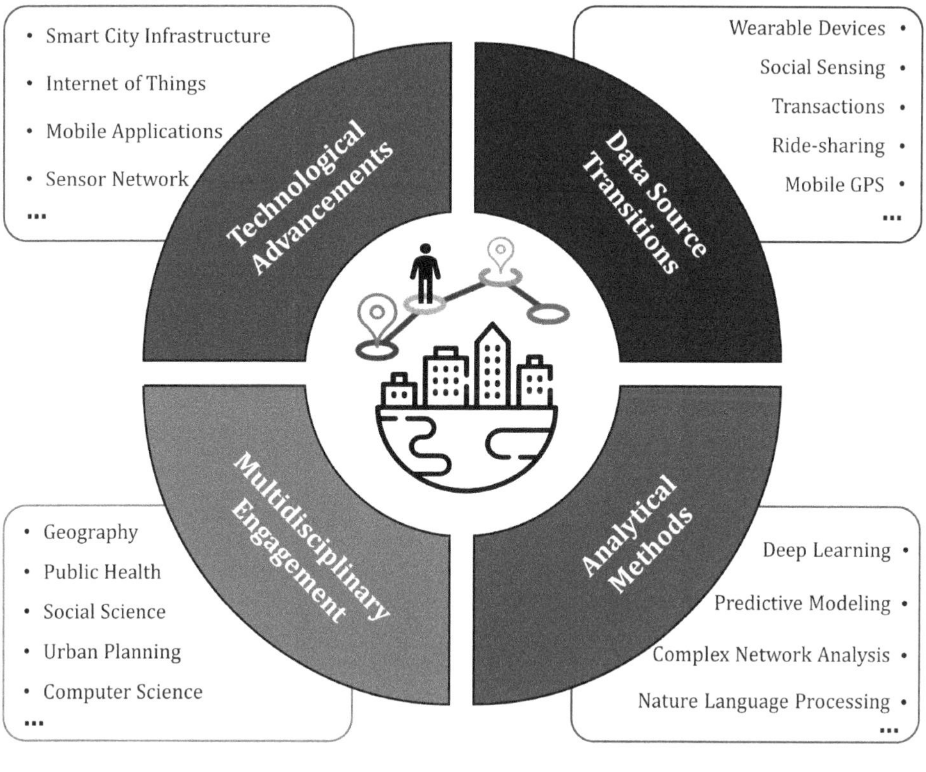

FIGURE 2.1 Pillars of modern urban mobility studies.

DOI: 10.1201/9781003503262-3

communication technologies, the evolving landscape of data sources, the cutting-edge analytical methods driving new insights, and the indispensable contributions of various disciplines to the field of urban mobility studies.

2.2 TECHNOLOGICAL ADVANCEMENTS

The transformative impact of information and communication technologies (ICTs) on urban mobility is profound and far-reaching. ICTs have revolutionized our approach to managing urban transportation systems, offering new insights and enhanced operational efficiencies (Giannopoulos 2004). Through the deployment of sensor networks, mobile applications, and advanced data processing technologies, we have shifted toward more dynamic and informed management of urban systems. These tools facilitate extensive data collection as well as enable real-time analysis and forecasting, significantly enhancing decision-making processes. Moreover, the integration of ICTs in urban mobility goes beyond mere data handling; it fosters a user-centric framework that is both efficient and adaptive to the needs of rapidly evolving urban landscapes. This integration helps to streamline traffic management, reduce congestion, and improve the overall user experience by providing timely information and personalized travel suggestions. As urban areas continue to grow and the demand for more sustainable and resilient transportation systems increases, the role of ICT becomes increasingly critical. It serves as a cornerstone for developing smarter, more connected cities that prioritize accessibility, sustainability, and the well-being of their inhabitants.

Mobile devices, equipped with advanced global positioning system (GPS) capabilities, have become pervasive across the globe, furnishing researchers and urban planners with a rich source of data regarding individual movement patterns. These data prove invaluable for analyzing a wide array of urban dynamics, including travel behaviors, peak travel times, and congestion zones (Schläpfer et al. 2021; Lu et al. 2024; Chukwu et al. 2024). Applications such as Google Maps and Waze epitomize the utility of these data. These platforms have largely facilitated urban navigation and compile extensive traffic data to offer real-time updates and optimized routing suggestions. By harnessing these massive datasets, these applications are able to predict traffic conditions with a high degree of accuracy, enhance route-planning efficiency, and provide users with timely alerts. This integration of geospatial technology into daily transportation logistics not only streamlines individual commutes but also contributes to broader urban planning and management strategies, making urban environments more navigable, less congested, and ultimately more sustainable.

The advent of the Internet of Things (IoT) has catalyzed transformative changes in urban mobility, characterizing smart cities through highly interconnected systems and devices that bolster transportation efficiency and sustainability (Rosa et al. 2021). Embedded sensors within road infrastructures, traffic lights, and public transit systems gather critical data on traffic dynamics, vehicular speeds, and mass transit utilization (Oladimeji et al. 2023). These data underpin the optimization of traffic signals, congestion alleviation, and the enhancement of public transportation schedules. The seamless interoperability among these devices fosters a synchronized and adaptive transportation network, adept at responding to instantaneous variations in traffic conditions and demands (Coleri et al. 2004). Exemplifying the application of IoT, cities such as Singapore have implemented smart traffic management systems that dynamically adjust traffic signals based on real-time data, thereby minimizing wait times at intersections and enhancing traffic circulation (Kumar Debnath et al. 2011). Furthermore, IoT-enabled public transit systems offer passengers up-to-the-minute updates on schedules, thereby elevating the reliability and convenience of public transportation. Such integration also augments operational efficiency, curtails delays, and improves commuter experience.

Technological advancements have notably spurred the growth of autonomous vehicles (AVs), which promise to transform urban mobility significantly. These vehicles, spearheaded by companies like Tesla, operate using a sophisticated blend of sensors, GPS technology, and advanced algorithms, enabling them to navigate without human intervention. Extensive testing and deployment in urban settings underscore the potential of AVs to reduce accidents, cut emissions, and enhance mobility for those who cannot drive, such as the elderly and disabled. AVs also have the potential to streamline traffic flow, diminish parking space requirements, and reshape the urban landscape, thereby fostering safer and more accessible environments (Duarte and Ratti 2018). Simultaneously, the progression of smart city infrastructure plays a crucial role in advancing urban mobility. This development encompasses the deployment of extensive sensor networks that capture a range of data, from environmental conditions to traffic and pedestrian dynamics. By leveraging mobile applications and advanced data processing technologies, cities can manage and interpret this wealth of data, gaining actionable insights that drive swift responses to evolving urban conditions. For instance, integrated city dashboards provide real-time data visualization, supporting city officials in making informed decisions regarding traffic management, emergency responses, and urban planning strategies (Kourtit and Nijkamp 2018). Together, these technologies interlink to revolutionize how cities function, offering more sustainable, efficient, and user-friendly urban environments.

2.3 DATA SOURCE TRANSITIONS

Urban mobility studies have traditionally relied on data derived from surveys and traffic sensors, such as household travel surveys (Stopher and Greaves 2007) and manual traffic counts (Zheng and Mike 2012). While these traditional sources are foundational, offering baseline insights into human travel patterns, they are often constrained by their limited scope and infrequent data collection intervals. These methods can be labor-intensive and costly and may not adequately capture the dynamic and rapidly changing nature of urban mobility. With the advent of cutting-edge technologies, a diverse array of modern data sources has dramatically broadened both the scope and depth of data available for urban mobility studies (Butler et al. 2020; Li et al. 2021; Hu et al. 2021). These include social media platforms (Huang et al. 2022a; Hu et al. 2023), mobile applications (Seliverstov et al. 2020), and IoT devices (Solmaz and Turgut 2019), which facilitate continuous, real-time data collection and offer a more nuanced view of urban dynamics. For instance, social media platforms like Twitter and Instagram provide immediate insights into public sentiment and mobility trends through geotagged posts, which can pinpoint popular destinations, peak travel times, and the impacts of specific events on urban mobility (Huang et al. 2022b).

Additionally, mobile applications such as Strava and Citymapper deliver precise tracking of individual movements, generating detailed datasets on walking, cycling, and public transit usage (Sun and Mobasheri 2017). These insights are invaluable for urban planning and transportation management, enabling more responsive and adaptive strategies. Expanding further, wearable devices and smart card transaction records are becoming increasingly important in urban mobility research. Wearables provide personal health and location data that help in understanding patterns in physical activity and mobility preferences (Seneviratne et al. 2017). Similarly, smart card data from public transit systems offer detailed information on passenger flows and transit network usage (Pelletier et al. 2011). Ride-sharing apps contribute data on travel routes, times, and user preferences, while GPS pins from mobile devices enrich our understanding of spatial dynamics across different times and conditions (Wang et al. 2024). Together, these diverse data sources create a comprehensive ecosystem of mobility data, enhancing our ability to understand and improve urban environments in real time, ensuring they are more responsive, sustainable, and aligned with the needs of their inhabitants.

However, the integration of big data into urban mobility studies introduces several challenges alongside its benefits. A significant concern is the protection of data privacy and security (Huang et al. 2024). As personal data collection volumes grow, the implementation of stringent data protection measures becomes critical to safeguard individual privacy and prevent breaches. This includes enforcing robust encryption methods, adhering to international data protection standards, and conducting regular security audits. Furthermore, the management and processing of extensive datasets necessitate substantial computational resources and specialized expertise. This requirement can be a barrier for many institutions, particularly those with limited technological infrastructure or access to skilled data scientists. Despite these hurdles, the opportunities presented by big data in urban mobility are substantial. Enhanced analytics powered by these large datasets can provide a more accurate and comprehensive understanding of urban mobility patterns, facilitating better-informed decision-making and policy development. Such insights can drive significant improvements in traffic management, urban planning, and the overall quality of urban life, making the challenges worthwhile to address.

2.4 ANALYTICAL METHODS

In modern urban mobility studies, analytical methods have evolved significantly, driven by the advent of big data and the proliferation of advanced computational techniques. This section summarizes the ongoing trends in urban human mobility studies, which include harnessing the power of data through transformative approaches, appreciating the power of data fusion for multimodal synergies, employing cutting-edge intelligent algorithms for smart solutions, and engaging time and space in the analysis of spatiotemporal dynamics.

2.4.1 Harnessing the Power of Data: Transformative Approaches

We have entered the big data era, where data-driven models are becoming the preferred choice for scholars in urban mobility studies. The abundance of data generated from various sources, such as social media, GPS, sensors, and mobile devices, provides unprecedented opportunities to analyze and understand complex urban mobility patterns (Wang et al. 2019). Scholars are harnessing such big data to develop sophisticated models that capture the intricacies of human movement within urban environments. Data-driven approaches enable the integration of diverse datasets, allowing for a more comprehensive analysis of mobility trends. These approaches leverage the vast amounts of data available to uncover patterns, predict future movements, and optimize transportation systems. The proliferation of big data facilitates real-time analysis and decision-making. By incorporating real-time data inputs, researchers can monitor current conditions, detect anomalies, and make dynamic adjustments to improve the efficiency of transportation networks. This capability is particularly valuable for managing traffic flow, optimizing public transit schedules, and enhancing overall urban mobility. Moreover, big data analytics empowers policymakers and urban planners to make evidence-based decisions. The insights derived from data-driven approaches inform the design and implementation of policies aimed at reducing congestion, improving accessibility, and promoting sustainable transportation options. These insights also support the evaluation of urban planning initiatives, enabling continuous improvement and adaptation to evolving mobility needs.

2.4.2 Multimodal Synergies: Appreciating the Power of Data Fusion

Today's urban systems leverage a vast array of diverse data sources. Individually, these sources may exhibit biases or fail to capture the full complexity of human mobility patterns (Huang et al. 2021). Against this limitation, scholars are increasingly recognizing the significance of data fusion, even when it involves data from different modalities. Integrating data from public transit systems, ride-sharing services, and personal mobile devices allows researchers to examine the interactions between various transportation modes. This holistic approach not only optimizes transportation networks but also boosts the overall efficiency of urban mobility. For instance, merging data from ride-sharing applications with public transit schedules helps identify service gaps and pinpoint areas that could benefit from additional transit options (Varone and Aissat 2015). Likewise, integrating GPS data from private vehicles with sensor data from road infrastructure enhances traffic management by providing real-time traffic updates and facilitating dynamic adjustments to traffic signals. Furthermore, data fusion underpins the development of integrated mobility platforms that ensure seamless access to various transportation modes, such as biking, walking, public transit, and car-sharing. These platforms deliver personalized travel recommendations based on real-time data, empowering users to make informed decisions about their travel choices.

2.4.3 Smart Solutions: Cutting-Edge Intelligent Algorithms

In the era of big data, deploying advanced, data-intensive models has become indispensable in urban mobility studies. These sophisticated intelligent algorithms offer unparalleled capabilities in processing and analyzing vast amounts of data to extract valuable insights and generate precise predictions (Luca et al. 2021). The integration of these advanced algorithms with big data analytics facilitates the development of adaptive and scalable intelligent urban mobility solutions. Continuous real-time data feeds from a multitude of sensors and devices provide essential inputs for these algorithms, enabling them to perpetually refine their models and predictions. This real-time adaptability is crucial for addressing the ever-evolving dynamics of urban environments, ensuring that mobility solutions remain responsive and efficient. These data-hungry models excel at handling high-dimensional data and uncovering complex patterns that conventional methods may overlook, facilitating the analysis of diverse data streams from various sources to understand and predict traffic flow, identify congestion points, and optimize route planning (Kim and Song 2018). Additionally, intelligent algorithms are well-suited for dynamic and interactive scenarios in urban mobility, as they can be employed to develop adaptive systems that learn optimal strategies based on real-time conditions.

2.4.4 Spatiotemporal Dynamics: Engaging Time and Space

Recent advancements in human mobility data capture with increasing spatiotemporal granularity have significantly benefited various fields that demand such detailed information. Scholars are developing models that engage more space and time components to better understand urban dynamics, facilitating deeper analysis of how spatial and temporal factors influence human behavior and mobility patterns. By incorporating both historical data and real-time inputs, these models can predict trends, assess the impact of urban planning decisions, and optimize resource allocation. Advanced machine learning techniques integrated with geospatial data enable dynamic visualization of changes over time, providing invaluable insights into cyclic and seasonal variations in urban mobility or mobility influenced by the pandemic (Zuo et al. 2020). This comprehensive approach enhances predictive accuracy and supports more effective policymaking and urban development strategies. Advanced spatial analysis techniques and temporal data modeling help identify trends, peak usage times, and high traffic density areas, while geographic information system (GIS) tools allow researchers to map and examine the spatial distribution of mobility patterns. Engaging with the interplay of time and space in urban mobility, researchers and planners can develop more effective strategies for managing transportation systems, enhancing accessibility, and improving the overall efficiency of urban mobility networks.

2.5 MULTIDISCIPLINARY ENGAGEMENT

In recent years, the study of urban mobility has become increasingly complex, necessitating a multidisciplinary approach to comprehensively address the various factors influencing movement within urban environments. This section highlights the transition to multidisciplinary engagement in urban mobility studies, particularly in the era of big data. It encompasses fields such as urban planning, engineering, data science, public policy, sociology, environmental science, transportation economics, geography, psychology, public health, computer science, and civil engineering to advance our understanding and management of urban mobility.

- *Urban Planning and Engineering*: Urban planners and engineers play pivotal roles in the development and execution of efficient transportation systems. Urban planners strategically design transportation networks that are aligned with overarching city development goals, land use policies, and sustainability initiatives, ensuring comprehensive accessibility and connectivity throughout urban regions. Engineers, on the other hand, concentrate on the technical elements, focusing on the construction and maintenance of transportation infrastructure. They employ advanced materials and technologies to improve the infrastructure's durability and efficiency. Together, these disciplines tackle both the strategic and practical facets of urban mobility, ranging from conceptual designs to operational implementations.

- *Data Science and Information Technology (IT)*: Data scientists and IT specialists harness the power of big data analytics and machine learning algorithms to decode complex transportation datasets. These experts devise predictive models that anticipate traffic flows, refine route planning, and enhance the efficacy of public transit systems. Technological advancements in IT enable the integration of diverse data sources, such as IoT devices and social media, offering a holistic perspective on urban mobility trends. This interdisciplinary effort significantly bolsters the ability to tackle real-time challenges and anticipate future developments in urban transportation.
- *Public Policy and Governance*: Public policy experts and government officials shape the regulatory frameworks that govern urban mobility. They develop policies that promote sustainable transportation, regulate emerging mobility services like ride-sharing and AVs, and ensure equitable access to transportation resources. Effective governance requires a deep understanding of both the technical and social dimensions of urban mobility, emphasizing the need for continuous dialogue between policymakers, researchers, and the public to create inclusive and forward-thinking transportation policies.
- *Sociology and Behavioral Studies*: Sociologists and behavioral scientists investigate the human factors that influence transportation behaviors. They analyze how variables such as socioeconomic status, cultural norms, and personal preferences shape travel choices and mobility patterns (Zhou et al. 2022). Understanding these human dimensions is crucial for designing transportation systems that cater to the diverse needs of urban populations and promote sustainable mobility practices. Insights gleaned from sociology are instrumental in developing targeted interventions aimed at encouraging the use of public transit, cycling, and walking, thus diminishing the dependency on private vehicles.
- *Environmental Science*: Environmental scientists assess the environmental impacts of urban transportation systems. They study the effects of transportation on air quality, greenhouse gas emissions, and urban heat islands, providing essential data to support sustainable mobility initiatives (Boschmann and Kwan 2008). Collaboration with environmental scientists ensures that urban mobility strategies align with broader environmental goals, such as reducing carbon footprints and enhancing urban resilience to climate change.
- *Transportation Economics*: Economists analyze the economic aspects of urban mobility, including the costs and benefits of different transportation modes, funding mechanisms for infrastructure projects, and the economic impacts of transportation policies. Their insights help optimize resource allocation, evaluate the economic viability of transportation projects, and develop strategies that maximize economic efficiency and sustainability.
- *Geography*: Geographers contribute spatial analysis techniques to understand the distribution of resources, travel patterns, and the spatial interactions between different urban areas (Roche 2014). GISs are particularly valuable for visualizing and analyzing spatial data, helping to identify areas of congestion, accessibility gaps, and opportunities for transportation improvements.
- *Psychology and Public Health*: Psychologists study the cognitive and emotional factors that influence human travel behaviors, such as risk perception, stress levels, and decision-making processes (McMichael 2020). Public health experts examine the health impacts of transportation systems, including physical activity levels, exposure to air pollution, and traffic-related injuries. Their research supports the development of transportation systems that promote mental and physical well-being.

2.6 CONCLUSION

In conclusion, the study of urban human mobility has evolved significantly, driven by technological advancements, diverse data sources, sophisticated analytical methods, and multidisciplinary approaches. As cities continue to expand and the demand for sustainable, resilient, and efficient transportation systems grows, the principles and approaches discussed in this chapter will remain crucial. The continued integration of technological innovations, diverse data sources, advanced analytical techniques, and multidisciplinary collaboration will be pivotal in creating smarter, more connected, and sustainable urban environments, ultimately enhancing the quality of life for urban residents.

REFERENCES

Boschmann, E. Eric, and Mei-Po Kwan. "Toward socially sustainable urban transportation: Progress and potentials." *International Journal of Sustainable Transportation* 2, no. 3 (2008): 138–157.

Butler, Luke, Tan Yigitcanlar, and Alexander Paz. "Smart urban mobility innovations: A comprehensive review and evaluation." *IEEE Access* 8 (2020): 196034–196049.

Chukwu, Michaelmary, Xiao Huang, Siqin Wang, Xiao Li, and Hanxue Wei. "Urban park accessibility assessment using human mobility data: A systematic review." *Annals of GIS* 30, no. 2 (2024): 181–198.

Coleri, Sinem, Sing Yiu Cheung, and Pravin Varaiya. "Sensor networks for monitoring traffic." In *Allerton Conference on Communication, Control and Computing*, pp. 32–40. 2004.

Duarte, Fábio, and Carlo Ratti. "The impact of autonomous vehicles on cities: A review." *Journal of Urban Technology* 25, no. 4 (2018): 3–18.

Giannopoulos, Georgios A. "The application of information and communication technologies in transport." *European Journal of Operational Research* 152, no. 2 (2004): 302–320.

Hu, Tao, Xiao Huang, and Siqin Wang. "Actionable Environmental Science Through Social Media Platforms." In Z. Sun (ed) Actionable Science of Global Environment Change: From Big Data to Practical Research, pp. 355–371. Cham: Springer International Publishing, 2023.

Hu, Tao, Siqin Wang, Bing She, Mengxi Zhang, Xiao Huang, Yunhe Cui, and Jacob Khuri et al. "Human mobility data in the COVID-19 pandemic: Characteristics, applications, and challenges." *International Journal of Digital Earth* 14, no. 9 (2021): 1126–1147.

Huang, Xiao, Zhenlong Li, Yuqin Jiang, Xinyue Ye, Chengbin Deng, Jiajia Zhang, and Xiaoming Li. "The characteristics of multi-source mobility datasets and how they reveal the luxury nature of social distancing in the US during the COVID-19 pandemic." *International Journal of Digital Earth* 14, no. 4 (2021): 424–442.

Huang, Xiao, Yago Martin, Siqin Wang, Mengxi Zhang, Xi Gong, Yue Ge, and Zhenlong Li. "The promise of excess mobility analysis: Measuring episodic-mobility with geotagged social media data." *Cartography and Geographic Information Science* 49, no. 5 (2022a): 464–478.

Huang, Xiao, Siqin Wang, Di Yang, Tao Hu, Meixu Chen, Mengxi Zhang, and Guiming Zhang et al. "Crowdsourcing geospatial data for earth and human observations: A review." *Journal of Remote Sensing* 4 (2024): 0105.

Huang, Xiao, Siqin Wang, Mengxi Zhang, Tao Hu, Alexander Hohl, Bing She, and Xi Gong et al. "Social media mining under the COVID-19 context: Progress, challenges, and opportunities." *International Journal of Applied Earth Observation and Geoinformation* 113 (2022b): 102967.

Kim, Dong Yup, and Ha Yoon Song. "Method of predicting human mobility patterns using deep learning." *Neurocomputing* 280 (2018): 56–64.

Kourtit, Karima, and Peter Nijkamp. "Big data dashboards as smart decision support tools for *i*-cities–An experiment on stockholm." *Land Use Policy* 71 (2018): 24–35.

Kumar Debnath, Ashim, M. Mazharul Haque, Hoong Chor Chin, and Belinda Yuen. "Sustainable urban transport: Smart technology initiatives in Singapore." *Transportation Research Record* 2243, no. 1 (2011): 38–45.

Li, Zhenlong, Xiao Huang, Xinyue Ye, Yuqin Jiang, Yago Martin, Huan Ning, Michael E. Hodgson, and Xiaoming Li. "Measuring global multi-scale place connectivity using geotagged social media data." *Scientific Reports* 11, no. 1 (2021): 14694.

Lu, Junyu, Xiao Xiao, Xiao Huang, Xiaowei Chuai, Zhenlong Li, Hanxue Wei, and Sicheng Wang. "Big data insights into urban park use in the pandemic: Changes in visitation patterns and exacerbated social inequalities in the US." *Cities* 152 (2024): 105204.

Luca, Massimiliano, Gianni Barlacchi, Bruno Lepri, and Luca Pappalardo. "A survey on deep learning for human mobility." *ACM Computing Surveys (CSUR)* 55, no. 1 (2021): 1–44.

McMichael, Celia. "Human mobility, climate change, and health: Unpacking the connections." *The Lancet Planetary Health* 4, no. 6 (2020): e217–e218.

Oladimeji, Damilola, Khushi Gupta, Nuri Alperen Kose, Kubra Gundogan, Linqiang Ge, and Fan Liang. "Smart transportation: An overview of technologies and applications." *Sensors* 23, no. 8 (2023): 3880.

Pelletier, Marie-Pier, Martin Trépanier, and Catherine Morency. "Smart card data use in public transit: A literature review." *Transportation Research Part C: Emerging Technologies* 19, no. 4 (2011): 557–568.

Roche, Stéphane. "Geographic information science I: Why does a smart city need to be spatially enabled?" *Progress in Human Geography* 38, no. 5 (2014): 703–711.

Rosa, Luís, Fábio Silva, and Cesar Analide. "Mobile networks and internet of things infrastructures to characterize smart human mobility." *Smart Cities* 4, no. 2 (2021): 894–918.

Schläpfer, Markus, Lei Dong, Kevin O'Keeffe, Paolo Santi, Michael Szell, Hadrien Salat, Samuel Anklesaria, Mohammad Vazifeh, Carlo Ratti, and Geoffrey B. West. "The universal visitation law of human mobility." *Nature* 593, no. 7860 (2021): 522–527.

Seliverstov, Yaroslav, Alexey Starichenkov, and Kirill Nikitin. "Using mobile applications to evaluate quality of road networks and transport mobility." *Transportation Research Procedia* 50 (2020): 636–646.

Seneviratne, Suranga, Yining Hu, Tham Nguyen, Guohao Lan, Sara Khalifa, Kanchana Thilakarathna, Mahbub Hassan, and Aruna Seneviratne. "A survey of wearable devices and challenges." *IEEE Communications Surveys & Tutorials* 19, no. 4 (2017): 2573–2620.

Solmaz, Gürkan, and Damla Turgut. "A survey of human mobility models." *IEEE Access* 7 (2019): 125711–125731.

Stopher, Peter R., and Stephen P. Greaves. "Household travel surveys: Where are we going?" *Transportation Research Part A: Policy and Practice* 41, no. 5 (2007): 367–381.

Sun, Yeran, and Amin Mobasheri. "Utilizing crowdsourced data for studies of cycling and air pollution exposure: A case study using strava data." *International Journal of Environmental Research and Public Health* 14, no. 3 (2017): 274.

Varone, Sacha, and Kamel Aissat. "Multi-modal transportation with public transport and ride-sharing-multi-modal transportation using a path-based method." In *International Conference on Enterprise Information Systems*, vol. 2, pp. 479–486. SCITEPRESS, 2015.

Wang, Jinzhong, Xiangjie Kong, Feng Xia, and Lijun Sun. "Urban human mobility: Data-driven modeling and prediction." *ACM SIGKDD Explorations Newsletter* 21, no. 1 (2019): 1–19.

Wang, Sicheng, Xiao Huang, and Qing Shen. "Disparities in resilience and recovery of ridesourcing usage during COVID-19." *Journal of Transport Geography* 114 (2024): 103745.

Zheng, Pengjun, and McDonad Mike. "An investigation on the manual traffic count accuracy." *Procedia-Social and Behavioral Sciences* 43 (2012): 226–231.

Zhou, Yang, Quan Yuan, Fangyi Ding, Mingyang Chen, Chao Yang, and Tangyi Guo. "Demand, mobility, and constraints: Exploring travel behaviors and mode choices of older adults using a facility-based framework." *Journal of Transport Geography* 102 (2022): 103368.

Zuo, Fan, Jingxing Wang, Jingqin Gao, Kaan Ozbay, Xuegang Jeff Ban, Yubin Shen, Hong Yang, and Shri Iyer. "An interactive data visualization and analytics tool to evaluate mobility and sociability trends during COVID-19." *arXiv preprint arXiv:2006.14882* (2020).

Section II

Aspects of Human Mobility in Modern Cities

3 Equity and Accessibility in Urban Mobility

B. van Wee

3.1 INTRODUCTION

Providing access to destinations is the core aim of the transport system and, to some extent, also the land use system. Although this notion is not at all new, it only recently has become generally recognized among many policymakers and others (Handy, 2020). This raises an important question: How should accessibility levels of societal trends and policies be assessed and interpreted? This is an interesting question in itself and also more specifically for the evaluation of policies that aim to improve accessibility. In the policy analysis literature, there is considerable agreement on the idea that "good" policies need to be (1) effective, (2) efficient, and (3) fair (Young and Tilley, 2006). However, until approximately 2010, fairness was only limitedly on the radar of people doing or asking for policy evaluations in the area of transport, the dominant exception being the literature on social exclusion (e.g., Lucas, 2012). A search in Scopus (on April 16, 2024) with the search string "accessibility AND equity AND travel" revealed 603 hits, of which only 49 were published before 2010.

That has rapidly changed over the past decade and a half: equity is now quite prominent on the radar of researchers and policymakers. So an important question is how to evaluate the equity or fairness (terms that are often used interchangeably) of societal trends and (candidate) policies. To answer this question, first an understanding of factors influencing equity levels of accessibility is needed. This might be helpful to generate policy options and to understand the mechanisms that influence equity levels of accessibility. What is needed next is an understanding of the normative judgements of fairness: Which ethical principles are available and most useful to assess the equity levels of accessibility, how do these principles translate to indicators, and which of those principles and indicators are most frequently used? A final topic relevant to answer the "how to" question presented earlier relates to the spatial scale of the analysis. A researcher can calculate levels of accessibility for all individuals, making use of their individual home addresses and the addresses of candidate destinations. But that is a lot of work, and data are not always available. So in practice, calculations are often made for groups of addresses, which can be labeled as zones (postal codes, neighborhoods, municipalities, regions, or other administrative boundaries, or based on grids of certain sizes, such as 1 × 1 km). Then it is important to realize that the level of aggregation can influence the outcomes of equity assessments.

This chapter aims to provide guidance for the assessment of equity of accessibility. Summarizing the topics just discussed, it aims to answer four questions:

1. Which factors influence equity of accessibility?
2. Which are the dominant ethical principles to assess equity of accessibility?
3. Which quantitative indicators are generally used to assess equity levels of accessibility?
4. How important are spatial aggregations for the outcomes of equity levels of accessibility?

Before answering these questions, I first provide the definition of accessibility as proposed by Geurs and van Wee (2004: 128): 'the extent to which land-use and transport systems enable (groups of) individuals to reach activities or destinations by means of a (combination of) transport mode(s)."

DOI: 10.1201/9781003503262-5

Next I provide a definition of equity from the World Health Organization (WHO): "The absence of avoidable or remediable differences among groups of people, whether those groups are defined socially, economically, demographically, or geographically" (Davis and Pilkington, 2019: 161).

These four questions are addressed in the next four subsections, respectively. Section 3.6 completes this chapter with some concluding remarks.

3.2 A CONCEPTUAL MODEL FOR EQUITY OF ACCESSIBILITY

Van Wee (2022) proposes a conceptual model for equity of accessibility; see Figure 3.1.

Figure 3.1 first shows that accessibility and differences in (perceived) levels of accessibility depend on the four factors determining levels of accessibility as proposed by Geurs and van Wee (2004): the land use system, the transport system, the temporal component, and the individual component. The land use system explains which activity locations (e.g., dwellings, jobs, schools, medical services, recreational facilities) are located where. The transport system describes the level of service of travel options between these locations and involves infrastructure, travel speeds (public transport) services provide, vehicles available, and the reliability of the transport system. The temporal component indicates which temporal constraints apply, such as opening hours of shops and kindergartens. The individual component describes the characteristics of individuals that are relevant for accessibility. This includes sociodemographic factors like income, driver's license holding, and age. For example, age can affect one's ability to get a driver's license, and therefore, access to destinations by car is irrelevant for people below the age at which they can get their driver's license (unless someone else drives that person to a destination). Next, attitudes and preferences matter: regardless of sociodemographic factors, people can have preferences for making use of specific transport modes and have different attitudes toward driving longer times or distances. Figure 3.1 also makes explicit that digital access to activities, like online shopping, working remotely, or e-consultation of doctors, influences levels of accessibility because they can be substitutes for physical accessibility. Next it explains that not only calculated levels of accessibility matter (the levels of accessibility calculated based on spatial data) but also perceptions of accessibility (see Pot et al., 2021). This contradicts that in a huge majority of accessibility studies, only calculated levels of accessibility are the topic of research. If perceived accessibility differs from calculated levels of accessibility, this can be the result of a person not being aware of destinations to travel to or of travel options. But it can also be that the person more accurately knows which options are available because they have time constraints that are ignored in calculated levels of accessibility (Pot et al., 2021).

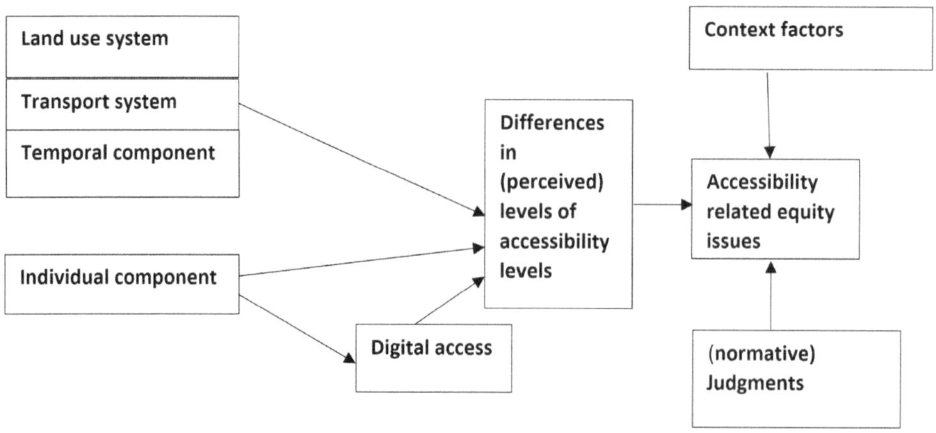

FIGURE 3.1 A conceptual model for equity of accessibility. (From van Wee 2023.)

Not all types of destinations are equally important for assessing the equity of accessibility (see Van Wee and Geurs, 2011, for a discussion): access to destinations like grocery stores, medical services, schools, social contacts, and jobs is more important than luxury destinations such as an expensive concert. So it makes sense to limit studies (and policies) on equity of accessibility to certain activity types.

Figure 3.1 next explains that to come to an assessment of equity of accessibility, first, normative judgments are needed: When are levels of accessibility or the distribution of those levels over groups of the population (e.g., by income level, vehicle ownership levels, or neighborhood/region) more or less inequitable? We discuss these normative judgments in the next section. Figure 3.1 also makes explicit that context factors are relevant for such assessments. A context factor could be the importance of religion: access to a mosque is more important in some Islamic countries than in countries with only a limited population of Muslims. Cultural factors are also context factors. For example, the occurrence and importance of specific sports vary considerably across countries, as does access to related sports events.

This conceptual model can be helpful to understand the mechanisms influencing assessments of equity of accessibility. Next, it is useful to systematically come to a list of candidate policy options to improve equity of accessibility. For researchers, it can be a source of inspiration for future research. The paper of Van Wee and de Jong (2023) proposes a related research agenda.

3.3 ETHICAL PRINCIPLES

As explained above, normative judgments are needed to say something about equity of accessibility. But equity is not a simple concept; there are many views on what is fair (e.g., Van Wee, 2011). Three ethical principles dominate the debate on equity and accessibility. One principle is utilitarianism. This principle states that an act is fair if it maximizes the sum of net utility of all people involved. This net utility is what is calculated in social cost-benefit analysis (SCBA), a dominant evaluation method to assess the pros and cons of (candidate) policy options. Equity is only included in an SCBA if one thinks utilitarianism is the (only) principle relevant to assess what is fair. But most researchers in the area of accessibility and equity disagree and thus focus on other principles. A second popular principle is egalitarianism, which indicates that an act should reduce differences between (groups of) people. A third principle, sufficientarianism, focuses only on those who are worse off and states that an act should reduce the number or share of people below a certain minimum level or that it should reduce the gap between the minimum and actual level of these people. Sufficientarianism is a principle underpinning a focus on social exclusion. Egalitarianism and sufficientarianism are two ethical principles often used in studies addressing equity and accessibility (see, for example, Lucas et al., 2016; Martens, 2016; Lucas et al., 2019). In other words, two views on equity and accessibility are dominant: (1) differences in levels of accessibility between groups of people should not be too large, and (2) society should primarily focus on those who have accessibility levels below a threshold.

A next question is: Which (groups of) people should be compared? Thomopoulos et al. (2009) give an overview of equity types relevant for accessibility. Here I limit myself to two important equity types. Horizontal equity states that accessibility levels of comparable (groups of) people or regions should be about equal, whereas vertical equity implies that disadvantaged individuals, groups, or regions should be treated with priority to increase their accessibility levels. Egalitarianism underpins the importance of horizontal equity, and vertical equity underpins the importance of sufficientarianism.

3.4 QUANTITATIVE ASSESSMENTS OF EQUITY OF ACCESSIBILITY

In some studies, metrics or quantitative indicators are used for the distribution of accessibility levels over (groups of) people. These metrics are mainly relevant for egalitarian views on what is fair. Van Wee and Mouter (2021) give an overview of these quantitative assessments of equity in the transport literature. The equity of accessibility is the most dominant topic covered by that literature; there is much more literature focusing on the equity of accessibility than the equity of (road) safety and the environment. They give an overview of several indices: the Gini, Suits, Palma, Theil, and Atkinson indices; the percentile ratio; and the coefficient of variation. Based on a literature review, they conclude that the Gini index and the coefficient of variation are probably the most used indices for evaluating equity of accessibility. The following sections on these two indices are taken from on that chapter (Van Wee and Mouter 2020: 110–114) and have been slightly adapted for this text.

3.4.1 Gini Index

The Gini index is a quantitative indicator expressing the level of (un)equality of a distribution. It expresses the index graphically. It sorts the unit of analyses (e.g., individuals, regions) on the x-axis, based on the unit of the variable for which the distribution is shown. That unit is expressed on the y-axis and shows the sum of that indicator up to any point on the x-axis. The graph also shows the Lorenz curve, which is the line representing a 100% equal distribution. Assuming an equal length of the x-axis and the y-axis, that line is the 45-degree line. The Gini index is the surface between the Lorenz curve and the equal distribution curve, divided by the triangle between the Lorenz curve, the x-axis, and the right-hand positioned y-axis. The larger the Gini index, the more unequal is the distribution. An often-used indicator for which distributions are visualized this way and expressed in terms of a Gini index is income. The indicator to be evaluated for the evaluation of distributions of accessibility could be any quantitative accessibility indicator. Groups of people are then positioned on the x-axis, ranging from the lowest level at the left-hand side to the highest level at the right-hand side. Figure 3.2 visualizes this graphically. The Gini index can directly be interpreted as an equity index of accessibility based on egalitarianism. Figure 3.2 can also be used if one prefers to base the equity analysis on sufficientarianism: those who have accessibility levels below a certain threshold value are positioned at the left of the x-axis. The people positioned at the left-hand side of the X_s line are assumed to have accessibility levels below the threshold level. Researchers can report the absolute number of people (or zones) below the threshold level or the share below that level. The surface A_s visualizes the inequalities area related to sufficientarianism, and the surface A_{ns} expresses inequalities not related to sufficientarianism. The area of A_s divided by the triangle

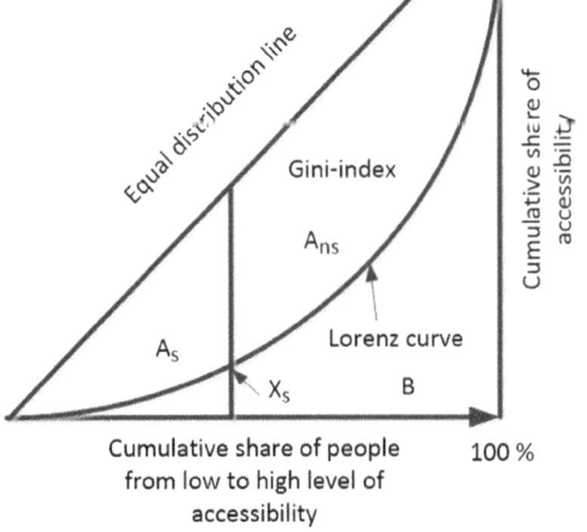

FIGURE 3.2 The Lorenz curve and the Gini coefficient applied to accessibility. (From Lucas et al. 2016.)

(A+B) can be interpreted as a metric for the severity of social exclusion. See Lucas et al. (2016) for an empirical example based on Figure 3.2.

3.4.2 Coefficient of Variation

This standard statistic can be interpreted as an indicator presenting the variability of any quantitative output indicator. A related metric is the standard deviation. In case of accessibility, the standard deviation of levels of accessibility (e.g., of people, the average of zones) can be calculated. The larger the standard deviation, the more unequally accessibility is distributed.

3.5 THE IMPACT OF SPATIAL AGGREGATION ON THE EQUITY OF ACCESSIBILITY

So far, this chapter has provided guidance for assessing equity of accessibility. Often, these levels are not assessed at the individual level, based on the exact location of origins and destinations, but on spatial aggregations. For example, average levels of accessibility per zone are calculated to assess inequalities in accessibility levels across these zones. An important question that then emerges is: What is the impact of spatial aggregation on these inequalities? Van Wee and de Jong (2023) aimed to answer this question, making used of a newly developed methodology of what they call dedicated accessibility points. They calculated Gini indices based on travel times by car and bicycle between places of residence and a selection of points of interest (POIs) at four different spatial scales with decreasing unit numbers and increasing spatial size, ranging from 8,131,689 individual addresses to 355 local municipalities covering the Netherlands. POIs include pharmacies, family doctors, hospitals, ambulance posts, fire stations, and policy stations. Note that the first three examples are destinations people might want to travel to, whereas the other three are services that might need to travel to one's dwelling. Table 3.1 shows the results of their analyses.

The text in the following section is largely based on the paper of Van Wee and de Jong (2023: 4–6).

3.5.1 Spatial Aggregation

A first topic to be discussed is the importance of spatial aggregation. Suppose a researcher or policymaker is interested in inequalities of accessibility at the individual level. Then any spatial aggregation implies a loss of information blurring the results. The ideal situation would be an assessment of inequalities in accessibility at the address level, as presented in the first row of each POI. The right-hand column in Table 3.1 divides the Gini index of spatial units larger than the address level by the address level, for a given POI and travel mode. This difference expresses the error in the calculated Gini indices due to the spatial aggregation. The results reveal that this problem does exist and is all but negligible. So, a very important conclusion is that the calculated magnitude of inequalities of accessibility heavily depends on the spatial scale at which travel times are calculated and averaged. And a related issue is that if one is interested in inequalities between individuals, problems due to spatial aggregation are significant.

But in many cases, the researcher, policymaker, or other "client" of research could have good reasons to not be interested in inequalities between individuals, but at a certain spatial scale. A policymaker could be interested in comparing accessibility levels of different neighborhoods, municipalities, or regions. For example, if a municipality wants to reduce differences in accessibility levels between neighborhoods or postal code zones, it makes sense to calculate the average level of accessibility for all people in each neighborhood or postal code zone based on a point of gravity for each zone and then next focus on the distribution of that aggregate accessibility level. For example, the municipality might be interested in differences in access to hospitals. Many smaller municipalities have only one hospital, and it makes sense to calculate, for example, travel times from a centroid

TABLE 3.1
Gini Coefficients per POI, Mode, and Area Type

POI	Travel Mode	Area Type	Gini Coefficient	Gini Addresses/Gini Area Type
Pharmacy (N = 1,672)	Car	Individual address	0.33	1.00
		Small neighborhood	0.27	1.25
		Four-digit postal code	0.23	1.45
		Municipality	0.18	1.86
	Bicycle	Individual address	0.42	1.00
		Small neighborhood	0.32	1.31
		Four-digit postal code	0.27	1.54
		Municipality	0.24	1.75
Family doctor (N = 4,705)	Car	Individual address	0.31	1.00
		Small neighborhood	0.23	1.38
		Four-digit postal code	0.19	1.67
		Municipality	0.14	2.28
	Bicycle	Individual address	0.43	1.00
		Small neighborhood	0.28	1.52
		Four-digit postal code	0.23	1.83
		Municipality	0.20	2.17
Hospital (N = 187)	Car	Individual address	0.32	1.00
		Small neighborhood	0.31	1.05
		Four-digit postal code	0.29	1.09
		Municipality	0.22	1.45
	Bicycle	Individual address	0.42	1.00
		Small neighborhood	0.40	1.06
		Four-digit postal code	0.38	1.12
		Municipality	0.27	1.56
Ambulance (N = 236)	Car	Individual address	0.30	1.00
		Small neighborhood	0.28	1.06
		Four-digit postal code	0.27	1.13
		Municipality	0.16	1.84
Fire station (N = 992)	Car	Individual address	0.28	1.00
		Small neighborhood	0.24	1.18
		Four-digit postal code	0.20	1.39
		Municipality	0.19	1.46
Police station (N = 337)	Car	Individual address	0.35	1.00
		Small neighborhood	0.32	1.09
		Four-digit postal code	0.29	1.18
		Municipality	0.21	1.65

(gravity point) of each postal code zone to the hospital or to a new hospital to be built. Theoretically, the comparison between spatial units could still be based on individual addresses, but that could be a lot of work, and spatial aggregation could reduce the workload. So the individual address level is not necessarily the preferred level to be used for the distributions of accessibility and related equity assessments.

Assuming there are good reasons to be interested in inequalities comparing averages at a certain spatial level, the authors next dig deeper in the results departing from this perspective. A first and very important conclusion, already presented earlier, is that the value of the Gini index can significantly differ, depending on the spatial aggregation level at which inequalities in the level of accessibility occur. The highest values of the Gini index for a given POI type and mode are 45%–128% higher (see right-hand column) than the lowest level, depending on the spatial scale. This conclusion applies to all POIs and modes evaluated in this study. So, if one is interested in comparing zones, it matters a lot how addresses are clustered in zones and thus how large zones are.

The conclusion that inequalities are higher at lower spatial scales also applies to all POIs and modes. In other words, levels of accessibility comparing small neighborhoods differ more across these neighborhoods than they differ comparing municipalities. This implies that if the aim is to reduce general inequalities across all people, regardless of the specific policy question at stake, the focus should primarily be on at least smaller geographical areas and to what extent policies could reduce differences in accessibility.

3.5.2 Differences Between POIs

The Gini index for POIs that are relatively nearby, at least in the Netherlands, such as family doctors and pharmacies, is influenced strongly by aggregating from the address level to larger areas. For other POIs for which people in general would need to travel farther because they are less nearby (in other words, there are fewer of those available in a city or town), as expected, the Gini index is less sensitive to spatial aggregation.

Looking at the three emergency services included in our analyses, we see that the differences at the municipality level are relatively small, but at the level of small neighborhoods, they become somewhat larger. Spatial scale matters most for ambulances, then for police stations, and finally for fire stations.

3.5.3 Car versus Bike

For the bicycle, the impact of spatial aggregation is larger than for the car for all POIs for which Gini indices for both modes are calculated. The authors see two possible explanations. The first is at the "within-city" level and the second is at the "city versus rural areas" level. First, a possible explanation is that the bicycle networks in the Netherlands are more uniform than those for cars. In central urban areas and neighborhoods around centers, several roads are closed for cars or are unidirectional for cars but not for bicycles. This means that the detours for bicycles in such areas are smaller than for cars. In addition, in such areas, the speeds of cars are lower than in the outskirts. This trend probably also applies to bicycles but to a lesser extent. The result of this pattern is that for cars the advantages of having POIs nearer by than farther away is partly compensated for by the lower average speeds on shorter distance and the longer detours, but for bicycles, this is less the case. In other words, distance matters more for bicycle trips than for car trips because of differences in car speeds across the network. Second, focusing on urban versus rural areas, an explanation can be that speeds for cars are much higher outside urban areas than in urban areas, but this hardly applies to bicycles. The higher speeds for cars reduce inequalities in travel times, whereas this does not apply for bicycles. Again, distance matters more for bicycle trips than for car trips because of car speed difference.

In sum, the authors conclude that spatial aggregation has major implications for the calculated Gini indices. In addition, the importance of spatial aggregation differs between POIs, and those differences are heavily influenced by the spatial distribution of those POIs. The authors also conclude that inequalities in travel times are smaller for the car than for the bike. This difference is probably mainly explained by network characteristics.

Because of the impact of spatial aggregation on the Gini index, it is important that researchers and policymakers are explicit about the aim of a related study. The aim should determine the preferred level of spatial aggregation (assuming the data needed are available). A "wrong" choice can severely influence the results and thus the preferences for policies.

3.6 CONCLUDING REMARKS

As made clear above, the equity impacts of transport policies in general and, more specifically, equity of accessibility are now more prominent areas of focus for researchers and policymakers compared with a decade and half ago. Sufficientarianism and egalitarianism are the dominant normative yardsticks to evaluate equity of accessibility. Despite this general pattern, at the time of writing this chapter in 2024, the literature in this area was still emerging. It could well be that other ethical principles will become more important in the future and that other metrics for the distribution of accessibility will become more important. It is also possible that the individual component will gain more attention and that researchers will focus increasingly on the capabilities, preferences, and perceptions of people. Social networks and digital access could also gain more attention in the future.

Although there is agreement in the literature that some destination types are more important for equity of accessibility than others, it is not yet clear which destinations should or should not be included in related analyses, and this selection will likely be context specific.

REFERENCES

Davis, A. and P. Pilkington. 2019. A public health approach to assessing road safety equity—the RoSE cycle. In: *Measuring Transport Equity* Ed. Lucas, K., Martens, K., F. Di Ciommo, A. Dupont-Kieffer. Amsterdam/Oxford (UK)/Cambridge (USA): Elsevier.

Geurs, K.T. and B. van Wee. 2004. Accessibility evaluation of land-use and transport strategies: Review and research directions, *Journal of Transport Geography*, 12(2), 127–140.

Handy, S. 2020. Is accessibility an idea whose time has finally come? *Transportation Research Part D: Transport and Environment*, 83, 102319.

Lucas, K. 2012. Transport and social exclusion: Where are we now? *Transport Policy*, 20, 105–113.

Lucas, K., K. Martens, F. Di Ciommo and A. Dupont-Kieffer. (Eds.). 2019. *Measuring Transport Equity*. Amsterdam/Oxford (UK)/Cambridge (USA): Elsevier.

Lucas, K., B. van Wee and K. Maat. 2016. A method to evaluate equitable accessibility: Combining ethical theories and accessibility-based approaches, *Transportation*, 43, 473–490.

Martens, K. 2016. *Transport Justice: Designing Fair Transportation Systems*. London: Routledge.

Pot, F.J., B. van Wee and T. Tillema. 2021. Perceived accessibility: What it is and why it differs from calculated accessibility measures based on spatial data, *Journal of Transport Geography*, 94, 103090.

Silber, J. (Ed.). 1999. *Handbook on Income Inequality Measurement*. Norwell, Massachusetts, Dordrecht, the Netherlands: Kluwer.

Thomopoulos, N., S. Grant-Muller and M.R. Tight. 2009. Incorporating equity considerations in transport infrastructure evaluation: Current practice and a proposed methodology, *Evaluation and Program Planning*, 32(4), 351–359.

Van Wee, B. 2011. *Transport and Ethics. Ethics and the Evaluation of Transport Policies and Projects*. Cheltenham: Edward Elgar.

Van Wee, B. 2022. Accessibility and equity: A conceptual model and research agenda, *Journal of Transport Geography*, 104, 103421.

Van Wee, B. and T. de Jong. 2023. Differences in levels of accessibility: The importance of spatial scale when measuring distributions of the accessibility of health and emergency services, *Journal of Transport Geography*, 106, 103511.

Van Wee, B. and K.T. Geurs. 2011. Discussing equity and social exclusion in accessibility evaluations, *European Journal of Transport and Infrastructure Research*, 11(4), 350–367.

Van Wee, B., N. Mouter. 2021. Evaluating transport equity. In: N. Mouter (ed.), New Methods, Reflections and Application Domains in Transport Appraisal. Elsevier Series, *Advances in Transport Policy and Planning*, 7, 103–126.

Young, W. and F. Tilley. 2006. Can businesses move beyond efficiency? The shift toward effectiveness and equity in the corporate sustainability debate, *Business Strategy and the Environment*, 15(6), 402–415.

4 Social and Cultural Dimensions of Urban Human Mobility

Xiao Huang and Malene Freudendal-Pedersen

4.1 INTRODUCTION

> *"... material transformations are remaking the 'social,' especially those diverse mobilities that, through multiple senses, imaginative travel, movements of images and information, virtuality and physical movement, are materially reconstructing the 'social as society' into the 'social as mobility.'"*
>
> **Urry (2000)**

In his book *Sociology Beyond Societies* (2000), John Urry emphasizes the importance of dynamic movements and flows as central to social structures and interactions, especially in increasingly connected modern cities. The social and cultural dimensions of urban human mobility encompass a wide array of factors influencing how people navigate urban environments. Studying these factors is crucial as they directly affect the accessibility, efficiency, and equity of urban transportation systems (Sheller 2018; Ryghaug et al. 2023). Understanding these dimensions allows for a deeper insight into the diverse needs and practices of urban populations, leading to more inclusive and effective mobility solutions.

Urban mobility extends beyond the physical movement of people to include how social interactions and cultural practices influence movement choices. The interplay between mobility and social dynamics is complex, involving numerous factors such as socioeconomic status, demographic characteristics, cultural backgrounds, and personal lifestyles, to list a few. This chapter explores these factors, highlighting their impact on urban mobility and providing insights into the challenges and opportunities they present.

This chapter provides a comprehensive examination of the social and cultural dimensions of urban mobility. It begins with an exploration of the social dimensions, emphasizing socioeconomic factors, demographic influences, and the role of social networks. The subsequent section addresses cultural influences, analyzing how cultural norms and lifestyle choices shape mobility patterns. Two case studies illustrate how human mobility records can enhance our understanding of community formation and social segregation. The chapter concludes with a discussion of strategies for promoting inclusive urban mobility.

By the end of this chapter, readers will gain a thorough understanding of the intricate relationship between urban mobility and the social and cultural fabric of cities. This knowledge is essential for developing mobility solutions that are efficient, sustainable, equitable, and responsive to the diverse needs of urban populations.

4.2 SOCIAL DIMENSIONS OF URBAN MOBILITY

Urban mobility is deeply intertwined with various social dimensions that significantly influence how individuals and communities navigate urban spaces. Understanding these social dimensions is crucial for developing inclusive and effective transportation systems that cater to the diverse needs of urban populations. This section delves into the key social dimensions of urban mobility, including socioeconomic factors, demographic influences, and the role of social networks.

DOI: 10.1201/9781003503262-6

4.2.1 Socioeconomic Factors

Socioeconomic status has certain influence on urban mobility patterns (Pucher and Renne 2003). Income, education, occupation, and access to resources impact individuals' transportation choices and opportunities. Higher-income individuals often have greater access to private vehicles, which are often seen as providing more flexibility and convenience (Krog Juvik and Halkier 2024). In contrast lower-income individuals, kids, and elderly people may rely heavily on public transportation, which in many places around the world can be less reliable and more time-consuming (Barbosa et al. 2021). Public transportation systems are crucial in providing mobility to those who choose not to own or cannot afford private vehicles. Socioeconomically strong neighborhoods often have high-quality transit options, whereas lower-income neighborhoods may have fewer public transportation routes, longer wait times, and less frequent service, exacerbating social inequalities. Moreover, transportation costs can be a significant burden for low-income households, affecting their ability to access education, employment, healthcare, and other essential services (Smith et al. 2012). Efforts to address these disparities include subsidized transportation programs, investments in improving public transit infrastructure, and policies aimed at making transportation more affordable and accessible. Ensuring that all socioeconomic groups have equitable access to transportation can promote social inclusion and is an important part of a sustainable mobility system.

4.2.2 Demographic Influences

Demographic characteristics such as age, gender, and ethnicity also play a role in shaping urban mobility patterns. Different demographic groups have varying mobility needs and face different challenges. The bundles of practices that need to be connected in everyday life are very different for families with children compared to young people in education (Plyushteva and Schwanen 2018).

Age impacts mobility needs across different life stages. Children and teenagers typically rely on their parents or guardians for transportation and may also use school buses or public transit. Young adults, particularly those in college or early in their careers, might favor public transportation, biking, or ride-sharing services. Middle-aged individuals often prefer private vehicles for their commuting and daily transportation needs. Older adults may require accessible transportation options due to mobility limitations. Providing age-appropriate transportation solutions, such as safe walking or cycling paths for children and accessible public transit for the elderly, is essential for fostering inclusive mobility across all age groups (Lin and Cui 2021).

Gender influences mobility patterns in several ways. Women often have different travel practices compared to men, influenced by factors such as caregiving responsibilities, safety concerns, and employment patterns (Hidayati et al. 2020). A study by Gauvin et al. (2020) concluded that women visit fewer locations than men and are more localized, tending to spend time within a few preferred areas. This reduced mobility among women may result from the interplay of cultural, infrastructure, resource, and safety constraints. Nobis (2007) found that men drive and cycle more than women but also noted that women are generally more likely to combine driving with other modes of transportation. In countries like Denmark and the Netherlands with strong cycling cultures and a prioritization of the cycling infrastructure, this is not the case though. Here, women often cycle more, and in families with one car, it is often the man who drives the car even if the woman is undertaking the transport of kids and grocery shopping (Freudendal-Pedersen 2015; Porrazzo et al. 2022). In line with this, Kuhnimhof et al. (2012) observed that men travel more by car than women but that the gender gap regarding licensure, car availability, and automobile travel is diminishing. Addressing these concerns requires gender-sensitive transportation planning, such as ensuring safe and well-lit transit stops, providing secure travel options, and considering the needs of caregivers in transportation design. It has been suggested that to achieve more gender-sensitive planning, it is paramount to change the composition of the sector when 80% of people working in the transport sector are male and an even smaller percentage of women work at the top level (Ng and Acker 2020).

Ethnic minorities often face unique mobility challenges related to systemic inequalities and discrimination. These groups may live in underserved neighborhoods with limited access to reliable public transportation, affecting their ability to access employment, education, and other essential services. Language barriers and cultural differences can also affect their ability to navigate the transportation system effectively. Policies aimed at improving transportation equity must consider the specific needs of ethnic minorities, ensuring that all community members have equal access to mobility opportunities.

4.2.3 Social Networks

Social networks and communities also play a role in shaping urban mobility practices. Individuals frequently rely on their social networks for information regarding transportation options, carpooling arrangements, and shared travel experiences. These social interactions can influence travel decisions, such as commuting with colleagues, traveling with family members, or engaging in community-based transportation initiatives (Rodríguez et al. 2020).

The advent of digital platforms and social media has revolutionized the way people interact with transportation systems. Online communities and mobile applications now provide real-time information on transit schedules, ride-sharing opportunities, and traffic conditions. These platforms enhance connectivity and convenience, facilitating travel planning and access to transportation services. Nonetheless, digital divides persist, with marginalized communities having limited access to technology and the Internet, thereby restricting their mobility options (Groth 2019). Enhancing urban mobility through strong social networks and digital tools involves promoting information sharing, community engagement, and collaborative transportation solutions. Initiatives such as community-based ride-sharing programs, neighborhood transit groups, and online mobility forums can empower individuals and communities to actively participate in shaping their transportation (Freudendal-Pedersen and Kesselring 2018).

4.3 CULTURAL INFLUENCES ON URBAN MOBILITY

4.3.1 Cultural Norms and Values

Cultural norms and values profoundly shape how people perceive and engage with transportation options in urban environments. In some cultures, there is a strong preference for private vehicle ownership, which is often associated with status, independence, and personal freedom. This cultural inclination can lead to high levels of car ownership and a corresponding demand for extensive road networks and parking facilities. For instance, in the United States, the car culture has historically dominated, influencing urban planning and resulting in sprawling cities with significant reliance on automobiles (Bohren 2016). Conversely, other cultures prioritize collective well-being and environmental sustainability, leading to greater acceptance and use of public transportation. In many European cities, for example, public transit systems are highly developed and widely used, reflecting a cultural value placed on environmental consciousness and community (Parent 2006). This cultural norm promotes investment in efficient, reliable, and comprehensive public transit networks, as well as cycling and walking infrastructure (Aldred and Jungnickel 2014).

Cultural norms also influence gender roles and expectations related to mobility. In some societies, cultural expectations may limit women's mobility by imposing restrictions on their use of certain modes of transportation or requiring male accompaniment (Uteng 2011). This can lead to disparities in access to education, employment, and other opportunities, perpetuating gender inequality. Moreover, cultural values impact the design and implementation of transportation policies. In cities with a strong cultural emphasis on inclusivity and accessibility, there is likely to be greater attention to ensuring that transportation systems are accessible to all individuals, including those with disabilities, the elderly, and other marginalized groups. This is evident in cities like

Tokyo, where public transportation is designed to accommodate a wide range of users, reflecting a cultural commitment to inclusivity (Chowdhury and McFarlane 2022).

4.3.2 Lifestyle and Personal Choices

Lifestyle choices also influence urban mobility patterns and shape how individuals navigate their daily environments. These choices dictate various preferences and needs for transportation, leading to diverse mobility demands within urban settings. The transportation choices are seldom reflected once they are incorporated into the organization of everyday life (Freudendal-Pedersen 2022). The planning of transportation systems and lifestyles are closely connected. For instance, individuals with environmentally conscious lifestyles may opt for sustainable modes of transportation, such as biking, walking, or using public transit. But if the infrastructure to facilitate sustainable transport is not in place, this complicates everyday life organization. As such, the commitment to reducing carbon footprints and promoting environmental stewardship does not have a direct impact on increased investments in green transportation infrastructure. The way we plan urban infrastructure has a very strong impact on people's lifestyle choices, and the "culture of automobility" (Dennis and Urry 2009) plays a strong role in cities around the world. This also promotes fast-paced lifestyles where speed and convenience are prioritized, leading to a preference for private cars or ride-sharing services. The demand for quick, flexible, and reliable transportation options drives the need for sophisticated road networks, efficient traffic management systems, and innovative mobility services like car-sharing and electric scooters. Again, in some urban areas like Copenhagen and Amsterdam, the highly developed cycle system places the bicycle as the mode of transportation that offers quick, flexible, and reliable transportation (Cox and Koglin 2020).

Urban mobility is also profoundly affected by lifestyle trends such as telecommuting, flexible work hours, and the gig economy. The rise of remote work, accelerated by technological advancements and the COVID-19 pandemic, has altered daily commuting patterns (Chatterjee and Crawford 2021). This shift has reduced the demand for traditional rush-hour transit services and increased the need for flexible, on-demand transportation options. Consequently, companies and cities are focusing on developing adaptable transit systems that cater to these new commuting practices, incorporating features like dynamic scheduling and real-time data analytics to optimize service delivery. Similarly, the gig economy has created a demand for more adaptable and varied transportation solutions to accommodate the dynamic schedules of gig workers (Wang 2022). This sector's growth underscores the need for transportation systems that are both reliable and flexible enough to meet the fluctuating demands of workers in industries such as ride-hailing, food delivery, and freelance contracting. Furthermore, lifestyle choices related to health and wellness are influencing urban mobility. As more people adopt active lifestyles, there is an increased demand for infrastructure that supports walking, cycling, and other forms of active transportation (Young et al. 2020). Cities are responding by expanding bike lanes, pedestrian zones, and greenways, thereby creating safer and more attractive environments for nonmotorized mobility.

4.4 TWO CASE STUDIES

4.4.1 Mobility-Informed Community Homophily

Spatially, homophily suggests that stronger interactions occur among locations with similar characteristics, such as demographics and socioeconomic status. This principle is a key factor in human spatial and social interactions and has been extensively researched across various fields. In a case study by Huang et al. (2022), the researchers extracted human mobility structures by detecting communities formed by strong spatial links and examined the homophily effect at the community level using information entropy, which measures the chaoticness of local settings within communities. The study leveraged mobility networks derived from approximately 45 million mobile devices in the United States and selected the top 20 most-populated U.S. Metropolitan Statistical Areas (MSAs) as study areas.

In their workflow, the researchers first detected communities characterized by strong spatial interactions using a network community detection algorithm via aggregated mobility networks (Figure 4.1). Further, they explored the distribution of selected sociodemographic variables as well as different aspects of social vulnerability using entropy calculation. The more chaotic (high entropy) social settings are across community space, the less homophily this community exhibits. Their study presented varying patterns of community formation across different MSAs (Figure 4.2).

For example, the Atlanta MSA exhibits the lowest average community entropy in terms of household income, percentage of Black residents, and percentage of residents with low education levels

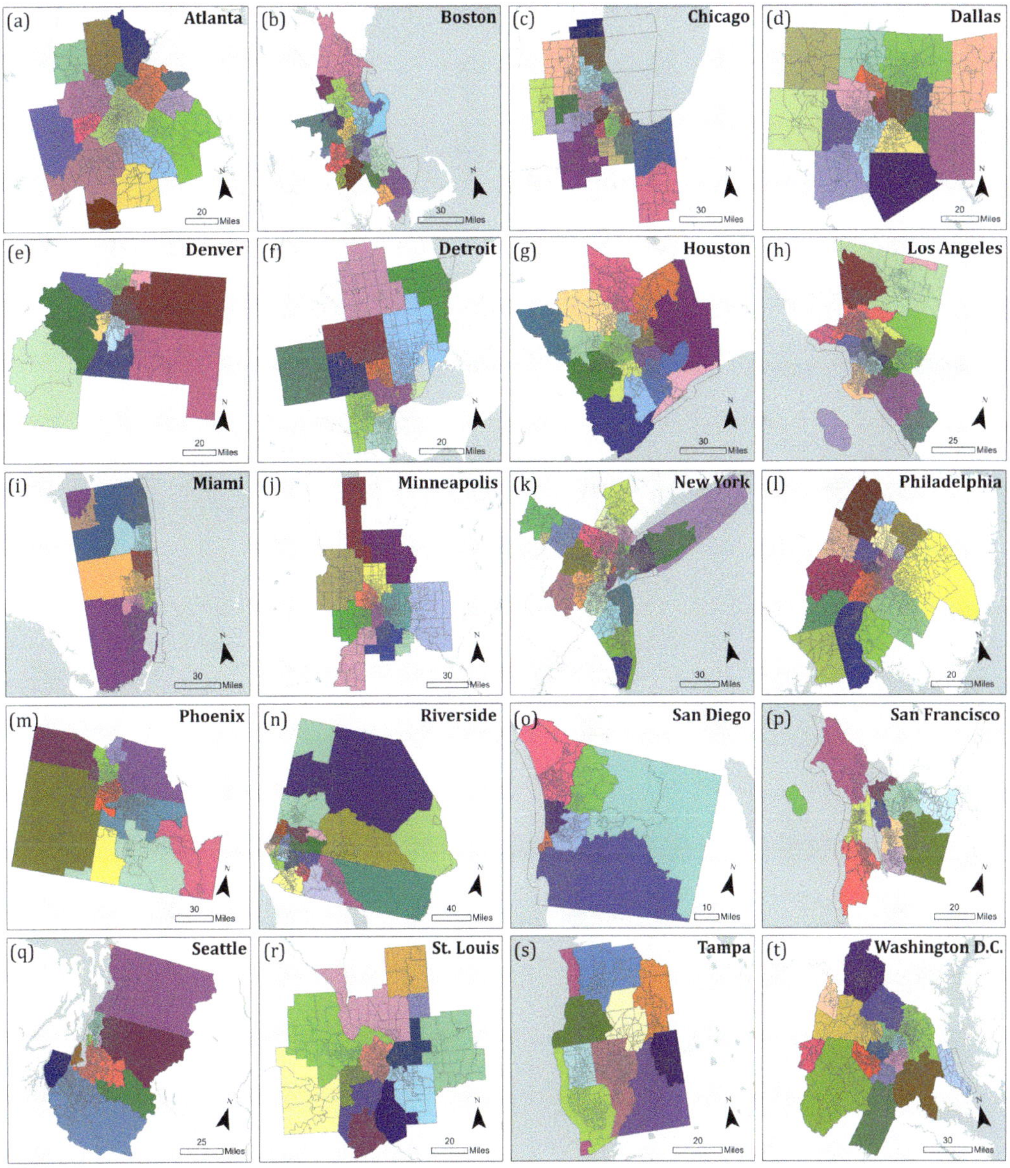

FIGURE 4.1 Detected communities in the top 20 most populated U.S. Metropolitan Statistical Areas. (From Huang et al. 2022.)

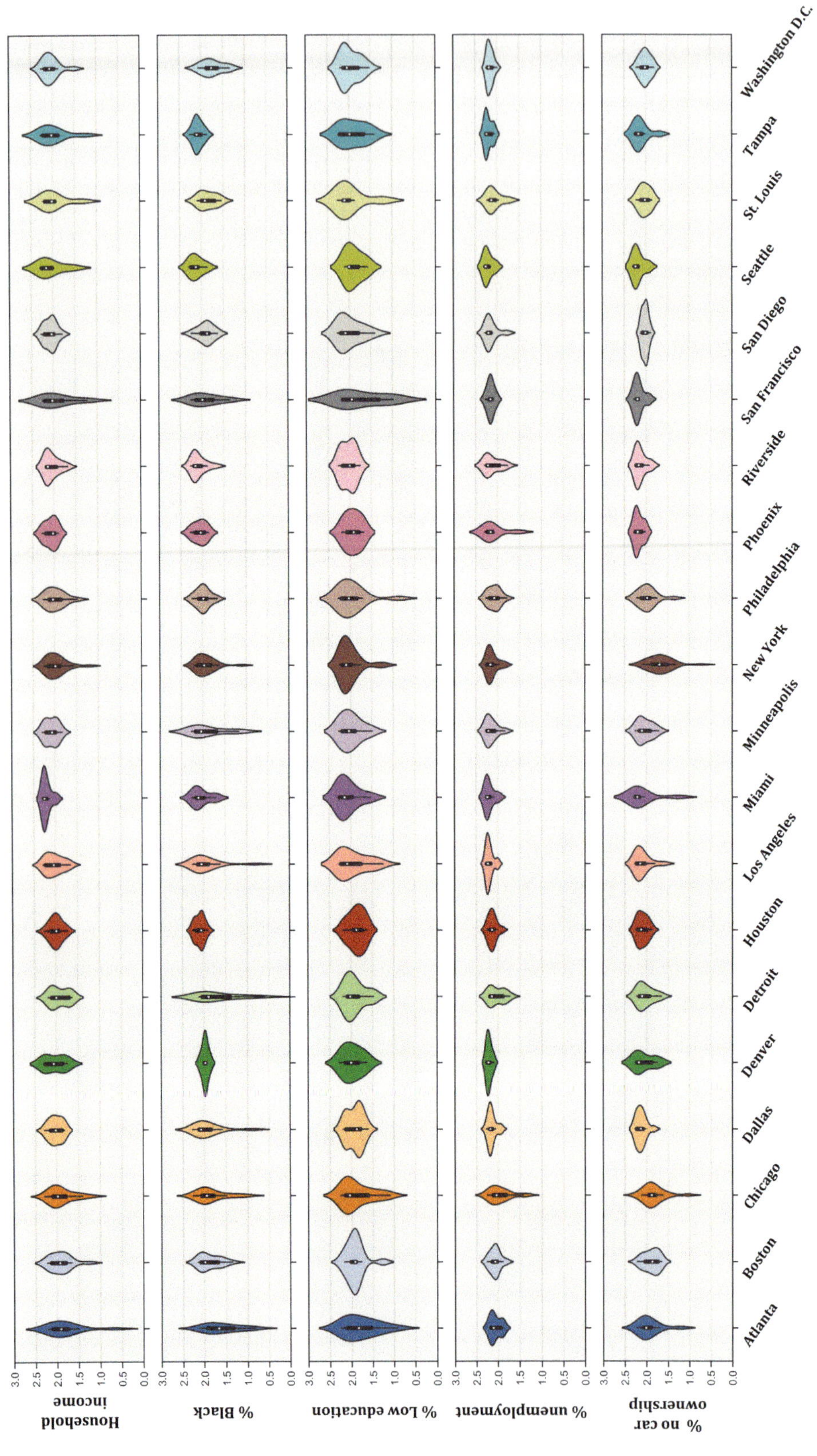

FIGURE 4.2 Community entropy distribution of selected demographic and socioeconomic variables in the top 20 most populated U.S. Metropolitan Statistical Areas. (From Huang et al. 2022.)

(Huang et al. 2022). Lower entropy signifies less variability, indicating that the Atlanta MSA displays the strongest homophily effect in these areas. This means that, compared to other MSAs, census tracts in the Atlanta MSA with similar socioeconomic characteristics are more likely to form communities with strong spatial interactions. The pronounced homophily effect in the Atlanta MSA aligns with existing research highlighting significant class and racial divisions and widening disparities in the region (e.g., Bullard et al. 1999). In contrast, the Miami MSA shows the highest entropy for household income and the third-highest entropy for the percentage of Black residents. This indicates that census tracts in the Miami MSA with varying income levels and racial compositions are more likely to intermingle and exhibit strong spatial interactions, leading to the formation of high-entropy communities (Huang et al. 2022).

4.4.2 Aggravated Social Segregation During the COVID-19 Pandemic

The COVID-19 pandemic significantly altered urban mobility patterns, exacerbating pre-existing social and spatial inequalities. In a study by Li et al. (2022), researchers assessed social segregation levels before and during the pandemic within the 12 most populous U.S. MSAs using large-scale individual-level mobility records.

They introduced a Social Segregation Index (SSI), which integrated mobility patterns with the Centers for Disease Control and Prevention's Social Vulnerability Index (SVI). Unlike previous studies that focused on racial or economic segregation, this research primarily investigated how socially vulnerable communities interacted with or became isolated from other communities through mobility flows.

The findings indicated that, since the onset of COVID-19, the aggregated SSI showed a statistically significant increase across all 12 MSAs, with New York experiencing the most notable rise (Li et al. 2022). The increase in the maximum value and standard deviation, coupled with a decrease in the minimum value of SSI in most MSAs, suggested varying levels of impact on segregation across different census tracts within each MSA. In Boston, San Francisco, and Washington, D.C., MSAs, more vulnerable groups (SVI levels 6 to 10) exhibited greater segregation compared to less vulnerable groups (SVI levels 1 to 5) (Figure 4.3), indicating that these communities faced higher levels of segregation overall. Furthermore, the study observed noticeable fluctuations in the local SSI of more vulnerable groups in nine of the 12 MSAs, including New York, Los Angeles, Chicago, Philadelphia, Houston, Miami, San Francisco, Atlanta, and Washington, D.C., suggesting that socially vulnerable communities were more adversely affected by the pandemic, leading to increased social segregation (Li et al. 2022).

4.5 TOWARD INCLUSIVE URBAN HUMAN MOBILITY

4.5.1 Equity-Focused Transportation Planning

Equity-focused transportation planning is essential for creating inclusive urban mobility systems that cater to the diverse needs of all residents, particularly marginalized and underserved communities. This approach prioritizes fairness in the distribution of transportation resources, ensuring everyone has access to safe, reliable, and affordable transportation options, regardless of socioeconomic status, race, gender, age, or ability. Addressing the spatial mismatch between where low-income individuals live and where jobs, services, and amenities are located is critical. Many marginalized communities reside in areas with limited public transit options, making it challenging to access employment, education, healthcare, and other essential services. To mitigate this, transit connectivity in underserved neighborhoods, expanded public transportation routes, and increased service frequency and reliability must be improved.

Affordability is another crucial component of equity in urban mobility. Transportation costs can be a significant burden for low-income households, limiting their mobility and access to

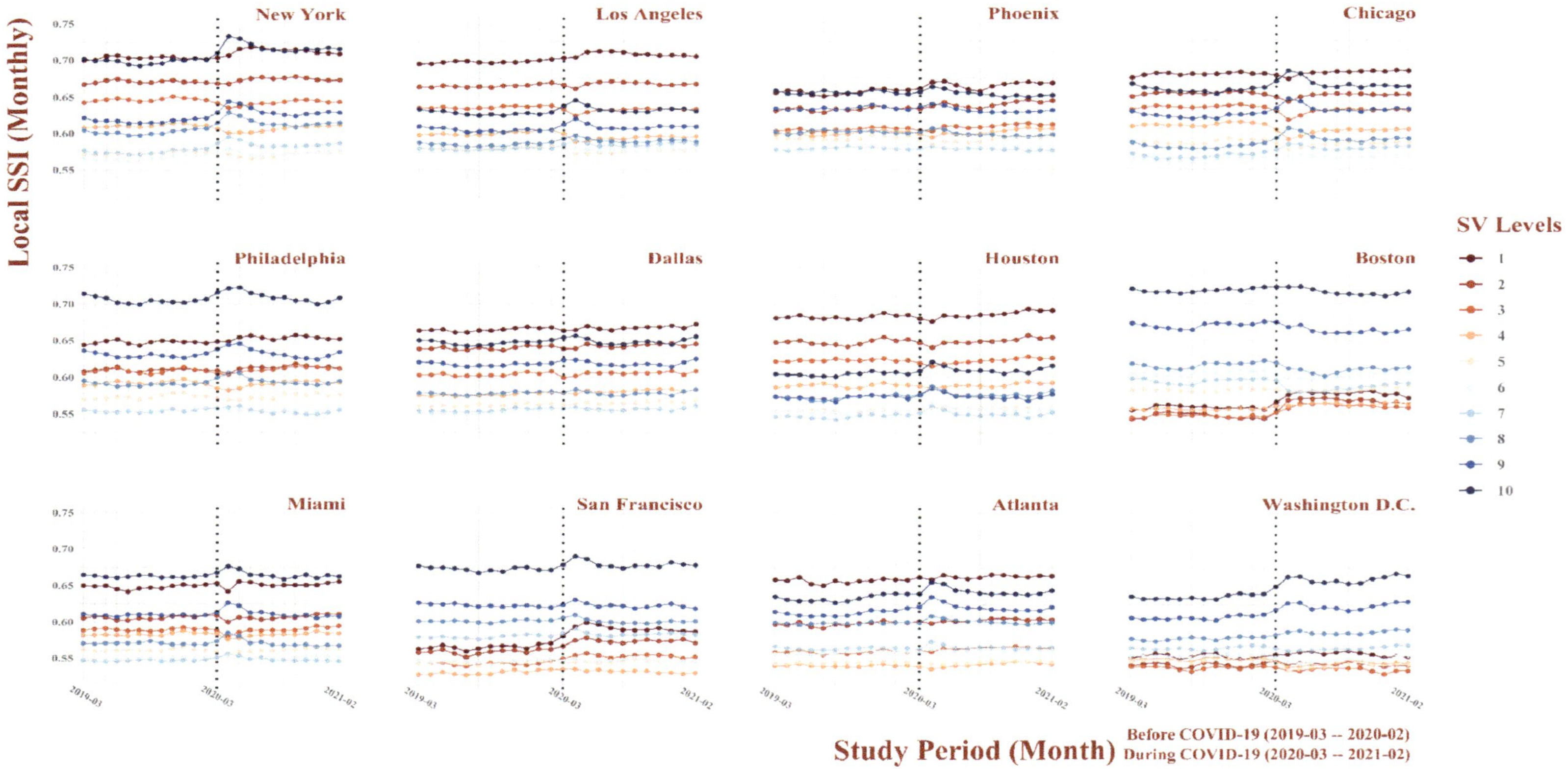

FIGURE 4.3 The mean value of monthly local Social Segregation Index (SSI) for census tracts at different social vulnerability (SV) levels before and during COVID-19. (From Li et al., 2022.)

opportunities (Guzman and Oviedo 2018). Implementing fare subsidies, offering reduced-cost transit passes for low-income individuals, and exploring alternative funding mechanisms can help keep transit services affordable without compromising quality. Additionally, investing in infrastructure that supports nonmotorized modes of transportation, such as walking and cycling, promotes equity. Safe and accessible pedestrian pathways, bike lanes, and greenways enhance mobility for those who cannot afford, or choose not to have, private vehicles, while also contributing to environmental sustainability and public health. In addition, engaging with communities to understand their specific needs and preferences ensures that marginalized groups are heard and considered in the planning process (Chatterjee et al. 2004). Community meetings, surveys, focus groups, and collaboration with local organizations facilitate this participatory approach. Addressing systemic biases and discrimination within transportation systems is also vital. Training transit staff to be sensitive to all passengers' needs, implementing policies to prevent harassment and discrimination, and designing accessible transit facilities are necessary steps.

4.5.2 Culturally Sensitive Mobility Planning

Culturally sensitive mobility planning respects diverse cultural practices, values, and preferences, ensuring transportation systems accommodate various needs and foster inclusivity. Understanding cultural contexts is crucial when it shapes mobility patterns and preferences. The role and significance of family and community can vary, requiring diverse transportation solutions. In some areas, this might necessitate options for group travel, while in others, the emphasis may be on creating walkable neighborhoods. In some cultures, personal freedom is related to the favoring of private vehicle ownership, and in others, it involves a transport system that offers many solutions like cycling, public transit, walking, mobility as a service, and so on (Freudendal-Pedersen et al. 2023). When planning transportation also involves engagement with cultural communities through outreach strategies, such as working with community leaders, conducting multilingual focus groups, and using relevant communication channels, transportation solutions that address specific cultural needs and values can be developed.

Recognizing historical and systemic factors is also vital. Minority communities often face limited access to transportation due to historical disinvestment (Inwood et al. 2015). Addressing these legacies involves targeted investments, equitable access policies, and building trust with affected communities. Incorporating cultural practices into transportation planning enhances acceptance. Transit stations and public spaces reflecting local cultural heritage can foster a sense of belonging. This can be achieved through architectural design, public art, multilingual signage, and cultural symbols.

Safety and security concerns of different cultural groups must be considered. Women and girls from certain cultural backgrounds may face specific safety challenges when using public transportation, particularly during nighttime or in isolated areas (Chowdhury et al. 2024). Addressing these concerns requires gender-sensitive planning, such as improving lighting, surveillance, and security measures in transit facilities and ensuring transportation services accommodate the needs of caregivers and families. Promoting cultural competency among transportation professionals is essential, including training staff to understand and respect cultural differences, recognize biases, and engage effectively with diverse communities. Fostering inclusivity within transportation agencies allows planners to develop more effective and inclusive mobility solutions.

4.6 CONCLUSION

The exploration of the social and cultural dimensions of urban human mobility highlights their significant impact on how individuals navigate urban environments. Understanding these factors is essential for comprehending the complex interplay between mobility and social dynamics, emphasizing the necessity for inclusive and equitable transportation planning. Recognizing and addressing the diverse needs and behaviors of urban populations enable planners to develop mobility solutions that enhance accessibility, efficiency, and fairness. Equity-focused transportation planning and culturally sensitive mobility initiatives underscore the importance of community involvement in the planning process. These strategies ensure that all residents, particularly marginalized groups, have access to reliable, safe, and affordable transportation options. Fostering an inclusive urban mobility system requires a holistic approach that considers the intricate relationships between social structures, cultural practices, and mobility patterns. This knowledge is crucial for creating sustainable, resilient, and just urban transportation systems that cater to the needs of all individuals, ultimately enhancing the overall quality of life in modern cities.

REFERENCES

Aldred, Rachel, and Katrina Jungnickel. "Why culture matters for transport policy: The case of cycling in the UK." *Journal of Transport Geography* 34 (2014): 78–87.

Barbosa, Hugo, Surendra Hazarie, Brian Dickinson, Aleix Bassolas, Adam Frank, Henry Kautz, Adam Sadilek, José J. Ramasco, and Gourab Ghoshal. "Uncovering the socioeconomic facets of human mobility." *Scientific Reports* 11, no. 1 (2021): 8616.

Bohren, Lenora. "Car culture and decision-making: Choice and climate change." In *Anthropology and Climate Change*, pp. 370–379. Routledge, 2016.

Bullard, Robert D., Glenn S. Johnson, and Angel O. Torres. "*Sprawl Atlanta: Social Equity Dimensions of Uneven Growth and Development*." Clark Atlanta University, The Environmental Justice Resource Center, Atlanta, GA, USA, 1999.

Chatterjee, Alina, Chung Tang, Cidalia Torres, Debbie Douglas, Duberlis Ramos, Raymond Micah, and Uzma Shakir. "Alternative social planning: A paradigm shift developing an inclusive, healthy Toronto." *Planning* 32 (2004): 32.

Chatterjee, Kiron, and Fiona Crawford. "Changing work and work-related travel and the impact of COVID-19." *Built Environment* 47, no. 3 (2021): 336–354.

Chowdhury, Romit, and Colin McFarlane. "The crowd and citylife: Materiality, negotiation and inclusivity at Tokyo's train stations." *Urban Studies* 59, no. 7 (2022): 1353–1371.

Chowdhury, Subeh, Priya Patel, Varun Giridharan, and Vania Ceccato. "Formation of fear and adaptive behavior in young ethnic minority women riding public transport." *Transportation Research Record* 2678, no. 3 (2024): 687–697.

Cox, Peter, and Till Koglin. *The Politics of Cycling Infrastructure*. Policy, Bristol, 2020.

Dennis, Kingsley, and John Urry. *After the Car*. Polity, 2009.

Freudendal-Pedersen, Malene. "Cyclists as part of the city's organism: Structural stories on cycling in Copenhagen." *City & Society* 27, no. 1 (2015): 30–50.

Freudendal-Pedersen, Malene. *Making Mobilities Matter*. Routledge, 2022.

Freudendal-Pedersen, Malene, and Sven Kesselring. "Sharing mobilities. Some propaedeutic considerations." *Applied Mobilities* 3, no. 1 (2018): 1–7.

Freudendal-Pedersen, Malene, Malene Rudolf Lindberg, Katrine Hartmann-Petersen, and Toke Haunstrup Christensen. "Outgrowing the private car—Learnings from a mobility-as-a-service intervention in Greater Copenhagen." *Sustainability* 15, no. 17 (2023): 13187.

Gauvin, Laetitia, Michele Tizzoni, Simone Piaggesi, Andrew Young, Natalia Adler, Stefaan Verhulst, Leo Ferres, and Ciro Cattuto. "Gender gaps in urban mobility." *Humanities and Social Sciences Communications* 7, no. 1 (2020): 1–13.

Groth, Sören. "Multimodal divide: Reproduction of transport poverty in smart mobility trends." *Transportation Research Part A: Policy and Practice* 125 (2019): 56–71.

Guzman, Luis A., and Daniel Oviedo. "Accessibility, affordability and equity: Assessing 'pro-poor'public transport subsidies in Bogotá." *Transport Policy* 68 (2018): 37–51.

Hidayati, Isti, Wendy Tan, and Claudia Yamu. "How gender differences and perceptions of safety shape urban mobility in Southeast Asia." *Transportation Research Part F: Traffic Psychology and Behaviour* 73 (2020): 155–173.

Huang, Xiao, Yuhui Zhao, Siqin Wang, Xiao Li, Di Yang, Yu Feng, Yang Xu, Liao Zhu, and Biyu Chen. "Unfolding community homophily in US metropolitans via human mobility." *Cities* 129 (2022): 103929.

Inwood, Joshua FJ, Derek Alderman, and Jill Williams. ""Where do we go from here?" Transportation justice and the struggle for equal access." *Southeastern Geographer* 55, no. 4 (2015): 417–433.

Krog Juvik, Amanda, and Bente Halkier. "Pathways to more resource-intensive consumption through convenient bundles and complexes of food, mobility and housing practices." *Consumption and Society* 3, no. 1 (2024): 2–20.

Kuhnimhof, Tobias, Ralph Buehler, Matthias Wirtz, and Dominika Kalinowska. "Travel trends among young adults in Germany: Increasing multimodality and declining car use for men." *Journal of Transport Geography* 24 (2012): 443–450.

Li, Xiao, Xiao Huang, Dongying Li, and Yang Xu. "Aggravated social segregation during the COVID-19 pandemic: Evidence from crowdsourced mobility data in twelve most populated US metropolitan areas." *Sustainable Cities and Society* 81 (2022): 103869.

Lin, Dong, and Jianqiang Cui. "Transport and mobility needs for an ageing society from a policy perspective: Review and implications." *International Journal of Environmental Research and Public Health* 18, no. 22 (2021): 11802.

Ng, Wei-Shiuen, and Ashley Acker. "The gender dimension of the transport workforce." International Transport Forum Discussion Papers, No. 2020/11, OECD Publishing, Paris, 2020. https://doi.org/10.1787/0610184a-en

Nobis, Claudia. "Multimodality: Facets and causes of sustainable mobility behavior." *Transportation Research Record* 2010, no. 1 (2007): 35–44.

Parent, Michel. "New technologies for sustainable urban transportation in Europe." *Transportation Research Record* 1986, no. 1 (2006): 78–80.

Plyushteva, Anna, and Tim Schwanen. "Care-related journeys over the life course: Thinking mobility biographies with gender, care and the household." *Geoforum* 97 (2018): 131–141.

Porrazzo, Alexandra, Caroline Samson, and Malene Freudendal-Pedersen. "Gender and mobility planning: The influence of national culture on planning processes." *Case Studies on Transport Policy* 10, no. 2 (2022): 1102–1109.

Pucher, John, and John L. Renne. "Socioeconomics of urban travel: Evidence from the 2001 NHTS." *Transportation Quarterly* 57, no. 3 (2003): 49–77.

Rodríguez, Lucas, Javier Palanca, Elena del Val, and Miguel Rebollo. "Analyzing urban mobility paths based on users' activity in social networks." *Future Generation Computer Systems* 102 (2020): 333–346.

Ryghaug, Marianne, Ivana Subotički, Emilia Smeds, Timo von Wirth, Aline Scherrer, Chris Foulds, Rosie Robison et al. "A social sciences and humanities research agenda for transport and mobility in Europe: Key themes and 100 research questions." *Transport Reviews* 43, no. 4 (2023): 755–779.

Sheller, Mimi. *Mobility Justice: The Politics of Movement in an Age of Extremes.* Verso Books, 2018.

Smith, Noel, Donald Hirsch, and Abigail Davis. "Accessibility and capability: The minimum transport needs and costs of rural households." *Journal of Transport Geography* 21 (2012): 93–101.

Urry, John. "Societies." In *Sociology Beyond Societies: Mobilities for the Twenty-First Century.* Routledge.

Uteng, Tanu P. Gender and Mobility in the Developing World. World Development Report 2012, Gender Equality and Development. Background Paper, 2011. http://siteresources.worldbank.org.

Wang, Hai. "Transportation-enabled urban services: A brief discussion." *Multimodal Transportation* 1, no. 2 (2022): 100007.

Young, Deborah R., Angie L. Cradock, Amy A. Eyler, Mark Fenton, Margo Pedroso, James F. Sallis, and Laurie P. Whitsel; American Heart Association Advocacy Coordinating Committee. "Creating built environments that expand active transportation and active living across the United States: A policy statement from the American heart association." *Circulation* 142, no. 11 (2020): e167–e183.

5 Natural Disasters and Urban Human Mobility

Xiang Li and Yi Qiang

5.1 INTRODUCTION

Over half of the world's population now resides in urban areas, and this number is projected to grow to 68% by 2050 (United Nations 2018). Urban areas, characterized by high population density and extensive mobility because of well-developed transportation networks and infrastructure, play a critical role in shaping daily human activity. However, as cities continue to expand, the associated risks of climate change and natural disasters become more pronounced. In recent decades, extreme weather events have caused large-scale population movements and widespread socioeconomic effects. For instance, Hurricane Katrina in 2005 resulted in the evacuation of over a million New Orleans residents, many of whom did not return, leading to lasting demographic shifts and profound impacts on urban development (Groen and Polivka 2010; Fussell, Curtis, and DeWaard 2014; Qiang and Lam 2016). In 2017, the Atlantic hurricane season experienced the "terrible trio" of Hurricanes Harvey, Irma, and Maria, causing millions of people to evacuate, making it the costliest season since 1851 in the United States, with economic losses reaching nearly $370 million (Halverson 2018). More recently, Hurricane Ian in 2022 struck Florida, causing hundreds of deaths and injuries, catastrophic property damage, and an economic loss exceeding $112 billion, while 2.5 million people were ordered to evacuate (Smith 2024). These extreme weather events highlight the critical role of human mobility in disaster scenarios, and understanding these mobility patterns is essential for improving disaster preparedness and enhancing postdisaster recovery efforts.

Research on urban human mobility in disasters is significant for prioritizing and optimizing disaster response strategies and promoting resilience. By analyzing population movements during disasters, researchers and policymakers can gain valuable insights into the socioeconomic and demographic factors that influence human mobility such as evacuation. One critical aspect of this is understanding compliance with evacuation orders, as failure to comply can significantly increase the risks of casualties and injuries (Apte 2010). Various factors, such as risk perception, trust in authorities, and access to resources, can affect order compliance, leading to disparities in evacuation behaviors across different socioeconomic groups (Roy et al. 2021; Karaye et al. 2022). This understanding is crucial for optimizing resource allocation, developing more effective evacuation orders, and ensuring that vulnerable populations are adequately supported during emergencies. Therefore, a comprehensive understanding of how different population groups respond to disasters allows for the development of more equitable and targeted disaster management policies.

Urban human mobility research during disasters is researched via a wide range of data sources and methodologies. The new mobility data sources, such as mobile phone data and social media data, can provide near real-time insights into population movements during disasters. These datasets are particularly useful for capturing the spatial and temporal patterns of evacuation behaviors. Beyond that, the emergence of geospatial artificial intelligence (GeoAI), which leverages advanced machine learning methods and multisource geospatial data to uncover patterns and trends in population movement, offers new opportunities and has become an increasingly powerful tool in human mobility research (Janowicz et al. 2020). By integrating data from diverse sources such as satellite imagery, mobility data, and survey data, GeoAI enables the analysis of mobility dynamics at a finer spatial and temporal resolution, which offers relatively accurate and rapid insights into

DOI: 10.1201/9781003503262-7

evacuation patterns or other disaster responses (Wang et al. 2024). Furthermore, GeoAI's ability to model complex relationships between socioeconomic factors and human mobility enhances our understanding of how socioeconomic and demographic characteristics influence movement decisions in disasters.

This chapter introduces an analytical workflow of utilizing human mobility data to analyze the evacuation pattern during a major disaster. First, it examines how different types of data can be utilized to capture and measure mobility patterns during or after disasters. Second, the chapter presents a case study to empirically assess the short-term impacts of a specific disaster on human mobility. Third, this chapter addresses the key challenges and highlights some potential future directions in the field. The content from this chapter demonstrates the value of mobility data in disaster response and resilience research. The analysis results contribute to a deeper understanding of the mechanism and inequalities of urban human movement in disasters, providing insights that can enhance disaster management and resilience planning.

5.2 MOBILITY DATA FOR DISASTER RESEARCH

Understanding human mobility in urban disaster contexts is crucial for effective disaster response and management. Over time, the research of human mobility has evolved with advancements in data collection and modeling techniques, enabling researchers to capture increasingly relatively accurate, near real-time, and complex information.

Traditionally, researchers often used surveys to gather data related to the evacuation decision-making of residents at disaster risk. These methods allow researchers to obtain first-hand information of mobility decisions and the personal factors influencing population movements during disasters. For example, Adeola (2009) employed paper-and-pencil surveys to explore the relationship between the duration of residency as well as prior experiences and evacuation in Hurricane Katrina. The results indicate that the duration of residency is positively affected by evacuation, while prior experiences are negatively associated with evacuation. Brown et al. (2016) discussed factors that may have influenced evacuation behavior and evacuation times in Hurricane Sandy based on in-person surveys. In addition, other types of surveys, including mail surveys (Lindell, Lu, and Carla 2005; Brackenridge et al. 2012; Zhang, Prater, and Lindell 2004), and mixed surveys (McClure et al. 2011) are frequently applied in postdisaster investigations.

While survey data remain essential for mobility analysis, the advent of big data technologies has revolutionized human mobility research, providing real-time or near real-time, high-resolution insights for disaster management and disaster resilience research (Yu, Yang, and Li 2018). Among these new data sources, mobile phone tracking data have emerged as a key data source for studying population movement at fine spatial and temporal resolutions. Given the widespread use of global positioning system (GPS)-enabled mobile devices and location-based services, human movements in geographical space create a wealth of digital traces in cyberspace, resulting in a variety of human mobility data. By analyzing these mobility data, researchers can monitor large-scale human mobility patterns (Yabe et al. 2022), detect traffic bottlenecks in road networks (Wang et al. 2024), and assess the effectiveness of evacuation orders during disasters (Yabe, Ukkusuri, and Rao 2019). During COVID-19, mobile phone data helped researchers and scientists to predict the spread of COVID-19 (Alessandretti 2022), evaluate the effectiveness and adherence to intervention policies (Levin et al. 2021), and analyze impacts on socioeconomic activities (de Palma, Vosough, and Liao 2022). Besides mobile phone GPS data, vehicles equipped with GPS data also add a precise geographic dimension to urban mobility research. It not only captures the direction and distance of population movements but also reveals specific routes taken during the disasters. This is particularly valuable in dense urban environments where transportation infrastructure plays a critical role in shaping mobility (Gong et al. 2023).

Another valuable source of information is remote sensing data, which uses satellite and aerial imagery to monitor changes in human activity and infrastructure in disaster-affected urban areas.

Remote sensing can directly detect the disaster extent, the building damages, power outages, and population movements, all of which help us understand how urban populations respond to and recover from disasters (Krichen et al. 2024). For example, Qiang, Huang, and Xu (2020) utilized night-time light (NTL) remote sensing images to explore the great economic impact of Katrina and the spatial variation of the disturbance and recovery pattern of economic activity. Ghaffarian et al. (2021) utilized remote sensing data to monitor and evaluate the recovery process after Typhoon Haiyan. Derdouri and Osaragi (2021) used geotagged photos shared online to examine the disaster-induced human movement patterns in Tokyo between 2008 and 2019.

Third, geotagged social media data are another common data source for studying human dynamics in disasters. Platforms such as Twitter, Facebook, Weibo, and Instagram have been increasingly used to analyze movement patterns during disasters, offering a real-time lens on population behavior. Wang and Taylor (2016) utilized Twitter data to examine the influence of natural disasters on human mobility patterns in urban populations. Martín, Li, and Cutter (2017) developed the approach to utilize geotagged Twitter data to assess evacuation responses and examine the spatiotemporal variability in Hurricane Matthew. Zou et al. (2018) analyzed the spatiotemporal patterns of Twitter data to understand and assess community resilience. Metaxa-Kakavouli, Maas, and Aldrich (2018) employed Facebook data to explore the influence of social ties on hurricane evacuation behaviors.

The development of data sources in urban mobility research reflects broader technological advancements and the increasing complexity of modern cities. From the foundational survey data of the past to today's high-resolution, dynamic data sources, the study of human mobility has become more detailed and actionable. By integrating these diverse datasets, researchers and policymakers can better understand the complex interactions of urban populations in disaster scenarios, leading to more effective and responsive disaster management strategies.

5.3 CASE STUDY: DETECTING EVACUATION PATTERNS IN HURRICANE IAN

This section demonstrates the use of human mobility data to detect the spatiotemporal patterns of population evacuation during Hurricane Ian in 2022. The study relied on SafeGraph Weekly Pattern data, which tracks anonymized visits to points of interest (POIs) and links them to home locations, allowing for an analysis of population movements over time. This dataset provides high-resolution, real-time mobility data that were crucial for detecting shifts in human mobility during the hurricane. The study focused on three key objectives: (1) understanding changes in population flows during different phases of the hurricane, (2) identifying evacuation routes and destinations, and (3) assessing geographic variation in compliance with evacuation orders.

5.3.1 Analytical Workflow

The analysis of evacuation patterns during Hurricane Ian employs a structured analytical workflow designed to capture and interpret large-scale human mobility data. The workflow begins by defining the study period, which is divided into three distinct phases: the pre-hurricane phases (3 weeks before the landfall), in-hurricane (the week right after the landfall), and post-hurricane (the following week). During these periods, population movements are analyzed both within Florida and beyond, with a particular focus on the areas most affected by the hurricane. The specific framework utilized for this study is depicted in Figure 5.1.

5.3.1.1 Data Preprocessing

The study primarily utilizes the SafeGraph Weekly Pattern dataset to track human mobility patterns. This dataset records visits to specific POIs, such as commercial areas, parks, and residences, providing a detailed view of population movement during the disaster. The origin-destination (O-D) flows were created by linking the zip codes of POIs (destinations) and the block groups of visitors'

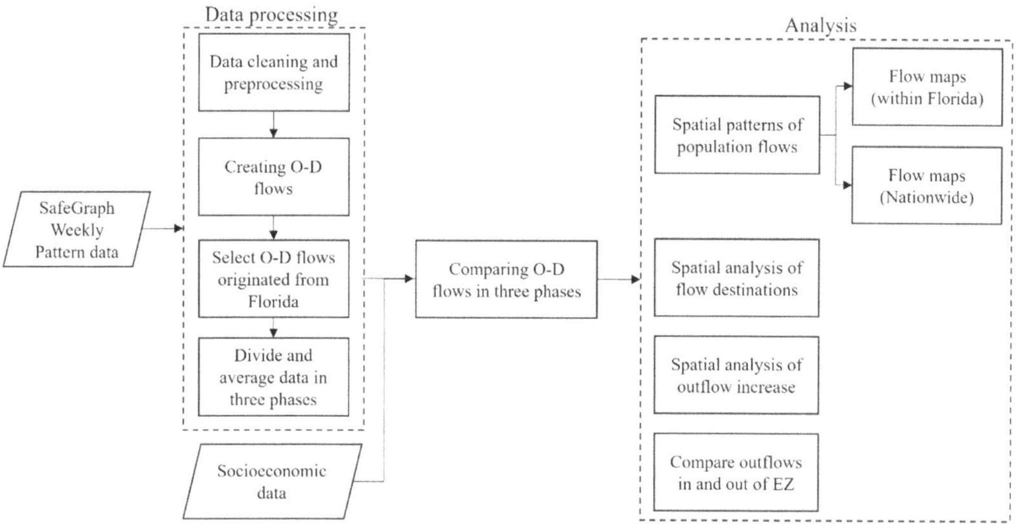

FIGURE 5.1 The analytical workflow. *Abbreviations*: EZ, evacuation zone; O-D, origin-destination. (From Li, Qiang, and Cervone 2024.)

origins (origins). To focus on intercity evacuations rather than short-distance travels like shopping or daily commuting, we only analyzed O-D flows with a distance exceeding 100 km. Then, we aggregated the destinations of the O-D flows into block groups and counties to be associated with socioeconomic data. As the visitor origins are aggregated in weekly intervals, the O-D flows represent population flows that occurred in different weeks. To detect the mobility changes caused by the hurricane, we selected the O-D flows that have an origin from Florida. The O-D flows are compared in the pre-, in-, and post-hurricane phases defined above to detect mobility changes in the different phases of the hurricane.

5.3.1.2 Spatial Analysis

Population movements are analyzed over the three phases to identify significant shifts in mobility. First, we created flow maps to visualize net population flows in O-D flows during the three phases of the hurricane. To calculate net population flows based on O-D flows, we established the total inflows and outflows between each O-D flow. Net population flow is determined by subtracting the total number of outflows from inflows for each destination. In addition, we mapped the net population flows for each of the three phases and displayed the change ratios in population flows during the in- and post-hurricane phases, as compared to the pre-hurricane phase (the baseline). Second, hot spot analysis was conducted to detect the popular destinations of population outflows from the affected areas. The study employs the local Getis-Ord Gi* statistics (Getis and Ord 1992) to detect local clusters of visitor change ratios. The Getis-Ord Gi* statistics conduct a statistical test at each spatial feature to assess whether values in the surrounding features are in complete spatial randomness (the null hypothesis). A high value surrounded by high values is a hot spot, and conversely, a low value surrounded by low values is a cold spot. The z-scores and p values indicate the statistical significance of the hot and cold spots.

5.3.2 Key Findings

The study revealed distinct spatial patterns of evacuation during Hurricane Ian. Compared to the baseline (pre-hurricane phase), population outflows increased significantly from Florida's west coast, particularly the Tampa Bay and Fort Myers areas, which were in the direct path of the hurricane.

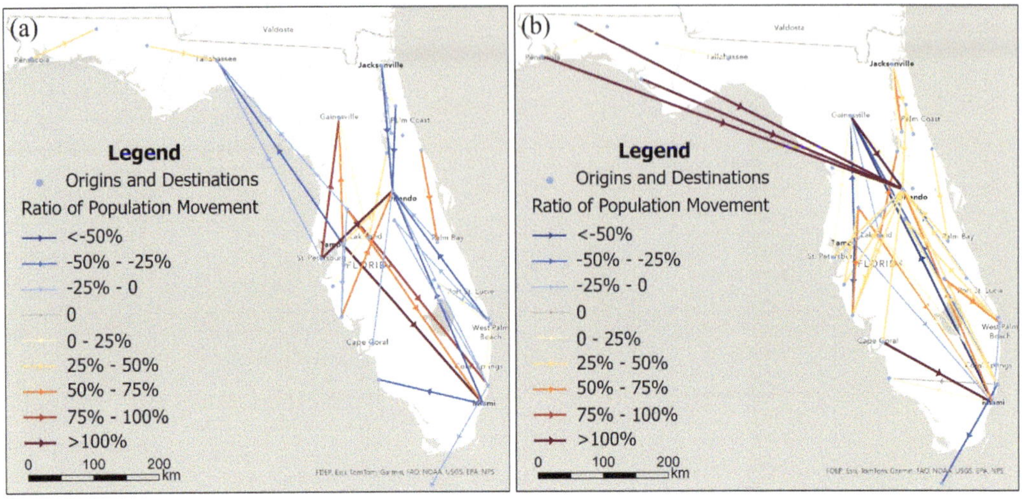

FIGURE 5.2 Percent changes of net population flows during the in-hurricane (a) and post-hurricane (b) phases in comparison to the baseline. The arrows represent the directions of the net outflows. (From Li, Qiang, and Cervone 2024.)

The number of populations traveling from the east coast to Orlando decreased significantly compared to the baseline condition (Figure 5.2a). Population flows from Tampa Bay to Tallahassee also decreased during this phase. These evacuees predominantly traveled eastward and northward, toward cities on Florida's east coast (such as Miami, Palm Beach, and Jacksonville). The post-hurricane phase shows a widespread increase in population movements in most areas (Figure 5.2b), particularly in the routes from the panhandle area and Gainesville to Orlando and from Cape Coral to Miami. This increase may be attributed to the release of pent-up travel demands and the returning evacuees after the hurricane. This evacuation pattern is consistent with media reports, which highlighted a mass exodus from the Gulf Coast toward safer, less affected regions.

In addition to analyzing net population flows within Florida, the study explored the popular destinations via hot spot analysis (Figure 5.3). Using hot spot analysis, we detected local clusters of high (hot spots) and low (cold spots) percent changes. Most travelers chose to move within Florida during the hurricane. In addition to counties within Florida, hot spots of Florida visitors tend to spread toward the northwest neighbor states including Alabama, Georgia, Mississippi, and even as far as Louisiana and Texas. However, a few long-distance outflows were also detected, with evacuees traveling to states as far as Montana and North Dakota. These long-distance travels suggest that some residents may have had pre-existing socioeconomic ties in these areas, such as friends, family, or accessible shelter options.

Compliance with evacuation orders varied across counties and regions, reflecting the complexity of human decision-making during disasters. The study analyzed population outflows from the 12 Florida counties under mandatory evacuation orders, revealing significant geographic variation (Figure 5.4). Lee County, which was hit hardest by the hurricane, exhibited a 46.49% increase in population outflow during the in-hurricane phase, indicating high compliance with the evacuation order. Other counties, such as Pinellas County (43.01%) and Hillsborough County (37.80%), also experienced substantial outflows as residents moved to safer areas. These counties had been forecasted as potential landfall sites early in the storm's development, likely prompting early evacuations. However, compliance was lower in counties farther from the storm's projected path, such as Levy County and St. Johns County, where outflows were lower, even within designated evacuation zones. This suggests that the timing and precision of evacuation orders are critical factors in influencing public response.

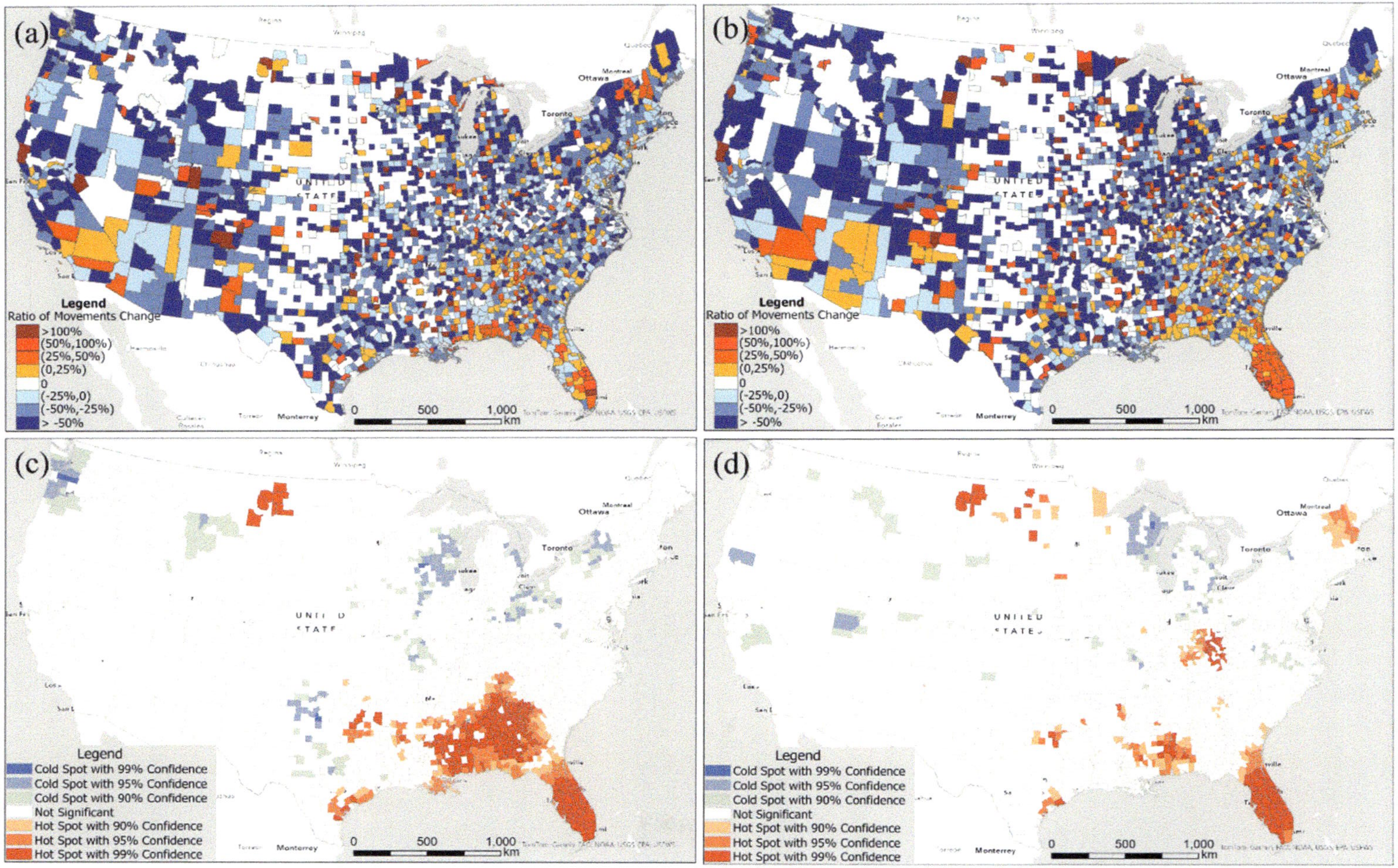

FIGURE 5.3 Percent changes of visitors from Florida in contrast to the baseline condition. (a) Percent changes of Florida visitors during the in-hurricane phase. (b) Percent changes of Florida visitors during the post-hurricane phase. (c) Hot spot analysis of (a). (d) Hot spot analysis of (b). (From Li, Qiang, and Cervone 2024.)

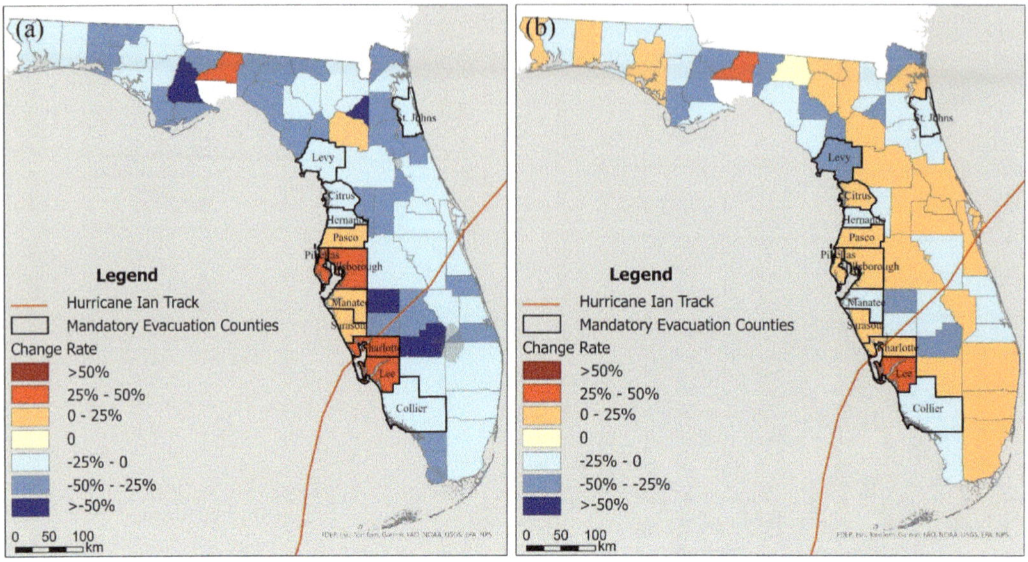

FIGURE 5.4 Percent changes of population outflows in Florida counties during the (a) in-hurricane and (b) post-hurricane phases. (From Li, Qiang, and Cervone 2024.)

In the post-hurricane phase, population outflows generally return to the baseline level. However, Lee County, where the hurricane made landfall, sustained a substantial rise (43.84%) in population outflow, indicating continuous evacuation from the hardest-hit area. Lee County issued its evacuation order just 27 hours before landfall, which may have contributed to delayed evacuations and increased casualties. The study's analysis highlights the need for timely and clear communication of evacuation risks to ensure that residents take appropriate action.

5.4 DISCUSSION

In the case study, we employed a human mobility dataset to analyze the spatiotemporal changes of population flows in different phases of Hurricane Ian. The main objective of the study was to evaluate the utility of the emerging mobility data in detecting evacuation patterns in natural hazards. The analyses show that, compared to the pre-hurricane phase (the baseline condition), the population movements changed during the in- and post-hurricane phases. During the in-hurricane phase, eastbound travels from the Gulf Coast in Florida toward central Florida and the Atlantic Coast increased. After the hurricane, the movement pattern largely resumed to the baseline, except for continued outflow from the most affected areas. The detected patterns of population movements are in line with the evacuation routes in our expectation or media reports, which confirms the effectiveness of the mobility dataset in detecting evacuation behaviors during the hurricane.

5.4.1 Evacuation Patterns and Effect of Evacuation Order

This study reveals several evacuation patterns that have not been documented in other studies or media reports. The destinations of the population outflow display a distance decay pattern, with most travels concentrated within Florida as well as a prominent overflow of Florida travelers to neighboring states during the hurricane. Our analysis reveals that population outflow from Florida avoided the East Coast and instead moved toward neighboring states to the north and west (e.g., Alabama and Mississippi).

The population outflows from Lee County, which was directly hit by the hurricane, continued into the post-hurricane phase, even a week after the landfall. The distant hot spots of increased Florida visitors in Texas, South Dakota, and Idaho reveal strong connections between Florida and these areas. The effect of mandatory evacuation orders is evident in the increased population outflows from the affected counties as well as the greater outflow increase within the designated evacuation zones compared to the outside areas. These spatiotemporal patterns detected in the analyses demonstrate the potential of human mobility data in monitoring large-scale evacuation patterns in natural hazards, which effectively overcomes the limitations of traditional data sources (e.g., questionnaires and surveys).

5.4.2 Challenges and Future Directions

Despite the valuable insights gained from human mobility data in disaster contexts, several challenges remain that need to be addressed to enhance the accuracy and applicability of such data for disaster response and resilience.

One of the primary challenges lies in balancing data accessibility and data privacy. While mobile phone, GPS, or social media data can offer near real-time information on population movements, the collection and use of such data often raise concerns about privacy and the potential misuse of personal information. In many cases, data are anonymized to protect individual identities, but even anonymized data can sometimes be re-identified or misused if not properly safeguarded (Kondor et al. 2018). Ensuring strict data privacy measures while maintaining high-quality access requires advanced techniques or possibly new legislative frameworks to protect individuals while enabling researchers to use data effectively for disaster response (Rao et al. 2020).

Another key challenge is data representativeness. Mobility data are often derived from smartphone users or social media activity, which may underrepresent certain demographic groups, such as the elderly, low-income populations, or those without access to digital technology (Lin et al. 2023). This can lead to biases in urban human mobility analysis, overlooking the unique challenges faced by vulnerable communities during disasters. In terms of future research directions, there is a strong need for improving data fusion techniques to combine multiple types of mobility data for comprehensive research.

In addition, while mobility data can capture movement patterns, they may not explain the reasons behind these movements. For instance, individuals may evacuate due to mandatory orders or socioeconomic pressures, and these motivations can vary widely across different population groups. A feasible approach is to combine multisource mobility data to explore human mobility in a more comprehensive and in-depth manner. For example, Huang et al. (2021) utilized four different datasets to study the mobility patterns of different income groups during the COVID-19 pandemic, revealing the mobility disparities across income levels during the crisis. The integration of multiple data sources can provide different dimensions of mobility information, helping to overcome the biases that may arise from relying on a single data source.

5.5 CONCLUSION

This chapter explored the critical role of human mobility data in disaster response, particularly through one case study of evacuation in Hurricane Ian. It presents an analytical framework for utilizing mobility data to detect evacuation patterns during natural disasters. The study provides a detailed analysis of population movements across different phases of the disaster, highlighting spatial variations in evacuation behavior and compliance with evacuation orders. The study reveals notable changes in population movements, both within Florida and nationwide, which are potentially linked to the hurricane-induced population evacuation. Furthermore, the increase in population outflows from the impacted areas suggests the effectiveness of mandatory evacuation orders. A more pronounced increase in outflows from designated mandatory evacuation areas points to the public awareness of the evacuation zone designation. However, several challenges remain in the practical application of mobility data in disaster contexts, including balancing data privacy and data accessibility, data representativeness, and the integration of multiple data sources. To address these challenges, future research should focus on refining data fusion techniques, enhancing the inclusivity of data collection, and employing advanced analytical frameworks to better understand and predict human mobility in response to disasters.

REFERENCES

Adeola, Francis O. 2009. "Katrina cataclysm: Does duration of residency and prior experience affect impacts, evacuation, and adaptation behavior among survivors?" *Environment and Behavior* 41 (4): 459–489. https://doi.org/10.1177/0013916508316651

Alessandretti, Laura. 2022. "What human mobility data tell us about COVID-19 spread." *Nature Reviews Physics* 4 (1): 12–13. https://doi.org/10.1038/s42254-021-00407-1

Apte, Aruna. 2010. "Humanitarian logistics: A new field of research and action." *Foundations and Trends® in Technology, Information and Operations Management* 3 (1): 1–100.

Brackenridge, Sandra, Lisa K. Zottarelli, Erin Rider, and Bev Carlsen-Landy. 2012. "Dimensions of the human–animal bond and evacuation decisions among pet owners during Hurricane Ike." *Anthrozoös* 25 (2): 229–238.

Brown, Shakara, Hilary Parton, Cynthia Driver, and Christina Norman. 2016. "Evacuation during hurricane Sandy: Data from a rapid community assessment." *PLOS Currents* 8. https://doi.org/10.1371/currents.dis.692664b92af52a3b506483b8550d6368

de Palma, André, Shaghayegh Vosough, and Feixiong Liao. 2022. "An overview of effects of COVID-19 on mobility and lifestyle: 18 months since the outbreak." *Transportation Research Part A: Policy and Practice* 159: 372–397.

Derdouri, Ahmed, and Toshihiro Osaragi. 2021. "How Various Natural Disasters Impact Urban Human Mobility Patterns: A Comparative Analysis Based on Geotagged Photos Taken in Tokyo." In Urban Informatics and Future Cities, edited by S. C. M. Geertman, Christopher Pettit, Robert Goodspeed, Aija Staffans, 151–171. Cham: Springer International Publishing.

Fussell, Elizabeth, Katherine J. Curtis, and Jack DeWaard. 2014. "Recovery migration to the City of New Orleans after Hurricane Katrina: A migration systems approach." *Population and Environment* 35: 305–322.

Getis, Arthur, and J. Keith Ord. 1992. "The analysis of spatial association by use of distance statistics." *Geographical Analysis* 24 (3): 189–206.

Ghaffarian, Saman, Debraj Roy, Tatiana Filatova, and Norman Kerle. 2021. "Agent-based modelling of post-disaster recovery with remote sensing data." *International Journal of Disaster Risk Reduction* 60: 102285. https://doi.org/10.1016/j.ijdrr.2021.102285

Gong, Shuhui, Xiangrui Dong, Kaiqi Wang, Bingli Lei, Zizhao Jia, Jiaxin Qin, Chris Roadknight, Yu Liu, and Rui Cao. 2023. "Agent-based modelling with geographically weighted calibration for intra-urban activities simulation using taxi GPS trajectories." *International Journal of Applied Earth Observation and Geoinformation* 122: 103368. https://doi.org/10.1016/j.jag.2023.103368

Groen, Jeffrey A., and Anne E. Polivka. 2010. "Going home after Hurricane Katrina: Determinants of return migration and changes in affected areas." *Demography* 47: 821–844.

Halverson, Jeffrey B. 2018. "The Costliest Hurricane Season in U.S. History." *Weatherwise* 71 (2): 20–27. https://doi.org/10.1080/00431672.2018.1416862

Huang, Xiao, Zhenlong Li, Yuqin Jiang, Xinyue Ye, Chengbin Deng, Jiajia Zhang, and Xiaoming Li. 2021. "The characteristics of multi-source mobility datasets and how they reveal the luxury nature of social distancing in the U.S. during the COVID-19 pandemic." *International Journal of Digital Earth* 14 (4): 424–442. https://doi.org/10.1080/17538947.2021.1886358

Janowicz, Krzysztof, Song Gao, Grant McKenzie, Yingjie Hu, and Budhendra Bhaduri. 2020. GeoAI: Spatially Explicit Artificial Intelligence Techniques for Geographic Knowledge Discovery and Beyond. Taylor & Francis.

Karaye, Ibraheem M, Nicholas Taylor, Maria Perez-Patron, Courtney Thompson, and Jennifer A. Horney. 2022. "Factors associated with hurricane evacuation: A statistical meta-analysis of studies,

1999–2018." *Disaster Medicine and Public Health Preparedness* 16 (3): 1064–1072. https://www.cambridge.org/core/journals/disaster-medicine-and-public-health-preparedness/article/abs/factors-associated-with-hurricane-evacuation-a-statistical-metaanalysis-of-studies-19992018/BDB4F8F767CB0A8DD9D1ED0DACFCBEB2

Kondor, Daniel, Behrooz Hashemian, Yves-Alexandre de Montjoye, and Carlo Ratti. 2018. "Towards matching user mobility traces in large-scale datasets." *IEEE Transactions on Big Data* 6 (4): 714–726.

Krichen, Moez, Mohamed S. Abdalzaher, Mohamed Elwekeil, and Mostafa M. Fouda. 2024. "Managing natural disasters: An analysis of technological advancements, opportunities, and challenges." *Internet of Things and Cyber-Physical Systems* 4: 99–109. https://doi.org/10.1016/j.iotcps.2023.09.002

Levin, Roman, Dennis L. Chao, Edward A. Wenger, and Joshua L. Proctor. 2021. "Insights into population behavior during the COVID-19 pandemic from cell phone mobility data and manifold learning." *Nature Computational Science* 1 (9): 588–597. https://doi.org/10.1038/s43588-021-00125-9

Li, Xiang, Yi Qiang, and Guido Cervone. 2024. "Using human mobility data to detect evacuation patterns in hurricane Ian." *Annals of GIS*: 30 (4): 493–511.

Lin, Binbin, Lei Zou, Heng Cai, Mingzheng Yang, and Bing Zhou. 2023. "Sensing the pulse of the pandemic: Geovisualizing the Demographic Disparities of Public Sentiment toward COVID-19 through Social Media." *arXiv preprint arXiv:2304.06120.*

Lindell, Michael K., Jing-Chein Lu, and S. Prater Carla. 2005. "Household decision making and evacuation in response to Hurricane Lili." *Natural Hazards Review* 6 (4): 171–179. https://doi.org/10.1061/(ASCE)1527-6988(2005)6:4(171)

Martín, Yago, Zhenlong Li, and Susan L. Cutter. 2017. "Leveraging twitter to gauge evacuation compliance: Spatiotemporal analysis of Hurricane Matthew." *PLoS One* 12 (7): e0181701. https://doi.org/10.1371/journal.pone.0181701

McClure, Laura A., Michael L. Boninger, Michelle L. Oyster, Mary Joan Roach, Jennifer Nagy, and Gregory Nemunaitis. 2011. "Emergency evacuation readiness of full-time wheelchair users with spinal cord injury." *Archives of Physical Medicine and Rehabilitation* 92 (3): 491–498.

Metaxa-Kakavouli, Danaë, Paige Maas, and Daniel P. Aldrich. 2018. "How social ties influence hurricane evacuation behavior." *Proceedings of the ACM on Human-Computer Interaction* 2 (CSCW): Article 122. https://doi.org/10.1145/3274391

Qiang, Yi, and Nina S.-N. Lam. 2016. "The impact of Hurricane Katrina on urban growth in Louisiana: An analysis using data mining and simulation approaches." *International Journal of Geographical Information Science* 30 (9): 1832–1852.

Qiang, Yi, Qingxu Huang, and Jinwen Xu. 2020. "Observing community resilience from space: Using nighttime lights to model economic disturbance and recovery pattern in natural disaster." *Sustainable Cities and Society* 57: 102115. https://doi.org/10.1016/j.scs.2020.102115

Rao, Jinmeng, Song Gao, Yuhao Kang, and Qunying Huang. 2020. "LSTM-TrajGAN: A deep learning approach to trajectory privacy protection." *arXiv preprint arXiv:2006.10521.*

Roy, Kamol Chandra, Samiul Hasan, Aron Culotta, and Naveen Eluru. 2021. "Predicting traffic demand during hurricane evacuation using Real-time data from transportation systems and social media." *Transportation Research Part C: Emerging Technologies* 131: 103339. https://doi.org/10.1016/j.trc.2021.103339

Smith, Adam B. 2024. "2023 US Billion-dollar weather and climate disasters in historical context." 104th AMS Annual Meeting.

United Nations. 2018. Revision of World Urbanization Prospects. New York, NY.

Wang, Qi, and John E. Taylor. 2016. "Patterns and limitations of urban human mobility resilience under the influence of multiple types of natural disaster." *PLoS One* 11 (1): e0147299.

Wang, Siqin, Xiao Huang, Pengyuan Liu, Mengxi Zhang, Filip Biljecki, Tao Hu, Xiaokang Fu, Lingbo Liu, Xintao Liu, Ruomei Wang, Yuanyuan Huang, Jingjing Yan, Jinghan Jiang, Michaelmary Chukwu, Seyed Reza Naghedi, Moein Hemmati, Yaxiong Shao, Nan Jia, Zhiyang Xiao, Tian Tian, Yaxin Hu, Lixiaona Yu, Winston Yap, Edgardo Macatulad, Zhuo Chen, Yunhe Cui, Koichi Ito, Mengbi Ye, Zicheng Fan, Binyu Lei, and Shuming Bao. 2024. "Mapping the landscape and roadmap of geospatial artificial intelligence (GeoAI) in quantitative human geography: An extensive systematic review." *International Journal of Applied Earth Observation and Geoinformation* 128: 103734. https://doi.org/10.1016/j.jag.2024.103734

Yabe, Takahiro, Nicholas K. W. Jones, P. Suresh C. Rao, Marta C. Gonzalez, and Satish V. Ukkusuri. 2022. "Mobile phone location data for disasters: A review from natural hazards and epidemics." *Computers, Environment and Urban Systems* 94: 101777.

Yabe, Takahiro, Satish V. Ukkusuri, and P. Suresh C. Rao. 2019. "Mobile phone data reveals the importance of pre-disaster inter-city social ties for recovery after Hurricane Maria." *Applied Network Science* 4 (1): 98. https://doi.org/10.1007/s41109-019-0221-5

Yu, Manzhu, Chaowei Yang, and Yun Li. 2018. "Big data in natural disaster management: A review." *Geosciences* 8 (5): 165.

Zhang, Yang, Carla S. Prater, and Michael K. Lindell. 2004. "Risk area accuracy and evacuation from Hurricane Bret." *Natural Hazards Review* 5 (3): 115–120.

Zou, Lei, Nina S. N. Lam, Heng Cai, and Yi Qiang. 2018. "Mining twitter data for improved understanding of disaster resilience." *Annals of the American Association of Geographers* 108 (5): 1422–1441. https://doi.org/10.1080/24694452.2017.1421897

6 Economics of Urban Mobility

Qianwen Guo, Xiao Huang, and Raj Ponnaluri

6.1 INTRODUCTION

Urban mobility is the lifeblood of modern cities, directly influencing their economic dynamics and shaping the daily lives of residents. As cities grow and develop, the efficient movement of people and goods becomes increasingly critical to economic prosperity, social mobility, and environmental sustainability. The challenges associated with urban mobility are multifaceted, involving complex interactions between demand and supply dynamics, pricing mechanisms, and infrastructure development. Addressing these challenges requires a deep understanding of the underlying factors driving urban transportation systems, the economic implications of mobility decisions, and the policies that can foster efficient, equitable, and sustainable mobility solutions.

This chapter explores various facets of urban mobility, focusing on the economic analysis of transportation systems, the factors shaping demand and supply, and the strategies that can help manage and optimize urban transportation networks. We begin by examining the fundamental relationship between urban mobility and economic development, highlighting the role of transportation systems in shaping labor markets, business competitiveness, and land values. Subsequently, we explore the dynamics of demand and supply in urban mobility, discussing how population density, income levels, urbanization, and land use patterns influence transportation demand, and how the provision of infrastructure and services meets the growing needs of urban populations. Finally, we delve into the pricing strategies used to manage congestion and promote equitable access to mobility, with a particular focus on congestion pricing and innovative approaches such as free fare transit.

6.2 URBAN MOBILITY AND ECONOMIC ANALYSIS

Urban areas thrive on the efficient movement of people and goods. Transportation systems directly influence labor markets, business competitiveness, and land value and urban development. Investments in urban mobility are often driven by their potential to yield substantial economic benefits, which, in turn, reshape the urban economic landscape. Residential household location serves as a critical driver of urban dynamics, influencing employment patterns, economic growth, social structures, spatial segregation, and transportation systems. Consequently, understanding and modeling residential location choice behavior are a key focus for urban planners, policymakers, and researchers.

The origins of residential location modeling can be traced back to von Thünen's (1826) pioneering work in land use modeling, where he examined how transport costs affect activity locations and the land market within an agricultural region centered on a single market. A foundational concept in this field is the bid-rent theory, which posits that landowners lease their properties to the highest bidder. Alonso (1964) extended this theory to urban contexts, focusing on residential locations within a monocentric city. He proposed that individuals and households select residential locations to maximize utility, which is influenced by expenditure on goods, land lot size, and proximity to the city center. Concurrently, Lowry (1964) introduced the gravity model to residential location analysis, assuming predefined employment centers in each zone. In his framework, households are distributed across zones based on a deterrence function that accounts for employment opportunities and residential attractiveness, with the latter determined by the availability of land for development in each zone.

DOI: 10.1201/9781003503262-8

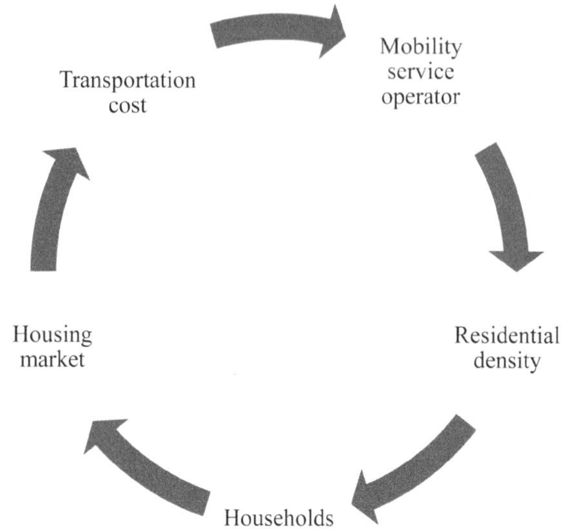

FIGURE 6.1 The interaction between mobility service operator and households.

However, many studies assumed that the distribution of residential areas within the planning region was exogenously fixed (Beckmann, 1969; Beckmann, 1974; Quigley, 1984; Anas & Xu; 1999; Brueckner, 2007; Li et al., 2012; Li et al., 2013). This approach overlooked the impact of transportation infrastructure design on residential location choices and the housing market. Empirical studies by McDonald and Osuji (1995) and Bowes and Ihlanfeldt (2001) revealed that introducing new transportation lines enhances accessibility, driving changes in urban land use, property values, and the housing market, including housing prices and space availability.

Figure 6.1 illustrates the interconnected relationship between mobility services and household residential location choices. Mobility services influence travel costs and trip accessibility, affecting housing rental prices and, ultimately, residential location decisions. Conversely, residential location choices shape residential density, influencing rail passenger demand, which in turn affects mobility service operations. This bidirectional relationship has spurred research incorporating the effects of mobility services on residential location choices, property values, and the housing market (Li et al., 2015; Li & Peng, 2016; Li and Guo, 2017).

6.3 DEMAND AND SUPPLY DYNAMICS IN URBAN MOBILITY

6.3.1 Demand for Urban Mobility

Urban mobility demand is shaped by population density, urbanization rates, income levels, and land use patterns. Effective demand management, encompassing sustainability, planning, and implementation, plays a vital role in enhancing service quality and environmental sustainability. A well-designed transportation system that aligns with societal needs and aspirations can drive long-term success and sustainability for communities and enterprises alike.

Existing literature highlights the challenges of transportation demand management (TDM) in regulating the mobility of people and goods within metropolitan areas (Hendricks and Georggi, 2007). However, critical yet underexplored factors contribute to rising transport demand, with transport congestion being a major symptom of urban transportation network overcapacity (Bylinko, 2020). Economic growth further exacerbates transportation demand in both advanced and emerging economies. Despite substantial investments in infrastructure, demand often outpaces capacity, creating an imbalance that underscores the importance of TDM.

TDM has emerged as a crucial approach to addressing urban transportation challenges, mitigating adverse effects such as air and noise pollution, traffic congestion, and accidents. By reallocating or reducing travel demand across time and space, TDM fosters a more efficient and sustainable urban mobility system (Bjorefeldt and Grunden, 2019). Treating transportation as a commodity with unique supply and demand characteristics, TDM shifts focus from expanding infrastructure to optimizing utilization.

Transport infrastructure managers play a pivotal role by overseeing road networks, transportation services, and vehicular traffic regulations. Their strategies often aim to limit unnecessary private car use while promoting efficient, eco-friendly public transportation.

Studies indicate that TDM strategies are not only cost-effective but also viable alternatives to car-centric infrastructure development (Litman, 2003). By encouraging active modes of travel and public transport, TDM seeks to shift current transportation behaviors, offering practical solutions to urban mobility challenges (Hasnine and Habib, 2020).

6.3.2 Supply of Urban Mobility

The supply of urban mobility encompasses a diverse range of infrastructure and services designed to meet the transportation needs of growing urban populations. These include public infrastructure such as roads, rail networks, and bus systems, as well as private services like ride-hailing, carsharing, and micromobility options (e.g., shared bikes and scooters). Together, these systems form the backbone of urban transportation, enabling efficient movement of people and goods. Among various public mobility services, public transit is a vital component of urban mobility supply, offering scalable solutions to address the increasing demand for transportation in densely populated areas. Its role is particularly significant in alleviating traffic congestion, reducing carbon emissions, and promoting equitable access to mobility (Sun et al., 2022). To tackle the challenges of urbanization and traffic congestion, many large cities have invested heavily in public transportation infrastructure in recent years. Cities worldwide have implemented ambitious projects to expand their mass transit networks, focusing on systems such as metro, light rail transit (LRT), and bus rapid transit (BRT) (Sun et al., 2018; Guo et al., 2023). These modes of transportation can accommodate high passenger volumes, providing a cost-effective and environmentally friendly alternative to private vehicles (Parajuli and Wirasinghe, 2001). However, the development of such systems requires significant capital investment, long-term planning, and political commitment (Beaudoin et al., 2015; Sun et al., 2016). For example, Second Avenue Subway in New York City cost $2.5 billion per mile, involving tunneling under dense urban areas, extensive utility relocation, and high labor costs (Killada and Raju, 2018).

Each type of transit technology has its advantages and disadvantages (Allport, 1981; Stutsman, 2002; Bruun, 2005; Tirachini et al., 2010). For instance, metro systems offer high capacity and faster operating speeds, making them suitable for densely populated urban cores. However, they require significantly higher capital and operating costs compared to bus systems (Sun et al., 2020). BRT systems are a cost-effective alternative to metros, combining the speed, reliability, and amenities of rail transit with the flexibility of bus services. They are less expensive to construct and maintain but have lower capacity and speed, which may limit their effectiveness in high-demand urban areas. LRT systems occupy a middle ground between metro and BRT systems in terms of speed, capacity, and cost. These above-ground rail systems are well-suited for medium-density corridors, offering greater capacity and reliability than buses while incurring lower costs than metros.

The decision to invest in a particular transit technology involves evaluating trade-offs between operating speed, capacity, and investment costs. Numerous studies in the literature have explored transit technology selection and investment timing (Guo et al., 2017). Parajuli and Wirasinghe (2001) developed a multi-attribute decision model to guide transit technology selection, using a combination of field surveys and empirical statistical approaches. Tirachini et al. (2010) proposed models for social welfare and private profit maximization to evaluate the investment benefits of different transit technologies along a linear transportation corridor. Chen et al. (2015) addressed the transit technology selection problem using a static, deterministic model, assuming time-independent travel demand. Chu and Polzin (2000) examined timing rules for major transportation investments through cost-benefit analysis, whereas Sivakumaran et al. (2014) explored how access modes (e.g., walking, cycling, or feeder buses) influence the choice of trunk-line transit technologies. These studies underscore the complexity of optimizing urban mobility supply, requiring careful consideration of economic, social, and operational factors to ensure sustainable and effective transportation systems.

6.4 PRICING STRATEGIES

This section examines pricing strategies that can help manage urban mobility challenges and improve the efficiency of transportation systems.

6.4.1 Congestion Pricing

With rapid urbanization and economic growth, traffic congestion has become a pressing issue in many densely populated U.S. cities, including New York City, Los Angeles, San Francisco, Chicago, and Houston (Duranton and Turner, 2011). This challenge primarily arises from an imbalance between transportation demand and supply. A commonly proposed solution to alleviate urban traffic congestion involves expanding transportation infrastructure capacity. However, this approach often leads to inefficient resource allocation, as it creates excessive capacity during nonpeak hours while failing to fully resolve congestion during morning and evening commuting periods. Additionally, expanding capacity can induce new traffic demand, further exacerbating congestion during peak times. Moreover, many urban areas face financial and spatial constraints that limit the feasibility of infrastructure expansion.

Cordon toll pricing has emerged as a widely recommended alternative to infrastructure expansion due to its relative ease of implementation and ability to internalize congestion externalities (Anas & Rhee, 2006, Baghestani et al., 2022), as shown in Figure 6.2. This strategy involves charging vehicles for entering specific urban areas during peak hours, thereby managing demand more effectively. Prominent examples of cordon toll pricing systems include those implemented in cities like Singapore, London, Hong Kong, Oslo, and Bergen (Ho et al., 2005; Yang and Huang, 2005; Rouwendal and Verhoef, 2006). These systems have demonstrated potential to reduce traffic congestion while promoting more sustainable urban mobility solutions.

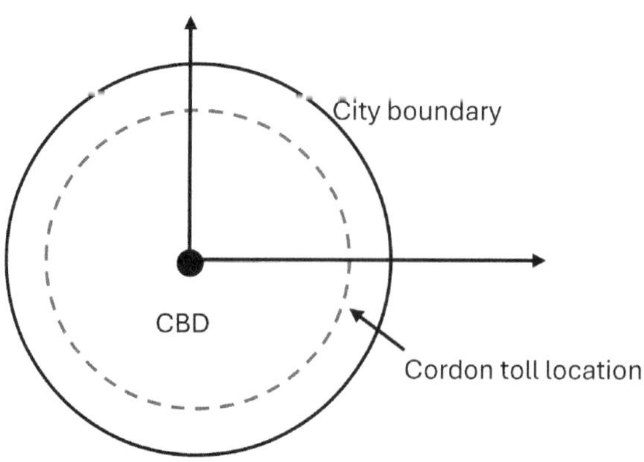

FIGURE 6.2 Cordon toll pricing system in a circular monocentric city. *Abbreviation:* CBD, Central Business District.

6.4.2 Transit Fare

Transit agencies routinely assess fare structures to ensure that their services remain efficient and accessible. One innovative approach to addressing urban mobility challenges is free fare transit, also known as fare-free or zero-fare transit. This progressive concept challenges conventional transit systems by eliminating passenger fares altogether (Studenmund and Connor, 1982). The goal of free fare transit is to create a public transportation network that offers unrestricted access to all individuals, fostering social equity and addressing various societal and environmental challenges (Štraub and Jaroš, 2019, Kirschen et al, 2022). This section examines the potential benefits, funding mechanisms, challenges, and transformative impacts of implementing free fare transit in urban communities.

6.4.2.1 Benefits of Free Fare Transit

Implementing free fare transit offers both immediate and long-term advantages. The most significant benefit is the promotion of social equity. By removing financial barriers, free fare transit ensures that all individuals have access to transportation, which is particularly important for marginalized populations, such as low-income individuals, students, and the elderly who rely on public transit for their daily commuting needs. Furthermore, it incentivizes a shift from private vehicle use to public transit, reducing traffic congestion, carbon emissions, and air pollution, which are critical factors in addressing climate change and improving urban air quality.

Free fare transit also improves public safety by reducing reliance on private vehicles, resulting in fewer traffic accidents and less road congestion. It promotes social cohesion by facilitating interactions between people from different socioeconomic backgrounds, thus contributing to the creation of more inclusive and vibrant urban environments. Additionally, free fare transit can stimulate local economies by increasing foot traffic to businesses near transit hubs and making cities more attractive to tourists due to enhanced accessibility.

6.4.2.2 Funding Mechanisms for Free Fare Transit

While the concept of free fare transit may initially appear financially challenging, there are several funding strategies that can help make it feasible. One approach is reallocating funds from existing transportation budgets. The money currently allocated to fare collection infrastructure, ticketing systems, and administrative costs can be redirected to enhance service quality, expand transit networks, and support the transition to free fare policies. Various schemes for implementing free fare transit exist, whether applied across the entire system, within specific service zones, during certain time periods, or for designated passenger groups, as summarized in Figure 6.3. These schemes are mutually exclusive, but in some cases, multiple options can be combined to suit the needs of different transit agencies (Table 6.1).

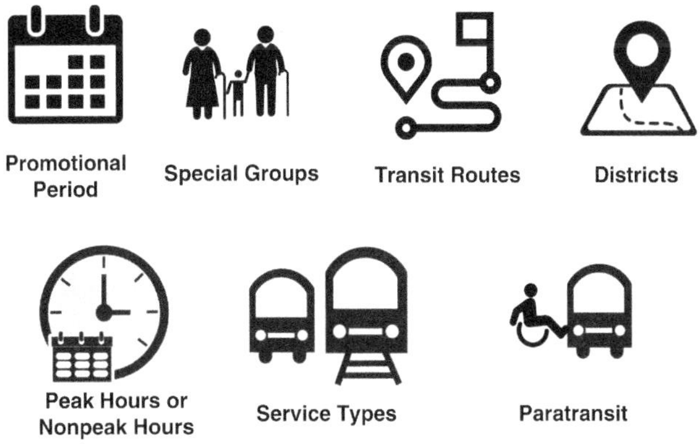

FIGURE 6.3 Fare-free public transit policy schemes.

TABLE 6.1
Typology of Zero-/Reduced-Fare Transit

Option	Description	Type	Case in United States
Customer groups	Zero- or reduced-fare programs for customers who meet eligibility requirements	Disadvantaged/low-income groups	RideRTA Fare Relief Program Transit Access Pass Lifeline Transportation Program
		K–12 students	Massachusetts Bay Transportation Authority (MBTA) buses (Massachusetts,) The Seattle Public Schools' ORCA (Seattle)
		University students and faculty	U-Pass Program Magic Bus system Youth Opportunity Pass
		Employees of specific organization	Google Bus Program Apple Shuttle Microsoft Connector
		Tourism	SunRunner Bus Rapid Transit (Las Vegas) The Las Vegas Strip
Service zones	Zero-fare rides for customers who travel within a predefined geography or zone	Resort community	SunTrolley (Fort Lauderdale) Breckenridge Free Ride (Breckenridge)
		Downtown circulation	Miami Metromover (Miami) Ride On program (Montgomery)
		Small urban/rural area	Area Regional Transit (St. Lucie) Atomic City Transit (Los Alamos)
		Mid-sized regional	Capital Area Transportation Authority (Lansing) Kansas City Area Transportation Authority
		Large urban area	Los Angeles Metro Denver Regional Transportation District Utah Transit Authority
		University-dominated community	Ellensburg Central Transit (Central Washington University) Bear Transit (University of California)
Service type	Zero- or reduced-fare rides on subsystem transit modes	Bus	Chapel Hill Transit The Jolley Trolley (Clearwater)
		Metro/subway	Washington Metro MARTA Rail (Atlanta) Los Angeles Metro Rail
		Light rail	Hudson-Bergen Light Rail (New Jersey) River Line (New Jersey)
		Ferry/boat	Staten Island Ferry Charleston Water Taxi Alameda/Oakland Ferry
		Streetcar	Seattle Streetcar; Portland Streetcar
		Trolley/shuttle	AirBART (Oakland) Newark AirTrain
		Bike-sharing	Boise GreenBike Holyoke Bike Share Santa Monica Bike Share
		Demand-responsive transit	RideKC Freedom (Kansas City) Capital MetroAccess (Austin)

(Continued)

TABLE 6.1 *(Continued)*
Typology of Zero-/Reduced-Fare Transit

Option	Description	Type	Case in United States
Ticket-free/reduced strategy	Implementation of different strategy on zero-fare transit system	Peak hours free	Downtown Spirit (Seattle)
			Route 17 Express (Santa Cruz)
		More rides, more discount	MARTA (Atlanta)
			Chicago Transit Authority
		Off-peak hours free	StarMetro (Tallahassee)
			Kansas City Streetcar
		Rest day free	City of Commerce Municipal Bus Lines (Los Angeles)
			Portland Streetcar

Furthermore, collaborations with local businesses, corporations, and philanthropic organizations can help generate additional financial support through sponsorships and advertising, creating an extra revenue stream to offset operational costs. Governments can also explore alternative funding models, such as congestion pricing, parking fees, or carbon emissions taxes, to generate supplementary income to sustain free fare transit systems.

6.4.2.3 Challenges in Implementing Free Fare Transit

Despite its many benefits, implementing free fare transit comes with challenges. One of the primary concerns is the potential strain on public transit budgets, especially when fare revenue is eliminated. However, studies indicate that increased ridership resulting from free fare policies can help offset the loss of fare revenue, making the financial burden more manageable. Additionally, effective cost management, resource allocation, and ongoing evaluation of funding strategies can help transit agencies address budgetary challenges. Another challenge is the risk of overcrowding and capacity constraints. To mitigate this issue, transit agencies must carefully plan and optimize routes, increase fleet size, and improve service frequency. Emerging technologies such as demand-responsive transit and real-time passenger data analytics can also help optimize operations and improve efficiency.

6.4.2.4 Farebox Recovery Ratio

A related financial metric used by transit agencies to evaluate public transit systems is the farebox recovery ratio. This ratio measures the extent to which fare revenue covers the operating costs of providing transit services. It is calculated by dividing fare revenue by total transit expenses. A higher farebox recovery ratio indicates that a larger portion of operating costs is covered by fare revenue, which signals greater financial sustainability. Conversely, a lower ratio suggests that the system relies more heavily on subsidies or other funding sources.

The farebox recovery ratio is an essential tool for transit agencies and policymakers to assess the financial health and efficiency of a transit system. It provides insight into the system's ability to generate revenue and manage costs effectively, guiding decisions on funding, fare structure adjustments, and service evaluations. Achieving a high farebox recovery ratio can be challenging, as it depends on factors such as fare levels, ridership, operating expenses, and service coverage. Different transit agencies and regions may set various targets based on their specific goals, funding needs, and regional conditions.

6.4.2.5 Cost-Effectiveness of Free Fare Transit

Economists often question the economic efficiency of introducing a fare-free public transit system in a city. Some argue that for transit agencies to succeed, they must operate like for-profit entities, responding to market-driven customer demands (Fearnley, 2013; Kębłowski, 2019). However, research has shown that the costs associated with fare collection can exceed the revenue generated by fares (Perone, 2002; Sasaki, 2014). These costs include capital costs for fareboxes and related infrastructure, as well as future technology updates; operating and maintenance costs for fare collection systems; fare policy planning costs, including staff time spent analyzing and implementing fare changes; and lost time and productivity due to the time spent collecting and explaining fares. Given these considerations, state agencies, such as those in Florida and Washington, should carefully weigh the net costs or revenues of free fare transit. In some cases, the costs of fare collection may be equal to or greater than the revenues generated, meaning there is no net benefit from fare collection. By eliminating fares, these costs could be eliminated, potentially offsetting the loss of revenue. When funding for transit systems is tied to ridership levels, eliminating fares or reducing fares can actually generate additional funding (Jenelius and Cebecauer, 2020). In fact, for zero-fare systems, increased ridership can lower the average subsidy per passenger, even without fare revenue (Saphores et al., 2020; Dai et al., 2021). However, it is important to consider that lost fare revenue and increased operating costs can also lead to greater financial losses. The potential costs of implementing a fare-free policy can be further understood by examining data from various transit systems, as shown in the Fare-Free Transit Evaluation Framework (Kirschen et al., 2022).

Free fare transit represents a shift toward more equitable, sustainable, and inclusive urban mobility. Although its implementation requires careful planning, innovative funding models, and effective strategies to address challenges, the potential benefits, including enhanced social equity, reduced congestion, and improved environmental sustainability, make it a compelling solution for many urban areas (Guo et al., 2021). By leveraging diverse funding mechanisms, efficiently managing costs and capacity, and ensuring long-term financial sustainability, cities can pave the way for a future where public transit is truly accessible to all, transforming urban communities for the better.

6.5 CONCLUSION

Urban mobility stands at the crossroads of economic growth, social equity, and environmental sustainability. The way in which cities manage transportation systems has profound implications not only for the daily lives of their residents but also for their long-term economic prosperity. Throughout this chapter, we have explored the intricate relationship between urban mobility and economic development, the factors influencing demand and supply dynamics, and the pricing strategies that cities can implement to optimize their transportation systems.

One of the key insights is that urban mobility is integral to the economic structure of cities. Efficient transportation systems act as catalysts for economic growth by improving access to labor markets, businesses, and residential areas, thus enhancing overall productivity and competitiveness. Investments in mobility infrastructure, such as public transit and road networks, yield substantial returns by boosting social equity, driving urban development, and shaping land values. Understanding the influence of residential location choices is crucial for designing transportation systems that align with the economic realities of cities, ensuring that mobility solutions contribute to, rather than hinder, sustainable urban development.

The analysis of demand and supply dynamics further underscores the complexity of urban mobility systems. The demand for transportation is closely tied to population growth, urbanization, and economic factors such as income levels and land use patterns. As cities expand, the strain on existing transportation infrastructure often leads to congestion, reduced service quality, and environmental degradation. Effective demand management strategies, such as TDM policies, are essential for addressing these challenges. On the supply side, the variety of transportation options, ranging from public transit to micromobility services, plays a vital role in meeting the growing demand. Public transit, in particular, remains a cornerstone of urban mobility, offering scalable solutions that can accommodate high passenger volumes and help mitigate the negative effects of congestion.

The need for innovative pricing strategies becomes apparent when examining the challenges of congestion and accessibility. Congestion pricing, which incentivizes the use of alternative transportation modes while reducing road congestion, has proven effective in cities worldwide. By charging vehicles to enter highly congested areas, cities can generate revenue that can be reinvested in public transportation infrastructure, further enhancing the overall transportation network. Additionally, pricing strategies that encourage the use of public transit, such as free fare systems, offer a transformative approach to urban mobility. While challenges exist in funding such initiatives, the potential benefits in terms of increased ridership, reduced inequalities, and environmental impact make free fare transit a promising avenue for future exploration.

In conclusion, urban mobility is not only a critical component of urban planning but also a central factor in shaping the economic and social fabric of cities. The interplay between transportation demand, infrastructure supply, and pricing strategies is complex, yet understanding and managing these elements is key to creating efficient, sustainable, and equitable urban mobility systems. As cities continue to grow and evolve, the policies and investments made in transportation will play a pivotal role in defining their future economic success and livability. By adopting comprehensive strategies that balance demand management, infrastructure investment, and innovative pricing models, cities can build transportation systems that support the needs of their residents while fostering long-term economic development and environmental sustainability.

REFERENCES

Allport, R. J. (1981). The costing of bus, light rail transit and metro public transport systems. *Traffic Engineering & Control*, *22*, 633–639.

Alonso, W. (1964). Location and land use: Toward a general theory of land rent. Harvard University Press.

Anas, A., & Rhee, H. J. (2006). Curbing excess sprawl with congestion tolls and urban boundaries. *Regional Science and Urban Economics*, *36*, 510–541.

Anas, A., & Xu, R. (1999). Congestion, land use, and job dispersion: A general equilibrium model. *Journal of Urban Economics*, *45*, 451–473.

Baghestani, A., Tayarani, M., Allahviranloo, M., Nadafianshahamabadi, R., Kucheva, Y., Mamdoohi, A. R., & Gao, H. O. (2022). New York City cordon pricing and its impacts on disparity, transit accessibility, air quality, and health. *Case Studies on Transport Policy*, *10*(1), 485–499.

Beaudoin, J., Farzin, Y. H., & Lawell, C. Y. C. L. (2015). Public transit investment and sustainable transportation: A review of studies of transit's impact on traffic congestion and air quality. *Research in Transportation Economics*, *52*, 15–22.

Beckmann, M. J. (1969). On the distribution of urban rent and residential density. *Journal of Economics Theory*, *1*, 60–67.

Beckmann, M. J. (1974). Spatial equilibrium in housing market. *Journal of Urban Economics*, *1*, 99–107.

Bjorefeldt, J., & Grunden, J. (2019). Demand and Capacity Asymmetry in Carsharing. A Comparative Case Study of Preventative Measures (Master's Thesis). Chalmers University of Technology Gothenburg, p. 103.

Bowes, D. R., & Ihlanfeldt, K. R. (2001). Identifying the impacts of rail transit stations on residential property values. *Journal of Urban Economics*, *50*(1), 1–25.

Brueckner, J. K. (2007). Urban growth boundaries: An effective second-best remedy for unpriced traffic congestion? *Journal of Housing Economics*, *16*(3–4), 263–273.

Bruun, E. (2005). Bus rapid transit and light rail: Comparing operating costs with a parametric cost model. Transportation Research Record Journal of the Transportation Research Board, *1927*, 11–21.

Bylinko, L. (2020). The practical importance of transport demand management policy. *Transport Problem*, *15*(4), 191–201.

Dai, J., Liu, Z., & Li, R. (2021). Improving the subway attraction for the post-COVID-19 era: The role of fare-free public transport policy. *Transport Policy*, *103*, 21–30.

Duranton, G., & Turner, M. A. (2011). The fundamental law of road congestion: Evidence from US cities. *American Economic Review*, *101*(6), 2616–2652.

Chen, Y. J., Li, Z. C., & Lam, W. H. (2015). Modeling transit technology selection in a linear transportation corridor. *Journal of Advanced Transportation*, *49*(1), 48–72.

Chu, X., & Polzin, S. E. (2000). Timing rules for major transportation investments. Transportation, *27*, 201–219.

Fearnley, N. (2013). Free fares policies: Impact on public transport mode share and other transport policy goals. *International Journal of Transportation*, *1*(1), 75–90.

Guo, Q., Chen, S., Sun, Y., & Schonfeld, P. (2023). Investment timing and length choice for a rail transit line under demand uncertainty. *Transportation Research Part B: Methodological*, *175*, 102800.

Guo, Q., Sun, Y., Schonfeld, P., & Li, Z. (2021). Time-dependent transit fare optimization with elastic and spatially distributed demand. *Transportation Research Part A: Policy and Practice*, *148*, 353–378.

Guo, Q. W., Chow, J. Y., & Schonfeld, P. (2017). Stochastic dynamic switching in fixed and flexible transit services as market entry-exit real options. *Transportation Research Procedia*, *23*, 380–399.

Hasnine, M. S., & Habib, K. N. (2020). Transportation demand management (TDM) and social justice: A case study of differential impacts of TDM strategies on various income groups. *Transport Policy*, *94*, 1–10.

Hendricks, S. J., & Georggi, N. L. (2007). Documented impact of transportation demand management programs through the case study method. *Journal of Public Transportation*, *10*(4), 79–98.

Ho, H. W., Wong, S. C., Yang, H., & Loo, B. P. (2005). Cordon-based congestion pricing in a continuum traffic equilibrium system. *Transportation Research Part A: Policy and Practice*, *39*(7–9), 813–834.

Jenelius, E., & Cebecauer, M. (2020). Impacts of COVID-19 on public transport ridership in Sweden: Analysis of ticket validations, sales and passenger counts. *Transportation Research Interdisciplinary Perspectives*, *8*, 100242.

Killada, M., & Raju, G. (2018). World's top economies and their metro systems' ridership and financial performance. *International Journal of Traffic and Transportation Engineering*, *7*, 91–97.

Kirschen, M., Pettine, A., Adams, M., & Persaud, H. (2022). Fare-free transit evaluation framework. *Transportation Research Board*. https://doi. org/10.17226/26732.

Kębłowski, W., Van Criekingen, M., & Bassens, D. (2019). Moving past the sustainable perspectives on transport: An attempt to mobilise critical urban transport studies with the right to the city. Transport Policy, *81*, 24–34.

Li, Z. C., Chen, Y. J., Wang, Y. D., Lam, W. H., & Wong, S. C. (2013). Optimal density of radial major roads in a two-dimensional monocentric city with endogenous residential distribution and housing prices. *Regional Science and Urban Economics*, *43*(6), 927–937.

Li, Z. C., & Guo, Q. W. (2017). Optimal time for implementing cordon toll pricing scheme in a monocentric city. *Papers in Regional Science*, *96*(1), 163–191.

Li, Z. C., Guo, Q. W., Lam, W. H., & Wong, S. C. (2015). Transit technology investment and selection under urban population volatility: A real option perspective. *Transportation Research Part B: Methodological*, *78*, 318–340.

Li, Z. C., Lam, W. H., Wong, S. C., & Sumalee, A. (2012). Design of a rail transit line for profit maximization in a linear transportation corridor. *Transportation Research Part E: Logistics and Transportation Review*, *48*(1), 50–70.

Li, Z. C., & Peng, Y. T. (2016). Modeling the effects of vehicle emission taxes on residential location choices of different-income households. *Transportation Research Part D: Transport and Environment*, *48*, 248–266.

Litman, T. (2003). The online TDM encyclopedia: Mobility management information gateway. *Transport Policy*, *10*(3), 245–249.

Lowry, I. S. (1964). A model of metropolis. The RAND Corporation.

McDonald, J. F, & Osuji, C. I. (1995). The effect of anticipated transportation improvement on residential land values. *Regional Science and Urban Economics*, *25*, 261–278.

Parajuli, P. M., & Wirasinghe, S. C. (2001). A line haul transit technology selection model. *Transportation Planning and Technology*, *24*(4), 271–308.

Perone, J. S. (2002). Advantages and disadvantages of fare-free transit policy. University of South Florida, National Center for Transportation Research. https://digitalcommons.usf.edu/cgi/viewcontent.cgi?article=1216&context=cutr_nctr

Quigley, J. M. (1984). The production of housing services and the derived demand for residential energy. *The RAND Journal of Economics*, 15, 555–567.

Rouwendal, J., & Verhoef, E. T. (2006). Basic economic principles of road pricing: From theory to applications. *Transport Policy*, *13*, 106–114.

Saphores, J. D., Shah, D., & Khatun, F. (2020). A Review of reduced and free transit fare programs in California. The University of California Institute of Transportation Studies. https://escholarship.org/uc/item/74m7f3rx

Sasaki, Y. (2014). Optimal choices of fare collection systems for public transportations: Barrier versus barrier-free. *Transportation Research Part B: Methodological*, *60*, 107–114.

Štraub, D., & Jaroš, V. (2019). Free fare policy as a tool for sustainable development of public transport services. *Human Geographies*, *13*(1), 45–59.

Studenmund, A. H., & Connor, D. (1982). The free-fare transit experiments. *Transportation Research Part A: General*, *16*(4), 261–269.

Sivakumaran, K., Li, Y., Cassidy, M., & Madanat, S. (2014). Access and the choice of transit technology. Transportation Research Part A: Policy and Practice, *59*, 204–221.

Stutsman, J. M. (2002). Bus rapid transit or light, rail transit: How to decide? Los Angeles case study. *Transportation Research Record*, *1793*, 55–61.

Sun, Y., Chen, S., & Guo, Q. (2022). Evaluating the environmental benefits of personalized travel incentives in dynamic carpooling. *KSCE Journal of Civil Engineering*, *26*(7), 3082–3093.

Sun, Y., Gong, H., Guo, Q., Schonfeld, P., & Li, Z. (2020). Regulating a public transit monopoly under asymmetric cost information. *Transportation Research Part B: Methodological*, *139*, 496–522.

Sun, Y., Guo, Q., Schonfeld, P., & Li, Z. (2016). Implications of the cost of public funds in public transit subsidization and regulation. *Transportation Research Part A: Policy and Practice*, *91*, 236–250.

Sun, Y., Schonfeld, P., & Guo, Q. (2018). Optimal extension of rail transit lines. *International Journal of Sustainable Transportation*, *12*(10), 753–769.

Tirachini, A., Hensher, D. A., & Jara-Diaz, S. R. (2010). Comparing operator and users costs of light rail, heavy rail and bus rapid transit over a radial public transport network. *Research in Transportation Economics*, *29*, 231–242.

Von Thünen, J. H. (1826). Derisolierte Statt in Belziehung auf Landwirtschaft und National ökonomie. Part I. Pethers.

Yang, H., & Huang, H. J. (2005). Mathematical and economic theory of road pricing. Emerald Group Publishing Limited.

7 Data Governance Policies for Human-Centric Urban Mobility Services in Smart Cities

Bokolo Anthony Jr.

7.1 INTRODUCTION

According to the United Nations, due to rapid urbanization worldwide, over 66% of the world's inhabitants will reside in cities by 2050 (Zhong and Wang 2022). As such, the concept of smart cities has emerged as a viable solution to address challenges arising from rapid urbanization (Choenni et al. 2022). Smart cities deploy digital technologies for urban development for a better quality of life, efficient resource usage, and sustainable economic development (Lis et al. 2022). Thus, cities need to enable data sharing and use of data to create value-added services such as intelligent mobility services to citizens (Früh et al. 2022; Anthony Jnr 2023c). Data governance is emphasized as a mechanism that influences the policy, dynamics, and even success of data-driven services in the urban context (Lis and Otto 2021). Recent contributions recognize the importance of data governance in smart cities and further examine data governance as an approach that facilitates data ecosystems in smart cities (Artyushina 2020; Choenni et al. 2022). In urban environments, data governance supports the strategic use of data to promote data-driven services (Meijer 2018; Anthony Jnr 2022a). In supporting cities to become smart cities, it is necessary to improve urban mobility services to achieve sustainable transportation development (Tan 2023).

Additionally, the provision of intelligent mobility services depends greatly on the availability of data, and the role of human-centric urban mobility services–based data governance in the context of smart cities has not yet been systematically investigated (Lis and Otto 2021). However, research in data governance has been examined in the literature (Lis and Otto 2021; Jansson et al. 2022). Studies on mobility data governance policies in smart cities are limited. Furthermore, researchers have investigated the importance of applications data governance, but little is known about how citizens can govern their mobility data generated from mobility applications. In addition, the contextual factors that influences the data governance adoption from the human-centric perspective have not been explored (De Prieëlle et al. 2020). Likewise, there is a noticeable lack of theoretical insights regarding drivers that influence the design and establishment of data governance policies in smart cities (Lis et al. 2022). As a result, this study addresses the following research questions:

- Which key mechanisms and procedures can improve human-centric data governance practices for urban mobility service in smart cities?
- What are the factors and challenges that influence human-centric data governance for urban mobility service in smart cities?

Thus, this study aims to create a theoretical and practical understanding of the application and characteristics of data governance policies to improve urban mobility services in smart cities. Building on secondary data, this study contributes to urban human mobility by enabling human-centered data procedures that support data quality for urban mobility services in smart cities by

DOI: 10.1201/9781003503262-9

proposing a data governance framework. The framework is conceptualized grounded in the literature to examine technical, social, regulatory, and organizational factors that influence smart cities in managing data governance issues for achieving human-centric urban mobility services. Findings from this study identify procedures and mechanisms that improve human-centric data governance policies. This chapter is arranged as follows. Section 7.2 discusses the methods. Section 7.3 presents the key findings. Section 7.4 provides the discussion and implications. Finally, Section 7.5 presents the conclusion.

7.2 METHODS

This study draws upon narrative literature review by examining suitable and relevant sources (Webster and Watson 2002; Snyder 2019). Online databases (Web of Science, Scopus, and Google Scholar) were used to retrieve scientific articles using different keyword search as search strategies. The primary keywords include "(human-centric OR citizen-centered OR urban mobility services OR urban mobility application OR smart cities) AND (data governance OR urban data governance)" (Anthony Jnr 2023d). The inclusion criteria include concepts, mechanisms, strategies, factors, challenges, and recommendations of data governance policies in the context of urban mobility services in smart cities from the perspective of citizens. Also, only studies published in English are included.

Sources not related to data governance from urban context were excluded. After identification, the retrieved articles were assessed by reading the abstracts, and relevant sources were selected that were later reviewed in more detail. Next, a full content review of the sources was done with emphasis on the introduction and conclusion. Finally, qualitative data were extracted from the selected 40 sources presented in the reference section of this chapter, and descriptive analysis was employed to present key findings as related to the research questions being explored in this study.

7.3 FINDINGS

This section presents descriptive analysis from the literature review. The results are presented in different thematic subsections that discuss human-centric governance policies for urban mobility service in smart cities.

7.3.1 Governance of Urban Mobility Services in Smart Cities

Cities are vibrant areas for any nation's economic, social, and environmental development. But the increase in human population and settlements has resulted in depletion of natural resources, environmental degradation, and urban congestion. To address the challenges faced by cities, in 2008–2009, IBM suggested the "smart earth" notion and later proposed the "smart city" concept. The concept of smart cities involves the deployment of advanced technologies to improve a city's livability, suitability (Tan 2023), and sustainability (Anthony Jnr 2023c). Smart cities provide inhabitants with better living conditions, contributing to a better natural environment by promoting efficient and equal public services, facilitating operational interaction, and encouraging green and low-carbon infrastructures. Smart cities aggregate and utilize data for better planning of a municipality's public transportation (Anthony Jnr 2022b; Tan 2023).

The same data can be exploited by transport service providers to develop novel mobility service for citizens, such as real-time route planners that recommend the safest and shortest direction to citizens (Tan 2023; Anthony Jnr 2024b). In the context of smart cities, governance refers to procedures, social activities, and practices performed by different actors within the society (Bevir 2013). Typically, governance from an urban perspective refers to organizational policies that guide citizens, companies, and municipalities in achieving tasks within their society (Anthony Jnr 2018). Thus, governance can be seen as the deployment of new organizational structures, instruments, and decision-making processes involving a multitude of actors (van den Hurk et al. 2021). Governance encompasses the rule sets that constitute decision-making for management between stakeholders (Docherty 2018).

7.3.2 Citizen-Centric Governance of Urban Mobility Applications

Urban mobility has become an important domain in cities to ensure seamless and effective travel of residents and flow of goods (Tan 2023; Anthony Jnr 2024a). Mobility applications are faced with data governance issues, which concern how to get the correct data and how to aggregate, utilize, and share the available data to the appropriate actors. This issue can be addressed by multiple actors, such as citizens, transport authorities, transport service providers, application developers, infrastructure managers, municipalities, and others, that are involved in the provision, processing, analyzing, and use of mobility-related data (Veeneman et al. 2018). Mobility applications adopted by citizens use and generate data, which are utilized by the transport service providers. These platforms facilitate data sharing and enable data-driven services such as data visualizations, analytics, and optimization. However, citizens do not have complete control over their data collected via mobility applications used by transport service providers (De Prieëlle et al. 2020). Also, defining the requirements for data quality and establishing procedures for citizen-centered services in the urban context are often ambiguous (Lis and Otto 2021).

Currently, transport service providers employ data-centric approaches for different purposes, such as for gaining insights on business performance, development of novel mobility business models, effective decision-making, and monitoring and reporting (Lis et al. 2022). Transport service providers collect and control citizen data using mobility applications, which are then used to provide urban mobility services. Simultaneously, transport service providers involve citizens to participate in governing their data by agreeing to predrafted terms and conditions. However, citizens do not have access and rights to their data. Hence, the distribution of decision rights is unbalanced between transport service providers and citizens (Zhong and Wang 2022). Therefore, to achieve urban mobility services, more attention needs to be centered on human-centric data governance with a focus on citizen involvement and collaborative participation (Tan 2023).

7.3.3 From Data Management to Data Governance in Smart Cities

As data have become a valuable asset, cities are now utilizing data to innovate novel business models to provide value-added services to citizens (Torre-Bastida et al. 2022). However, in smart cities, the use and sharing of data face several challenges, such as whether the data quality is good enough to improve mobility operations, access rights, and usage and ownership policies, which are mandatory to prevent security breaches and privacy issues while using and sharing mobility-related data (Choenni et al. 2022). Therefore, there is need for an approach on how to govern and manage data use and sharing in smart cities to address the aforementioned challenges when using data to provide data-driven solutions such as urban mobility services (Zorrilla and Yebenes 2022). Data governance ensures that data are managed in the appropriate way, whereas data management is focused on the responsibility of managing the use of data in order to achieve domain objectives (FHWA 2015). Data governance can be defined as how decision rights, responsibilities, and access as related to data are defined within a domain. Thus, data governance involves the definition of who and what are needed to make decision related to data (Nokkala et al. 2019).

Data governance typically captures the exercise of control and authority over the management of data assets (Lis et al. 2022). Data governance offers a framework to ethically and strategically exploit data (Lis et al. 2022). In general, data governance entails the efficient management and maintenance of data assets by overseeing aspects, such as data privacy, data rights, and data security, among others (Torre-Bastida et al. 2022). Thus, it enables decision-making processes and responsibilities needed to manage data creation, collection, processing, manipulation, storage, and dissemination for use/exploitation within the whole data lifecycle (Torre-Bastida et al. 2022). It can be modeled based on the responsible, accountable, consulted, informed (RACI) matrix (Wende 2007). Data governance within the urban context denotes who is accountable and holds decision rights for using and sharing data (Weill and Ross 2004). Data governance provides a compliance-based structure (Lis et al. 2022) of accountabilities, roles, and decision rights for the use of data (De Prieëlle et al. 2020). It promotes desirable procedures for using data, improving data quality, and managing data across a domain (Anthony Jnr 2018; Lis and Otto 2021).

In this regard, data governance offers a medium to exploit value from data resources (Lis et al. 2022). Within the context of urban mobility services in smart cities, data governance involves who owns data (data producers and data owners), how data can be processed (data consumers and data users), and who can access data (other actors and third parties) and under what usage conditions (De Prieëlle et al. 2020). Additionally, the Federal Highway Administration (FHWA) stated that data governance offers an approach for data collection and management that involves specifying clear standards (FHWA 2015). In the transportation domain, FHWA (2015), Green and Lucivero (2018), and Sooch (2023) specified that data governance denotes the established requirements and criteria for data that includes quality, usage policies, management, control process, and risk management for handling mobility data. Green and Lucivero (2018) added that data governance refers to a high-level policy that ensures that data are effectively used across an organization. Therefore, in the context of this study, data governance thereby refines mobility application efficiency and decision-making processes (Sooch 2023). Data governance aims to foster desirable behavior and regulation regarding the use of data and to increase the value of data assets by considering data quality, risks, and possible data-related costs (Lis et al. 2022). Evidence from the literature highlighted that data governance policies are captured across society at three levels, as seen in Figure 7.1.

Figure 7.1 depicts that the goals and scopes of data governance vary across the micro, meso, or macro levels within society. The *micro* level mainly involves intraorganizational policies aimed at maximizing the full potential and value from available data assets (Torre-Bastida et al. 2022). The *meso* level includes interorganizational policies focused on common rules and principles that regulate the accessibility or usage of data within a group of participating organizations or consortium within a trusted data community or in a data space. Data governance policies at the meso level also provide requirements for orchestrating data control access, ownership, and management for stakeholders based on their roles assigned within the data-sharing ecosystem. The meso level is deployed based on a set of overlying technical, administrative, legal, business, and organizational procedures that specify the roles, rights, functions, attributions, obligations, and responsibilities of each actor (Torre-Bastida et al. 2022). Finally, the *macro* level comprises a set of measures employed to support societal policies, regulations, and strategies regarding the management of data. An example

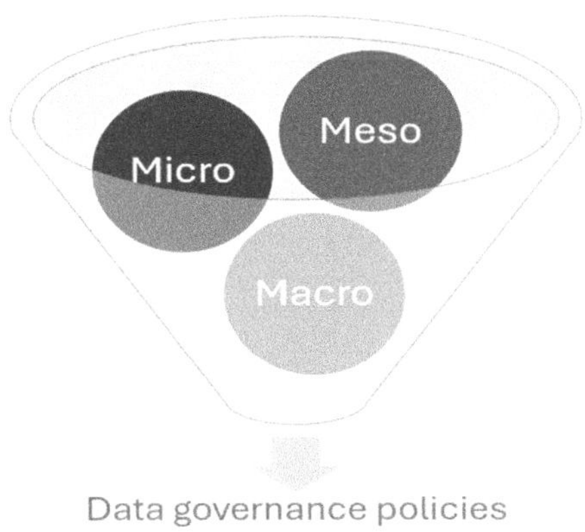

FIGURE 7.1 Data governance levels within society.

of the macro level in society is the General Data Protection Regulation (GDPR), which comprises policies that govern the exploitation and sharing of personal data for European member states. The GDPR also governs the free movement of personal data at the micro level internally across organizations (Torre-Bastida et al. 2022).

7.3.4 Data Governance of Human-Centric Mobility Services in Smart Cities

In smart cities, data governance focuses on what activities and what key aspects linked to data should be managed and which roles are to be involved in the decision-making processes, such as data producers, data stewards, data controllers, data owners, data consumers, and data committees, among others (Zorrilla and Yebenes 2022). In addition, in smart cities, the governance of data provides a data ecosystem that enables all stakeholders, including citizens, to shape the ownership and access control of data associated with them within a predefined environment system. This may include assigning, defining, and allocating responsibilities based on the values, norms, and goals of the cities (Anthony Jnr 2023d).

The responsibilities can include strategizing on how to guarantee secure data storage and exchange, ensuring trustworthy and quality data, fostering data privacy by safeguarding data usage, and fostering the operationalization of the findable, accessible, interoperable, and reusable (FAIR) guideline for the data (Choenni et al. 2022). According to the literature (FHWA 2015), the goals of data governance include supporting leadership roles needed to govern data, improving data quality, prioritizing data needs, promoting cooperation with different actors, fostering flexibility in developing data-driven services, and improving the utilization of available data, as presented in Figure 7.2.

Figure 7.2 depicts a data governance procedure that includes data ownership and access, adherence to regulatory policies, assessment of datasets, cataloguing of datasets, contractual usage schemes, monitoring of data usage, and data provenance (tracking and tracing) to be employed by the transport service providers and citizens. Figure 7.3 depicts the mechanisms to be employed to support a human-centric mobility service in smart cities as adapted from Lee et al. (2018). In

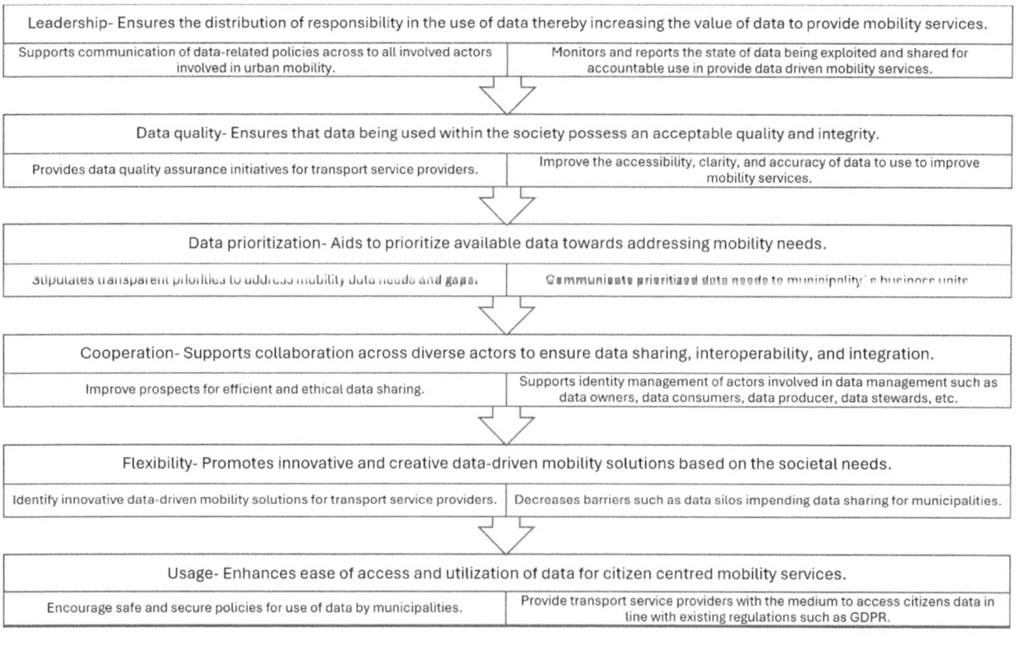

FIGURE 7.2 Data governance procedures for mobility services.

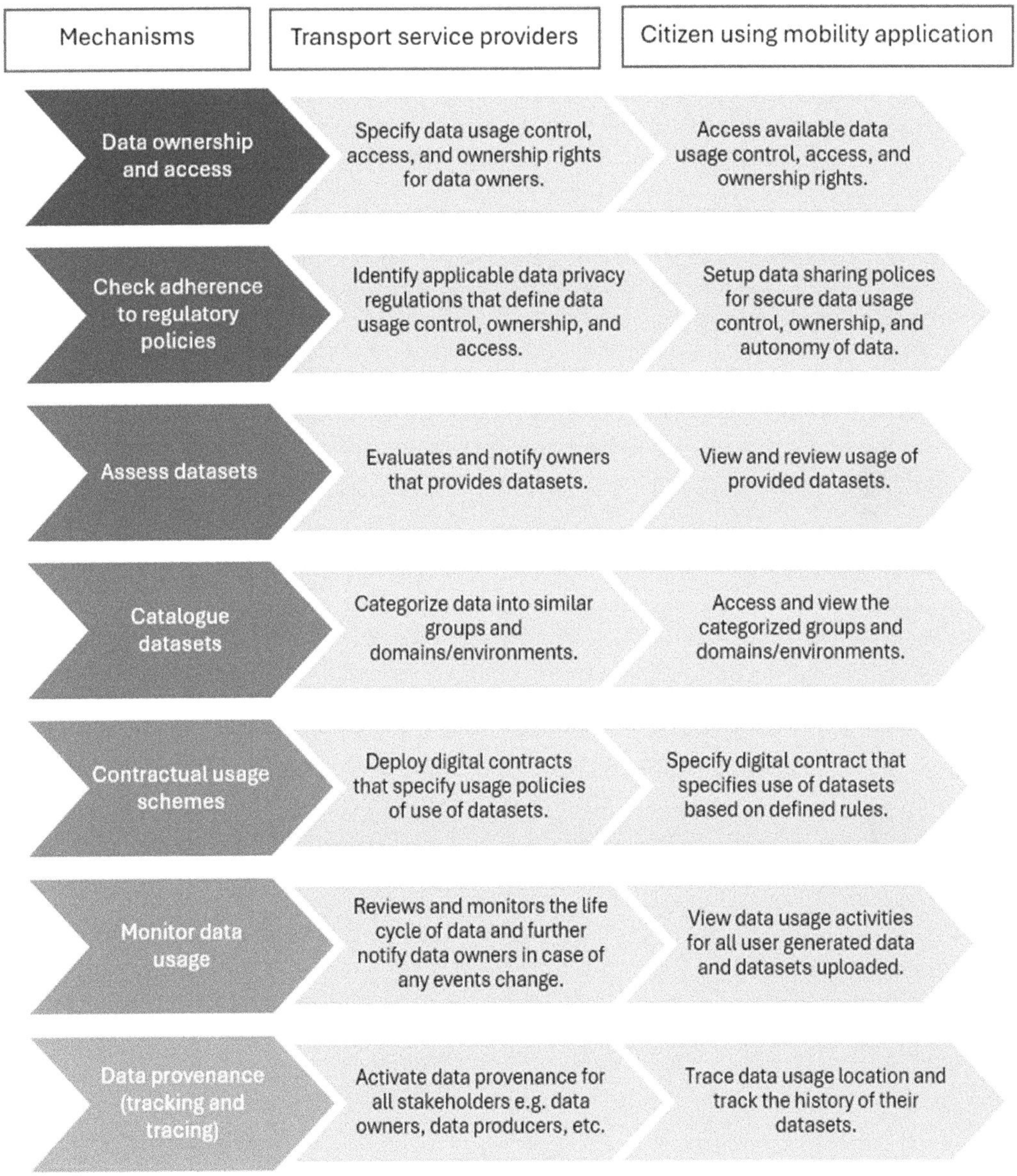

FIGURE 7.3 Data governance mechanisms.

addition, Figure 7.3 highlights the interaction between the transport service providers who provide the mobility application and the citizens that use the mobility application.

Hence, data governance initiated in mobility applications enables the establishment of necessary policies to enable citizens to exercise autonomy, control, and authority for making decisions regarding the management of their mobility-related data. This enables a citizen-centric approach for defining who has the responsibilities and decision-making rights for orchestrating data-related operations. For transport service providers who collect data via mobility applications used by citizens, data governance is initiated based on the strategy, standards, culture, and mission of the organization as related to the management of its data, which is often seen as a strategic asset. Accordingly, data governance for the transport service providers aims to guide procedures related to data management and orchestrate transversality across the technologies, people, and processes within the organization.

7.3.5 Factors That Impact Human-Centric Data Governance Policies

Data governance in the context of mobility services in smart cities refers to the complete set of rules that are pertinent for gathering, processing, analyzing, exploiting, sharing, and monetizing data in a specific sector or a data ecosystem (Murati 2023b). However, for data to be effectively governed, measures need to be addressed. As reviewed in the literature, several studies have explored smart city concepts, but research centered on human-centric data governance for urban mobility services in smart cities is limited. Moreover, the key factors required to improve human-centric data governance practices need to be established based on an operational set of initiatives (policies, procedures, and standards) and are needed for managing and monitoring urban mobility data generation, acquisition, processing, exploitation, and dissemination. Therefore, this section is dedicated to exploring the contextual technical, social, regulatory, and organizational factors impacting human-centric data governance policies for urban mobility services that have not been thoroughly investigated yet. Figure 7.4 depicts the proposed human-centric data governance framework identified in the literature to support urban mobility services in smart cities. Each of the factors are discussed below.

- *Technical Factors*: Citizens are challenged regarding how to fully guarantee that their privacy is not violated when they share data (Zhong and Wang 2022). Citizens' mobility-related data must be safe and not compromised at any point throughout the data lifecycle (Torre-Bastida et al. 2022). Hence, *data security and privacy* are issues that should be considered for successful data governance of urban mobility. *Data quality,* which refers to the techniques, processes, operations, and algorithms aimed at improving the trustworthiness and quality of the dataset provided by citizens, is an important factor (Torre-Bastida et al. 2022). There is also a need to improve clear *data provenance* practices (traceability

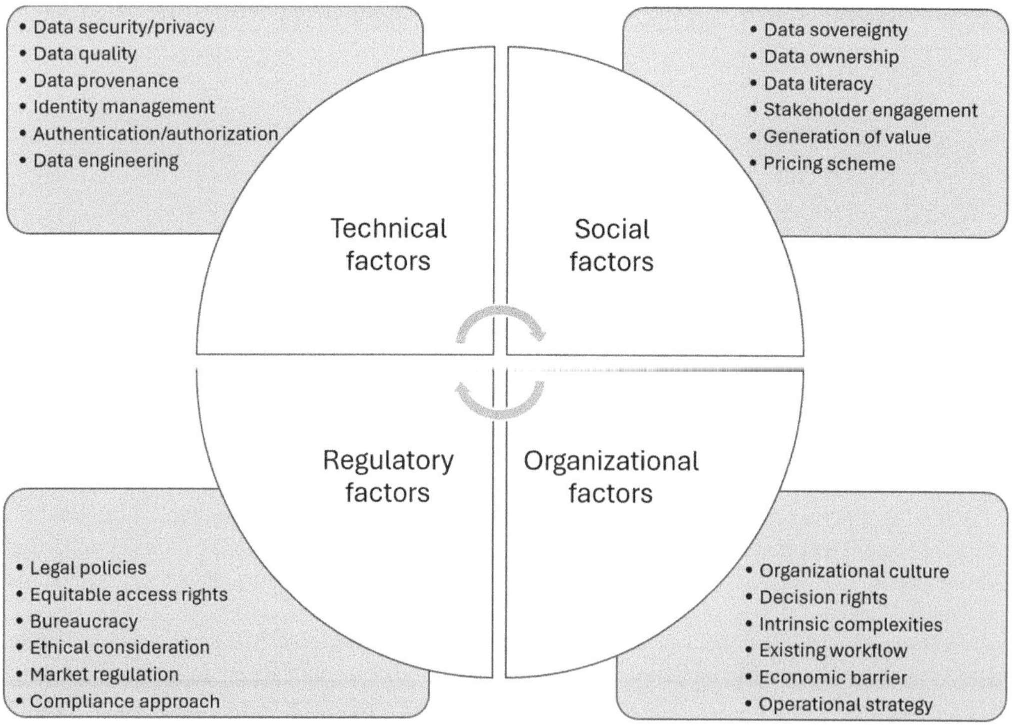

FIGURE 7.4 Proposed human-centric data governance framework.

and tracking), because the recording or logging of data can be used to monitor and detect data manipulation or denial of data usage by an unauthorized actor, including mobility application owners. This is important because it is necessary to avoid data misuse/abuse and privacy violations (Lee et al. 2018).

An *identity management* module and *authentication and authorization* schemes are employed to ensure data security and privacy (Anthony Jnr 2023b). Authentication schemes verify which users have access to particular mobility data using asymmetric encryption, which includes the generation of a private and public key pair (Hoffmann et al. 2021). This encryption scheme enables citizens to have control over their data using their private key signature and/or matching public key. Likewise, authorization proves the identity or permission rights of a user by asserting whether the user is allowed to use a particular data-driven service (Hoffmann et al. 2021). The administrators' emphasis on *data engineering* can limit them from benefiting from the value of data governance. This is because most data administrators are mostly focused on data production, operations, and maintenance. It is also difficult for them to understand how data governance policies fit into their daily operations. They also strive for communication and do not entirely see data as a product (Green and Lucivero 2018).

- *Social Factors*: From a societal perspective, access to citizens' data such as open data will depend on jurisdictions, i.e., a municipality's policies concerning standard procedures and mechanisms for who "owns" or "controls" the data to be exploited and shared (Murati 2023b). *Data sovereignty* is another factor to be considered to govern data. It entails giving the decision rights, duties, and responsibilities back to citizens or stakeholders who are the data owners and data producers (Torre-Bastida et al. 2022). *Stakeholder engagement,* which involves practices that involve key actors who have an interest or stake in societal activities, is another social factor. Stakeholder engagement helps to get stakeholder response, participation, and support and to resolve their concerns (Sooch 2023). As such, it is important to consider stakeholders' needs, particularities, and contexts (Anthony Jnr 2024b; de Souza et al. 2024). The *generation of value* from the provided data based on a business model is another important determinant because data are now seen as an asset for organizations. The monetization of data aims for data owners and data providers to obtain revenue from sharing their data (Torre-Bastida et al. 2022). Similarly, it is challenging for citizens to define a stable *pricing scheme* for data usage and to govern access rules and data sharing between the transport service providers, municipalities, and third parties (Murati 2023a). Regarding *data literacy,* most citizens do not understand how their data can be used or how meaningful insights can be derived from their personal data (Torre-Bastida et al. 2022).

  Citizens as "data owners" are faced with the lack of *data ownership* of mobility-related data (even though it is their data). This essentially involves them taking responsibility or being involved in data production (via mobility application) and data exploitation and sharing procedures carried out solely by data producers (transport service providers) who become responsible for managing citizen mobility-related data (Green and Lucivero 2018). However, retaining ownership of data is crucial when data are to be negotiated and shared with third parties via digital contracts (Torre-Bastida et al. 2022). Data ownership equips citizens with privileges, autonomy, and rights to execute control over access or usage of their mobility-related data. Data ownership entails assigning responsibilities and tasks to data owners/data producers and other users assigned to the data (Anthony Jnr 2023a). In smart cities, context data ownership specifies whether the rights to the data are set at personal, organizational, or shared access. Data ownership from the personal context involves rights relating to private or individual data (Lis and Otto 2021; Zhong and Wang 2022). From the organizational perspective, data ownership relates to data created or acquired by an enterprise to be shared through an economic medium. Such data are seen as commercial data, and rights can only be purchased or licensed to stake a claim of data ownership (Lis and Otto 2021). Shared ownership depicts data without any restrictions regarding use and ownership of the data. This is common with the open data standard, where data is made freely available and is accessible by everyone (Lis and Otto 2021; Anthony Jnr 2022a).

- *Regulatory Factors*: Theses are *legal policies* used to restrict and govern data sharing (Murati 2023a). The regulatory factors are concerned with processing and use of citizen-related mobility data in line with needs and legal requirements (i.e., GDPR), as related to when, who, why, and how to share. In the context of this study, this includes legal jurisdiction frameworks on decision-making adopted by the metropolitan authority on mobility and urban development (Docherty 2018). This also involves approaches in the context of business-to-business (B2B) data sharing by explicitly using contractual agreements as a basis to selectively grant data access to data consumers or data users. The contract is then used to check the legal data-sharing agreements to administer control as related to terms of use (Murati 2023a). The legal perspective ensures that citizens have *equitable access rights* to their data and that transport service providers' use of citizen data is constrained by existing law or is influenced in practice by data-sharing agreements (Murati 2023b; Sooch 2023).

  Ethical consideration is another factor that refers to the notion of correct and incorrect conduct in relation to exploitation of data, especially personal data (Torre-Bastida et al. 2022). *Bureaucracy* between management and the government can result in miscommunication. The value of data governance policies to upper executives can be diminished by bureaucracy in setting policies. This is because most upper management does not fully understand the need for governing data because the importance is not well conveyed, resulting in miscommunication that then affects the data sharing within the organization (Green and Lucivero 2018). The extent of *market regulation* is a factor based on the existing data model employed by the transport service providers and municipality (Weber et al. 2009). Some institutions employ a regulated environment based on a centralized data governance *approach for compliance*, whereas others use a less (or no) regulated approach where decision rights are distributed across all actors through a decentralized approach (Lee et al. 2018).
- *Organizational Factors*: These depend mostly on the general mindset of the city in relation to the data, processes, and participants. They also include how to coordinate data management between the various needs of different stakeholders (Docherty 2018) and further involve the necessary economic and administrative actions to be employed for effective use of data to support mobility services (Anthony Jnr 2023a; Murati 2023b). Organizational factors identified from the literature include, among others, the culture of the people involved in data management. The *culture of an organization* regarding the value of data will influence how they govern data (Lis et al. 2022). For example, some organizations consider data management a task for all employees, rather than relying on data specialists who are involved in data-driven operations (Sooch 2023). The current *decision rights* within the organization in relation to how decisions are implemented as related to data-driven innovation can impact data governance policies (Lis et al. 2022). The decision rights determine who has power over decision-making for data-related resources such as determining usage conditions or data standards (Lis and Otto 2021).

  The *intrinsic complexities* that data owners and data producers are faced with when introducing data governance initiatives were also emphasized as an issue. This is because the *existing workflow* for transport service providers is usually altered when they incorporate citizens' needs. For these reasons, data governance procedures and mechanisms require teams capable of steering resistance to change, advancing strong team dynamics, and ensuring seamless data management practices across society (Sooch 2023). Moreover, *economic barriers* are an important driver for any data-sharing model because they enable actors such as citizens to create and capture value because data sharing typically entails maintenance costs for transport service providers and data owners. Thus, if transport service providers cannot recoup a satisfactory level of profit on their data-related investments, e.g., through revenues obtained from granting data access and sharing, there is a possibility that data access and sharing will be inhibited. The *operational strategy,* as mentioned by Weber et al. (2009), stated that if an organization is more focused on profit, the efficiency of their policies will be based on the firm's investment. This will be evaluated in relation to their operating costs because higher operational efficiency results in greater profitability (Sooch 2023).

7.4 DISCUSSION AND IMPLICATIONS

Data governance aims to improve data through appropriate structure, methods, and implementation, ensuring that all stakeholders such as transport service providers have ready access to data for better decision-making and optimization of their business processes. Data governance commonly seeks to address data-related issues such as maintaining valuable data that can support decision-making, locating data, identifying data quality issues, eliminating redundancies, standardizing data collection, and supporting data access (Green and Lucivero 2018). In smart cities, transport service providers increasingly depend on data to make important decisions concerning the deployment, operation, and maintenance of safe, effective, and sustainable transportation systems. Municipalities and transport service providers have recognized the importance of managing data as an asset, including establishing policies, standards, mechanisms, and procedures for data usage, development, and organization, as well as creating an appropriate managerial structure and technological infrastructure to support effective data governance.

However, for cities to effectively implement human-centric data governance policies, the factors that may motivate or inhibit the exploitation and sharing of data need to be explored. Findings from this study focus on exploring the key mechanisms and procedures that improve human-centric data governance practices for urban mobility service in smart cities. The key mechanisms and procedures, as seen in Figures 7.2 and 7.3, offer recommendations as a comprehensive guide for deploying effective data governance policies. Building on the secondary data, a data governance framework is proposed to offer practical experiences and theoretical insights. The framework contributes to social change by equipping public transit agencies, transport service providers, and similar organizations with actionable initiatives for enhancing data governance and management, benefiting not only the transport service providers but also citizens.

This research aims to facilitate positive social change by empowering citizens to make informed decisions, promote operational efficiency, enhance transparency and accountability, and foster collaboration and innovation for urban mobility services in smart cities. The findings discuss possible policy approaches that contribute to the current knowledge on policies and strategies needed for human-centric urban mobility services in smart cities. Evidence from this study provides valuable strategies and insights needed to enhance data governance for urban mobility services in smart cities. The identified data governance mechanisms and procedures could be adopted to improve data utilization, sharing, and management, ultimately enhancing data quality, integrity, accuracy, and trustworthiness. The outcomes of this study could help transport service providers comply with data regulations, ethics, and policies to minimize data mismanagement or misuse.

7.5 CONCLUSION

In the data-driven and increasingly digitalized landscape of urban mobility services in smart cities, the efficiency of data governance policies has emerged as a tool for operational and strategic decision-making. This chapter proposes a human-centric data governance framework grounded in the literature to support urban mobility services in smart cities. To accomplish this goal, this study describes citizen-centric governance of urban mobility applications and data governance of human-centric mobility services in smart cities. More importantly, findings from this study present the mechanisms, strategies, and factors that impact human-centric data governance policies. Further work will be dedicated to evaluating the proposed framework based on real mobility use cases in smart cities.

REFERENCES

Anthony Jnr, Bokolo. "Using green IT governance as a catalyst to improve sustainable practices adoption: A contingency theory perspective." *International Journal of Business Continuity and Risk Management* 8, no. 2 (2018): 124–157.

Anthony Jnr, Bokolo. "Exploring data driven initiatives for smart city development: Empirical evidence from techno-stakeholders' perspective." *Urban Research & Practice* 15, no. 4 (2022a): 529–560.

Anthony Jnr, Bokolo. "Toward a collaborative governance model for distributed ledger technology adoption in organizations." *Environment Systems and Decisions* 42, no. 2 (2022b): 276–294.

Anthony Jnr, Bokolo. "Sustainable mobility governance in smart cities for urban policy development–A scoping review and conceptual model." *Smart and Sustainable Built Environment* (2023c). https://doi.org/10.1108/SASBE-05-2023-0109

Anthony Jnr, Bokolo. "Data driven approaches for smart city planning and design: A case scenario on urban data management." *Digital Policy, Regulation and Governance* 25, no. 4 (2023d): 351–367.

Anthony Jnr, Bokolo. "A developed distributed ledger technology architectural layer framework for decentralized governance implementation in virtual enterprise." *Information Systems and e-Business Management* 21, no. 3 (2023a): 437–470.

Anthony Jnr, Bokolo. "Investigating the decentralized governance of distributed ledger infrastructure implementation in extended enterprises." *Journal of the Knowledge Economy* 14, no. 4 (2023b): 5003–5032.

Anthony Jnr, Bokolo. "Developing green urban mobility policies for sustainable public transportation in local communities: A Norwegian perspective." *Journal of Place Management and Development* 17, no. 1 (2024a): 136–155.

Anthony Jnr, Bokolo. "Developing a decentralized community of practice-based model for on-demand electric car-pooling towards sustainable shared mobility." *Case Studies on Transport Policy* 15 (2024b): 101136.

Anthony Jnr, Bokolo. "The role of community engagement in urban innovation towards the co-creation of smart sustainable cities." *Journal of the Knowledge Economy* 15, no. 1 (2024c): 1592–1624.

Artyushina, Anna. "Is civic data governance the key to democratic smart cities? The role of the urban data trust in Sidewalk Toronto." *Telematics and Informatics* 55 (2020): 101456.

Bevir, Mark. A theory of governance. University of California Press, Berkeley CA. (2013).

Choenni, Sunil, Mortaza S. Bargh, Tony Busker, and Niels Netten. "Data governance in smart cities: Challenges and solution directions." *Journal of Smart Cities and Society* 1, no. 1 (2022): 31–51.

De Prieëlle, Fabian, Mark De Reuver, and Jafar Rezaei. "The role of ecosystem data governance in adoption of data platforms by Internet-of-Things data providers: Case of Dutch horticulture industry." *IEEE Transactions on Engineering Management* 69, no. 4 (2020): 940–950.

de Souza, Thais Assis, Kelly Carvalho Vieira, Isabelle Nicolaï, and André Grützmann. "Can urban mobility be responsible? A governance perspective." *Technology Analysis & Strategic Management* 36, no. 11 (2024): 3507–3523.

Docherty, Iain. "New governance challenges in the era of 'smart'mobility." In Governance of the Smart Mobility Transition, pp. 19–32. (2018). Emerald Publishing Limited.

FHWA. Data Governance Plan Volume 1 Data Governance Primer -U.S. Department of Transportation: The Federal Highway Administration (FHWA). (2015).

Früh, Alfred, Nadja Braun Binder, and Robert Schibli. "Data Governance für Smart Mobility aus rechtlicher Perspektive." *Sui Generis* (2022): 15–24. https://sui-generis.ch/article/download/sg.200/2455

Green, Michael, and Anthony Lucivero. "Data governance & data management case studies of select transportation agencies." United States Department of Transportation (2018). https://rosap.ntl.bts.gov/view/dot/53783/dot_53783_DS2.pdf

Hoffmann, Ilko, Nils Jensen, and Anamaria Cristescu. "Decentralized governance for digital platforms-architecture proposal for the mobility market to enhance data privacy and market diversity." IEEE 18th Annual Consumer Communications & Networking Conference (CCNC) (2021), pp. 1–6.

Jansson, Jonas, Joakim Dahlman, and Hillevi Ternström. "National data governance in the transport sector: An international outlook and literature review." VTI rapport 1115A (2022). https://www.diva-portal.org/smash/get/diva2:1643419/FULLTEXT01.pdf

Lee, Sung Une, Liming Zhu, and Ross Jeffery. "Designing data governance in platform ecosystems." (2018).

Lis, Dominik, Michael Arbter, Markus Spindler, and Boris Otto. "An investigation of antecedents for data governance adoption in the rail industry—Findings from a case study at Thales." *IEEE Transactions on Engineering Management* 70, no. 7 (2022): 2528–2545.

Lis, Dominik, and Boris Otto. "Towards a taxonomy of ecosystem data governance." (2021).

Meijer, Albert. "Datapolis: A public governance perspective on "smart cities." *Perspectives on Public Management and Governance* 1, no. 3 (2018): 195–206.

Murati, Erion. "Establishing a Three-Dimensional MaaS Data Sharing Governance." In Regulating Mobility as a Service (MaaS) in European Union: A Legal Analysis of Digitalization, Competition, and Multimodality, pp. 47–76. (2023b). Springer Nature Switzerland.

Murati, Erion. "Adoption of a MaaS data governance decision model to facilitate voluntary sharing of transport service data." In Regulating Mobility as a Service (MaaS) in European Union: A Legal Analysis of Digitalization, Competition, and Multimodality, pp. 77–107. (2023a). Springer Nature Switzerland.

Nokkala, Tiina, Hannu Salmela, and Jouko Toivonen. "Data governance in digital platforms." (2019).

Snyder, Hannah. "Literature review as a research methodology: An overview and guidelines." *Journal of Business Research* 104 (2019): 333–339.

Sooch, Manjit K. "Implementing data governance for a data-driven culture in public transit agencies." PhD diss., University of the Southwest, 2023.

Tan, Hongze. "Smarter mobility for better urban development: An exploration of urban smart transportation governance based on international comparison." *Studies in Social Science & Humanities* 2, no. 5 (2023): 1–11.

Torre-Bastida, Ana I., Guillermo Gil, Raúl Miñón, and Josu Díaz-de-Arcaya. "Technological perspective of data governance in data space ecosystems." In Data Spaces: Design, Deployment and Future Directions, pp. 65–87. (2022).

van den Hurk, Martijn, Peter Pelzer, and Rianne Riemens. "Governance challenges of mobility platforms: The case of Merwede, Utrecht." *European Transport Research Review* 13 (2021): 1–12.

Veeneman, Wijnand, Haiko van der Voort, Fabio Hirschhorn, Bauke Steenhuisen, and Bram Klievink. "PETRA: Governance as a key success factor for big data solutions in mobility." *Research in Transportation Economics* 69 (2018): 420–429.

Weber, Kristin, Boris Otto, and Hubert Österle. "One size does not fit all—a contingency approach to data governance." *Journal of Data and Information Quality (JDIQ)* 1, no. 1 (2009): 1–27.

Webster, Jane, and Richard T Watson. "Analyzing the past to prepare for the future: Writing a literature review." *MIS Quarterly* 26, no. 2 (2002): 13–23.

Weill, Peter, and Jeanne W. Ross. "IT governance on one page." *SSRN Electronic Journal* (2004): 349. https://doi.org/10.2139/ssrn.664612

Wende, Kristin. "A model for data governance–Organising accountabilities for data quality management." (2007).

Zhong, Ziyan, and Ying Wang. "Open data governance in smart mobility: An empirical investigation of Shanghai." Master's thesis, Nord Universitet, 2022.

Zorrilla, Marta, and Juan Yebenes. "A reference framework for the implementation of data governance systems for industry 4.0." *Computer Standards & Interfaces* 81 (2022): 103595.

8 Urban Human Mobility and Energy

Ehsan Dorostkar, Mahsa Najarsadeghi, and Xiao Huang

8.1 INTRODUCTION

Rapid urbanization and uncontrolled city growth have led to numerous challenges that adversely affect the quality of life (QOL) for urban residents. The World Health Organization defines QOL as individuals' perceptions of their position in life within the context of the culture and value systems in which they live. Key challenges to QOL in urban environments include issues related to energy consumption and distribution, human mobility, environmental degradation, and socioeconomic disparities.

Urban human mobility is intricately linked to energy consumption and distribution, significantly influencing the development of complex and hierarchical transportation networks. As cities evolve, there is a pressing need to transition toward smart cities—urban areas that leverage data and technology to enhance political, social, and environmental decision-making processes (Allam and Dhunny 2019; Dorostkar et al. 2023). Smart cities aim to address the unique needs and challenges posed by rapid urbanization, making it essential to evaluate and understand the fundamental data that inform decision-makers (De Guimarães et al. 2020; Dorostkar and Najarsadeghi 2024).

Technological innovations such as electric and hybrid vehicles, the integration of renewable energy sources in transportation, smart grids, and the proliferation of the Internet of Things (IoT) are transforming the landscape of urban mobility and energy consumption. These advancements offer promising solutions to mitigate the environmental impact of human mobility, reduce greenhouse gas emissions, and enhance energy efficiency within cities. However, the adoption of these technologies is influenced by various factors, including urban design and infrastructure, population density, social and economic dynamics, policy and regulations, and technological advancements in vehicles and fuels. Understanding these factors is crucial for developing effective strategies to manage energy consumption and promote sustainable urban mobility.

This chapter explores how energy consumption and distribution are influenced by human mobility and examines the consequent effects on urban environments. It delves into the technological innovations shaping the future of urban transportation and energy management, discusses the factors affecting energy consumption, and outlines strategies to achieve sustainable urban mobility. By addressing these critical issues, the chapter aims to contribute to the global effort to mitigate adverse environmental impacts and promote sustainability in the context of urban development and human mobility.

8.2 TECHNOLOGICAL INNOVATIONS

Human mobility within urban environments and its connection with energy involve various factors. These factors can often be categorized into specific groups. The main categories in this sector include electric and hybrid vehicles, renewable energy sources in transportation, smart grids and energy management systems, and the role of the IoT and smart city technologies. Technological innovations are regarded as the driving force behind human mobility and energy.

 DOI: 10.1201/9781003503262-10

8.2.1 Electric and Hybrid Cars

Electric and hybrid vehicles are at the forefront of transforming urban transportation and play a critical role in managing energy consumption within cities. Their adoption is directly linked to human mobility patterns and is significantly influenced by a country's electricity production methods and the integration of renewable energy sources. As cities strive to become smarter and more sustainable, electric and hybrid vehicles represent a vital first link in the innovation value chain at the urban level. Optimizing the energy conversion performance of electric and hybrid vehicles is essential for maximizing their efficiency and environmental benefits. According to Du et al. (2024), energy management strategies are key to enhancing the overall performance of a vehicle's power system. These strategies focus on factors such as vehicle speed, battery state of charge, and environmental conditions like temperature and road gradients. By effectively managing these variables, the energy consumption and emissions of vehicles can be significantly reduced.

Proper planning of vehicle speed is another critical aspect that influences energy efficiency. Li et al. (2024) emphasize that optimizing speed profiles not only enhances safety, comfort, and traffic flow but also improves fuel consumption. Leveraging the characteristics of a vehicle's power system allows for intelligent control of acceleration and deceleration, reducing unnecessary energy expenditure and extending the lifespan of the vehicle's battery. In addition to individual vehicle optimization, the integration of electric and hybrid vehicles with autonomous microgrid systems offers a holistic approach to energy management. Er et al. (2024) identify cost minimization, emission reduction, and reliability maximization as the most widely pursued objectives for these systems. By connecting vehicles to smart grids and energy management systems, cities can optimize energy distribution, utilize renewable energy sources more effectively, and enhance the reliability of the power supply for transportation needs.

Electric and hybrid vehicles also contribute to the development of sustainable urban environments by reducing reliance on fossil fuels and decreasing greenhouse gas emissions. Their widespread adoption requires supportive infrastructure, such as readily available charging stations powered by renewable energy, and policies that encourage the use of clean transportation alternatives.

8.2.2 Renewable Energy Sources in Transportation

The utilization of renewable energy gained significant momentum in the late 1970s following the major oil crisis. During that period, economic consequences and costs were paramount, leading to heightened interest in alternative energy sources. However, as oil prices declined, the enthusiasm for renewable energy waned due to the restored economic viability of fossil fuels. In recent years, the renewed focus on renewable energy is largely attributable to the severe environmental impacts associated with fossil fuel consumption, such as climate change, air pollution, and ecological degradation. One of the most significant challenges in the field of electricity is the inability to store electrical energy on a large scale efficiently and economically. This limitation hampers the widespread adoption of renewable energy sources like solar and wind, which are intermittent by nature. As oil and fossil fuel resources continue to diminish and their prices, along with environmental consequences, escalate, there is a growing global impetus to transition toward renewable and clean energy sources in the coming decades. This shift underscores the critical importance of investing in and acquiring the technical and technological knowledge necessary for their effective utilization.

Humanity's extensive need for energy has always influenced fundamental aspects of societal development and quality of life. The environmental problems caused by fossil fuel consumption, coupled with the increasing global energy demand, make the transition toward renewable energies—and the development and deployment of these resources—more urgent each day. In this context, the use of renewable energy in transportation emerges as a key component of optimal energy use and sustainability, positioning it as a primary option in efforts to reduce environmental impact within the transportation sector.

In the realm of transportation, renewable energy sources such as solar energy, kinetic energy generated by vehicle movement, and biomass energy can be effectively harnessed:

- *Solar energy*: Solar energy can be captured through photovoltaic cells installed on vehicles or integrated into transportation infrastructure. For example, solar panels on electric buses or cars can provide auxiliary power or extend the driving range by charging the battery while the vehicle is in use or parked. Solar-powered charging stations for electric vehicles are also becoming increasingly prevalent, reducing reliance on grid electricity generated from fossil fuels (Yap et al. 2022).
- *Kinetic energy recovery systems (KERS)*: KERS capture the kinetic energy typically lost during braking and deceleration. This energy is converted into electrical energy and stored in batteries or supercapacitors for later use, improving vehicle efficiency and reducing overall energy consumption. Regenerative braking systems are now standard in many hybrid and electric vehicles, contributing to significant energy savings (Qiu and Wang 2016).
- *Biomass energy*: Biomass energy involves converting organic materials—such as biodegradable components of products, agricultural residues and wastes, plant and animal materials, forestry residues, and industrial and urban biodegradable wastes—into biofuels like biodiesel, bioethanol, and biogas. These biofuels can be used in compatible internal combustion engines, reducing reliance on petroleum-based fuels and decreasing greenhouse gas emissions (Awogbemi et al. 2021).

8.2.3 Smart Grid and Energy Management Systems

Smart grids represent a significant evolution in the management and distribution of electrical energy. While various definitions exist, they generally converge on the concept of an advanced electricity network that utilizes digital communications technology to detect and respond to local changes in usage, thereby improving efficiency, reliability, and sustainability (El-Hawary 2014). Two primary characterizations of smart grids are commonly cited. First, a smart grid is often described as a "network of networks." This perspective emphasizes that a smart grid is an interconnected system comprising multiple smaller networks, such as telecommunications networks, information technology networks, and control networks. These integrated networks enable seamless communication and coordination across the entire electrical grid, facilitating real-time monitoring and management. Second, smart grids are characterized by "using megabytes of data to move megawatts of power." This description highlights the critical role of data in optimizing power flow within the grid. By leveraging vast amounts of data collected from sensors and measuring devices, smart grids enhance decision-making processes, improve demand response, and ensure efficient energy distribution.

The objectives for implementing smart grids vary by region due to differing priorities and challenges. In the United States, the primary goal is to increase the reliability of the electricity supply. Smart grids aim to reduce outages, improve grid resilience, and enhance service quality by quickly responding to disturbances and preventing widespread blackouts. In Europe, the focus is on environmental sustainability and the integration of renewable energy sources. Smart grids facilitate the accommodation of variable renewable generation, such as wind and solar power, by providing flexibility and advanced control mechanisms to balance supply and demand. In China, the main objective is to improve grid stability, thereby enhancing power system security.

The implementation of smart grids introduces fundamental changes to the existing electrical network structure. These changes include the development of a highly reliable network with inherent security at all levels, as well as decentralized and comprehensive control enabled by extensive use of sensors and measuring devices (Tuballa and Abundo 2016). This allows for a self-healing distribution network capable of automatically detecting, isolating, and repairing faults, thereby minimizing the impact of outages on consumers.

8.2.4 Role of the Internet of Things and Smart City Technologies

The IoT and smart city technologies are pivotal in advancing urban human mobility and achieving the overarching goals of smart cities. By embedding IoT devices within urban infrastructure, cities can enhance the efficiency of transportation systems, increase security and public safety, and promote health, sustainability, and overall quality of life. These technologies enable real-time data collection and analysis, facilitating more responsive and adaptive urban environments. For example, smart floor tiles have been developed to harvest energy from the pressure of pedestrian footsteps, converting kinetic energy into electricity that can power streetlights or charge devices (Visconti et al. 2022). In residential settings, similar technologies monitor physical activity to support health and safety, particularly for the elderly or those with mobility challenges.

In urban settings, interconnected devices communicate through Wi-Fi and cloud computing systems, interacting with each other, the environment, and people to optimize mobility. Barcelona's implementation of intelligent light-emitting diode (LED) streetlights exemplifies this integration. These streetlights receive real-time data on humidity, temperature, pollution levels, pedestrian locations, and street noise, allowing them to adjust light intensity to improve visibility and safety while conserving energy. Equipped with a central computing unit, each lighting fixture communicates with others to manage electric vehicle charging stations and provide public Wi-Fi access, contributing to a more connected and efficient transportation network. Moreover, IoT technologies facilitate advanced traffic management systems that reduce congestion and improve public transit services. Sensors and cameras installed at intersections monitor traffic flow, enabling dynamic adjustment of traffic signals to optimize vehicle movement and reduce delays. Public transportation vehicles equipped with IoT devices offer real-time location and arrival information to commuters via mobile apps, enhancing user experience and encouraging the use of public transit.

Bicycle-sharing programs also benefit from IoT integration, with smart docking stations and global positioning system (GPS)-enabled bikes providing data on availability and usage patterns (Tekouabou 2021). This information helps in redistributing bikes to meet demand and in planning future infrastructure investments. Autonomous vehicles, another emerging facet of IoT in urban mobility, rely on a network of sensors and communication systems to navigate streets safely, promising further reductions in traffic congestion and emissions.

8.3 FACTORS AFFECTING ENERGY CONSUMPTION

8.3.1 Urban Design and Infrastructure

Urban design and infrastructure have a profound impact on energy consumption within cities, both historically and in contemporary contexts. The way cities are planned and developed influences patterns of energy use, as infrastructure provides access to facilities and services that are managed based on prevailing energy consumption models. Modern urban design increasingly incorporates new knowledge and technologies to optimize energy use and improve design paradigms. For instance, the integration of renewable energy sources into urban environments is becoming more prevalent. Solar panels can be seamlessly incorporated into building surfaces, sometimes replacing traditional electricity generation methods entirely (Kovács et al. 2016). This technology is also applicable to public amenities, such as street lighting and traffic signals, reducing reliance on grid electricity. While wind energy presents challenges due to typically low wind speeds in urban areas and difficulties in integrating micro wind turbines, advancements continue to be made. The effectiveness of micro-generation from renewable sources is heavily dependent on the city's climatic conditions and morphological characteristics. Therefore, thoughtful urban design that considers these factors is crucial for promoting sustainable energy consumption.

8.3.2 Population Density and Urban Expansion

The rapid growth of urban populations necessitates development models that can accommodate large numbers of residents in a managed and sustainable manner. Two prevalent models of smart city development address this need. The first model adopts a technological approach, evolving from the information and communication technology sector to enhance the efficiency and effectiveness of urban services and processes. In this framework, data collected from various sources are transformed into valuable urban intelligence through powerful analytics, thereby improving urban management. This model aligns with the concepts of smart industries and digital cities. The second model offers a more holistic approach, striving for a balance among socioeconomic, environmental, human, cultural, and technological factors. This comprehensive perspective is often the focus of academic research, aiming to address new urban challenges such as climate change, economic development, and the efficient use of resources. Both models underscore the importance of integrating technology and sustainability principles to manage urban expansion and population density effectively.

8.3.3 Social and Economic Factors

By 2050, it is projected that 70% of the world's population will reside in cities (World Bank 2024), underscoring the significant role urban life plays in developing human, cultural, economic, environmental, and social capital. This concentration of people intensifies the consumption of resources, increases waste production, attracts capital, and elevates the demand for services. Similar to the industrial revolution of the 19th century, which introduced challenges related to health and housing, the current era of knowledge and information technology presents new opportunities and issues. Advancements in information and communication technology offer solutions to contemporary urban problems such as energy resource utilization, equitable access to urban services, and efficient mobility. The COVID-19 pandemic in 2020 further highlighted the necessity for remote services, smart banking, digital networking, and advanced healthcare services. These developments create new avenues for cities to address natural and human-made disasters, emphasizing the need for resilient and adaptable urban systems that can respond to multifaceted social and economic challenges.

8.3.4 Technological Innovations in Vehicles and Fuel

The emergence of self-driving cars represents a significant advancement in smart transportation, incorporating a series of innovations and cutting-edge technologies aimed at enhancing safety, productivity, and the overall travel experience. These vehicles utilize a combination of lasers, radars, sensors, and cameras to perceive their environment, supported by algorithms and artificial intelligence (AI) models that analyze sensor data, predict environmental behavior, and make real-time decisions without human intervention. Key technologies in this field include LiDAR (light detection and ranging) systems for high-precision environmental mapping through laser beams and reflections; image recognition systems for detecting obstacles, other vehicles, and road signs; sensors and radars for measuring distances and identifying objects on the road; AI for processing data and decision-making; advanced GPS systems for accurate positioning and routing; and V2X (vehicle-to-everything) communication networks that enable direct interaction between the vehicle and other environmental elements like traffic signals and road infrastructure. These technologies are continually evolving, promising even greater advancements in autonomous vehicle capabilities and their integration into smart transportation networks.

8.3.5 Effects of Policy and Regulations

The global energy crisis manifests in several critical ways: the impending depletion of fossil fuel reserves, the substantial financial investment required to transition to new and alternative energies—which poses significant challenges for developing countries—and the intensified impact of the energy crisis on these nations, as they may incur higher energy costs for economic development. Developing and implementing appropriate policies and regulations that consider the environmental, economic, cultural, and social conditions of cities are essential for influencing energy consumption both directly and indirectly. Human mobility within cities initiates energy consumption across various frameworks, systematically linked to factors such as the type and purpose of mobility. Contemporary issues are complex and multifaceted, necessitating a comprehensive approach to urban policy. For example, implementing population decentralization policies in densely populated cities can address multiple urban challenges and significantly affect energy consumption patterns. By adopting strategic policies and regulations, cities can manage energy resources more effectively, promote sustainable development, and mitigate the adverse effects of the energy crisis on economic growth.

8.4 FUTURE PATHS

8.4.1 Predictions for Future Urban Mobility and Energy Trends

High levels of uncertainty regarding future energy demands impose limitations on the continuous growth of cities worldwide. Urban areas, which have become the most complex built environments in human history, are rapidly expanding in both size and population (Mohammadi and Taylor 2017). In the future, urban areas will face significant challenges related to energy consumption, particularly in sectors that are difficult to predict accurately. A key aspect of these challenges is the integration of robotics, AI, and intelligent systems into urban environments. The advancement of robotics and AI will introduce practical urban functionalities that currently do not exist, transforming infrastructure and services to better meet the needs of citizens. Urban mobility will be profoundly impacted by the incorporation of robotics and AI. The integration of these technologies into urban systems will lead to a major transformation in how people and goods move within cities, introducing new activities and functions. This increased mobility and access to more activities will, in turn, create a specific and complete cycle of heightened energy consumption. Current predictions indicate that human and urban mobility—alongside the expansion of robotics and AI, the development of platforms for smart cities, and the continuous increase in big data transmission—will lead to the formation of new systems of mobility and energy governance at the urban level. This evolution requires thorough preparation across various departments of urban management, from macro to micro levels.

8.4.2 Innovations on the Horizon: Self-Driving Vehicles and Green Hydrogen

As discussed in the previous section, the future of urban environments will be shaped by robotics, AI, and the development of smart cities compatible with these technologies. Consequently, innovations in areas such as self-driving vehicles and green hydrogen are inevitable. Future self-driving vehicles must incorporate a variety of advanced features. These include integration with big data analytics systems, enabling communication with comprehensive and diverse networks, and establishing real-time interactions with both active and nearby systems. Such vehicles should possess the ability to learn and adapt, using existing knowledge and acquiring new insights in harmony with their environment. They will utilize multidimensional energy sources efficiently, adapt to temporal and spatial changes through intelligent processing, and coordinate with environmental conditions to reduce energy consumption. The use of real-time data will allow these vehicles to adjust energy usage based on available facilities, and remote sensing systems will evaluate optimal mobility conditions across different territories.

Innovation in robotics, AI, and adaptive smart city technologies is highly dynamic and diverse. These advancements will transform cities into "thinking cities" that interact seamlessly with their inhabitants. The integration of self-driving vehicles and sustainable energy sources like green hydrogen will play a critical role in achieving this vision, promoting more efficient, sustainable, and intelligent urban mobility systems.

8.4.3 Strategies to Achieve Sustainable Urban Mobility

Sustainable development is a global paradigm endorsed by the United Nations, introduced in 1987 through the Brundtland Commission Report. It is defined as development that meets the needs of the present generation without compromising the ability of future generations to meet their own needs. Sustainable development encompasses four intertwined dimensions: society, environment, culture, and economy. Therefore, the primary goals and missions of smart cities should align with their foundational concepts and ideal performance, which are intrinsically linked to sustainability across economic, social, environmental, and cultural dimensions. Additionally, the political context of governments where smart cities evolve plays a crucial role. In neoliberal economies, cities offer diverse facilities and services to citizens through private sector participation within competitive environments. This concept has rapidly gained prominence in industry, politics, and municipal discourse regarding urban development.

To achieve sustainable urban mobility, social and economic approaches can effectively influence energy consumption by considering the following strategies:

- *Engaging public organizations and citizens*: Involve city residents and public organizations in energy consumption management to foster collective responsibility toward energy efficiency.
- *Economic incentives*: Provide economic savings and incentives for residents who manage and reduce their energy consumption, encouraging sustainable practices.
- *Creating value chains*: Establish value chains that link social and economic activities in energy consumption management, promoting collaboration and shared benefits.
- *Regulatory frameworks*: Implement regulations that enhance energy management and improve social and economic efficiency within urban areas.
- *Social activities as catalysts*: Recognize social activities as the initial link in energy consumption management, leveraging community engagement to drive sustainable practices.

By integrating these strategies, cities can progress toward sustainable urban mobility, ensuring that development aligns with the overarching goals of sustainability and meets the needs of both current and future generations.

8.5 CONCLUSION

Urban human mobility and energy are deeply interconnected, serving as critical determinants in the evolution of modern cities. This chapter has explored how rapid urbanization and the complexities of energy consumption present both challenges and opportunities for urban environments. Technological innovations—such as electric and hybrid vehicles, renewable energy integration in transportation, smart grids, and the deployment of IoT and smart city technologies—are pivotal in addressing these challenges. Additionally, factors like urban design, population density, social and economic dynamics, technological advancements in vehicles and fuels, and policy regulations significantly influence energy consumption patterns. Looking toward the future, the integration of robotics, AI, self-driving vehicles, and sustainable energy sources like green hydrogen is expected to revolutionize urban mobility and energy governance. These advancements will transform cities into intelligent ecosystems that not only meet the evolving needs of their citizens but also promote sustainability and environmental stewardship. Achieving sustainable urban mobility requires strategic approaches that engage public organizations and citizens, provide economic incentives, establish value chains in energy management, implement robust regulatory frameworks, and leverage social activities to drive energy efficiency.

REFERENCES

Allam, Zaheer, and Zaynah A. Dhunny. "On big data, artificial intelligence and smart cities." *Cities* 89, no. 89 (2019): 80–91.

Awogbemi, Omojola, Daramy Vandi Von Kallon, Emmanuel Idoko Onuh, and Victor Sunday Aigbodion. "An overview of the classification, production and utilization of biofuels for internal combustion engine applications." *Energies* 14, no. 18 (2021): 5687.

De Guimarães, Julio Cesar Ferro, Eliana Andréa Severo, Luiz Antonio Felix Júnior, Wênyka Preston Leite Batista Da Costa, and Fernanda Tasso Salmoria. "Governance and quality of life in smart cities: Towards sustainable development goals." *Journal of Cleaner Production* 253 (2020): 119926.

Dorostkar, Ehsan, and Mahsa Najarsadeghi. "Hidden urban patterns: Existential discovery of urban patterns based on traffic and virtual space." *Transportation Research Interdisciplinary Perspectives* 23 (2024): 100997.

Dorostkar, Ehsan, Mahsa Najarsadeghi, Mehrnaz Molavi, and Nader Zali. "Human mobility and energy: How do human mobility and energy affect urban policy and planning?" *Journal of Urban Management* 12, no. 4 (2023): 413–419.

Du, Aimin, Yeyang Han, Zhongpan Zhu, and Chuanchuan Chu. "Multi-objective optimization of safety, comfort and economy of hybrid electric vehicle in car-following scenario." *International Journal of Green Energy* 21, no. 5 (2024): 1060–1071.

El-Hawary, Mohamed E. "The smart grid—State-of-the-art and future trends." *Electric Power Components and Systems* 42, no. 3–4 (2014): 239–250.

Er, Gulfem, Gurkan Soykan, and Ethem Canakoglu. "Stochastic optimal design of a rural microgrid with hybrid storage system including hydrogen and electric cars using vehicle-to-grid technology." *Journal of Energy Storage* 75 (2024): 109747.

Kovács, András, Roland Bátai, Balázs Csanád Csáji, Péter Dudás, Borbála Háy, Gianfranco Pedone, Tibor Révész, and József Váncza. "Intelligent control for energy-positive street lighting." *Energy* 114 (2016): 40–51.

Li, Cheng, Xiangyang Xu, Helong Zhu, Jiongpeng Gan, Zhige Chen, and Xiaolin Tang. "Research on car-following control and energy management strategy of hybrid electric vehicles in connected scene." *Energy* 293 (2024): 130586.

Mohammadi, Neda, and John E. Taylor. "Urban energy flux: Spatiotemporal fluctuations of building energy consumption and human mobility-driven prediction." *Applied Energy* 195 (2017): 810–818.

Qiu, Chengqun, and Guolin Wang. "New evaluation methodology of regenerative braking contribution to energy efficiency improvement of electric vehicles." *Energy Conversion and Management* 119 (2016): 389–398.

Tekouabou, Stephane Cedric Koumetio. "Intelligent management of bike sharing in smart cities using machine learning and internet of things." *Sustainable Cities and Society* 67 (2021): 102702.

Tuballa, Maria Lorena, and Michael Lochinvar Abundo. "A review of the development of smart grid technologies." *Renewable and Sustainable Energy Reviews* 59 (2016): 710–725.

Visconti, Paolo, Laura Bagordo, Ramiro Velázquez, Donato Cafagna, and Roberto De Fazio. "Available technologies and commercial devices to harvest energy by human trampling in smart flooring systems: A review." *Energies* 15, no. 2 (2022): 432.

World Bank—"Overview". Accessed September 28, 2024. https://www.worldbank.org/en/topic/urbandevelopment/overview.

Yap, Kah Yung, Hon Huin Chin, and Jiří Jaromír Klemeš. "Solar energy-powered battery electric vehicle charging stations: Current development and future prospect review." *Renewable and Sustainable Energy Reviews* 169 (2022): 112862.

Section III

Technologies and Techniques Shaping Urban Human Mobility Studies

9 Prediction and Analysis of Urban Mobility Based on Attention Mechanism and Geographic Information Embedding

Yao Yao and Zijin Guo

9.1 INTRODUCTION: BACKGROUND AND DRIVING FORCES

The expansion of cities has led to a growing contradiction between the increasing human mobility and the finite urban resource space, resulting in urban issues such as traffic congestion and regional development imbalances (Cobbinah et al. 2022).To address these challenges, geospatial data generated by communication base stations or smartphones have been used to understand the dynamic operation of cities (Liu et al. 2015; Wang et al. 2023). Geospatial data with large sample size, wide coverage, and high timeliness can effectively support urban study (Liu et al. 2020).

An in-depth analysis of the predictability of human mobility makes it possible to determine the activity patterns of users at different times of the day, such as the differences between weekdays and weekends (Wang, Zhang and Li 2022). Periodic patterns in the trajectories not only reflect habits at the individual level, but also reveal the rhythms and patterns of urban operation at the macro level (Zhao et al. 2023). The use of geospatial data to monitor urban dynamics has important reference value for urban planning and helps to improve the efficiency of urban operations and the quality of life of residents.

Next, location prediction refers to the prediction of one or more steps forward based on existing historical trajectories (Yao et al. 2023, 102009). Existing location prediction studies in a city can be categorized by subject as either human activity or vehicle. Vehicle trajectories are recorded by vehicle-mounted global positioning system (GPS) devices and represent the continuous driving paths of vehicles (such as cars, buses, and bicycles) on the road network. Restricted by road network structure and traffic rules, vehicle trajectories exhibit certain regularity. However, human activity trajectories and vehicle trajectories differ significantly in predictability and positioning accuracy (Qian et al. 2021). The human trajectories are less continuous than vehicle trajectories and are more influenced by personal preferences and activity purposes, making them more random and complex. Human activity trajectories can typically exhibit certain periodic characteristics (Deschaintres, Morency and Trépanier 2022). Consequently, the development of a method applicable to human activity trajectories is necessary to mine the periodic feature information in trajectories.

Using machine learning techniques and considering the complex mechanism information in multisource spatial data, the extraction of human activity purposes and behavioral preferences can become accurate. Existing researchers have made efforts on location prediction using the temporal features of trajectories and employed three main location prediction techniques: pattern-based, model-based, and neural network–based methods. The pattern-based method is sensitive to the selection of parameters and requires a long time to find frequent patterns (Guo et al. 2010). Model-based approaches determine the most likely locations by calculating the transfer probabilities

DOI: 10.1201/9781003503262-12

between all locations and utilizing dynamic programming. For example, Gambs, Killijian and del Prado Cortez (2012) utilize Markov chains to compute movement probabilities between locations based on historical visit data, while Qiao et al. (2014) capture key parameters of realistic scenarios by improving hidden Markov models (HMMs). However, these models assume trajectory independence, i.e., the current location is only related to previous locations, an assumption that limits the predictive accuracy of the models and leads to poor performance.

Complex and effective deep learning techniques have also been applied to the mining of human activity patterns. Neural network–based methods such as recurrent neural network (RNN) and long short-term memory (LSTM) have been used to mine long- and short-term dependencies from trajectories (Solomon et al. 2021). LSTM can efficiently solve gradient boosting and explosion problems, but it can only exploit historical contextual information (Solomon et al. 2021). The attention mechanism is parameterized by a feedforward neural network and can be integrated into the training of deep neural networks (Miao et al. 2020). Since LSTM models can capture contextual information in trajectories, the combination of attention mechanism and LSTM models further improves the performance of location prediction (Bao et al. 2021). However, an accurate description of trajectories or points of interest in urban space is usually required before applying neural network methods. Neural networks in general have difficulty accounting for the spatial properties and topological relationships of geospatial data, which limits their ability to understand human activity patterns and behavioral preferences.

Urban entity representation is a technique for describing spatial entities such as human activities and urban objects in mathematical vector spaces (Yin et al. 2021). Existing research approaches typically use Word2Vec models to extract spatial information from trajectories or points of interest. Yao et al. (2017a) first combined points of interest with Google's open-source machine learning word embedding model to establish a framework for sensing the spatial structure of urban land use. Zhang et al. (2021) proposed a Traj2Vec model that embeds the context of locations in trajectories to extract semantic information about how people use urban space. For the semantic information in the trajectory, it refers to the spatiotemporal transition relationship indicating the user's intent. The word-embedded representation vectors reflect human activity purposes, which are helpful to improve the performance of land-use classification.

Multisource spatiotemporal data can reveal social attributes in cities, such as functional semantics or population distribution (Yao et al. 2017b). Existing studies indicate that complex bidirectional feedback mechanisms exist between human activity trajectories and urban space (Liu et al. 2022). Considering the geographic information in different spatiotemporal data can improve the expression and understanding of geographic space semantics. However, multisource spatiotemporal data have complex modal features, which makes the fusion of multisource spatiotemporal data difficult and requires full consideration of spatial interconnections.

In response to the above issues, this chapter presents an innovative location prediction model that combines geographic information embedding and attention mechanisms. First, the model encodes location information in trajectories, transforming location information into vectors that capture their spatial characteristics. Then, the model employs a neural network structure based on attention mechanisms, which effectively integrates the spatial features of trajectories and the contextual relationships between locations. In this way, the model is able to capture the complexity and dynamics of human activities more accurately. To verify the effectiveness of the proposed model, this chapter introduces prediction experiments on the Chinese human trajectory dataset. These experiments can deepen the understanding of the relationship between urban spatial area functions and human activities.

9.2 METHOD

This chapter presents a location prediction model based on geographic information embedding and multisource data fusion. The model implements location prediction based on a bidirectional LSTM (Bi-LSTM) neural network and also incorporates the geographic information embedding method

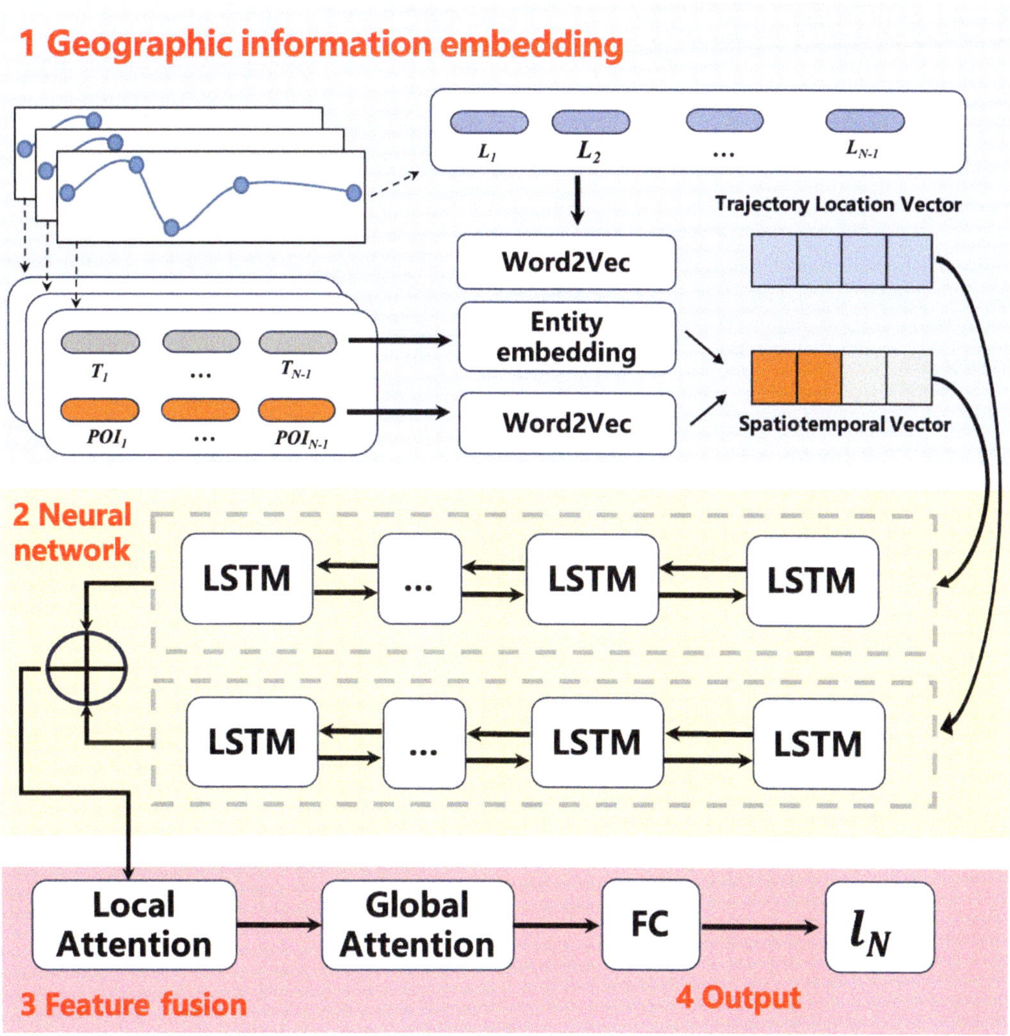

FIGURE 9.1 GEMA-BiLSTM model framework.

and multilayer attention mechanism, which we call GEMA-BiLSTM. Figure 9.1 shows the three core modules of the location prediction model:

1. In the geographic information embedding module, trajectories will be represented based on a word embedding model to extract contextual information from human activities. In addition, the encoding of points of interest and temporal features is also included to enable the acquisition of semantic information in the trajectory.
2. In the neural network module, long-term historical information in trajectories will be captured based on Bi-LSTM. The transfer of long- and short-term dependent information is considered through a series of layers, thus enhancing the model's representation ability and generalization performance.
3. In the feature fusion module, the attention layer is mainly employed to fully fuse the embedding vectors of trajectory and functional features, in which the associations of the features can all be taken into account and dealt with in the location prediction process.

9.2.1 Trajectory and Spatiotemporal Feature Embedding Based on Word Embedding Model

Human movement has certain regularities (Gonzalez, Hidalgo and Barabasi 2008). For example, the next step at home on a weekday morning is likely to be the office, while in the evening, the next step after work is usually to return to the residence. The contextual information provided by location sequence helps to infer the intent of human mobility. Word2Vec is one of the most popular models for learning word embeddings through the use of shallow neural networks (Mikolov 2013). The use of the Word2Vec model is motivated by the fact that the representational vectors of location capture contextual information (Zhang et al. 2021). The neural network used in Word2Vec learns the distribution of the words in the corpus in order to obtain representational vectors for the words. When the representative vectors are similar in distance, the contextual information of the location is also similar.

In generating location representation vectors with rich contextual information based on the word embedding model, a training dataset is first established. The sequence of the kth user's trajectory can be represented as $U_k = \{X_1,\ldots,X_i,\ldots,X_M\}$, where X_i is the ith step of the user's trajectory and M is the total number of locations in the trajectory. Notably, each trajectory in the collection of trajectory sequences needs to be sampled. For example, if sampling is performed at 1-hour intervals, then the trajectories in each user contain at most 24 locations, and the time intervals between neighboring locations are all 1 hour. The reason for extracting a fixed-length sequence of locations is to avoid learning inconsistent human travel patterns, which gives the model accuracy and reliability in predicting users' future locations or behaviors. All the users' trajectories formed the training dataset for the word embedding model.

Skip-Gram and CBOW (Continuous Bag of Words) architectures are the two main architectures in Word2Vec (Mikolov 2013). To efficiently process and capture the contextual information of each location in the trajectory, the Word2Vec architecture used in this study is CBOW. The location is first represented by one-hot coding, e.g., $X_i = (x_{i1}, x_{i2}, x_{i3},\ldots,x_{iv})$, where V is the total number of grids. Then it is fed into the training process of the word embedding model. Figure 9.2 shows the

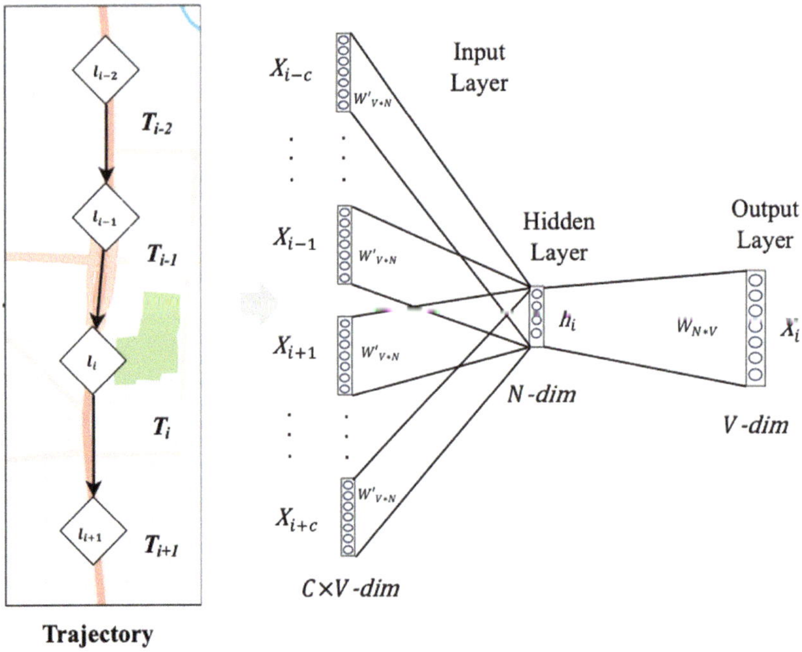

FIGURE 9.2 The CBOW framework of the Word2Vec model.

CBOW architecture training process. The input vectors are the location information in the context, and the output is the vector of the current location.

When the size of the sliding window is set to c, contextual information about X_i in the trajectory including $\{X_{i-c},\ldots,X_{i-1},X_{i+1},\ldots,X_{i+c}\}$ can be obtained. During the training process, the model will learn the contextual relationships between locations to generate vector representation of the locations. Then, the CBOW architecture is trained to obtain the weight matrix $W'_{V\times N}$, and the output matrix W_{V*N}. N is the dimension of the embedded vector. $W'_{V\times N}$ is used for mapping from input layer to hidden layer. By taking the average of the intermediate vectors in the window, the output vector h_i can be obtained. Each location encoded with one-hot can be converted into an N-dimensional word vector using the transposition of output matrix $W_{N\times V}$, and each user's trajectory is represented as an $M \times N$ matrix.

Spatiotemporal feature vectors are formed by concatenating temporal embeddings with the functional semantic embedding of the location. The embedding information of time is usually obtained by linear layer projection. The functional semantic embedding of location is then based on Word2Vec to obtain the embedding vectors of functional attributes of location (Yao et al. 2017a). The first step is to construct functional category sequences based on Delaunay triangular network and random wandering strategy. In the next step, the functional category sequences are trained using Word2Vec's Skip-Gram architecture. During training, the model learns a representation of the functional categories by maximizing predictive accuracy.

9.2.2 Capture Long-Term Historical Information in Trajectories Based on Bi-LSTM

In location prediction problems, utilizing the long- and short-term dependencies of sequences is essential to improve the accuracy and usefulness of the prediction (Chen et al. 2017). The LSTM model transfers information between different locations in a sequence based on a neural network structure of input gates, forget gates, and output gates. However, unidirectional LSTM can only pass information in one direction. The current location is related to the past state and may be related to the future state (Chen et al. 2017).

Bi-LSTM is a combination of forward LSTM and reverse LSTM. The application of Bi-LSTM in location prediction can effectively deal with the long- and short-term dependencies of sequences (Bao et al. 2021). The Bi-LSTM model needs to be inputted with the embedding vectors of the trajectories or the embedding vectors of the spatiotemporal features before training. The trajectory representation vectors and the spatiotemporal feature sequence vectors are independently used as inputs to the Bi-LSTM model. In the training layer, each sequence needs to be cut into some fixed-length segments in the neural network. The length of sequence is set to M. We then use the previous (M-1) location information and the previous (M-1) spatiotemporal feature information to predict the final location information. The last location is used to evaluate the performance of the model. When the input sequence of trajectories is not long enough, the last element of the sequence is usually repeated until the desired length is reached. Through training, the Bi-LSTM model is able to learn the long-term history information of the sequence, which helps to improve the accuracy of location prediction.

9.2.3 Attention Mechanism–Based Feature Fusion Module

Different features in trajectories are essential for understanding the spatiotemporal patterns of human mobility (Miao et al. 2020). Location coordinates are the most basic feature of a trajectory, which is directly associated with the geospatial distribution. Timestamps, on the other hand, provide information about when the trajectory occurred. Activity types can reflect the behavioral intentions and activity patterns of individuals, such as work, leisure, or transportation trips. Single features may not be sufficient to reflect complex patterns. In real human activity scenarios, location transfer during individual movement is closely related to timestamps and activity types (Chen, Li and Liu 2019, 30–41). An attention mechanism–based feature fusion module will be introduced in this section to integrate location sequences with spatiotemporal features. The feature fusion module based on the attention mechanism can integrate the complex relationship between features such as location index, timestamp, and activity type to accurately describe human activity behavior.

The attention mechanism captures the long-term dependencies in the sequence by calculating the similarity between each element in the input sequence and all other elements (Lyu et al. 2019), which avoids the problem of decreasing the efficiency of information transfer due to the increase in the length of the sequence in LSTM. For each element in the sequence, three different weight matrices and a linear transformation (matrix multiplication) are used to compute its corresponding Query, Key, and Value vectors. Given the input sequence element *i*, its Query, Key, and Value vectors are:

$$Query_i = X_i \cdot W^Q \tag{9.1}$$

$$Key_i = X_i \cdot W^K \tag{9.2}$$

$$Value_i = X_i \cdot W^V \tag{9.3}$$

The attention score obtained by computing the dot product between the Query and all the Keys of the current element is then computed, and the score is normalized using the softmax function in order to convert the score into attention weights. Each element in the sequence is assigned different attention weights, which reflect the correlation between the point and other elements in the sequence. Multiply the Value vector at each position by the corresponding attention weight, and perform a weighted sum to obtain the context vector. This context vector contains information for each element in the sequence. By concatenating the weighted sum results along the length dimension of the sequence, the final feature vector for each element can be obtained.

Figure 9.3 illustrates that the attention mechanism is initially employed to fuse the location and spatiotemporal features output by the Bi-LSTM model. The hidden state vector of each location and

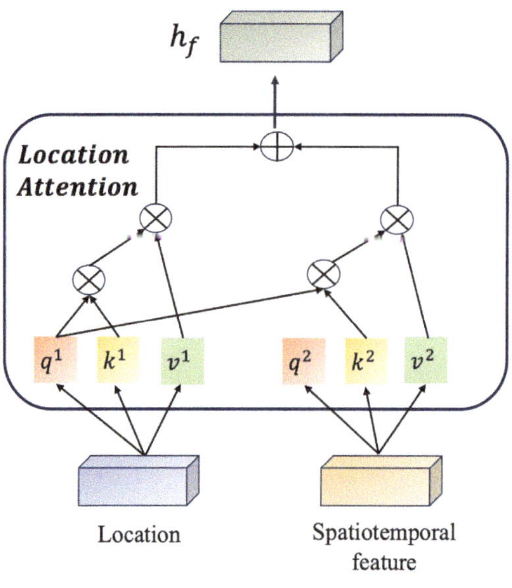

FIGURE 9.3 Integrating contextual features and corresponding spatiotemporal features of positions in trajectories based on attention mechanism.

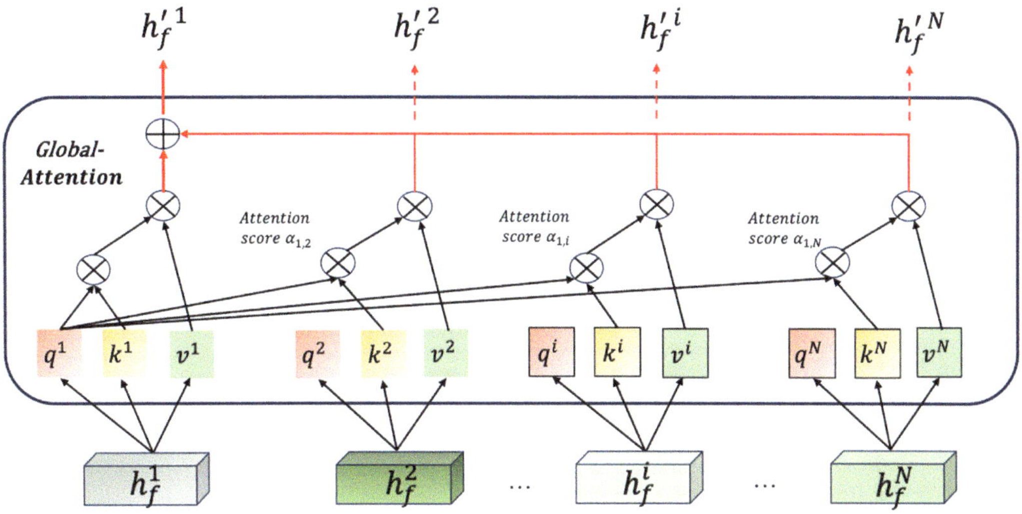

FIGURE 9.4 Integrating global information based on attention mechanism.

the hidden state vector of the corresponding spatiotemporal feature can be obtained after Bi-LSTM training. The hidden state vectors of locations and features are concatenated on the feature dimension to obtain h_f. At this stage, the location attention layer would initially assign weights to the location and spatiotemporal features and obtain the fused representation vectors for each location feature by weighted summation.

Human movement exhibits irregular patterns of transformation (Feng et al. 2020). The prediction of location requires attention to the impact of global location information on predicting the next location. In this chapter, we construct a global attention layer to explore the impact of each location information on prediction. Figure 9.4 shows the process of integrating global information based on attention mechanism. The input to the self-attention layer is the output of the feature fusion layer. The feature fusion vectors at each location are given initial weight vectors to obtain the hidden vectors of the combined feature and location information. The hidden vectors of the combined feature and location information at all moments are weighted and summed, and the probability of each candidate location is obtained by a simple fully connected layer.

9.2.4 Model Prediction and Validation

In the next location prediction task, the previous (M-1) position information and the previous (M-1) spatiotemporal feature information are mainly used to predict the last location. The last location is then used to evaluate the performance of the model. The models that are usually used in the next location prediction task include Markov, convolutional neural networks (CNNs), LSTM, and Bi-LSTM.

To examine the performance in the location prediction task, this chapter applies Accuracy as the key evaluation metric. Accuracy is the ratio of the number of correctly predicted samples to the total number of samples, which is calculated as follows:

$$\mathrm{Hit}(y_i, \hat{y}_i) = \begin{matrix} 1, & Dis\left(y_i, \hat{y}_i\right) \le \gamma \\ 0, & Dis\left(y_i, \hat{y}_i\right) > \gamma \end{matrix} \tag{9.4}$$

$$\text{Accuracy}@k = \frac{\sum_{i=1}^{n} Hit(y_i, \hat{y}_i)@k}{n} \quad (9.5)$$

where n is the number of trajectories participating in the evaluation, $\hat{y}_i$ is the predicted label of user *i*, y_i is the true label of user *i*, and $\text{Hit}(y_i, \hat{y}_i)@k$ is the top *K* labels with the highest probability in the predicted results. $\Delta Dis(y_i, \hat{y}_i)$ is the spherical distance between the predicted location and the true location. $Distance_i$ is the sum of the moving distances of the user *i*, and γ is the error threshold. Considering the range of human activities, γ is set to 500 m.

In addition, in order to evaluate the performance of location prediction in terms of spatial error, the mean absolute error (MAE), root mean square error (RMSE), and mean relative error (MRE) are also used as evaluation metrics as shown in Equations 9.6 to 9.8:

$$\text{MAE} = \frac{1}{n}\sum_{i=1}^{n}\left|\Delta Dis(y_i, \hat{y}_i)\right| \quad (9.6)$$

$$\text{RMSE} = \sqrt{\frac{1}{n}\sum_{i=1}^{n}\left(\Delta Dis(y_i, \hat{y}_i)\right)^2} \quad (9.7)$$

$$\text{MRE} = \frac{1}{n}\sum_{i=1}^{n}\frac{\left|\Delta Dis(y_i, \hat{y}_i)\right|}{\text{Distance}_i} \quad (9.8)$$

Travel semantic prediction performance measures whether the predicted location matches the travel semantics of the actual location. Travel semantic is the type of point of interest (POI) that humans are most likely to visit in the neighborhood of the travel destination. To evaluate travel semantic prediction performance, we first identify the functional semantics of each land in the city using POI data. Then, the POI type with the highest attractiveness in the neighborhood of the travel destination is used as the travel semantic. Finally, the confusion matrix is used to assess the travel semantic overall accuracy, which is the proportion of correctly identified samples to the total number of samples in the matrix.

9.3 EXPERIMENTS

9.3.1 Study Area and Study Data

The experiments on location prediction are conducted on mobile phone signaling data (MPSD) recorded by a Shenzhen-based communication operator. A total of approximately 16.3 million tracks are included. The number of users in the MPSD is similar to the year-end permanent population data from the 2012 Shenzhen Statistical Yearbook (10.46 million), indicating that the MPSD is representative of the city's population. The MPSD utilized in this study encompasses crucial information, including user identification (ID), record times, and geographical location. Figure 9.5 shows that there are 5,943 base stations in Shenzhen. This study drew Voronoi diagrams based on the base station to get Thiessen polygon and divided and counted the service area of Shenzhen's mobile phone base station.

Shenzhen's Autonavi POI data in 2012 were used to calculate spatial semantic features. In order to determine the travel semantics of the user, we reclassified the POI categories as done by Li et al. (2021). Table 9.1 shows the classification rules and the proportion of the corresponding

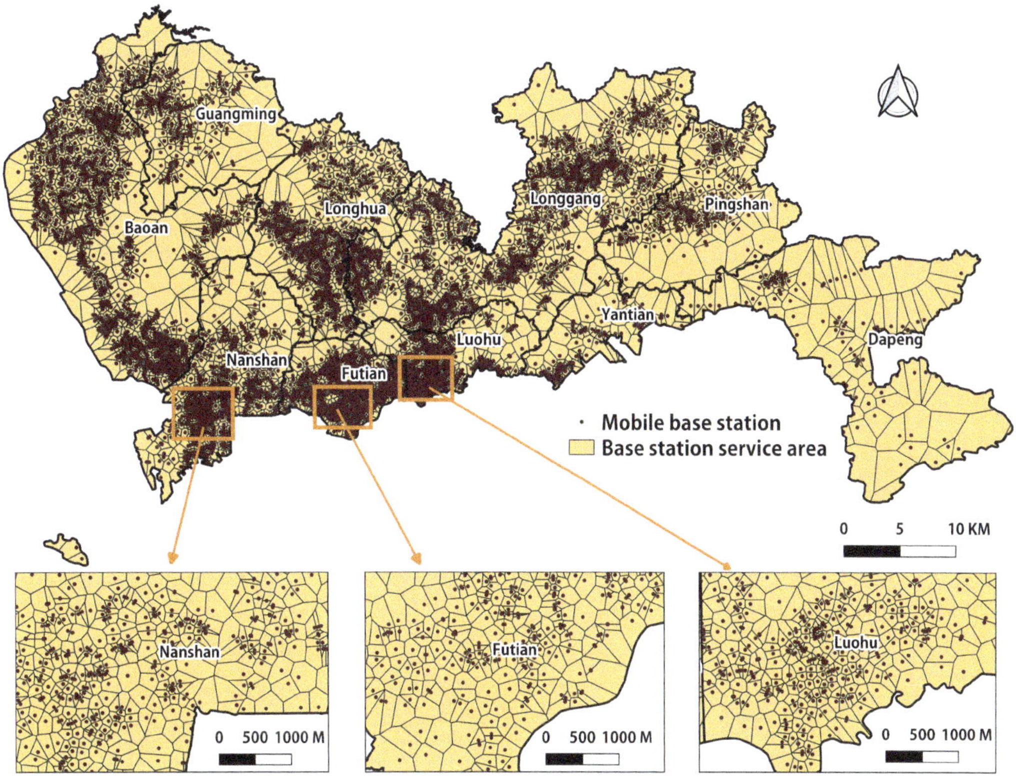

FIGURE 9.5 Distribution of mobile base stations in Shenzhen.

TABLE 9.1
POI and Urban Land Use Type Mapping Relationship

POI Types	Urban Land Use Types	Percentage (%)
Residential Communities	Residential	5.33%
Medical Institutions	Public Services	26.64%
Education and Culture		
Administrative Landmarks		
Government Agencies		
Catering Services	Commercial Services	49.81%
Shopping Stores		
Automobile Services		
Hotels		
Financial Institutions		
Business Buildings		
Entertainment Venues		
Transportation Facilities	Transportation	2.54%
Factories	Industrial	14.40%
Tourist Attractions	Scenery	1.28%

functional semantics. Similar to the results of the proportion of urban function classification (Li et al. 2021), the commercial service type is more distributed, followed by public service and residential areas.

9.3.2 Experimental Settings

For effective embedding using the Word2Vec model, the embedding vector size is set to 300 and the window size c is set to 5. We perform location prediction at 1-hour intervals. The analysis focused on trajectories exhibiting at least one geolocation record per 1-hour temporal interval. When there are multiple locations in a time interval, only one location is retained as the anchor point, and the retention principle is the longest dwell time. By preprocessing the original data, we obtained about 6 million trajectory data and divided the human trajectory dataset into training sets and test sets in a 7:3 ratio for experiments.

9.3.3 Results

9.3.3.1 Next-Location Prediction Results

Table 9.2 delineates the performance of various models in location prediction. Compared to the traditional Markov model, better results are indicated in all metrics for LSTM, CNN, BiLSTM-CNN, and GEMA-BiLSTM, which shows the superiority of deep learning models in location prediction. The GEMA-BiLSTM model proposed performs the best in all metrics, with an improvement in prediction accuracy@1 of 6.21% and 2.28% compared to the LSTM and BiLSTM-CNN models, respectively. Furthermore, we observed a gradual improvement in the performance of deep neural network models with the incorporation of an attention module. The addition of a local attention layer and a global attention layer improved performance by 4.58% and 5.69%, respectively, compared to the LSTM model. This improvement may be due to the fact that the attention module can extract the most relevant contextual information to enhance the performance of the model.

Considering the temporal characteristics of human activities, a day was divided into four parts: morning rush hour (6:00–9:59 a.m.), daytime slow hour (10:00 a.m.–15:59 p.m.), evening rush hour (16:00–20:59 p.m.), and night slow hour (21:00 p.m.–5:59 a.m. the next day). Figure 9.6 clearly demonstrates a significant difference in the accuracy performance among the models across the four periods. In the morning rush hour, the accuracy of the proposed GEMA-BiLSTM model (84.66%)

TABLE 9.2
Comparison of the Predicted Performance

Model	MAE	RMSE	MRE	Accuracy@1
Markov	1,344.90	4,157.06	0.22	69.99
CNN	967.32	3,162.40	0.14	76.10
LSTM	833.98	2,819.13	0.11	77.91
GEMA-BiLSTM (no attention model)	758.19	2,698.33	0.11	79.09
GEMA-BiLSTM (only local attention layer)	677.52	2,478.21	0.08	83.67
GEMA-BiLSTM (only global attention layer)	616.44	2,405.71	0.07	84.78
BiLSTM-CNN	579.17	2,286.45	0.07	85.35
GEMA-BiLSTM	498.23	2,081.89	0.06	87.63

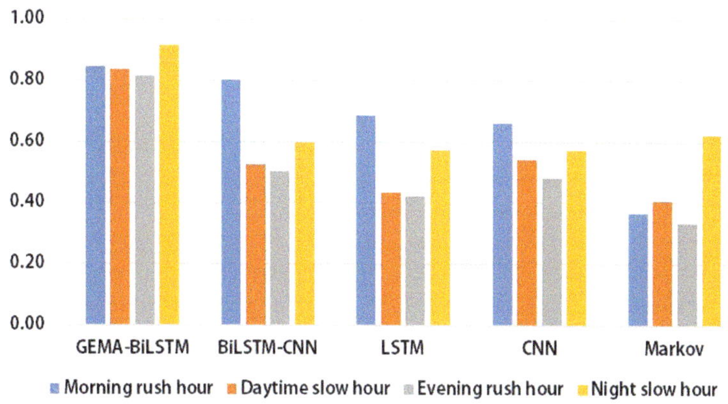

FIGURE 9.6 Comparison of the accuracy of different models in four periods.

is the highest, while the accuracy of the Markov model (36.25%) is the lowest. Meanwhile, the GEMA-BiLSTM model accuracy also has a significant advantage in all other periods. Moreover, our prediction model's performance is associated with the time period. All models always have lowest accuracy in predicting the location at the evening rush hour than in other periods, which indicates the more diverse purpose of human activity at the evening rush hour.

9.3.3.2 Uncertainty Analysis of Location Prediction Results at Different Spatial Scales

To validate the effect of temporal features and spatial-functional semantic features in trajectories and to further understand their importance in trajectory representation, ablation experiments were designed on the next location prediction and the uncertainty of the location prediction model with deeply fused geo-semantics was explored. The experiments were based on the number of spatial points to divide the study area into clustered regions ranging from 100 to 800, representing different scales.

As observed in Table 9.3, the location prediction model incorporating trajectory, temporal, and POI features consistently exhibits the highest accuracy on all scales. Comparing the performance of the model considering only trajectories (average accuracy 71.97%) and considering both trajectories and spatial-functional semantic features (average accuracy 74.25%), the prediction of trajectories after sequentially incorporating POI features and temporal features (average accuracy 75.89%) is improved by 2.28% and 1.64%, respectively. In addition, the incorporation of spatial-functional semantic features in the trajectory enhances the model more significantly compared to

TABLE 9.3
Prediction Performance of the Next Location at Different Regional Scales

		100	200	300	400	500	600	700	800
Accuracy@1	GEMA-BiLSTM	0.8413	0.8129	0.7531	0.7492	0.7533	0.7408	0.7164	0.7049
	GEMA-BiLSTM (Trajectory + POI)	0.8207	0.8015	0.7507	0.7350	0.7334	0.7071	0.7107	0.6851
	GEMA-BiLSTM (Trajectory)	0.8066	0.7789	0.7092	0.7156	0.7151	0.6912	0.6595	0.6817

the incorporation of temporal information, which indicates that spatial-functional semantic features are particularly important to enable the model to better make sense of human activity patterns in the trajectory data. This indicates that the trajectory prediction framework proposed in this chapter has good generalization ability and robustness.

By observing the performance of the next location prediction at different scales in Table 9.3, it can be seen that the prediction accuracy as a whole shows a stable decreasing trend. As the scale is refined, the proportion of functional attributes within the study area becomes more complex and the human activities reflected in the trajectory data become more complex and diverse, thus leading to a decreasing trend in the accuracy of the user's location prediction. However, the location prediction framework proposed still maintains a relatively stable prediction accuracy, which means that the model is able to capture general patterns and regularities in the data, rather than just performing well at specific scales. This indicates that the trajectory prediction framework proposed in this chapter has good generalization ability and robustness.

9.3.3.3 Travel Semantic Prediction Accuracy

To further analyze the proposed algorithm's performance, we evaluated its travel semantic prediction accuracy. Figure 9.7 shows that the GEMA-BiLSTM can effectively tap into users' travel semantics by combining functional semantics of locations. Overall, GEMA-BiLSTM outperforms the LSTM (63.27%), Markov (67.04%), CNN (71.70%), and BiLSTM-CNN (72.96%) models with the highest overall accuracy (75.35%) for identifying travel semantics. The model performs well in identifying various functional areas, including residential, public services, scenery, industrial, commercial services, and transportation areas. Notably, the proposed model significantly reduces the percentage of misclassification into diverse functional lands. For example, for inferring the travel semantics to residential area, GEMA-BiLSTM (19.61%) has less error compared to LSTM (29.96%), Markov (27.40%), BiLSTM-CNN (24.20%), and CNN (22.96%) for identifying into commercial areas. Moreover, the error rate of GEMA-BiLSTM misclassification into commercial is also the lowest in all the inferences of other semantics.

This study found that residential activities are frequently classified as commercial services, with GEMA-BiLSTM misclassifying almost 20% of residential activities. This suggests that the travel characteristics of commercial activities are complex and have similar characteristics as other travel semantics. The complexity may be due to mixed functions of urban land use, where commercial activities are frequently combined with other functions, such as entertainment, tourism, or cultural education (Cheng et al. 2018).

9.3.3.4 Human Activity Pattern Analysis

To analyze the pattern of human activity, we measured the proportion of functional area in population activities during the four periods. Figure 9.8 shows the proportion trend of different functional land use types in different periods. The study identified the "residence-work-residence" activity pattern using location prediction on weekdays. The morning and evening rush hour periods had a higher proportion of transportation, as office workers predominantly use public transport to commute. The proportion of activities in industrial and public service areas and scenery function areas increased initially until the daytime slow hour, then decreased, whereas the residential areas exhibited an inverse trend. These findings suggest that geospatial big data can reflect activity preferences, with the weekday activity patterns of users aligning with those of residents

Nighttime attraction to commercial services is a unique phenomenon, as suggested by the changing trend in the proportion of activities from the daytime slow hour to the evening rush hour. The proportion of residential and transportation activities increases, reflecting people's commuting and rest needs. Furthermore, the proportion of commercial service areas has increased, potentially resulting from government policies promoting the development of the nighttime economy. The land for commercial services has launched night leisure projects to promote economic development (Seijas and Gelders 2021).

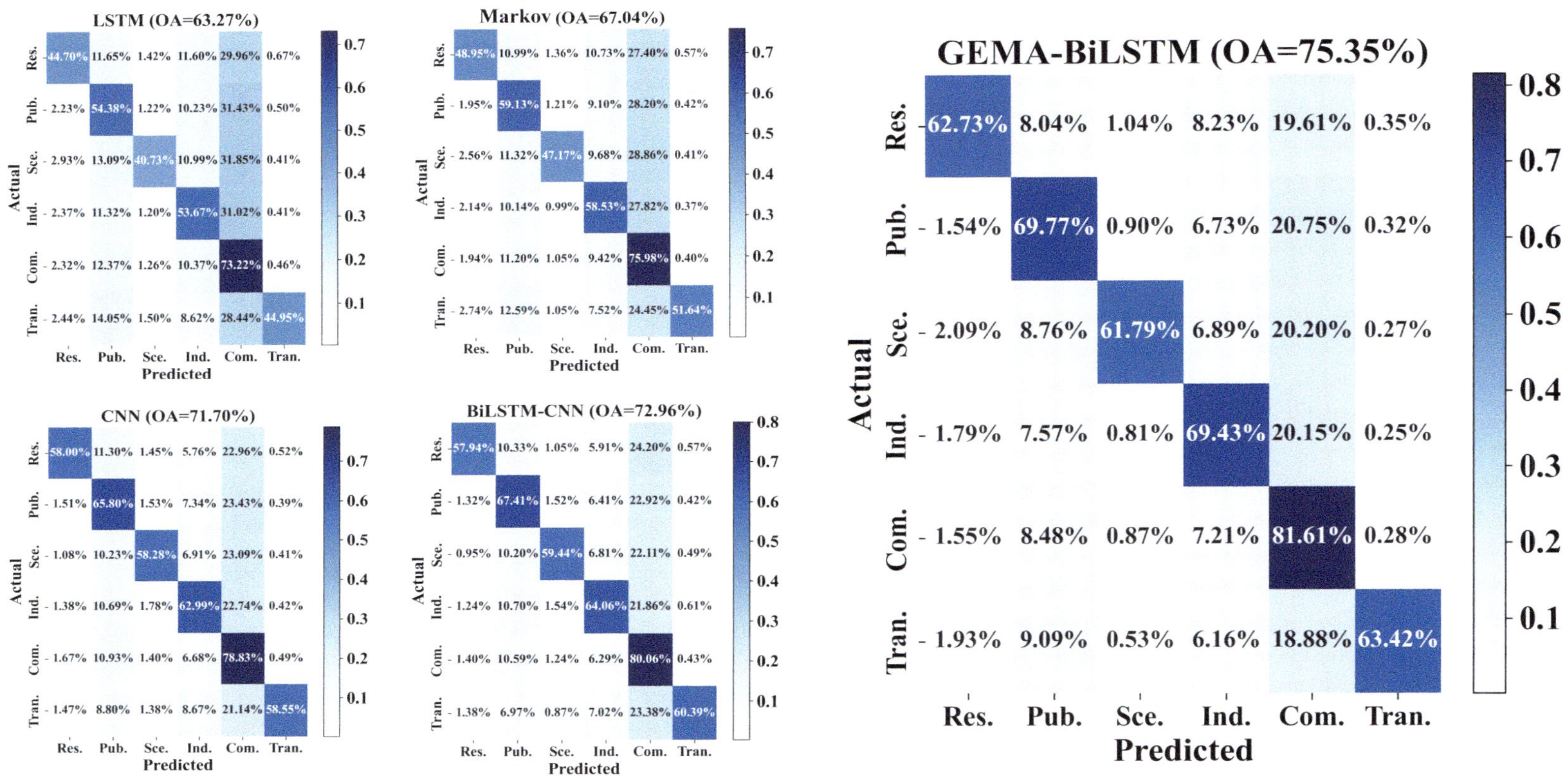

FIGURE 9.7 Confusion matrix for semantic prediction of location by seven models. The semantic categories include residential (Res), public services (Pub), scenery (Sce), industrial (Ind), commercial services (Com), and transportation (Tran).

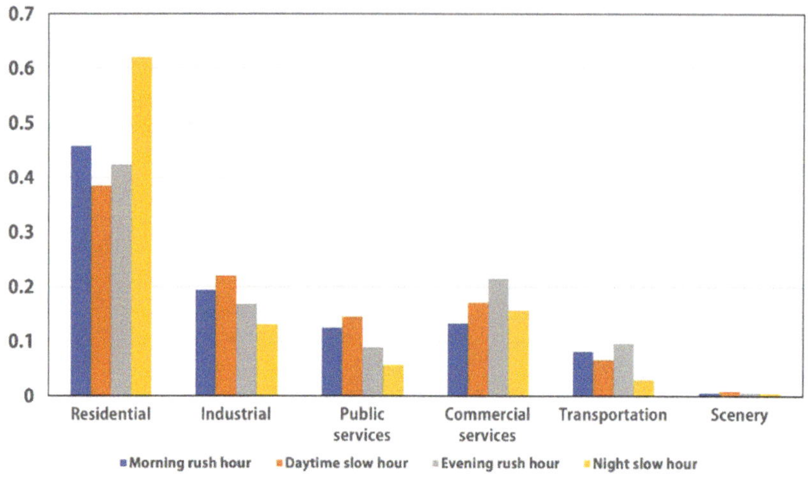

FIGURE 9.8 Proportion trend of different functional land use types in different periods.

9.4 DISCUSSION AND CONCLUSION

The location prediction of human mobility helps discover complex human behavior and travel patterns. Previous studies have not fully integrated human trajectories with spatiotemporal feature information, which limits the ability of models to extract deep semantic information from trajectories and, consequently, impacts the application of location prediction. In this chapter, we proposed a new location prediction framework based on geographic embedding and a multilayer attention mechanism that combines MPSD and POI to extract deep semantic information from human trajectories. Our case study based on the trajectory data of Shenzhen city demonstrates the effectiveness of our proposed model.

The advantage of the proposed model is that the GEMA-BiLSTM model can accurately predict the next location in users' activities in terms of accuracy. With the consideration of spatiotemporal semantic features, the GEMA-BiLSTM (no attention model) model generally outperforms both neural network–based (LSTM, CNN) prediction models and model-based prediction models (Markov). Furthermore, with the addition of a multilayer attention module, the GEMA-BiLSTM model is better than the advanced hybrid neural network model (BiLSTM-CNN) in accuracy, RMSE, MAE, and MRE. Specifically, the GEMA-BiLSTM model's accuracy improved by 17.64%, 11.53%, 9.72%, and 2.28% compared with Markov, CNN, LSTM, and BiLSTM-CNN, respectively. Ablation experiments with the GEMA-BiLSTM model reveal that both the local and global attention layers improve the neural network model's performance. The addition of the attention layer is more conducive to the mining of the trajectory long- and short-term dependence patterns.

Another benefit is that the travel semantics of users can be effectively mined by combining the spatiotemporal semantic features of locations. GEMA-BiLSTM significantly improved the accuracy of the next location in the residential, scenery, and transportation functional areas, with maximum improvements of 18.03%, 21.06%, and 18.47%, respectively. This indicates that the model is effective in capturing more travel features of human commuting and leisure behaviors. The proposed model produces prediction results that are closer to real human activity, which guarantees more interpretable and reasonable predictions of the next location. In addition, the location prediction model proposed in this study has strong generalization capabilities. Stable accuracy is demonstrated in quantitative experiments on multiple research unit scales, and it also shows that incorporating spatial functional semantics is more critical to capture user travel semantics than temporal information.

From the perspective of the temporal characteristics of human mobility, the performance of the model highlights the complexity of human activity in different periods. It outperforms other comparison models in predicting human mobility, but accuracy varies across periods. Active periods such as the morning and evening rush hours are predicted with lower accuracy. Conversely, stationary periods such as the daytime and night slow hours are relatively easy to predict, as people tend to remain in one place. At the night slow hour, most people rest at home, and urban residents go out for a few activities and travel for a single purpose (Dai et al. 2017). These findings reinforce the importance of considering temporal characteristics when modeling human mobility.

In addition, our location prediction models successfully revealed weekday activity patterns consistent with actual human behavior in developed cities, demonstrating a "residence-work-residence" pattern. However, our study also found abnormal activity patterns during evening rush hour periods, especially with an increase in commercial services. This highlights the diverse patterns of movement related to the development of the nighttime economy. This study suggests that the government optimizes the allocation of public transport resources to the land for commercial services, thus facilitating transport for urban residents.

However, there are still aspects that need to be further studied. While our prediction accuracy remains high, further research is needed to consider additional factors such as weekends and holidays. Additionally, limited by the computational and time costs, our proposed model only explores the uncertainty on the spatial scale, and different time scales are also important to verify the generalization performance of the model (Huang et al. 2023,), which will be further verified in the new and improved model in the future.

REFERENCES

Bao, Yi, Zhou Huang, Linna Li, Yaoli Wang, and Yu Liu. 2021. "A BiLSTM-CNN model for predicting users' next locations based on geotagged social media." *International Journal of Geographical Information Science* 35 (4): 639–660.

Chen, Gengxiang, Yingguang Li, and Xu Liu. 2019. "Pose-dependent tool tip dynamics prediction using transfer learning." *International Journal of Machine Tools and Manufacture* 137: 30–41.

Chen, Tao, Ruifeng Xu, Yulan He, and Xuan Wang. 2017. "Improving sentiment analysis via sentence type classification using BiLSTM-CRF and CNN." *Expert Systems With Applications* 72: 221–230.

Cheng, Zhe, Huanming Wang, Ling Wang, and Jing Lin. 2018. "Mix leading to success? Exploring the innovative development model in peri-urban China." *Habitat International* 82: 1–8.

Cobbinah, Patrick Brandful, Prosper Issahaku Korah, Joachim Boadi Bardoe, Rhoda Mensah Darkwah, and Abraham Marshall Nunbogu. 2022. "Contested urban spaces in unplanned urbanization: Wetlands under siege." *Cities* 121: 103489.

Dai, Linlin, Tao Xue, Bihu Wu, Xiao Rong, and Bixia Xu. 2017. "Spatiotemporal structure features of network check-in activities of urban residents and their impacting factors: A case study in six urban districts of Beijing." *Journal of Asian Architecture and Building Engineering* 16 (1): 131–138.

Deschaintres, Elodie, Catherine Morency, and Martin Trépanier. 2022. "Cross-analysis of the variability of travel behaviors using one-day trip diaries and longitudinal data." *Transportation Research Part A: Policy and Practice* 163: 228–246.

Feng, Jie, Yong Li, Zeyu Yang, Qiang Qiu, and Depeng Jin. 2020. "Predicting human mobility with semantic motivation via multi-task attentional recurrent networks." *IEEE Transactions on Knowledge and Data Engineering* 34 (5): 2360–2374.

Gambs, Sébastien, Marc-Olivier Killijian, and Miguel Núñez del Prado Cortez. 2012. "Next place prediction using mobility Markov chains." *Proceedings of the first Workshop on Measurement, Privacy, and Mobility.*

Gonzalez, Marta C, Cesar A Hidalgo, and Albert-Laszlo Barabasi. 2008. "Understanding individual human mobility patterns." *Nature* 453 (7196): 779–782.

Guo, L, Z Ding, Z Hu, and C Chen. 2010. "Uncertain path prediction of moving objects on road networks." *Journal of Computer Research and Development* 47 (1): 104–112.

Huang, Weiming, Daokun Zhang, Gengchen Mai, Xu Guo, and Lizhen Cui. 2023. "Learning urban region representations with POIs and hierarchical graph infomax." *ISPRS Journal of Photogrammetry and Remote Sensing* 196: 134–145.

Li, Zehui, Limin Jiao, Boen Zhang, Gang Xu, and Jiafeng Liu. 2021. "Understanding the pattern and mechanism of spatial concentration of urban land use, population and economic activities: A case study in Wuhan, China." *Geo-Spatial Information Science* 24 (4): 678–694.

Liu, Shichun, Libang Ma, Yao Yao, and Xijuan Cui. 2022. "Man-land relationship based on the spatial coupling of population and residential land–A case study of Yuzhong County in Longzhong Loess Hilly Region, China." *Land Use Policy* 116: 106059.

Liu, Wei, Wenjie Wu, Piyushimita Thakuriah, and Jianghao Wang. 2020. "The geography of human activity and land use: A big data approach." *Cities* 97: 102523.

Liu, Yu, Xi Liu, Song Gao, Li Gong, Chaogui Kang, Ye Zhi, Guanghua Chi, and Li Shi. 2015. "Social sensing: A new approach to understanding our socioeconomic environments." *Annals of the Association of American Geographers* 105 (3): 512–530.

Lyu, He, Ningyu Sha, Shuyang Qin, Ming Yan, Yuying Xie, and Rongrong Wang. 2019. "Advances in neural information processing systems." *Proceedings of the 2019 International Conference on Neural Information Processing Systems.*

Miao, Congcong, Ziyan Luo, Fengzhu Zeng, and Jilong Wang. 2020. "Predicting human mobility via attentive convolutional network." *Proceedings of the 13th International Conference on Web Search and Data Mining.*

Mikolov, T. 2013. "Efficient estimation of word representations in vector space." *arXiv preprint* arXiv:1301.3781.

Qian, Jiale, Zhang Liu, Yunyan Du, Fuyuan Liang, Jiawei Yi, Ting Ma, and Tao Pei. 2021. "Quantify city-level dynamic functions across China using social media and POIs data." *Computers, Environment and Urban Systems* 85: 101552.

Qiao, Shaojie, Dayong Shen, Xiaoteng Wang, Nan Han, and William Zhu. 2014. "A self-adaptive parameter selection trajectory prediction approach via hidden Markov models." *IEEE Transactions on Intelligent Transportation Systems* 16 (1): 284–296.

Seijas, Andreina, and Mirik Milan Gelders. 2021. "Governing the night-time city: The rise of night mayors as a new form of urban governance after dark." *Urban Studies* 58 (2): 316–334.

Solomon, Adir, Amit Livne, Gilad Katz, Bracha Shapira, and Lior Rokach. 2021. "Analyzing movement predictability using human attributes and behavioral patterns." *Computers, Environment and Urban Systems* 87: 101596.

Wang, Kuo, LingBo Liu, Yang Liu, GuanBin Li, Fan Zhou, and Liang Lin. 2023. "Urban regional function guided traffic flow prediction." *Information Sciences* 634: 308–320.

Wang, Ruoxi, Xinyuan Zhang, and Nan Li. 2022. "Zooming into mobility to understand cities: A review of mobility-driven urban studies." *Cities* 130: 103939.

Yao, Yao, Xia Li, Xiaoping Liu, Penghua Liu, Zhaotang Liang, Jinbao Zhang, and Ke Mai. 2017a. "Sensing spatial distribution of urban land use by integrating points-of-interest and google Word2Vec model." *International Journal of Geographical Information Science* 31 (4): 825–848.

Yao, Yao, Xiaoping Liu, Xia Li, Jinbao Zhang, Zhaotang Liang, Ke Mai, and Yatao Zhang. 2017b. "Mapping fine-scale population distributions at the building level by integrating multisource geospatial big data." *International Journal of Geographical Information Science* 31 (6): 1220–1244.

Yao, Yao, Zijin Guo, Chen Dou, Minghui Jia, Ye Hong, Qingfeng Guan, and Peng Luo. 2023. "Predicting mobile users' next location using the semantically enriched geo-embedding model and the multilayer attention mechanism." *Computers, Environment and Urban Systems* 104: 102009.

Yin, Jiadi, Jinwei Dong, Nicholas AS Hamm, Zhichao Li, Jianghao Wang, Hanfa Xing, and Ping Fu. 2021. "Integrating remote sensing and geospatial big data for urban land use mapping: A review." *International Journal of Applied Earth Observation and Geoinformation* 103: 102514.

Zhang, Jinbao, Xia Li, Yao Yao, Ye Hong, Jialyu He, Zhangwei Jiang, and Jianchao Sun. 2021. "The Traj2Vec model to quantify residents' spatial trajectories and estimate the proportions of urban land-use types." *International Journal of Geographical Information Science* 35 (1): 193–211.

Zhao, Yuhui, Bi Yu Chen, Fei Gao, and Xinyan Zhu. 2023. "Dynamic community detection considering daily rhythms of human mobility." *Travel Behaviour and Society* 31: 209–222.

10 Human Mobility Clustering Analysis

A Case Study of Ride-Hailing Demand Analysis

Sudipta Roy, Jiechao Zhang, and Samiul Hasan

10.1 INTRODUCTION

As cities continue to grow and evolve, understanding urban mobility behavior has become important to address the challenges of rapid urbanization. Analyzing human movement within urban environments is crucial to design efficient transportation systems, improve urban planning, and enhance the overall quality of life for residents. One of the most significant and transformative developments in the transportation sector in recent years has been the advent of ride-hailing services, which are reshaping how people move within cities by providing flexible and on-demand mobility solutions.

Ride-hailing services such as Uber and Lyft have gained substantial popularity in U.S. cities. According to the "New York City Mobility Report," 2.5% of the total number of trips used for-hire vehicles as their medium of movement ("New York City Mobility Report" 2019). The global ride-hailing service market had an estimated value of \$31.95 billion in 2022 and is projected to have a yearly growth rate of 15.7%, reaching \$104.93 billion by 2030 ("Research and Markets" 2023). As these services continue to grow, a deeper understanding of ride-hailing services and their impacts is essential, not only to efficiently manage the demand, but also to effectively integrate them with existing transportation facilities and services.

Despite significant advancements in ride-hailing mobility research over the last two decades, comparative studies across multiple cities in this domain are limited. Such cross-city analyses are crucial for addressing two key needs in urban mobility management: generalizability and customization (Canitez 2020; Lyu et al. 2021; Wang et al. 2024a). Generalizability or universal applicability ensures the development of sustainable transportation policies, innovative infrastructure guidelines, and robust travel demand models that can be applied across different urban contexts (González, Hidalgo, and Barabási 2008; Alessandretti, Aslak, and Lehmann 2020; Lyu et al. 2021). On the other hand, customization is critical for adapting strategies to the unique mobility patterns and challenges of individual cities (Mazza 2011). For instance, understanding how New York's transit-heavy system affects its ride-hailing demand compared to Chicago's auto-innovative model offers key insights on how these services interact with diverse urban transportation systems emphasizing both the universal and customized nature of ride-hailing across cities (Oke et al. 2019). Such a comparative analysis enables policymakers to create customized solutions that reflect the specific needs of each city while leveraging common mobility patterns for broader applications.

In this chapter, we investigate the dynamic and complex nature of ride-hailing mobility patterns in two of the large and densely populated cities in the United States: New York and Chicago. First, we explore the spatial and temporal variations in ride-hailing demand patterns for both cities, which provide us insights on how demand fluctuates across different times and regions. Building on this, we perform a clustering analysis of urban regions based on typological attributes to classify

DOI: 10.1201/9781003503262-13

neighborhoods according to shared spatial characteristics. This urban typology-based clustering helps explain variations in ride-hailing demand observed across the two cities.

Through this case study, we aim to contribute to the growing body of literature on urban mobility by providing a comparative analysis of ride-hailing demand patterns and associated urban typologies in two major U.S. cities. Our findings have implications for transportation policy, urban planning, and smart city initiatives, offering valuable insights that can help shape the future of urban mobility.

10.2 LITERATURE REVIEW

Over the past decade, advances in data collection technologies—such as global positioning systems (GPS), mobile phone data, and ride-hailing service logs—have enabled researchers to analyze human mobility with unprecedented detail and scale. In one of the seminal works in the field, analyzing mobile phone data, Gonzalez et al. discovered that human mobility follows highly regular patterns, with most individuals repeating a small set of daily routines (González, Hidalgo, and Barabási 2008). Understanding human mobility patterns will be beneficial for transportation planning, traffic management, and resource allocation (Zhang et al. 2020; Alessandretti, Aslak, and Lehmann 2020).

Recent studies have emphasized the role of ride-hailing services in reshaping urban mobility. Ride-hailing data provide a granular view of human movement, allowing for more precise clustering of mobility patterns. Previous studies on ride-hailing demand analysis have explored different spatial and temporal variations in trip demand and different factors contributing to these variations within cities (Du, Rakha, and Breuer 2022). Past research discovered that spatial attributes such as sociodemographic attributes, land use characteristics, and transportation infrastructure contribute most for ride-hailing demand variation (Dean and Kockelman 2021; Dey et al. 2021; Du, Rakha, and Breuer 2022; Roy, Nahmias-Biran, and Hasan 2025; Zhang, Huang, and Zhu 2020; Wang et al. 2024b). Li et al. concluded the travel distance between origin-destination pairs is significant for trip demand generation (Li et al. 2023). Liu et al. showed that spatial-temporal clustering of origin-destination pairs can efficiently reveal the demand variation (Liu, Krishnakumari, and Cats 2019). Alonso-González et al. concluded that efficient ride-hailing services can significantly reduce vehicle miles traveled (VMT) and emission (Alonso-González et al. 2021). Several studies also revealed that ride-hailing services may increase VMT and increase congestion, especially during peak hours and without sufficient pooling (Tirachini and Gomez-Lobo 2020; Liang et al. 2022; Naumov and Keith 2023; Lotze et al. 2023). Congestion pricing has been found to be an effective measure to improve ride-sharing ratio (Alonso-González et al. 2021; Zheng et al. 2023). Kontou et al. found that using accurate demand forecasting, deadheading miles can be reduced and hence congestion can be alleviated (Kontou, Garikapati, and Hou 2020). As such, analyzing ride-hailing service data allowed discovering various aspects concerning the impact of ride-hailing on urban mobility demand and supply.

Among various mobility analysis techniques, clustering has seen significant advancements to uncover the patterns that shape cities (Calabrese, Ferrari, and Blondel 2014; Fan et al. 2015; Liu, Krishnakumari, and Cats 2019; Dean and Kockelman 2021; Dey, Tirtha, and Eluru 2021; Du, Rakha, and Breuer 2022; Lam, Liu, and Hui 2021; Zhang and Zhao 2022; Li et al. 2023; Wang et al. 2024b). Researchers have explored various clustering methods such as factor analysis (FA), principal component analysis (PCA), and other techniques due to their ability to reduce the dimensionality of data and uncover underlying patterns. For example, Oke et al. used FA first to reduce the dimension of data and then applied a hierarchical clustering technique to form a cluster of cities and classified them based on different urban typologies (Oke et al. 2019). Popovich et al. introduced a geospatial typology framework that also used FA to identify the critical indicators and then deployed k-medoids clustering to classify U.S. regions into distinct categories based on commuting patterns, density, and polycentricity (Popovich et al. 2021). Zheng et al. applied PCA to extract key features from mobility data, followed by k-means clustering to identify distinct mobility patterns in urban environments (Zheng et al. 2009). This approach has been instrumental in identifying how different urban areas exhibit unique mobility behaviors, influenced by factors such as land use, population density, and socioeconomic status. Calabrese et al. utilized mobile phone data to discover urban clusters based on human activity, revealing the functional structure of cities and informing urban planning efforts (Calabrese, Ferrari, and Blondel 2014). Jiang et al. emphasized the temporal dimension of urban clustering, identifying daily activity patterns in different city areas (Jiang, Ferreira, and González 2012). Several studies further expanded this field by using geotagged social media data to detect regions of interest and individual destination choices (Hasan and Ukkusuri 2015; Hasnat and Hasan 2018; Wang et al. 2016). From the literature review, a comparative analysis of different clusters from two distinct cities is rare; such an analysis can provide a deeper understanding of ride-hailing demand patterns and the associated factors.

In summary, studying human mobility behavior has evolved significantly, with advancements in data collection and analysis techniques driving new insights. Analyzing ride-hailing data presents a promising frontier to understand spatiotemporal patterns of urban mobility. As cities continue to grow and evolve, empirical research on ride-hailing mobility patterns will be crucial for developing realistic urban mobility models. Despite the significant progress in urban clustering research based on human mobility behavior, several research gaps remain. The lack of extensive ride-hailing data in many studies limits the understanding of how services like Uber and Lyft influence urban mobility patterns. Furthermore, there is a need for a comparative analysis across different cities to identify commonalities and differences in urban clustering patterns. Existing models are tailored to specific cities. Without understanding the generalizability of the insights from these models, it will be challenging to apply them across diverse urban contexts. Addressing these gaps—particularly by incorporating ride-hailing data and conducting cross-city comparisons—will be crucial for advancing the field. These steps will enhance the effectiveness of urban planning, transportation policy, and smart city initiatives.

10.3 STUDY AREAS

In this study, we analyze ride-hailing mobility patterns of two major U.S. cities: New York and Chicago (Figure 10.1). New York City represents the case of a densely populated, mass transit–oriented metropolis, characterized by extensive public transportation options and higher levels of traffic congestion (Oke et al. 2019). On the other hand, Chicago represents an auto-centric city, known for its more expansive layout, relative affluence, high transit mode share, and lower congestion levels (Oke et al. 2019). Despite these differences, both cities share a high development index, including elevated costs of living, housing rents, and grocery and restaurant prices. Additionally, both cities boast strong innovation, Internet, and smartphone penetration, all of which are significant drivers of ride-hailing demand.

For this study, taxi zones were used as the spatial units for trip aggregation in New York, while community areas were used for Chicago. New York City is divided into 263 taxi zones across its five boroughs, covering a total area of 306 square miles. Our analysis was done using data from 222 of these zones, based on data availability. In contrast, Chicago's city area is divided into 77 community areas, spanning a total of 231 square miles. We used data from 74 of these areas for our analysis. Spatial units containing airports were excluded from the models due to their unique characteristics.

10.4 SPATIAL-TEMPORAL RIDE-HAILING MOBILITY BEHAVIOR PATTERNS

We considered 12 months of ride-hailing data for both New York (February 2019 to January 2020) and Chicago (January 2019 to December 2019), aggregating trip data by hour of the day. The spatial and temporal ride-hailing mobility pattern analysis of the aggregated ride-hailing data are presented in this section. The temporal variation is demonstrated over monthly and weekly contexts. Three key parameters including mean of total hourly ride-hailing demand, shared ride authorization ratio, and shared ride matching ratio are used to explain ride-hailing mobility patterns.

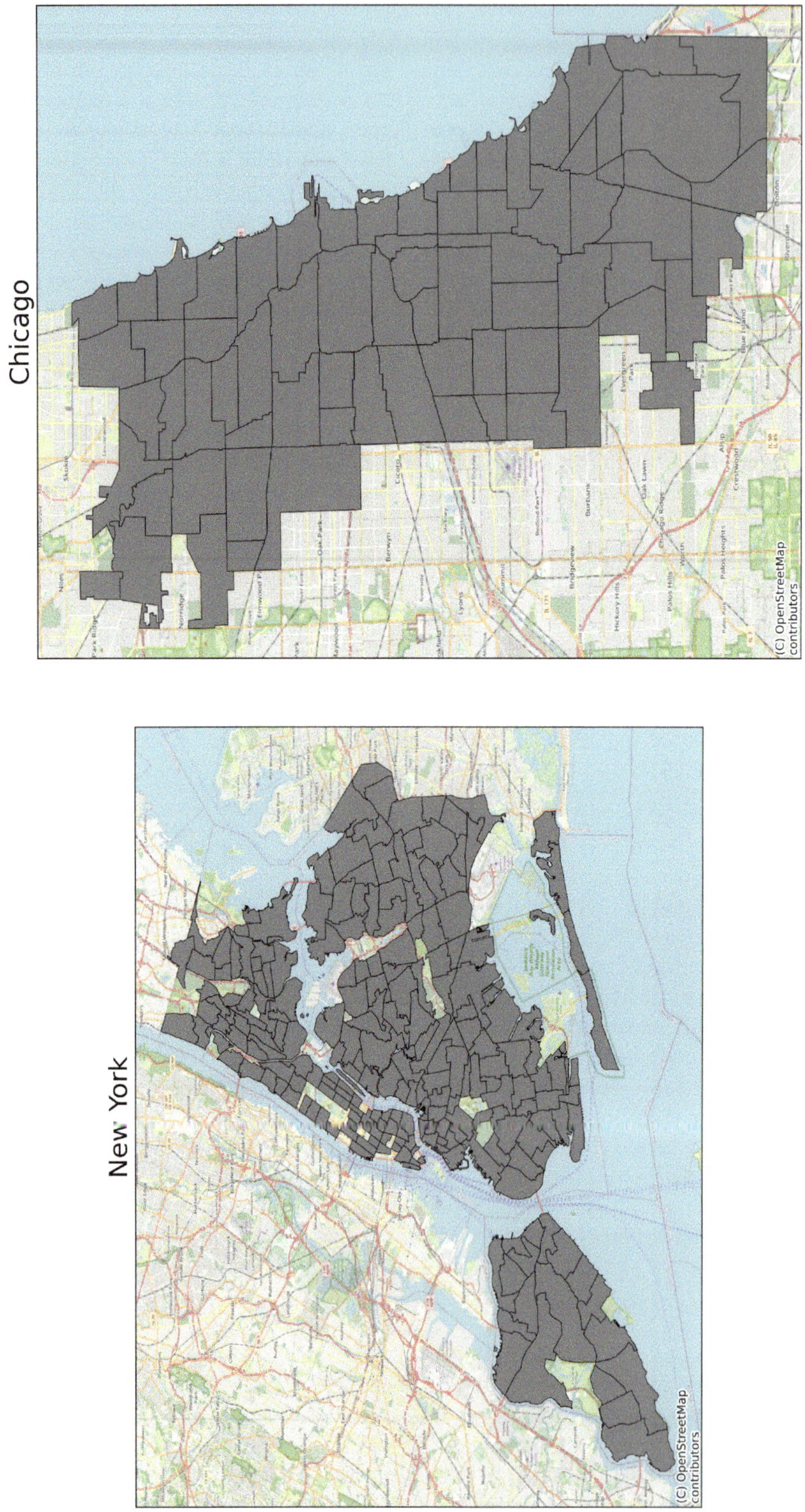

FIGURE 10.1 Study areas.

10.4.1 Temporal Variation

Analyzing the temporal variation of ride-hailing trips shows some common patterns in both cities. Figure 10.2 demonstrates that the monthly variation in ride-hailing demand for both New York and Chicago is relatively negligible. In both cities, while there is a slight increase in demand during the first few months of the year, the overall pattern remains consistent throughout the year.

Figure 10.3 shows the hourly ride-hailing demand variation throughout the days of the week. We observe significant variations in ride-hailing demand across different days of the week in both New York and Chicago. For both cities, the demand increases progressively from Monday to Saturday, peaking on Fridays and Saturdays, particularly during the evening hours. Sundays experience a notable drop in demand. This comparison highlights that the variation in demand is much more prominent across weekdays than across months in both cities. On weekdays, especially from Monday to Friday, the peak in ride-hailing demand during morning and evening hours corresponds to typical work commute times. The increased demand on Fridays and Saturdays, especially during the evening and late-night hours, is likely driven by trips to participate in social and entertainment activities.

Shared trip authorization indicates the willingness of a passenger to share the trip with someone else in case a similar or overlapped trip route exists. It is defined as the ratio of the number of trips authorized by a passenger to be shared to the total number of generated trips for each city. Figure 10.4 shows a similar trend in shared trip authorization ratio for both cities. This metric is important because it indicates passengers' openness to share a ride, which can lead to more efficient use of vehicles, reduced costs, and potentially lower environmental impact due to fewer vehicles being on the road. The shared trip authorization ratio is relatively higher during the weekdays. This suggests that passengers are more willing to share their rides during weekdays, likely due to the consistent and predictable commuting patterns associated with work and school. The ratio slightly decreases on Saturdays and Sundays, indicating that passengers may prefer private rides for leisure activities or weekend plans because of privacy or convenience.

Shared trip matching refers to the shared rides where multiple passengers with similar or overlapping routes are matched together in the same vehicle, i.e., when the authorized shared trips are actually matched. Figure 10.5 presents the shared trip matching ratios, i.e., the ratio of the number of shared trips matched to the total number of generated trips, for both cities. New York exhibits a more effective shared trip matching scenario compared to Chicago, with higher overall matching ratios, particularly during weekdays and peak commuting hours. This suggests a better performance to pair passengers with similar routes in New York, likely due to higher passenger density and demand. In contrast, Chicago's shared trip matching ratios have consistently lower and less pronounced variations throughout the week. Even during peak hours, the shared trip matching ratio is lower in Chicago, indicating challenges in promoting shared rides in this city.

The temporal analysis reveals consistent ride-hailing demand across seasons, with strong hourly fluctuations reflecting daily human activity patterns. This indicates that ride-hailing has become integral to urban mobility, aligning with the rhythms of city life. New York's higher ride-sharing ratio suggests more efficient utilization of resources compared to Chicago, likely driven by its denser urban landscape. These findings underscore the need for city-specific and time-sensitive shared mobility policies to ensure efficient, sustainable, and adaptive ride-hailing services.

10.4.2 Spatial Variation

The spatial variation analysis is done for two different periods, weekday morning peak demand (from 8:00 a.m. to 9:00 a.m.) and weekend night demand (from 11:00 p.m. to 12:00 a.m.), demonstrated in Figure 10.6. These periods were chosen because they are characterized by significantly high demand, as highlighted in Figure 10.3. The morning peak on weekdays captures the rush

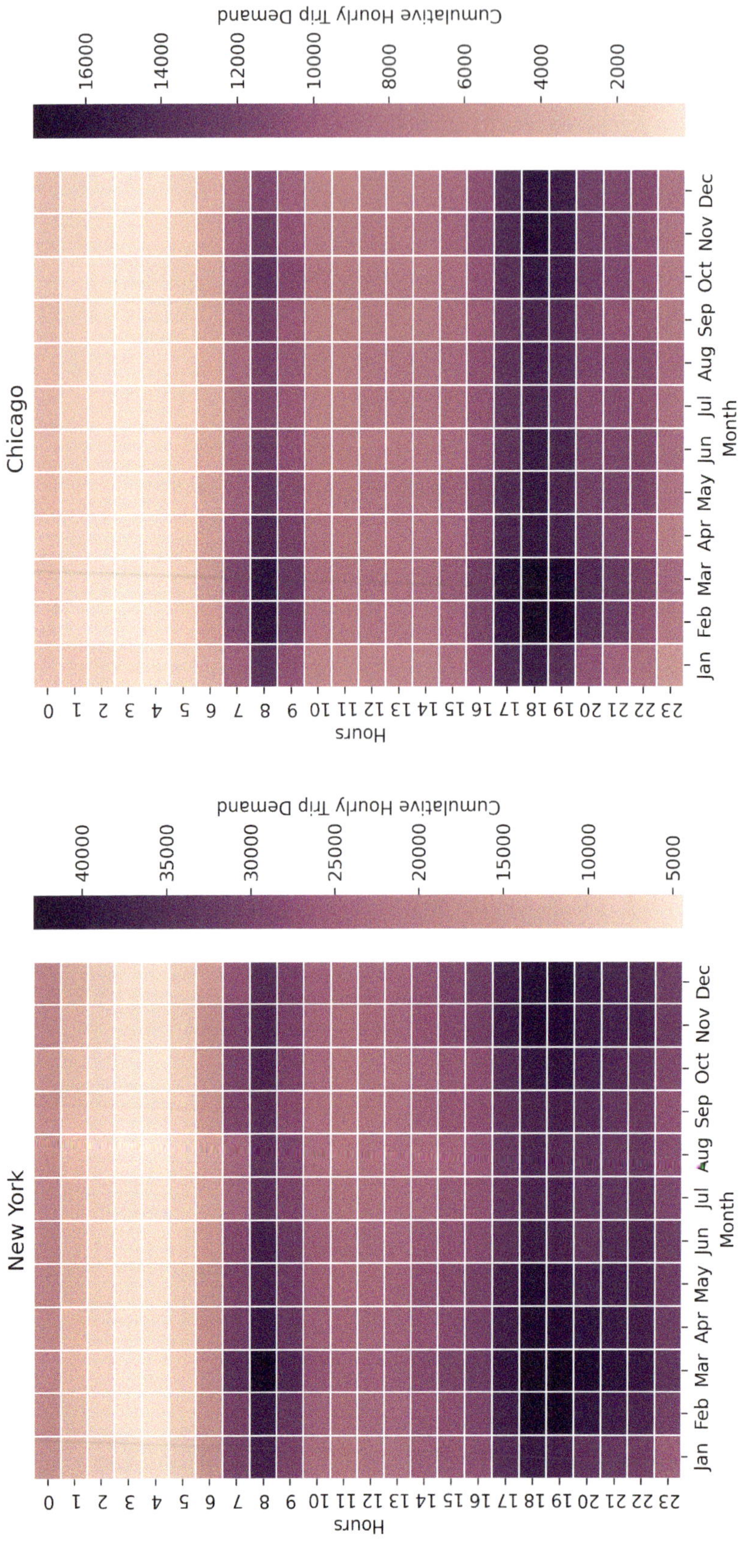

FIGURE 10.2 Ride-hailing trip demand monthly variation in New York and Chicago.

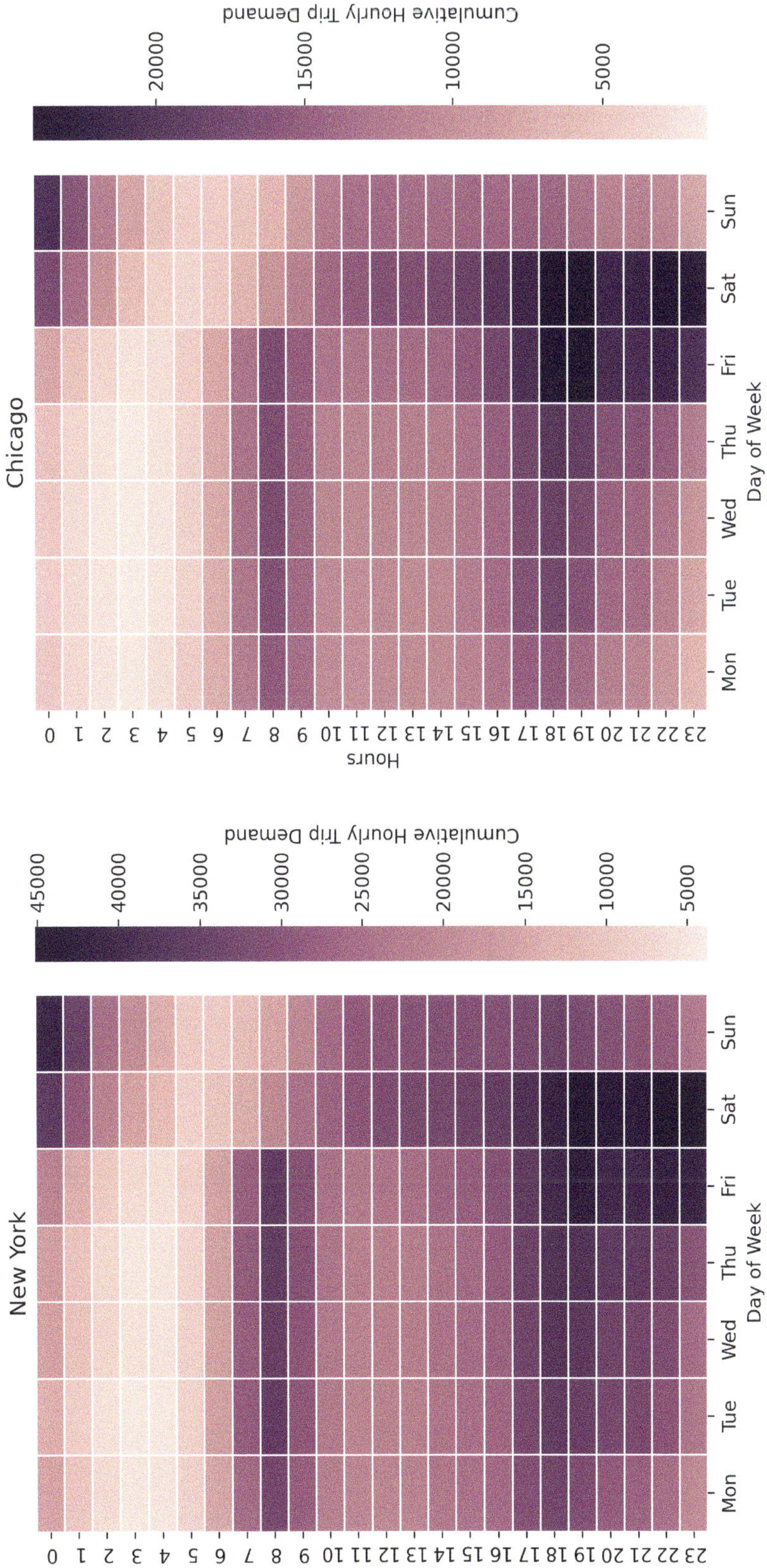

FIGURE 10.3 Ride-hailing trip demand weekly variation in New York and Chicago.

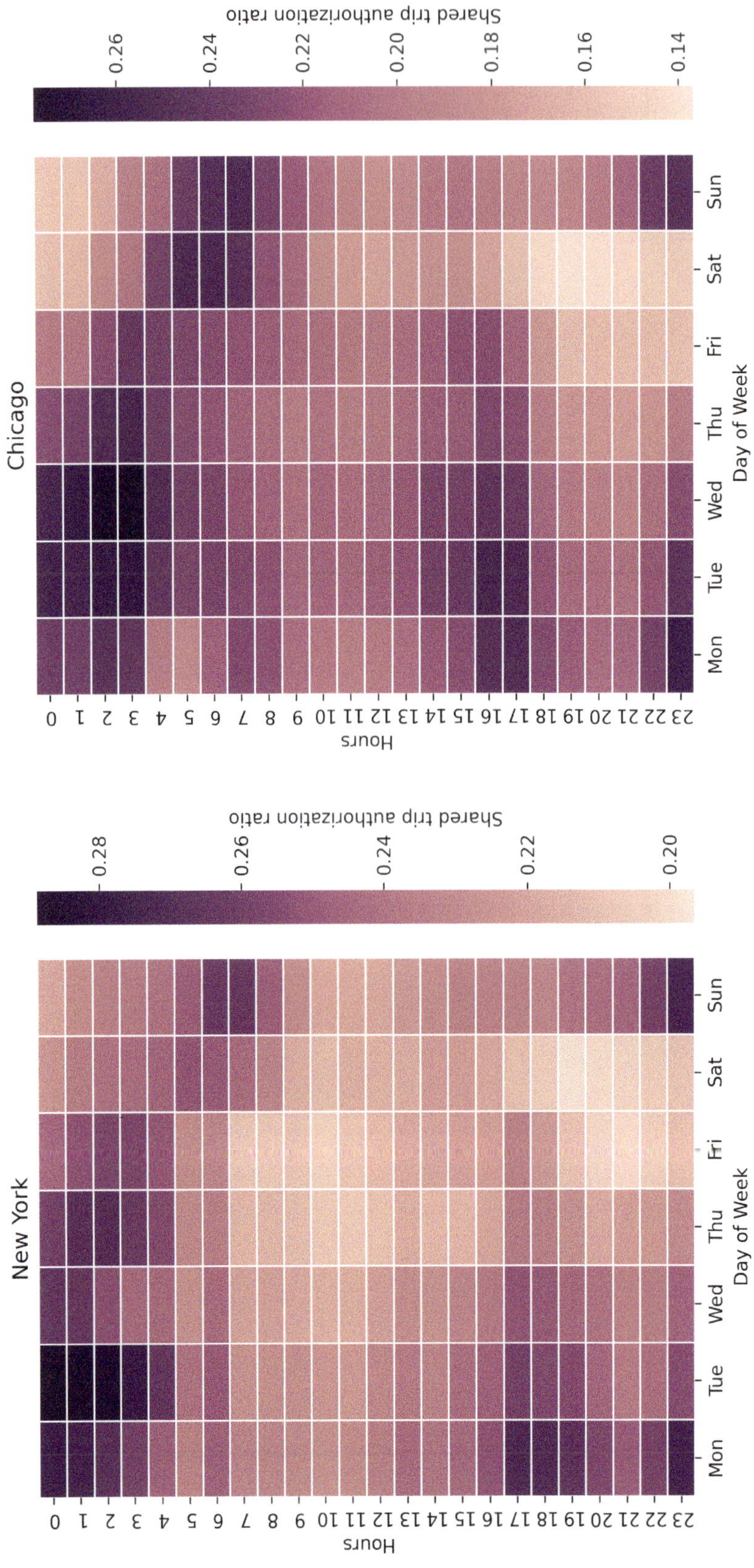

FIGURE 10.4 Shared trip authorization weekly variation in New York and Chicago.

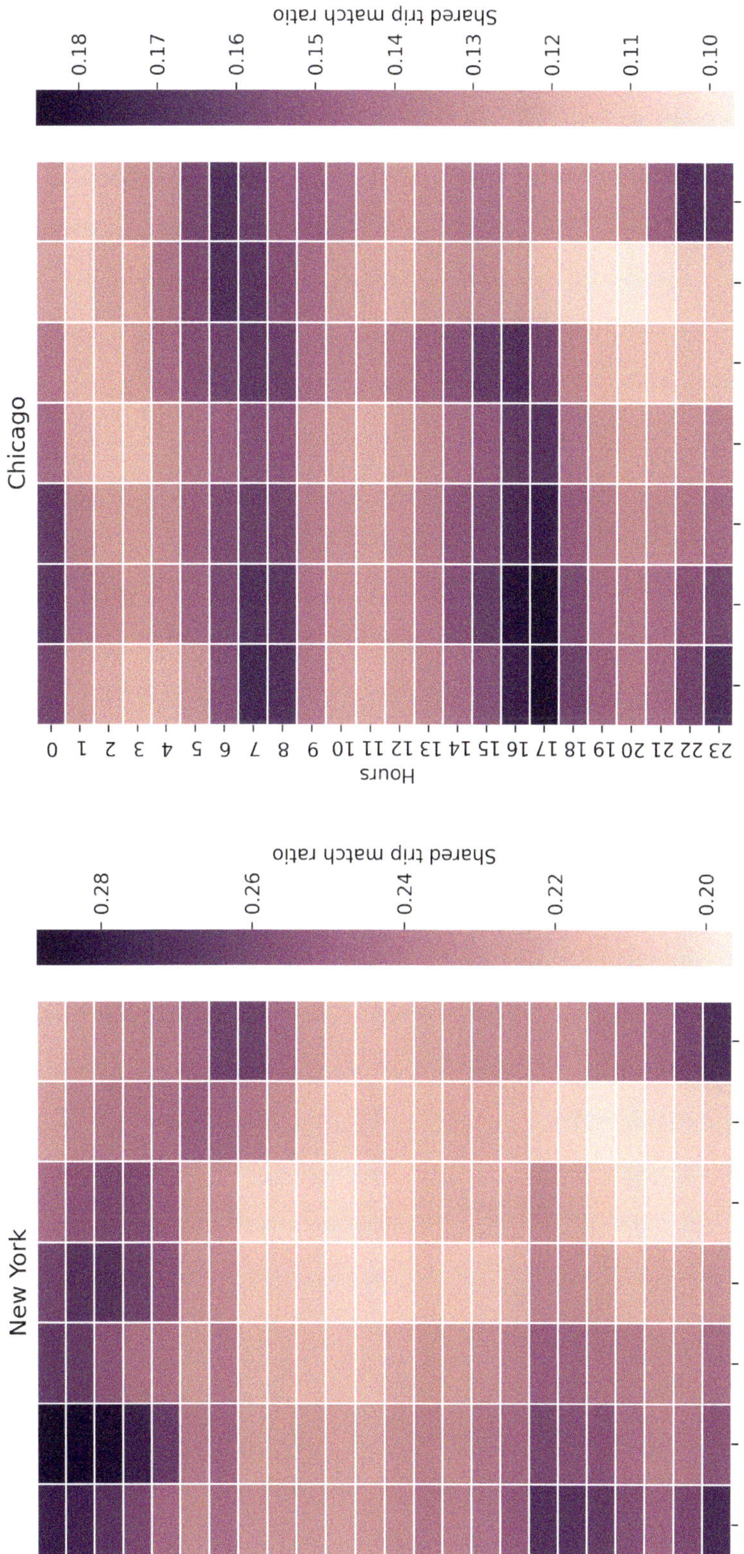

FIGURE 10.5 Shared trip matching weekly variation in New York and Chicago.

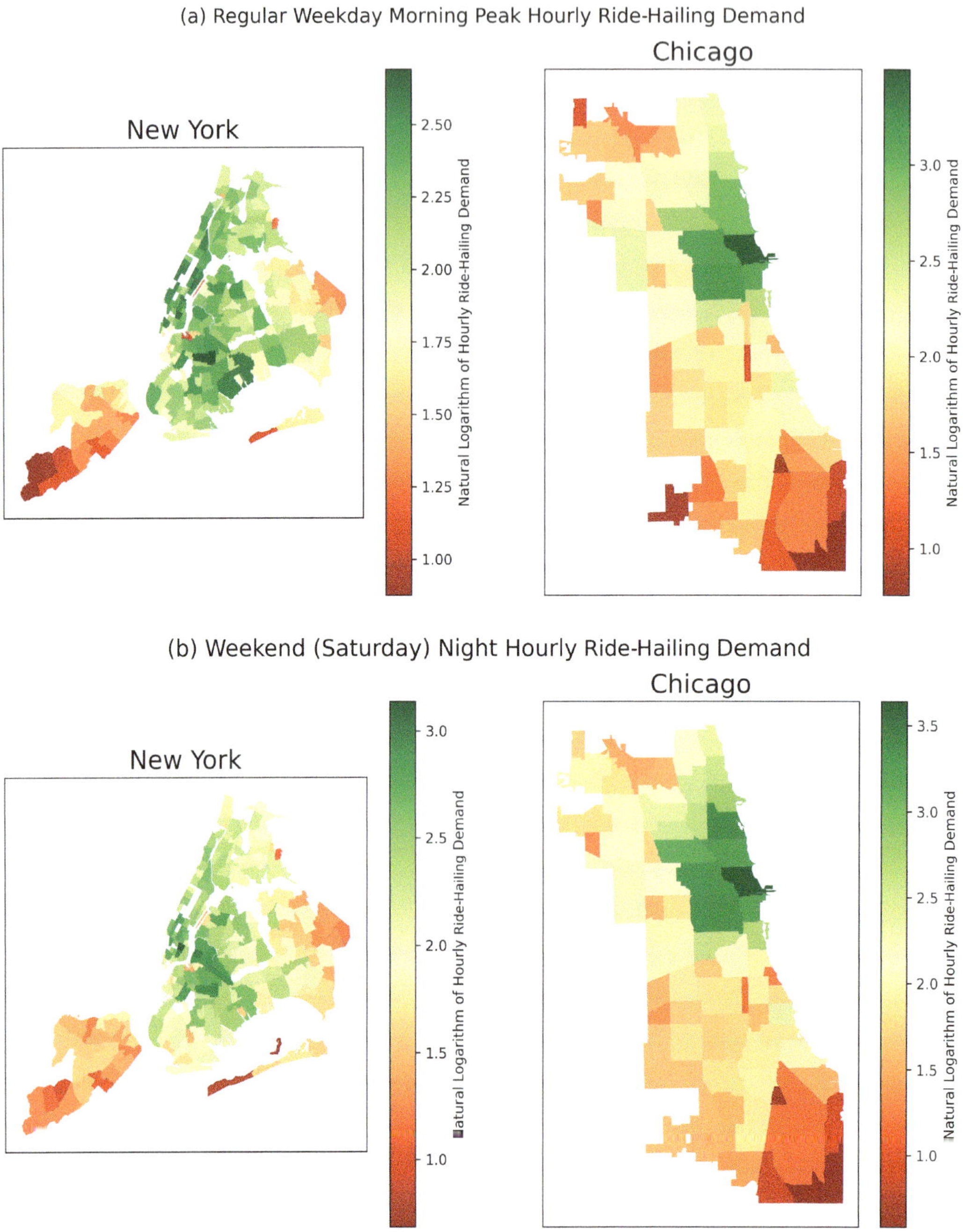

FIGURE 10.6 Spatial distributions of ride-hailing demand in (a) regular weekday morning peak and (b) weekend (Saturday) night.

hour, where a high volume of commuters requires ride-hailing services to reach their workplaces, schools, or other destinations. In contrast, the weekend night demand reflects the social and entertainment-oriented trips.

During the weekday morning peak (8:00 a.m. to 9:00 a.m.), ride-hailing demand in both New York City and Chicago is heavily concentrated in the central business districts. In New York, Manhattan emerges as the primary hub, with additional high-demand zones in Brooklyn. This pattern reflects the daily commute of workers heading into the city for work. Similarly, in Chicago, the highest demand is observed in the downtown area, particularly around the Loop and the Near North Side, indicating a similar commuter-driven demand focused on the city's core.

In contrast, the weekend night demand (11:00 p.m. to 12:00 a.m.) shows a broader distribution of ride-hailing activity in both cities. In New York, demand spreads beyond Manhattan into Brooklyn, especially in areas known for social activities and entertainment. Chicago exhibits a similar trend, with high demand in downtown and the North Side and an increase in activity in the southern community areas.

This shift from concentrated weekday morning peaks to more dispersed weekend night demand highlights the difference in travel behaviors associated with work-related versus leisure activities in urban environments. Spatially, however, there are significant differences between the two cities. In New York, ride-hailing demand is concentrated heavily in Manhattan and parts of Brooklyn, with lesser demand in outer boroughs like Queens and Staten Island. Chicago, on the other hand, exhibits a more centralized demand pattern focused around its downtown and northern neighborhoods.

10.5 ANALYSIS OF URBAN TYPOLOGY FOR BOTH CITIES

One of the crucial strategies for gaining a deeper understanding of mobility behavior within a city, particularly in the context of ride-hailing demand, is to analyze urban typologies. This approach helps predict changes and future trends in ride-hailing demand, enabling more efficient resource allocation and service provision. Additionally, understanding urban typologies aids in recognizing mobility patterns, assessing infrastructure needs, and designing customized transportation policies that can address the specific requirements of different urban zones.

We create a typological framework for both New York and Chicago cities, where the main output will be the clustered spatial units (i.e., taxi zones for New York and community areas for Chicago) based on different regional attributes. This framework consists of several components: arranging spatial attribute dataset, dimension reduction of variables, hierarchical clustering, and understanding of the clustered zones, which are discussed in the following subsections.

10.5.1 Dataset Preparation

The clustering analysis uses a set of independent variables that are classified in different categories, i.e., point of interest and distance from central business district (CBD) using different application programming interfaces (API), demographic and socioeconomic properties, transportation infrastructure, and longitudinal origin destination employment statistics (LODES). The details about each variable are presented in Table 10.1.

10.5.1.1 Point of Interest and Distance from CBD

Using the Google Places API and direction API, variables such as the total number of eateries/cafés, points of interest in each spatial unit, and distance from CBD are determined ("Google Places API" 2023). These variables can reflect the number of ride-hailing trips needed to participate in different types of activities.

10.5.1.2 Demographic and Socioeconomic Properties

For the study period (2019), data from the American Community Survey (ACS) were used to compile the demographic and socioeconomic information about the community areas in Chicago and the taxi zones in New York City ("Social Explorer" 2023). This dataset can be found in ACS at the census tract or census block level. After geoprocessing, the data are generated for each of the two study regions at the appropriate spatial unit level.

10.5.1.3 Transportation Infrastructure

The walk score, bike score, and transit score of the study areas—all of which were derived using the walk score API—were among the transportation infrastructure-related data ("Walkscore" 2023).

TABLE 10.1
Variables Used for Spatial Clustering

		New York				Chicago			
	Units	Mean	Std	Min	Max	Mean	Std	Min	Max
Zones, no.		222				74			
Land Use Properties									
Number of points of interest	Count	5.74	5.05	0	20	11.15	10.36	0	61
Number of restaurants	Count	5.35	4.48	0	19	10.65	8.51	0	52
Shortest distance (miles) from CBD	Miles	11.14	6.5	0.49	32.22	9.95	4.56	0.01	21.41
Demographic and Socioeconomic Properties									
Population density	*1,000 people per square mile	48.8	33.	0.7	185.2	13.5	7.4	1.6	37.4
Percentage of workers using public transport, taxi, and carpool	%	0.53	0.13	0.15	0.78	0.31	0.08	0.08	0.45
Median household income	*1,000 USD	82.3	38.9	20.9	221.3	63.1	29.7	16.5	141.4
Transportation Infrastructure Attributes									
Walk score	Score	83.8	15.0	13	100	71	17.9	4	97.5
Bike score	Score	69.2	16.5	31	98.5	67.61	14.3	30	95.5
Transit score	Score	86.09	16.88	0	100	60.72	11.24	31.5	100

* This indicates the Mean, Minimum, Maximum or Std values will be multiplied by 1000 people per square mile, Suppose, the Mean for New York is 48.8, It will be actually 48800 people per square mile.

10.5.2 Dimensionality Reduction Using Factor Analysis

In this study, we use an exploratory FA algorithm for dimension reduction, a statistical method to condense a large set of observed variables to latent/unobserved variables/factors. Given the complexity and high dimensionality of the data, encompassing too many variables for relatively few observations or spatial units may lead to overfitting, computational inefficiency, and reduced generalizability. FA helps mitigate these issues by identifying underlying relationships between the variables, reducing the data's dimensionality while preserving as much variance as possible (Johnson and Wichern 2007; Hair et al. 2019; Oke et al. 2019).

For a set of observed variables expressed as $X_1, X_2, \ldots X_n$, the theory of FA assumes these variables can be expressed as a linear combination of some latent factors $F_1, F_2, \ldots. F_m$, where $n \gg m$.

$$X_i = \lambda_{i1}F_1 + \lambda_{i2}F_2 + \cdots + \lambda_{im}F_m + \epsilon_i \tag{10.1}$$

where, X_i are the observed variables, $\lambda_{i1}, \lambda_{i2}, \ldots \lambda_{im}$ are the factor loadings, and ϵ_i is the error term for each variable.

The matrix form of the FA model can be expressed as (Johnson and Wichern 2007):

$$X = \Lambda F + E \tag{10.2}$$

where X is the observed variable vector, Λ is the factor loadings matrix, F is the vector of latent factors, and E is the vector of unique error terms.

10.5.2.1 Determining the Optimal Number of Factors

An optimal number of factors is identified through a scree test, where eigen values of the covariance matrix of the observed dataset are calculated first to get the amount of variance covered by each factor (Hair et al. 2019) (Figure 10.7). The factors with eigen values higher than 1 are assumed to

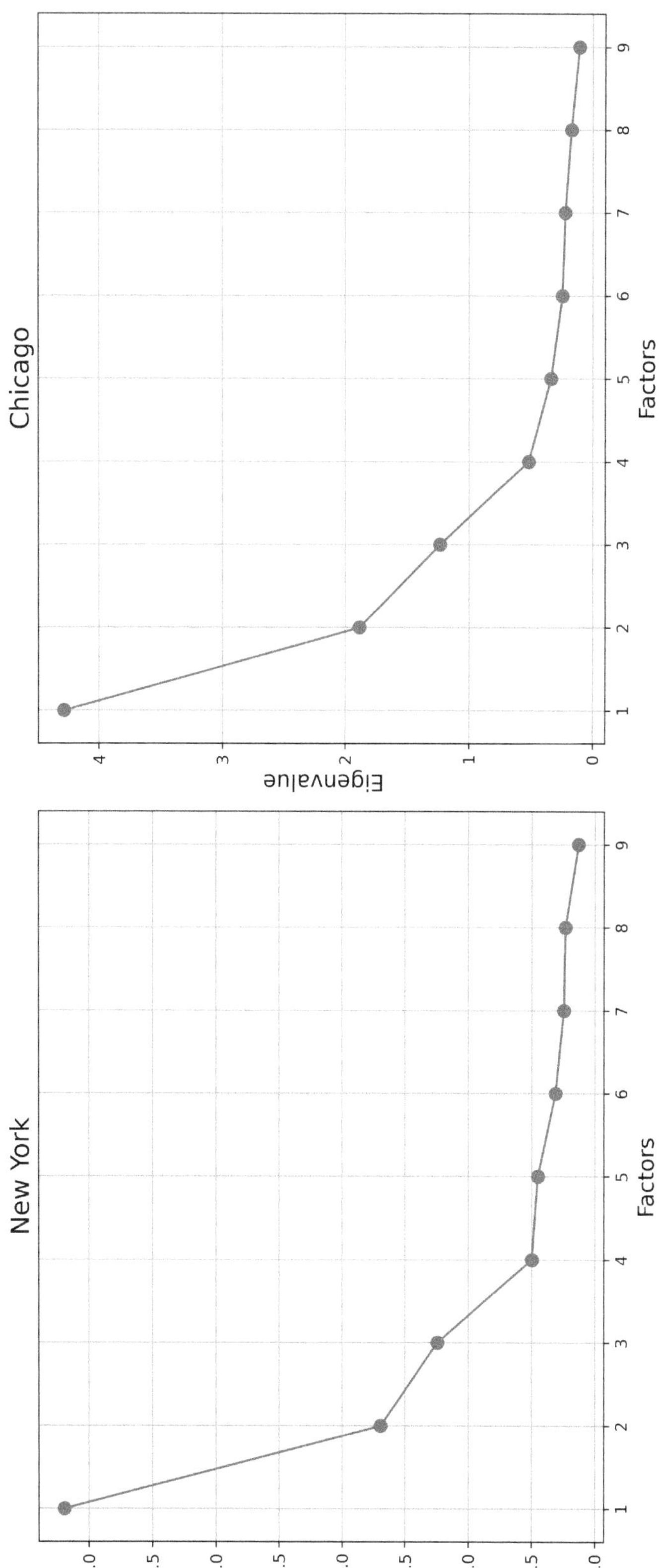

FIGURE 10.7 Cumulative variance explained by the factors.

TABLE 10.2
Factor Loadings for Both New York and Chicago Cities

	New York			Chicago		
Variables/Factors	**Factor 1**	**Factor 2**	**Factor 3**	**Factor 1**	**Factor 2**	**Factor 3**
Number of points of interest	0.05	**0.74**	−0.02	0.28	**0.91**	0.09
Number of restaurants	−0.03	**0.75**	0.06	0.14	**0.88**	0.05
Shortest distance from CBD	**−0.89**	−0.06	0.15	**−0.81**	−0.11	0.24
Population density	**0.65**	−0.14	0.14	**0.74**	0.07	0.1
Percentage of workers using public transport, taxi, and carpool	0.45	−0.05	**0.66**	0.42	−0.12	**−0.67**
Median household income	0.28	0.03	**−0.71**	0.31	0.19	**0.73**
Walk score	**0.89**	0.07	0.07	**0.85**	0.14	0.16
Bike score	**0.84**	−0.07	−0.22	**0.89**	0.05	0.18
Transit score	**0.87**	−0.07	0.20	**0.81**	0.17	−0.26

be optimum or appropriate number of factors. For both cities, three is the number of appropriate factors in this analysis.

10.5.2.2 Calculate Factor Loadings

Factor loadings reveal the strength of association between original variables and the derived factors, helping to assign meaningful names to each factor (Hair et al. 2019). The raw factor loadings from FA are rotated on a trial-and-error basis to get a unique correlated variable set with each factor. High positive or negative loadings indicate strong correlations, allowing us to interpret factors. The factors are named accordingly based on their associated observed variables. The factor loadings are shown in Table 10.2, where the loading absolute values higher than 0.50 are in bold. The factor names are commercial and economic factor, residential factor, socioeconomic factor, and urban mobility factor. The highly corelated variables with each factor are listed in Table 10.3.

10.5.3 Agglomerative Hierarchical Clustering

In this study, we have applied an agglomerative clustering strategy. The factors derived from the FA for different zones in both New York and Chicago cities are utilized as input for the clustering analysis.

Agglomerative hierarchical clustering is a commonly used approach for this kind of typology framework-building problem (Oke et al. 2019). This is a bottom-to-top approach where the individual zonal observations start to form clusters with neighbors of similar attributes and gradually expand the cluster. Two properties define the nature of agglomerative clustering. The first one is the distance calculation metric, and the second one is the linkage method.

TABLE 10.3
Highly Correlated Observed Variables with Each Factor for Both Cities

Factor	Observed Variables	Factor Name
Factor 1	Population density, walk score, bike score, transit score, shortest distance from CBD	Accessibility factor (AF)
Factor 2	Number of point of interests and restaurants	Point of interest factor (PF)
Factor 3	Percentage of workers using public transport, taxi, and carpool and median household income	Socioeconomic factor (SF)

In this study, we are applying agglomerative clustering using the Euclidean distance metric and Ward's method for linkage. The Euclidean distance metric is employed to measure the similarity between data points, ensuring that zones with similar factor profiles are grouped together (Hair et al. 2019). For two zones A and B with k factors or variables in both, the distance d_{AB} between them in the Euclidean method is:

$$d_{AB} = \sqrt{\sum_{i=1}^{k} (F_{Ai} - F_{Bi})^2} \tag{10.3}$$

where F_{Ai} and F_{Bi} are the ith factor of observations A and B.

The Ward's linkage method aims to minimize the variance within each cluster. In this method, the selection of which two elements will be combined to form a new cluster depends on minimizing the within-cluster sum of squares across all variables or factors (Hair et al. 2019).

To determine the optimal number of clusters, we experimented with different cluster number configurations. Although various numbers of clusters were tested, four clusters were ultimately selected for both New York and Chicago based on highest silhouette score (Figure 10.8). This decision was guided by the interpretability of the clusters, ensuring that each cluster represents a distinct and meaningful grouping of zones with similar characteristics.

If we cut the hierarchical dendrogram for New York at height 2 and for Chicago at height 1.25, we would get the desired clusters (Figure 10.9).

10.5.4 Understanding of Clustered Zones

To understand the properties of each cluster, we calculate the mean of the factor scores for all the zones within each cluster. By averaging the factor scores, we obtain a representative profile for each cluster, highlighting the dominant characteristics of the zones grouped together. This method allows us to interpret the clusters by identifying the prominent attributes of clustered zones. The attributes of the clusters are presented in Figure 10.10, and the maps of the clusters from New York and Chicago are displayed in Figure 10.11. We can also observe the percentage of ride-hailing demand generated from or attracted to each cluster, as presented in Table 10.4.

For New York, the cluster zones are sporadically distributed. Cluster 1 stands out with high accessibility and socioeconomic conditions, suggesting zones that are densely populated with good accessibility, have a high number of workers who directly use transit, carpool, or taxicabs to commute, and have a median household income that is high, though they have only a moderate concentration of points of interest. The attributes suggest high regular weekday peak period ride-hailing demands in these zones for home to work or work to home commute. Table 10.4 also shows that Cluster 1 generates the maximum percentages of ride-hailing trips. Cluster 2 represents areas with moderate levels across all factors, indicating a balanced but less affluent profile compared to Cluster 1. Cluster 3 shows similar profile to cluster 1, but with lower socioeconomic attributes and high accessibility, reflected by high ride-hailing trip generation and attraction in Table 10.4. Cluster 4 has high accessibility and a high number of points of interest but lower socioeconomic attributes. The high number of points of interest indicates the likelihood of having a high volume of ride-hailing traffic for social and entertainment activities. The cluster analysis shows that for New York, the ride-hailing demand volume is scattered across multiple clusters, reflected in Table 10.4.

In Chicago, the cluster analysis shows distinct spatial patterns that differ from New York. Cluster 1, with high accessibility and moderate socioeconomic status, is predominantly located in the central and northern parts of the city and also contains the CBD, indicating heavy ride-hailing traffic for both work–home and social gatherings, which is prominently shown in Table 10.4 as the highest percentage of ride-hailing volume is generated and attracted by a few zones of this cluster. Cluster 2 has low accessibility but moderate socioeconomic status, suggesting zones that are less

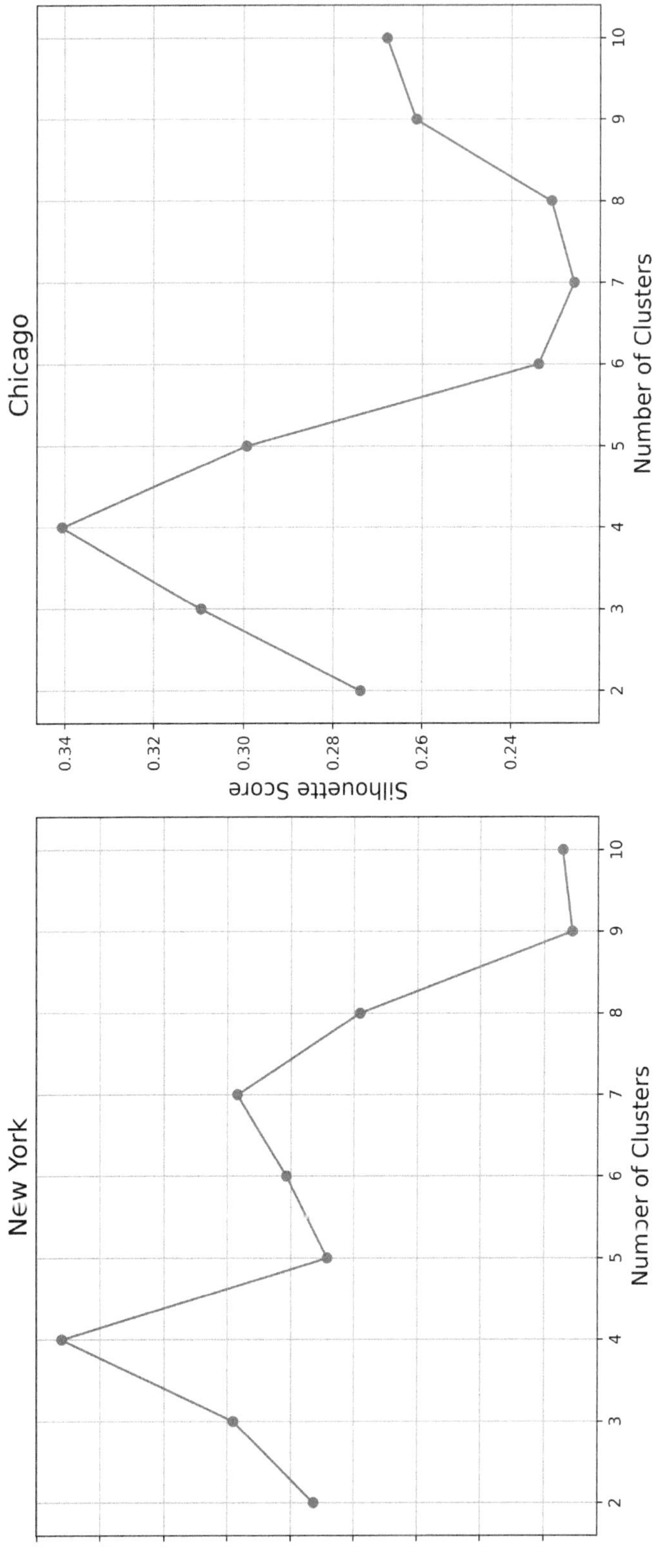

FIGURE 10.8 Silhouette scores for both New York and Chicago city datasets.

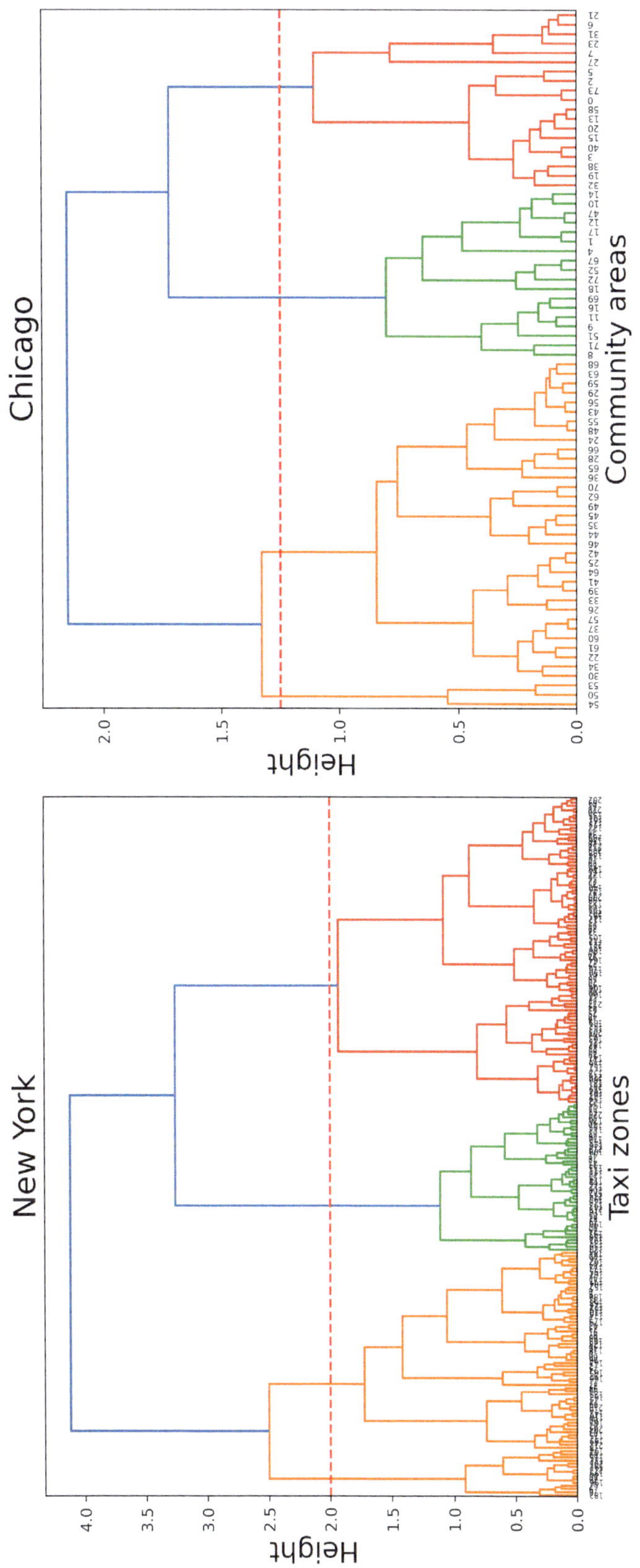

FIGURE 10.9 Hierarchical clusters of zones for both New York and Chicago city datasets.

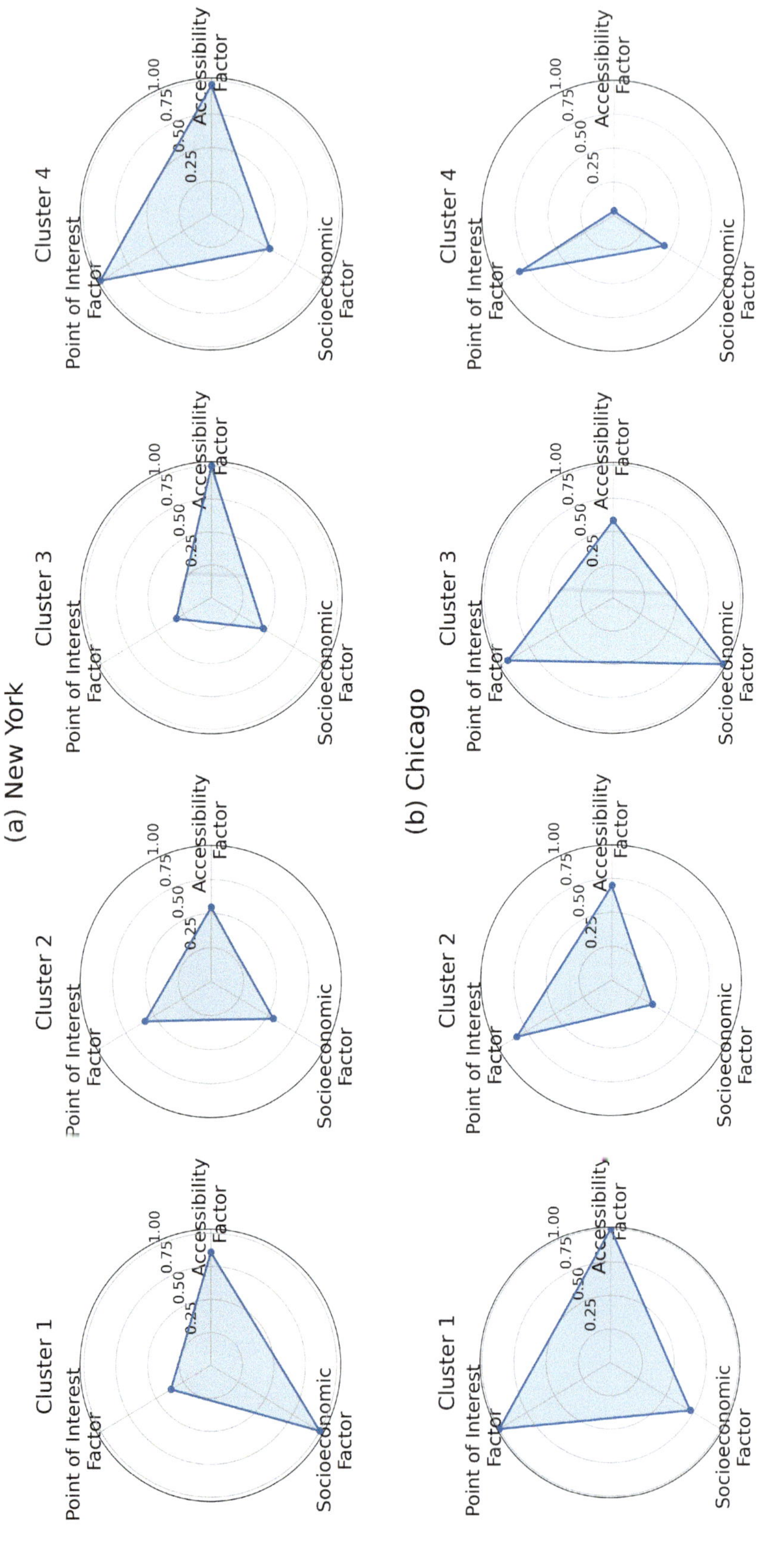

FIGURE 10.10 Comparison of cluster attributes for both cities.

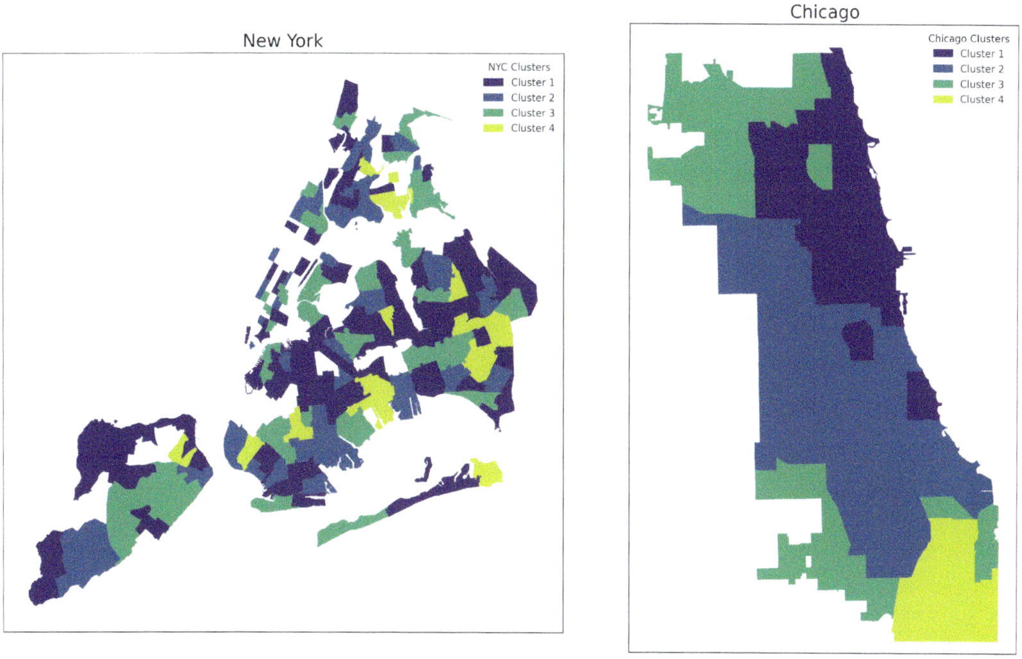

FIGURE 10.11 Clusters in both New York and Chicago.

affluent and less connected to the city's core, but the high point of interest factor indicates a good number of ride-hailing traffic for social gathering purposes, as shown in Table 10.4. Cluster 3, with moderate accessibility yet high socioeconomic status, is in the northwest and southwest parts of the city, suggesting affluent areas that are not necessarily centrally located. Finally, Cluster 4, characterized by very low accessibility and a low presence of points of interest, is found in the outer and less developed areas of the city, likely representing suburban or peripheral zones with lower connectivity and fewer amenities; these areas should experience a lower amount of ride-hailing traffic. The clustering activity in Chicago shows that the ride-hailing demand is mostly concentrated on the cluster 1 zones, as reflected in Table 10.4.

TABLE 10.4
Ride-Hailing Trip Generation and Attraction from Different Clusters

City	Cluster	Number of Zones	Morning Peak Hourly Trip Generation (%)	Morning Peak Hourly Trip Attraction (%)	Weekend Night Hourly Trip Generation (%)	Weekend Night Hourly Trip Attraction (%)
New York	1	85	49.69	45.78	43.10	47.88
	2	44	10.96	9.21	6.64	7.18
	3	40	28.60	31.69	34.60	30.99
	4	13	10.75	13.33	15.66	13.94
Chicago	1	19	73.47	81.15	83.41	81.41
	2	34	19.34	13.65	10.44	11.62
	3	18	6.92	5.01	6.03	6.82
	4	3	0.27	0.19	0.11	0.15

The clustering analysis presented here extends the spatial and temporal visualizations of ride-hailing demand analysis, providing several insights on potential factors that influence the demand generation process. First, for both New York and Chicago, the ride-hailing demand is shaped by the combination of urban accessibility, socioeconomic status, and presence of points of interest. Zones with high accessibility and more points of interest tend to generate higher demand during different periods. Second, the clustering also highlights that while New York ride-hailing demand is more decentralized, as it is dispersed across multiple clusters with different attributes, Chicago ride-hailing demand is more concentrated on the cluster containing the CBD. Third, this clustering analysis also allows us to anticipate how changes in different attributes such as development in transportation infrastructure, shift in socioeconomic condition, or having new points of interest might influence future ride-hailing trip generation or attractions. Therefore, this analysis not only enhances our understanding of current demand but also provides a foundation for predicting and managing future ride-hailing demand.

10.6 CONCLUSIONS

This chapter presents a comprehensive analysis of ride-hailing demand of two major cities in the United States including New York and Chicago, by examining both spatial and temporal patterns alongside a clustering of demand generation zones based on different spatial attributes. The temporal and spatial visualizations highlight the present demand scenarios, revealing peak periods and zones of high activity. However, to better understand the underlying factors for these demand patterns, FA is employed, followed by hierarchical clustering. This approach allowed us to group zones with similar accessibility, points of interest, and socioeconomic characteristics—offering insights on the reasons underlying ride-hailing demand. By understanding these clusters, a clearer picture of the present demand landscape has emerged; this will allow us to develop models for forecasting ride-hailing demand as cities evolve. Thus, this chapter provides valuable insights for urban planners and policymakers aiming to optimize ride-haling services and manage future demand.

REFERENCES

Alessandretti, Laura, Ulf Aslak, and Sune Lehmann. 2020. "The Scales of Human Mobility." *Nature* 587: 402–7. doi:10.1038/s41586-020-2909-1.

Alonso-González, María J., Oded Cats, Niels van Oort, Sascha Hoogendoorn-Lanser, and Serge Hoogendoorn. 2021. "What Are the Determinants of the Willingness to Share Rides in Pooled On-Demand Services?" *Transportation* 48 (4): 1733–65. doi:10.1007/s11116-020-10110-2.

Calabrese, Francesco, Laura Ferrari, and Vincent D. Blondel. 2014. "Urban Sensing Using Mobile Phone Network Data: A Survey of Research." *ACM Computing Surveys* 47 (2). doi:10.1145/2655691.

Canitez, Fatih. 2020. "Transferring Sustainable Urban Mobility Policies: An Institutional Perspective." *Transport Policy* 90: 1–12. doi:10.1016/j.tranpol.2020.02.005.

Dean, Matthew D., and Kara M. Kockelman. 2021. "Spatial Variation in Shared Ride Hail Trip Demand and Factors Contributing to Sharing: Lessons from Chicago." *Journal of Transport Geography* 91. doi:10.1016/j.jtrangeo.2020.102944.

Dey, Bibhas Kumar, Sudipta Dey Tirtha, and Naveen Eluru. 2021. "Transport Networking Companies Demand and Flow Estimation in New York City." *Transportation Research Record* 2675:139–53. doi:10.1177/03611981211000752.

Dey, Bibhas Kumar, Sudipta Dey Tirtha, Naveen Eluru, and Karthik C. Konduri. 2021. "Transformation of Ridehailing in New York City: A Quantitative Assessment." *Transportation Research Part C: Emerging Technologies* 129. doi:10.1016/j.trc.2021.103235.

Du, Jianhe, Hesham A. Rakha, and Helena Breuer. 2022. "An In-Depth Spatiotemporal Analysis of Ride-Hailing Travel: The Chicago Case Study." *Case Studies on Transport Policy* 10 (1): 118–29. doi:10.1016/j.cstp.2021.11.010.

Fan, Zipei, Xuan Song, Ryosuke Shibasaki, and Ryutaro Adachi. 2015. "CityMomentum: An Online Approach for Crowd Behavior Prediction at a Citywide Level." In *UbiComp 2015 - Proceedings of the 2015 ACM International Joint Conference on Pervasive and Ubiquitous Computing*, 559–69. Association for Computing Machinery, Inc. doi:10.1145/2750858.2804277.

González, Marta C., César A. Hidalgo, and Albert László Barabási. 2008. "Understanding Individual Human Mobility Patterns." *Nature* 453: 779–82. doi:10.1038/nature06958.

"Google Places API." 2023. https://developers.google.com/maps/documentation/places/web-service/search.

Hair, Joseph F., Barry J. Babin, William C. Black, and Rolph E. Anderson. 2019. *Multivariate Data Analysis*. 8th ed. Cengage.

Hasan, Samiul, and Satish V. Ukkusuri. 2015. "Location Contexts of User Check-Ins to Model Urban Geo Life-Style Patterns." *PLoS One* 10 (5). doi:10.1371/journal.pone.0124819.

Hasnat, Md Mehedi, and Samiul Hasan. 2018. "Identifying Tourists and Analyzing Spatial Patterns of Their Destinations from Location-Based Social Media Data." *Transportation Research Part C: Emerging Technologies* 96: 38–54. doi:10.1016/j.trc.2018.09.006.

Jiang, Shan, Joseph Ferreira, and Marta C. González. 2012. "Clustering Daily Patterns of Human Activities in the City." *Data Mining and Knowledge Discovery* 25: 478–510. doi:10.1007/s10618-012-0264-z.

Johnson, Richard A., and Dean W. Wichern. 2007. *Applied Multivariate Statistical Analysis*. 6. Pearson Education, Inc.

Kontou, Eleftheria, Venu Garikapati, and Yi Hou. 2020. "Reducing Ridesourcing Empty Vehicle Travel With Future Travel Demand Prediction." *Transportation Research Part C: Emerging Technologies* 121. doi:10.1016/j.trc.2020.102826.

Lam, Chungsang Tom, Meng Liu, and Xiang Hui. 2021. "The Geography of Ridesharing: A Case Study on New York City." *Information Economics and Policy* 57. doi:10.1016/j.infoecopol.2021.100941.

Li, Xuefeng, Jiacong Xu, Mingyang Du, Dong Liu, and Mei Po Kwan. 2023. "Understanding the Spatiotemporal Variation of Ride-Hailing Orders Under Different Travel Distances." *Travel Behaviour and Society* 32. doi:10.1016/j.tbs.2023.100581.

Liang, Zijun, Wei Kong, Xuejuan Zhan, and Yun Xiao. 2022. "Analysis of the Impact of Ride-Hailing on Urban Road Network Traffic by Using Vehicle Trajectory Data." *Journal of Advanced Transportation* 2022. doi:10.1155/2022/6940850.

Liu, T. L. K., P. Krishnakumari, and O. Cats. 2019. "Exploring Demand Patterns of a Ride-Sourcing Service Using Spatial and Temporal Clustering." In *2019 6th International Conference on Models and Technologies for Intelligent Transportation Systems (MT-ITS)*, 1–9. IEEE. doi:10.1109/MTITS.2019.8883312.

Lotze, Charlotte, Philip Marszal, Felix Jung, Debsankha Manik, Marc Timme, and Malte Schröder. June 2023. "Identifying the Threshold to Sustainable Ridepooling." http://arxiv.org/abs/2306.05851.

Lyu, Wenjun, Guang Wang, Yu Yang, and Desheng Zhang. 2021. "Mover: Generalizability Verification of Human Mobility Models via Heterogeneous Use Cases." *Proceedings of the ACM on Interactive, Mobile, Wearable and Ubiquitous Technologies* 5 (4). doi:10.1145/3494997.

Mazza, Mario. 2011. "Global Markets and Supply Customization in Urban Transportation Systems." *Anno Accademico*. Doctoral Thesis, Milan: University of Milano-Bicocca.

Naumov, Sergey, and David Keith. 2023. "Optimizing the Economic and Environmental Benefits of Ride-Hailing and Pooling." *Production and Operations Management* 32 (3): 904–29. doi:10.1111/poms.13905.

"New York City Mobility Report." 2019. https://www.nyc.gov/html/dot/downloads/pdf/mobility-report-singlepage-2019.pdf.

Oke, Jimi B, Youssef M Aboutaleb, Arun Akkinepally, Carlos Lima Azevedo, Yafei Han, P Christopher Zegras, Joseph Ferreira, and Moshe E Ben-Akiva. 2019. "A Novel Global Urban Typology Framework for Sustainable Mobility Futures." *Environmental Research Letters* 14 (9): 095006. doi:10.1088/1748-9326/ab22c7.

Popovich, Natalie, C. Anna Spurlock, Zachary Needell, Ling Jin, Tom Wenzel, Colin Sheppard, and Mona Asudegi. 2021. "A Methodology to Develop a Geospatial Transportation Typology." *Journal of Transport Geography* 93. doi:10.1016/j.jtrangeo.2021.103061.

"Research and Markets." 2023. www.researchandmarkets.com/r/sv7odc.

Roy, Sudipta, Bat-hen Nahmias-Biran, and Samiul Hasan. 2025. "Spatial transferability of machine learning based models for ride-hailing demand prediction." *Transportation Research Part A: Policy and Practice* 193. doi: 10.1016/j.tra.2025.104413.

"Social Explorer." 2023. https://www.socialexplorer.com/.

Tirachini, Alejandro, and Andres Gomez-Lobo. 2020. "Does Ride-Hailing Increase or Decrease Vehicle Kilometers Traveled (VKT)? A Simulation Approach for Santiago de Chile." *International Journal of Sustainable Transportation* 14 (3): 187–204. doi:10.1080/15568318.2018.1539146.

"Walkscore." 2023. https://www.walkscore.com/.

Wang, Yandong, Teng Wang, Ming Hsiang Tsou, Hao Li, Wei Jiang, and Fengqin Guo. 2016. "Mapping Dynamic Urban Land Use Patterns With Crowdsourced Geo-Tagged Social Media (Sina-Weibo) and Commercial Points of Interest Collections in Beijing, China." *Sustainability (Switzerland)* 8 (11). doi:10.3390/su8111202.

Wang, Yu, Tongya Zheng, Yuxuan Liang, Shunyu Liu, and Mingli Song. 2024a. "COLA: Cross-City Mobility Transformer for Human Trajectory Simulation." In *WWW 2024 - Proceedings of the ACM Web Conference*, 3509–20. Association for Computing Machinery, Inc. doi:10.1145/3589334.3645469.

Wang, Zhiqi, Yufeng Zhang, Bin Jia, and Ziyou Gao. 2024b. "Comparative Analysis of Usage Patterns and Underlying Determinants for Ride-Hailing and Traditional Taxi Services: A Chicago Case Study." *Transportation Research Part A: Policy and Practice* 179. doi:10.1016/j.tra.2023.103912.

Zhang, Jiechao, Samiul Hasan, Xuedong Yan, and Xiaobing Liu. 2021. "Spatio-temporal mobility patterns of on-demand ride-hailing service users." *The International Journal of Transportation Research* 14 (9). doi:10.1080/19427867.2021.1988439. doi:10.48550/arXiv.2010.10249.

Zhang, Xiaojian, and Xilei Zhao. 2022. "Machine Learning Approach for Spatial Modeling of Ridesourcing Demand." *Journal of Transport Geography* 100. doi:10.1016/j.jtrangeo.2022.103310.

Zhang, Xinxin, Bo Huang, and Shunzhi Zhu. 2020. "Spatiotemporal Varying Effects of Built Environment on Taxi and Ride-Hailing Ridership in New York City." *ISPRS International Journal of Geo-Information* 9 (8). doi:10.3390/ijgi9080475.

Zheng, Yu, Lizhu Zhang, Xing Xie, and Wei-Ying Ma. 2009. "Mining Interesting Locations and Travel Sequences from GPS Trajectories." In *Proceedings of the 18th International Conference on World Wide Web*, 791–800. New York, NY, USA: ACM. doi:10.1145/1526709.1526816.

Zheng, Yunhan, Patrick Meredith-Karam, Anson Stewart, Hui Kong, and Jinhua Zhao. 2023. "Impacts of Congestion Pricing on Ride-Hailing Ridership: Evidence from Chicago." *Transportation Research Part A: Policy and Practice* 170. doi:10.1016/j.tra.2023.103639.

11 Generative Artificial Intelligence–Powered Multi-Agent Paradigm for Smart Urban Mobility

Opportunities and Challenges for Integrating Large Language Models and Retrieval-Augmented Generation with Intelligent Transportation Systems

Haowen Xu, Jinghui Yuan, Anye Zhou, Guanhao Xu, Wan Li, Xuegang (Jeff) Ban, and Xinyue Ye

11.1 INTRODUCTION

Urbanization has led to an estimated 40% of the population spending over an hour commuting daily, making efficient urban mobility management essential (Schafer and Victor, 2000). The growth in urban populations and vehicle numbers has strained transportation systems, causing congestion, accidents, energy waste, and pollution (Xu et al., 2023a). Cities are now leveraging advancements in artificial intelligence (AI), the Internet of Things (IoT), 5G, and computing technologies to create more efficient, safe, and sustainable transportation systems (Xu et al., 2021a; Moriano et al., 2024). These technologies are central to intelligent transportation systems (ITS), which use data from traffic sensors, connected vehicles (CVs), simulations, and crowdsourcing for real-time situational awareness and predictive analytics (Li et al., 2021; Xu et al., 2023a). ITS optimizes infrastructure, traffic control, parking, and vehicle operations while providing intelligent advisory services for travelers and operators (Berres et al., 2021a, 2021b; Shao et al., 2022; Zhou et al., 2022c). Globally, cities are adopting ITS to enhance safety, mobility, and environmental performance, as seen in Japan's VICS and Smartway, Europe's CVIS and COOPERS, and the United States's ITS Strategic Plan.

Despite global successes, challenges persist in making ITS scalable, accessible, and interoperable, especially for diverse urban populations (Li et al., 2021). Scaling smart city technologies to meet growing urban demands requires extensive data processing, traffic simulations, and software development (Xu et al., 2022a, 2023b, 2023b, 2024b). Additionally, the inaccessibility of smart mobility services and transportation data limits their effective use by city residents (Cledou et al., 2018). There is a need for digital assistants that help users intuitively access ITS tools and information. As smart city planning shifts toward human-centric approaches, next-generation ITS and CVs

DOI: 10.1201/9781003503262-14

should promote public participation through user-friendly technologies that don't require specialized skills, allowing residents to customize services through simple interactions (Vasilieva et al., 2023).

Given AI's role in transportation research, there is a growing interest in using generative AI (GenAI) to enhance ITS through advanced language understanding, content generation, and reasoning. This paper explores integrating large language models (LLMs) and retrieval-augmented generation (RAG) technologies to transform ITS and CVs into intelligent, human-centric multi-agent systems (MASs). We review current mobility data and technologies in ITS and CV sectors and propose a framework for integrating GenAI to (1) promote science-based advisory for reducing congestion, accidents, and emissions; (2) enhance public engagement in mobility management; and (3) automate tasks like data analytics and traffic simulations. The target users include drivers, urban commuters, transportation planners, and traffic operators responsible for emergency response, traffic operations, and asset maintenance. The concept of our proposed framework is illustrated in Figure 11.1.

Compared to conventional chatbots using general-purpose LLMs, our LLM-RAG-powered MAS offers unique advantages:

1. Responses are tailored to user-defined scenarios, addressing the specific transportation and mobility needs of distinct user groups, ensuring relevance to each user's context.
2. By integrating real-time and historical data (e.g., traffic, weather, infrastructure, public transit, global positioning systems [GPS]), responses are timely and situationally aware, enhancing decision-making in urban mobility.
3. Responses are personalized, using advanced data-driven and simulation-based methods, providing context-specific insights rather than generic information.

Our GenAI-powered MAS combines ITS advancements with LLMs and RAG technologies to create more intuitive, efficient, and interoperable solutions for urban mobility management.

11.2 LITERATURE REVIEW

Over the past decades, ITS have advanced with information communication technology (ICT) innovations, transforming urban transportation into optimized smart mobility. ITS and CV applications reduce congestion, enhance safety, identify bottlenecks, and improve transportation planning. Research has focused on integrating AI, simulations, and communication technologies to create more efficient, safer, and sustainable urban transport systems. This chapter reviews key areas such as traffic state forecasting and vehicle behavior prediction, advanced driver assistance and intelligent speed adaptation, traffic safety and crash risk modeling, and network traffic control and optimization.

In the following subsections, we explore the data generated by modern intelligent transportation technologies, which are essential for advanced traffic analytics and optimization—key to achieving smart mobility systems. We will also review current research efforts in each area. This sets the stage for our vision of developing GenAI-powered MAS for next-generation urban smart mobility, focusing on the planning, management, and optimization of road transportation and urban mobility systems.

11.2.1 Popular Intelligent Transportation Applications

Driven by vast and diverse traffic data, novel methodologies have been integrated into ITS to enhance urban mobility from multiple perspectives, which are reviewed in the following subsections. Advances in ICT and computing now allow these ITS algorithms to be accessed via wireless communication protocols and application programming interface (API) endpoints. This enables AI

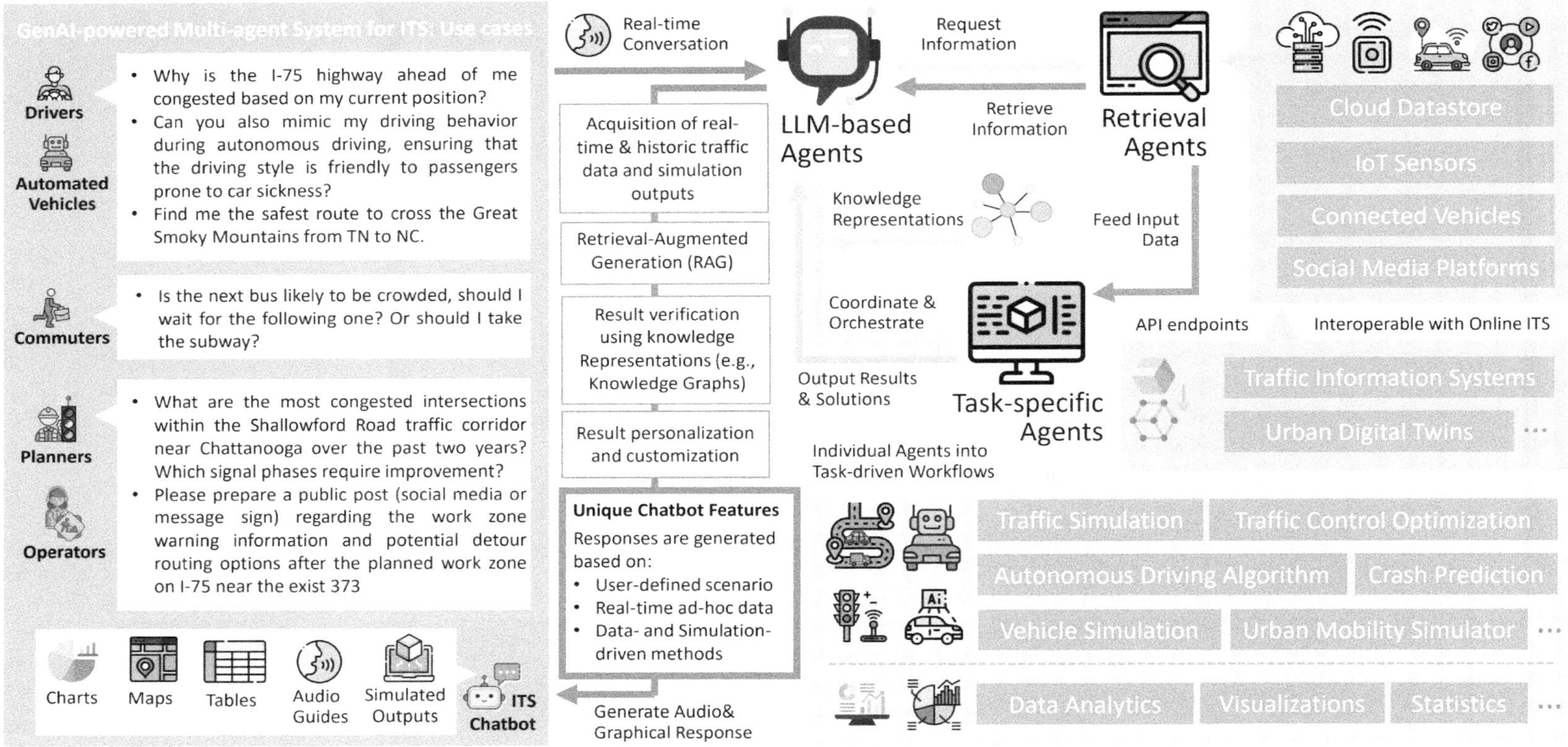

FIGURE 11.1 A GenAI-powered MAS for intelligent transportation and smart mobility.

agents to utilize these methods as web or cloud services, facilitating automated analytics and optimizations for smart mobility services.

11.2.1.1 Traffic State Forecasting and Vehicle Behavior Prediction

Traffic state forecasting and behavior prediction are essential for enhancing road safety, reducing congestion, and improving traffic efficiency. Traffic state forecasting predicts future conditions like flow, density, and speed using historical and real-time data from sources such as loop detectors, cameras, probing devices, and social media (Kumar and Raubal, 2021). Vehicle behavior prediction focuses on forecasting the trajectories of surrounding vehicles using data from historical trajectories and traffic environments gathered through sensors and communication networks.

Algorithms for traffic forecasting and behavior prediction fall into two categories: model-based and learning-based methods. Traditional model-based approaches, such as statistical time series models, rule-based models, and probabilistic models, perform well in specific scenarios but are limited by rigid structures and assumptions, making them less effective under uncertainty (Wang et al., 2014; Li et al., 2020b; Zhang et al., 2020; Wang et al., 2022b).

In contrast, model-free, learning-based methods, particularly those using deep neural networks (DNNs), offer greater accuracy and robustness under uncertain conditions. DNNs can handle multimodal inputs, generate high-dimensional outputs, and learn spatial and temporal dependencies, making them well-suited for traffic forecasting and behavior prediction (Do et al., 2018; Hosseini and Talebpour, 2019; Lin et al., 2022). Advanced techniques like convolutional neural networks (CNNs), graph neural networks (GNNs), and recurrent neural networks (RNNs) further enhance their capabilities by capturing social interactions and time-series dependencies (Cui et al., 2020a, 2020b; Mo et al., 2021). Incorporating physical principles into these models improves their generalizability and reliability (Mo et al., 2020; Geng et al., 2023).

The emergence of LLMs offers additional opportunities to refine forecasting and prediction by integrating expert feedback, improving DNN training, and enhancing the interpretability and trustworthiness of the overall process.

11.2.1.2 Driving Advisory and Intelligent Speed Adaptation

Advanced driving assistance systems (ADAS) enhance driver decision-making and road safety by providing real-time guidance using data from GPS, traffic systems, and vehicle sensors. ADAS offers advice on lane usage, merging, speed adjustments, and route selection, reducing driver workload and improving safety (Antony and Whenish, 2021; Ameta et al., 2023).

Intelligent speed adaptation (ISA) helps drivers maintain appropriate speeds by either advising on speed limits or automatically adjusting the vehicle's speed to comply with legal limits. ISA uses GPS, digital maps, and onboard cameras to determine speed limits, aiming to reduce speed-related accidents and improve road safety and fuel efficiency (Jamson et al., 2006, Ryan, 2019, Newsome and WSP, 2024).

ADAS and ISA are widely adopted in connected automated vehicles (CAVs), which vary in levels of cooperation and automation. In connected but nonautomated scenarios, advisory systems compute energy-saving or congestion-mitigating trajectories for human drivers. In automated but nonconnected scenarios, ADAS ensures smooth vehicle operations using onboard sensors (Zhou et al., 2022d, 2023). Fully connected and automated systems, based on the Society of Automobile Engineers standards, include Status sharing (Class A), Intent sharing (Class B), Agreement seeking (Class C), and Prescriptive (Class D) (Zhou et al., 2022a; Wang et al., 2024).

Challenges remain in personalizing autonomous systems, interpreting AI decisions, and building trust in human–machine interactions. Multimodal LLMs offer potential solutions by integrating human input and articulating AI and human rationales in a bidirectional manner.

11.2.1.3 Traffic Safety and Crash Risk Modeling and Prediction

Traffic safety applications utilize statistical and machine learning techniques to assess and predict the likelihood of traffic crashes (Chand et al., 2021). These models identify patterns and factors contributing to crashes, such as driver behavior, road conditions, weather, and traffic flow, with the aim of enhancing road safety (Theofilatos and Yannis, 2014). By incorporating historical crash data and real-time information, real-time crash risk prediction models have been developed to identify high-risk situations and locations. Additionally, traffic conflict-based safety models address the limitations of the rarity and randomness of crash data.

There are three primary types of traffic safety models:

1. *Crash frequency prediction models (safety performance functions [SPF])*: These models estimate the expected number of crashes at specific locations, helping planners identify high-risk areas and allocate resources effectively. They often consider factors like traffic volume, road geometry, and traffic control characteristics (Khattak et al., 2024).
2. *Crash severity models*: These models predict the severity of crashes by assessing factors such as vehicle speed, driver characteristics, collision type, and vehicle safety features. Understanding crash severity aids in developing targeted safety interventions and infrastructure improvements (Qin et al., 2013; Khattak and Fontaine, 2020).
3. *Real-time crash risk prediction models*: These models predict crash likelihood based on real-time traffic flow characteristics, enabling dynamic warnings for drivers and providing traffic operators with insights to improve emergency response and reduce incident clearance time (Yuan and Abdel-Aty, 2018; Yuan et al., 2020; Yu et al., 2020).

Additionally, surrogate safety models use measures like time to collision and deceleration rates to predict near-crashes before they occur, offering a scalable and implementable approach to traffic safety (Xing et al., 2020; Wang et al., 2021; Abdel-Aty et al., 2022).

These models are widely applied in urban planning, traffic management systems, ADAS, and autonomous driving, contributing to safer road networks and reduced accident rates.

11.2.1.4 Network Traffic Control and Optimization

While driving advisory systems and intelligent vehicle control enhance individual vehicle mobility, network traffic control strategies are crucial for optimizing entire transportation systems. These strategies improve travel times, reduce fuel consumption, and lower emissions, supporting sustainable urban mobility (Ugirumurera et al., 2020; Xu et al., 2021b; Wang et al., 2022a). Network traffic control involves managing and regulating traffic flows using real-time data and advanced algorithms to optimize signals, reroute vehicles, and adjust traffic patterns.

Traffic signal control is the most widely implemented strategy. Traditional fixed-time control often fails to adapt to real-time traffic, leading to inefficiencies. Adaptive signal control systems address this by using real-time data to adjust signal timings dynamically, optimizing flow and reducing delays (Wang et al., 2018; Li et al., 2022b). Signal coordination synchronizes signals along major corridors, minimizing stops and maintaining continuous traffic flow (Bazzan, 2005; Li and Ban, 2020). Transit signal priority systems further enhance efficiency by reducing delays for public transit vehicles (Smith et al., 2005; He et al., 2014).

Ramp metering controls vehicle entry onto freeways to maintain optimal flow and prevent congestion, improving throughput, speeds, and safety (Papageorgiou et al., 1991; Ma et al., 2020). It relies on real-time data, predictive modeling, and adaptive algorithms.

The macroscopic fundamental diagram (MFD), which describes the relationship between flow, density, and speed at a network level, is a valuable tool for managing congestion across networks. MFD-based strategies like perimeter flow control, congestion pricing, and area-wide signal control effectively optimize traffic flow (Aboudolas and Geroliminis, 2013; Xu and Gayah, 2024). These strategies require robust data collection and advanced analytics to interpret MFD data effectively.

Other methods include variable speed limits, which adjust speed limits based on real-time conditions (Lin et al., 2004), and integrated corridor management, which optimizes corridor performance by efficiently managing infrastructure and resources (Hashemi and Abdelghany, 2016). Integrating these strategies with LLMs can further enhance their effectiveness.

11.3 THE RAISE OF MULTI-AGENT SYSTEMS

A MAS is a framework where multiple autonomous agents, each with specific capabilities, collaborate to achieve a common goal or solve complex problems (Vogel-Heuser et al., 2020). These agents interact, communicate, and emulate collaborative behaviors seen in human and natural systems. Enhanced by AI and knowledge engineering, they now perform increasingly specialized tasks. MASs are gaining traction in fields like smart manufacturing and healthcare, where complex problem-solving is essential (Yahouni et al., 2021). The integration of GenAI technologies, including LLMs and RAG, has further evolved these systems, enabling them to address complex problems that require deep domain knowledge more effectively.

A typical GenAI-powered MAS includes key components: (1) LLM agents for natural language understanding and generation, (2) retrieval agents that gather relevant data from external sources, and (3) task-specific agents for specialized functions like sentiment analysis, optimization, or decision-making. These agents work together, guided by domain knowledge (e.g., ontologies and knowledge graphs), forming a cohesive unit that addresses intricate problems more effectively than a single model.

11.3.1 LLM Agents: Origin and Capabilities

In recent years, pretrained LLMs have greatly expanded AI's capabilities in natural language understanding, reasoning, and generation (Yang et al., 2024). Models like GPT-3, BERT, T5, and RoBERTa, built on the Transformer architecture, have become increasingly popular in industry and academia (Raiaan et al., 2024). These models excel in applications such as conversational agents and content generation due to their ability to leverage vast amounts of textual data, capturing nuanced linguistic patterns and contextual relationships. This makes them more versatile than traditional deep learning methods, which often require task-specific training and struggle with broader context comprehension.

LLMs' flexibility and robustness have led to their growing use in urban science research, where they analyze large volumes of textual data, extract insights, and support decision-making in areas like building management, natural hazard mitigation, energy system management, and freight system optimization (Choi et al., 2024; Tupayachi et al., 2024). Their ability to process and interpret complex datasets with minimal supervision is invaluable for addressing big data challenges in these fields.

In a GenAI-powered MAS, an LLM agent serves as a specialized entity for processing natural language data and generating content. Its roles include:

- *Natural language understanding*: Interpreting human language inputs to facilitate effective communication with users and other agents
- *Task automation*: Automating language-based tasks like report drafting, response generation, and information summarization, reducing workload on human agents
- *Mediation and coordination*: Facilitating communication and coordination between agents by translating, summarizing, or contextualizing information
- *User interaction*: Serving as the primary interface for human users, enabling intuitive and conversational interactions, enhancing user experience in complex systems

Overall, the LLM agent enhances a MAS's cognitive capability, enabling more sophisticated, context-aware, and human-centric interactions.

11.3.2 Retrieval Agent for RAG

Retrieval agents were introduced to overcome the limitations of early LLM agents, particularly in handling domain-specific queries and accessing up-to-date information (Hu et al., 2023). General-purpose LLMs, trained on static text, often struggle with specialized questions, leading to issues like hallucinations (Zaki and Krishnan, 2024). Instead of relying on the costly and data-intensive fine-tuning of LLMs, RAG provides an efficient solution by integrating external information retrieval systems, enhancing the LLMs' knowledge base with domain-specific information.

RAG retrieves relevant documents from a vector database, which indexes and searches data using vector embeddings. The retrieved information is then combined with the model's internal knowledge to generate responses (Cakaloglu et al., 2020). Vector databases are crucial for efficiently retrieving semantically similar data, ensuring that the most relevant information is used. Integrating retrieval mechanisms with generative models has been shown to improve factual accuracy and reduce hallucinations (Karpukhin et al., 2020).

Retrieval agents play a key role by sourcing, filtering, and organizing information from various repositories. Their main components include (1) an interface layer for communication, (2) a query processor that interprets queries, and (3) a data retrieval engine that interacts with databases to fetch the required information.

Different database technologies are used based on the data type and retrieval needs. Relational databases are suited for structured data with well-defined relationships, graph databases excel in scenarios involving complex relationships, and vector databases handle unstructured or high-dimensional data, such as natural language processing embeddings. By integrating these technologies, a retrieval agent can efficiently manage diverse data types, ensuring robust information retrieval in an MAS.

Recently, frameworks like LangChain, Hugging Face, and LlamaIndex (formerly GPT Index) have emerged, offering robust pipelines that connect LLM-based agents with various external data sources, including databases and APIs (Topsakal and Akinci, 2023; Shen et al., 2024). These frameworks enable the development of powerful retrieval agents capable of effectively processing and interacting with data from diverse sources.

11.3.3 Task-Specific Agents

In MAS, task-specific agents are specialized entities designed to perform distinct roles tailored to specific tasks. These agents are equipped with the necessary capabilities, knowledge, and behaviors to fulfill their roles, operating autonomously or in coordination with other agents to achieve localized goals that support the overall objectives of the system.

In the context of ITS and smart urban mobility, task-specific agents integrate specialized transportation tools and methodologies, such as agent-based traffic simulations and predictive models for traffic flow, accidents, and driver behavior. These agents can be developed using core ITS components, which are discussed as examples in Section 11.2. With advancements in cloud and edge computing, features like traffic simulations, vehicle simulators, and predictive analytics can be hosted as web services and delivered via wireless networks. This enables the creation of task-specific agents using computing paradigms like service-oriented architecture and microservices architecture, facilitating seamless and interoperable integration within a multi-agent framework.

11.4 MULTI-AGENT FRAMEWORK FOR ITS

This section reviews the current advancements of GenAI-powered MASs in research, highlighting their transformative potential. We then explore opportunities enabled by integrating LLM and RAG into MASs to address the challenges and limitations of traditional ITS applications, as discussed in Section 11.1.

11.4.1 MAS for Operations Research

While LLMs and RAG have gained traction in urban research, their use in intelligent transportation and smart mobility is still limited, with most developments focused on energy systems and smart grids.

Gamage et al. (2024) introduced a multi-agent chatbot within the La Trobe Energy AI Platform (LEAP) to enhance decision-making in net-zero emissions energy systems. This system integrates LLMs and RAG, with agents responsible for anomaly detection, data visualization, behavioral analysis, and information retrieval. By leveraging RAG, the chatbot provides contextually relevant responses to optimize decision support.

Choi et al. (2024) developed eGridGPT, a GenAI-powered platform by the National Renewable Energy Laboratory for power grid control rooms. eGridGPT uses LLMs and RAG to enhance situational awareness, simulate scenarios, and generate actionable recommendations, integrating digital twins for accurate, reliable outputs. This system streamlines operations, reduces decision-making time, and improves energy system management, which are crucial for future grid operations.

11.4.2 LLM and RAG in MAS: Opportunities for ITS

Traditional MASs were designed to consist of autonomous agents that performed specific tasks through rule-based reasoning and predefined communication protocols. They played key roles in distributed problem-solving across various domains like traffic management and robotics. However, these systems were limited by their fixed knowledge bases, lack of advanced learning capabilities, and inability to process or generate natural language, making them less adaptable and context aware (Wooldridge, 2009). The recent integration of LLMs and RAG has significantly expanded MAS capabilities by enabling advanced capabilities for natural language processing, reasoning, and dynamic knowledge retrieval, thus making these systems more versatile and autonomous in complex, real-world scenarios.

11.4.2.1 Enhancing System Scalability

In a GenAI-powered MAS, LLM-based agents, equipped with advanced language reasoning and dynamic knowledge capabilities, can autonomously manage repetitive, time-consuming tasks crucial for developing and maintaining ITS. These agents are particularly effective in handling the complexities of expansive urban areas, ensuring scalability and efficiency.

11.4.2.1.1 Autonomous Traffic Data Analytics and Interpretation

Building on studies that explore LLM capabilities in automating data analytics and interpretation (Li and Ning, 2023; Zheng et al., 2023), LLM-based agents integrated with a knowledge graph and RAG can autonomously process vast amounts of traffic data. These agents analyze real time and historical data, identify patterns, and generate insights to optimize traffic flow, reduce congestion, and enhance safety. By leveraging a knowledge graph and ontological learning, they can extract and interpret domain-specific patterns, continuously adapting to changing traffic conditions without requiring constant human intervention.

11.4.2.1.2 Automated Software and Simulation Development

Leveraging LLMs' potential for automating complex software engineering tasks, AI-powered agents can also automate the development of essential ITS components, including data stores, software modules, and traffic simulation scenarios (Xu et al., 2024a). By automatically generating and refining these components based on data and requirements, the system can rapidly deploy ITS solutions across large urban areas. This automation accelerates deployment and ensures that the ITS remains updated with the latest advancements in traffic management technologies.

11.4.2.2 Promoting Smart Mobility Service Accessibility

LLM-based agents enhanced with RAG can significantly improve user interaction and accessibility in ITS by leveraging their advanced natural language understanding. These agents can engage users in natural, human-like conversations, making ITS more user-friendly and eliminating the need for technical expertise. By interpreting users' mobility needs, such as reducing commute times, alleviating traffic congestion, or enhancing safety, LLM-based agents can map these requirements to the appropriate task-specific agents, such as those responsible for traffic simulations, flow optimizations, or crash predictions.

- *Intelligent task coordination*: In MAS, LLM-based agents act as orchestrators, coordinating various task-specific agents to complete complex decision-support tasks. For example, if a user wants to optimize their daily commute, the LLM-based agent interprets the request and delegates tasks to agents specializing in traffic data analysis, route optimization, and traffic simulation, delivering a comprehensive, intelligent solution for smart mobility.
- *Chatbot for intelligent driver assistance*: These agents offer conversational interaction, guiding drivers on optimal routes, warning of potential hazards, and advising on fuel efficiency. With advancements in ontological learning and RAG, these agents can handle sophisticated queries by leveraging task-specific agents and real-time traffic data to identify high-congestion areas and safety risks. This conversational interface enhances the driving experience and makes ITS features accessible to a broader audience, including those unfamiliar with advanced technology.

11.4.2.3 Improving Human-Centric Mobility Management

In a generative AI-powered MAS, LLM-based agents enhanced with RAG can revolutionize human-centric mobility management within ITS. These agents, with their advanced language understanding, can interpret diverse user needs—whether from urban commuters, drivers, or transportation planners—by processing everyday language and casual queries to extract valuable insights reflecting users' concerns, preferences, and expectations.

Leveraging their language capabilities, LLM-based agents can automatically identify relevant data sources and knowledge bases from various online resources. They generate precise queries using text-to-query conversions or self-query retrievers, delivering accurate traffic data, insights, and explanations for specialized urban mobility and transportation management tasks.

- *Facilitating crowd-sensing and crowdsourcing through conversations*: LLM-based agents enable crowd-sensing and crowdsourcing by engaging in conversational interactions with users through chatbots. They gather real-time feedback, complaints, and suggestions from urban residents about the transportation system, ensuring ITS remains responsive to public needs. This real-time feedback loop allows transportation planners to make data-driven decisions that reflect the community's actual experiences and concerns.
- *Expert system for public education and engagement*: These agents query versatile traffic data and knowledge bases available through public web services, providing clear, easy-to-understand explanations to urban residents. This empowers the public with knowledge about traffic patterns, safety measures, and mobility options, fostering a more informed and engaged community. By making complex transportation issues accessible to those without technical backgrounds, these expert systems promote greater public participation in urban mobility planning.

11.5 CONCEPTUAL FRAMEWORK: DESIGN AND CHALLENGES

We propose a conceptual framework for designing a GenAI-powered MAS to enhance smart urban mobility (Figure 11.2), with its components detailed in Section 11.3.

11.5.1 Overall System Design

Conceptually, in this framework, LLM-based agents are central to user interaction, reasoning, and task formulation using knowledge representations like ontologies and knowledge graphs. They coordinate tasks, orchestrate agents, and deliver personalized, science-based results through RAG. Retrieval agents work with LLM-based agents to fetch essential transportation data from distributed resources, including government portals, third-party data stores, and transportation digital twins. Task-specific agents, equipped with domain models for simulations and analytics, then generate scientific outputs to meet user requests, such as optimizing routes to reduce carbon emissions, finding efficient multimodal transportation, and optimizing IoT-based traffic controls.

This flexible framework is adaptable to various software development environments and industry-standard stacks, providing a plug-and-play solution for intelligent transportation by connecting LLM-based agents with task-specific agents for simulation and data analytics. Leveraging modern DevOps technologies, these modular agents can be deployed and scaled across edge and cloud architectures. LLM-based agents can be implemented using APIs like ChatGPT, Hugging Face Transformers, and Google's Bard, enabling chatbots on both server-side and client-side platforms. While external knowledge sources often eliminate the need for fine-tuning, it may still be beneficial for specific outputs or user interactions.

Incorporating transportation knowledge representations, such as ontologies and knowledge graphs, helps guide LLMs in task sequencing and querying the correct datastores and APIs. Retrieval agents, built with frameworks like LangChain and LlamaIndex, connect LLMs to data sources, while task-specific agents host simulations and analytical models as web services, all documented in the knowledge representations for seamless querying and access.

11.5.2 Potential Challenges

In this chapter, we highlight three key methodological challenges in designing effective MASs for ITS, deliberately omitting the technical challenges related to computational load and AI model limitations.

11.5.2.1 Optimized Task Coordination and Orchestration

Coordinating multiple agents is challenging, particularly when optimizing domain workflows and computing resources. Balancing diverse tasks while managing computational load can lead to conflicts, especially when agents compete for shared resources or task dependencies create bottlenecks. A potential solution is applying game theory to model agent interactions as strategic games. Concepts like Nash equilibrium can guide agents to cooperate or compete in ways that optimize overall system performance, ensuring efficient task allocation and resource management (Wang et al., 2011; Lownes et al., 2011).

11.5.2.2 Data Sovereignty

Data sovereignty issues, including data ownership, colonialism, and residency, pose significant obstacles to developing effective MASs for ITS. Data ownership laws may restrict cross-border information sharing, while concerns over data colonialism may lead regions to withhold data to protect their interests. Additionally, data residency requirements that mandate data stay within specific geographic boundaries complicate information retrieval and processing across distributed systems. These challenges limit the system's ability to access comprehensive and timely data, reducing its effectiveness in optimizing transportation solutions (Xu et al., 2024a).

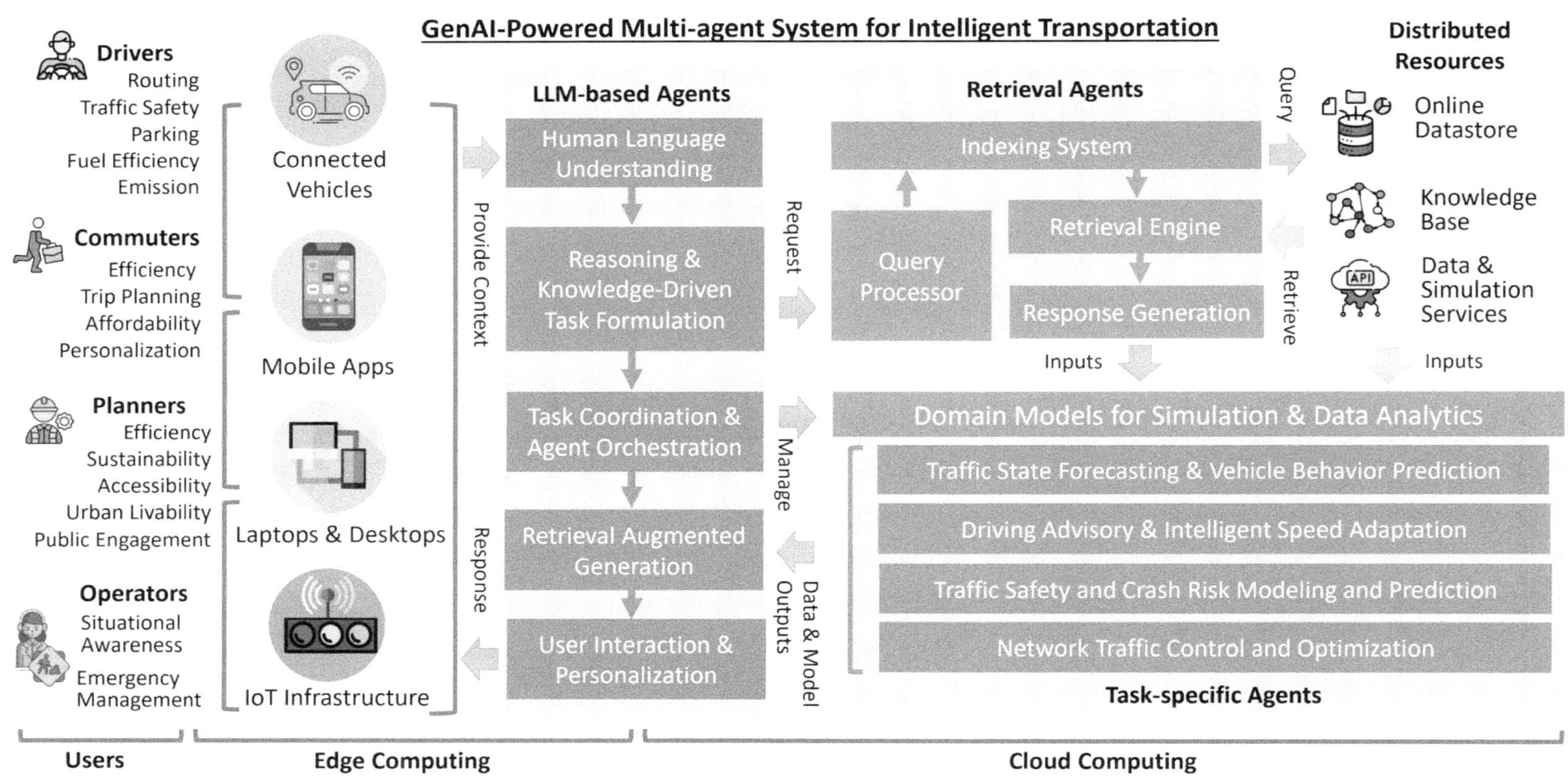

FIGURE 11.2 Overall system design to fulfill the needs for various system users in the urban mobility systems. The system is designed to fulfill the needs for four types of targeted users, including drivers (in both traditional and automated vehicles), commuters of public transit, transportation planners, and traffic operators from the government bodies.

11.5.2.3 AI Accountability

AI accountability is crucial in building MASs for ITS. Ensuring that decisions made by autonomous agents are traceable and that responsibility can be attributed is essential for addressing errors, biases, and unintended outcomes. Without clear accountability mechanisms, the reliability, safety, and ethical operation of ITS could be compromised. Implementing robust accountability measures is critical to building trust and ensuring system transparency.

11.6 CONCLUDING REMARKS

This chapter has explored the transformative potential of integrating LLMs and RAG into MASs for ITS. Our proposed framework demonstrates how these technologies can enhance scalability, accessibility, and human-centric management in urban mobility. LLMs, combined with RAG and task-specific agents, offer new capabilities for autonomous task management, natural language interaction, and efficient coordination, paving the way for advanced smart city applications.

However, challenges such as task coordination, data sovereignty, and AI accountability must be addressed to fully realize these benefits. Overcoming these hurdles will be critical for leveraging AI-powered MASs to create safer, more efficient, and sustainable urban transportation systems. Future research should focus on refining these technologies and expanding their applications across broader smart city contexts.

REFERENCES

Abdel-Aty, M., Wu, Y., Zheng, O., Yuan, J., 2022. Using closed-circuit television cameras to analyze traffic safety at intersections based on vehicle key points detection. Accident Analysis & Prevention, 176, 106794.

Aboudolas, K., Geroliminis, N., 2013. Perimeter and boundary flow control in multi-reservoir heterogeneous networks. Transportation Research Part B: Methodological, 55, 265–281.

Ameta, B., Mathur, V., 2023. Advanced driving assistance systems. PRATIBODH, 3(2).

Antony, M.M., Whenish, R., 2021. Advanced driver assistance systems (ADAS), in: Automotive Embedded Systems: Key Technologies, Innovations, and Applications. Springer, pp. 165–181.

Bazzan, A.L., 2005. A distributed approach for coordination of traffic signal agents. Autonomous Agents and MASs, 10, 131–164.

Berres, A.S., LaClair, T.J., Wang, C., Xu, H., Ravulaparthy, S., Todd, A., Tennille, S.A., Sanyal, J., 2021a. Multiscale and multivariate transportation system visualization for shopping district traffic and regional traffic. Transportation Research Record, 2675, 23–37.

Berres, A.S., Xu, H., Tennille, S.A., Severino, J., Ravulaparthy, S., Sanyal, J., 2021b. Explorative visualization for traffic safety using adaptive study areas. Transportation Research Record, 2675, 51–69.

Cakaloglu, T., Szegedy, C., Xu, X., 2020. Text embeddings for retrieval from a large knowledge base, in: International Conference on Research Challenges in Information Science, Springer, 338–351.

Chand, A., Jayesh, S., Bhasi, A., 2021. Road traffic accidents: An overview of data sources, analysis techniques and contributing factors. Materials Today: Proceedings, 47, 5135–5141.

Choi, S.L., Jain, R., Emami, P., Wadsack, K., Ding, F., Sun, H., Gruchalla, K., Hong, J., Zhang, H., Zhu, X., Kroposki, B., 2024. eGridGPT: Trustworthy AI in the Control Room. Technical Report NREL/TP-5D00-87440. National Renewable Energy Laboratory. Golden, CO.

Cledou, G., Estevez, E., Barbosa, L.S., 2018. A taxonomy for planning and designing smart mobility services. Government Information Quarterly, 35, 61–76.

Cui, Z., Ke, R., Pu, Z., Ma, X., Wang, Y., 2020a. Learning traffic as a graph: A gated graph wavelet recurrent neural network for network-scale traffic prediction. Transportation Research Part C: Emerging Technologies, 115, 102620.

Cui, Z., Ke, R., Pu, Z., Wang, Y., 2020b. Stacked bidirectional and unidirectional LSTM recurrent neural network for forecasting network-wide traffic state with missing values. arXiv abs/2005.11627.

Do, L.N., Taherifar, N., Vu, H.L., 2019. Survey of neural network-based models for short-term traffic state prediction. Wiley Interdisciplinary Reviews: Data Mining and Knowledge Discovery, 9(1), e1285.

Gamage, G., Mills, N., De Silva, D., Manic, M., Moraliyage, H., Jennings, A., Alahakoon, D., 2024. Multi-agent RAG chatbot architecture for decision support in net-zero emission energy systems, in: 2024 IEEE International Conference on Industrial Technology (ICIT), IEEE, 1–6.

Geng, M., Li, J., Xia, Y., Chen, X.M., 2023. A physics-informed transformer model for vehicle trajectory prediction on highways. Research Part C: Emerging Technologies, 154, 104272. https://api.semanticscholar.org/CorpusID:260764281.

Hashemi, H., Abdelghany, K.F., 2016. Real-time traffic network state estimation and prediction with decision support capabilities: Application to integrated corridor management. Transportation Research Part C: Emerging Technologies, 73, 128–146.

He, Q., Head, K.L., Ding, J., 2014. Multi-modal traffic signal control with priority, signal actuation and coordination. Transportation Research Part C: Emerging Technologies, 46, 65–82.

Hu, L., Liu, Z., Zhao, Z., Hou, L., Nie, L., Li, J., 2023. A survey of knowledge-enhanced pre-trained language models. IEEE Transactions on Knowledge and Data Engineering, 36(4), 1413–1430.

Jamson, S., Carsten, O., Cholrton, K., & Fowkes, M. (2006). Literature Review and Scoping Study. *The University of Leeds and MIRA Ltd.: Leeds, UK.* https://content.tfl.gov.uk/intelligent-speed-adaptation-literature-review-and-scoping-study-jan-2006.pdf

Karpukhin, V., Oğuz, B., Min, S., Lewis, P., Wu, L., Edunov, S., Chen, D., Yih, W.T., 2020. Dense passage retrieval for open-domain question answering. arXiv preprint arXiv:2004.04906.

Khajeh Hosseini, M., Talebpour, A., 2019. Traffic prediction using time-space diagram: a convolutional neural network approach. Transportation Research Record, 2673(7), 425–435.

Khattak, M.W., De Backer, H., De Winne, P., Brijs, T., Pirdavani, A., 2024. Analysis of road infrastructure and traffic factors influencing crash frequency: Insights from generalized Poisson models. Infrastructures, 9, 47.

Khattak, Z.H., Fontaine, M.D., 2020. A Bayesian modeling framework for crash severity effects of active traffic management systems. Accident Analysis & Prevention, 145, 105544.

Kumar, N., Raubal, M., 2021. Applications of deep learning in congestion detection, prediction and alleviation: A survey. Transportation Research Part C: Emerging Technologies, 133, 103432.

Li, W., Ban, X., 2020. Connected vehicle-based traffic signal coordination. Engineering, 6, 1463–1472.

Li, W., Wang, B., Khattak, Z.H., Deng, X., 2022a. Network-level traffic signal cooperation: A higher-order conflict graph approach. IEEE Transactions on Intelligent Transportation Systems, 24, 990–999.

Li, W., Wang, B., Liu, Z., Li, Q., Qi, G.J., 2022b. Point: Partially observable imitation network for traffic signal control. Sustainable Cities and Society, 76, 103461.

Li, W., Wang, J., Fan, R., Zhang, Y., Guo, Q., Siddique, C., Ban, X.J., 2020b. Short-term traffic state prediction from latent structures: Accuracy vs. efficiency. Transportation Research Part C: Emerging Technologies, 111, 72–90.

Li, X., Xu, H., Huang, X., Guo, C., Kang, Y., Ye, X., 2021. Emerging geo-data sources to reveal human mobility dynamics during COVID-19 pandemic: Opportunities and challenges. Computational Urban Science, 1, 1–9.

Li, Z., Ning, H., 2023. Autonomous GIS: The next-generation AI-powered GIS. International Journal of Digital Earth, 16, 4668–4686.

Lin, L., Li, J., Chen, F., Ye, J., Huai, J., 2017b. Road traffic speed prediction: A probabilistic model fusing multi-source data. IEEE Transactions on Knowledge and Data Engineering, 30, 1310–1323.

Lin, L., Li, W., Bi, H., Qin, L., 2022. Vehicle trajectory prediction using LSTMs with spatial–temporal attention mechanisms. IEEE Intelligent Transportation Systems Magazine, 14, 197–208.

Lin, P.W., Kang, K.P., Chang, G.L., 2004. Exploring the effectiveness of variable speed limit controls on highway work-zone operations, in: Intelligent Transportation Systems. Taylor & Francis, pp. 155–168.

Lownes, N.E., Wang, Q., Ibrahim, S., Ammar, R.A., Rajasekaran, S., Sharma, D., 2011. Many-to-many game-theoretic approach for the measurement of transportation network vulnerability. Transportation Research Record, 2263, 1–8.

Ma, X., Karimpour, A., Wu, Y.J., 2020. Statistical evaluation of data requirement for ramp metering performance assessment. Transportation Research Part A: Policy and Practice, 141, 248–261.

Mo, X., Xing, Y., Lv, C., 2021. Graph and recurrent neural network-based vehicle trajectory prediction for highway driving, in: 2021 IEEE International Intelligent Transportation Systems Conference (ITSC), 1934–1939.

Mo, Z., Di, X., Shi, R., 2020. A physics-informed deep learning paradigm for car-following models. arXiv abs/2012.13376.

Moriano, P., Berres, A., Xu, H., Sanyal, J., 2024. Spatiotemporal features of traffic help reduce automatic accident detection time. Expert Systems With Applications, 244, 122813.

Newsome, M., WSP, A.T.M., 2024. The potential for intelligent speed adaptation (ISA) to assist with road safety.

Papageorgiou, M., Hadj-Salem, H., Blosseville, J.M., et al., 1991. Alinea: A local feedback control law for on-ramp metering. Transportation Research Record, 1320, 58–67.

Qin, X., Wang, K., Cutler, C.E., 2013. Analysis of crash severity based on vehicle damage and occupant injuries. Transportation Research Record, 2386, 95–102.

Raiaan, M.A.K., Mukta, M.S.H., Fatema, K., Fahad, N.M., Sakib, S., Mim, M.M.J., Ahmad, J., Ali, M.E., Azam, S., 2024. A review on large language models: Architectures, applications, taxonomies, open issues and challenges. IEEE Access, 12, 26839–26874.

Ryan, M. (2019). Intelligent speed assistance technologies: A review. *Proceedings of the ITRN2019.*

Schafer, A., Victor, D.G., 2000. The future mobility of the world population. Transportation Research Part A: Policy and Practice, 34, 171–205.

Shao, Y., Wang, C., Berres, A., Yoshioka, J., Cook, A., Xu, H., 2022. Computer vision-enabled smart traffic monitoring for sustainable transportation management, in: International Conference on Transportation and Development 2022. ASCE, pp. 34–45.

Shen, Y., Song, K., Tan, X., Li, D., Lu, W., Zhuang, Y., 2024. HuggingGPT: Solving AI tasks with ChatGPT and its friends in Hugging Face. Advances in Neural Information Processing Systems, 36.

Smith, H.R., Hemily, B., Ivanovic, M., 2005. Transit signal priority (TSP): A planning and implementation handbook.

Theofilatos, A., Yannis, G., 2014. A review of the effect of traffic and weather characteristics on road safety. Accident Analysis & Prevention, 72, 244–256.

Topsakal, O., Akinci, T.C., 2023. Creating large language model applications utilizing LangChain: A primer on developing LLM apps fast, in: International Conference on Applied Engineering and Natural Sciences, 1050–1056.

Tupayachi, J., Xu, H., Omitaomu, O.A., Camur, M.C., Sharmin, A., Li, X., 2024. Towards next-generation urban decision support systems through AI-powered generation of scientific ontology using large language models–A case in optimizing intermodal freight transportation. arXiv preprint arXiv:2405.19255.

Ugirumurera, J., Severino, J.A., Wang, Q., Ravulaparthy, S., Berres, A., Nugent, P., Sorensen, H., Moore, A., Todd, A., Nag, A., et al., 2020. High performance computing traffic simulations for real-time traffic control of mobility in Chattanooga region. Technical Report. National Renewable Energy Lab (NREL), Golden, CO (United States).

Vasilieva, E., Artamonova, Y.S., Vasiliev, E., Brega, A., 2023. Participatory design as a tool for optimizing an intelligent transport system, in: 2023 Systems of Signals Generating and Processing in the Field of on Board Communications, IEEE, 1–5.

Vogel-Heuser, B., Seitz, M., Cruz Salazar, L.A., Gehlhoff, F., Dogan, A., Fay, A., 2020. MASs to enable Industry 4.0.

Wang, C., Xie, Y., Huang, H., Liu, P., 2021. A review of surrogate safety measures and their applications in connected and automated vehicles safety modeling. Accident Analysis & Prevention, 157, 106157.

Wang, H., Zhu, M., Hong, W., Wang, C., Li, W., Tao, G., Wang, Y., 2022a. Network-wide traffic signal control using bilinear system modeling and adaptive optimization. IEEE Transactions on Intelligent Transportation Systems, 24, 79–91.

Wang, J., Deng, W., Guo, Y., 2014. New Bayesian combination method for short-term traffic flow forecasting. Transportation Research Part C: Emerging Technologies, 43, 79–94.

Wang, J., Zhou, A., Liu, Z., Peeta, S., 2024. Robust cooperative control strategy for a platoon of connected and autonomous vehicles against sensor errors and control errors simultaneously in a real-world driving environment. Transportation Research Part B: Methodological.

Wang, Q., Fiondella, L., Lownes, N., Ivan, J., Ammar, R., Rajasekaran, S., Tolba, S., 2011. Integrating equilibrium assignment in game-theoretic approach to measure many-to-many transportation network vulnerability, in: 2011 IEEE International Conference on Technologies for Homeland Security (HST), IEEE, 351–357.

Wang, Y., Yang, X., Liang, H., Liu, Y., 2018. A review of the self-adaptive traffic signal control system based on future traffic environment. Journal of Advanced Transportation, 2018, 1096123.

Wang, Y., Yu, C., Hou, J., Chu, S., Zhang, Y., Zhu, Y., 2022b. ARIMA Model and few-shot learning for vehicle speed time series analysis and prediction. Computational Intelligence and Neuroscience, 2022.

Wooldridge, M., 2009. An Introduction to Multiagent Systems. John Wiley & Sons.

Xing, L., He, J., Abdel-Aty, M., Wu, Y., Yuan, J., 2020. Time-varying analysis of traffic conflicts at the upstream approach of toll plaza. Accident Analysis & Prevention, 141, 105539.

Xu, G., Gayah, V.V., 2024. On the impact of bus dwelling on macroscopic fundamental diagrams, in: International Conference on Transportation and Development 2024, 599–612.

Xu, H., Berres, A., Tennille, S.A., Ravulaparthy, S.K., Wang, C., Sanyal, J., 2021a. Continuous emulation and multiscale visualization of traffic flow using stationary roadside sensor data. IEEE Transactions on Intelligent Transportation Systems, 23, 10530–10541.

Xu, H., Berres, A., Wang, C.R., LaClair, T.J., Sanyal, J., 2021b. Visualizing vehicle acceleration and braking energy at intersections along a major traffic corridor, in: Proceedings of the Twelfth ACM International Conference on Future Energy Systems, 401–405.

Xu, H., Berres, A., Yoginath, S.B., Sorensen, H., Nugent, P.J., Severino, J., Tennille, S.A., Moore, A., Jones, W., Sanyal, J., 2023a. Smart mobility in the cloud: Enabling real-time situational awareness and cyber-physical control through a digital twin for traffic. IEEE Transactions on Intelligent Transportation Systems, 24, 3145–3156.

Xu, H., Omitaomu, F., Sabri, S., Li, X., Song, Y., 2024a. Leveraging generative AI for smart city digital twins: A survey on the autonomous generation of data, scenarios, 3D city models, and urban designs. arXiv preprint arXiv:2405.19464.

Xu, H., Shao, Y., Chen, J., Wang, C., Berres, A., 2024b. Semi-automatic geographic information system framework for creating photo-realistic digital twin cities to support autonomous driving research. Transportation Research Record. https://doi.org/10.1177/03611981231205884

Xu, H., Wang, C., Berres, A., LaClair, T., Sanyal, J., 2022a. Interactive web application for traffic simulation data management and visualization. Transportation Research Record, 2676, 274–292.

Xu, H., Yuan, J., Berres, A., Shao, Y., Wang, C., Li, W., LaClair, T.J., Sanyal, J., Wang, H., 2023b. A mobile edge computing framework for traffic optimization at urban intersections through cyber-physical integration. IEEE Transactions on Intelligent Vehicles.

Yahouni, Z., Ladj, A., Belkadi, F., Meski, O., Ritou, M., 2021. A smart reporting framework as an application of MAS in machining industry. International Journal of Computer Integrated Manufacturing, 34, 470–486.

Yang, J., Jin, H., Tang, R., Han, X., Feng, Q., Jiang, H., Zhong, S., Yin, B., Hu, X., 2024. Harnessing the power of LLMs in practice: A survey on ChatGPT and beyond. ACM Transactions on Knowledge Discovery from Data, 18, 1–32.

Yu, Z., Xu, G., Gayah, V.V., Christofa, E., 2020. Incorporating phase rotation into a person-based signal timing optimization algorithm. IEEE Transactions on Intelligent Transportation Systems, 23, 513–521.

Yuan, J., Abdel-Aty, M., 2018. Approach-level real-time crash risk analysis for signalized intersections. Accident Analysis & Prevention, 119, 274–289.

Yuan, J., Abdel-Aty, M.A., Yue, L., Cai, Q., 2020. Modeling real-time cycle-level crash risk at signalized intersections based on high-resolution event-based data. IEEE Transactions on Intelligent Transportation Systems, 22, 6700–6715.

Zaki, M., Krishnan, N.A., et al., 2024. MASCQA: Investigating materials science knowledge of large language models. Digital Discovery, 3, 313–327.

Zhang, S., Zhi, Y., He, R., Li, J., 2020. Research on traffic vehicle behavior prediction method based on game theory and HMM. IEEE Access, 8, 30210–30222.

Zheng, O., Abdel-Aty, M., Wang, D., Wang, Z., Ding, S., 2023. ChatGPT is on the horizon: Could a large language model be suitable for intelligent traffic safety research and applications? arXiv preprint arXiv:2303.05382.

Zhou, A., Peeta, S., Yang, M., Wang, J., 2022a. Cooperative signal-free intersection control using virtual platooning and traffic flow regulation. Transportation Research Part C: Emerging Technologies.

Zhou, A., Wang, Z., Cook, A., 2023. Model predictive control-based trajectory shaper for safe and efficient adaptive cruise control, in: 2023 IEEE International Automated Vehicle Validation Conference (IAVVC), 1–7.

Zhou, H., Zhou, A., Laval, J.A., Liu, Y., Peeta, S., 2022b. Incorporating driver relaxation into factory adaptive cruise control to reduce lane-change disruptions. Transportation Research Record, 2676, 13–27.

Zhou, H., Zhou, A., Li, T., Chen, D., Peeta, S., Laval, J.A., 2022c. Congestion-mitigating MPC design for adaptive cruise control based on Newell's car following model: History outperforms prediction. Transportation Research Part C: Emerging Technologies.

Zhou, H., Zhou, A., Li, T., Chen, D., Peeta, S., Laval, J.A., 2022d. Significance of low-level control to string stability under adaptive cruise control: Algorithms, theory and experiments. Transportation Research Part C: Emerging Technologies. https://api.semanticscholar.org/CorpusID:248931658.

12 Agent-Based Modeling and Simulation for Smart Cities

Lingbo Liu

12.1 INTRODUCTION

Agent-based modeling (ABM) has emerged as a powerful tool in the development of smart cities, providing the ability to simulate individual behaviors and collective dynamics, which are crucial for urban planning and management (Crooks et al., 2021). By focusing on individual agents and their interactions, ABM allows for a granular understanding of complex systems, helping cities become more efficient and adaptive to change. The importance of ABM lies in its capability to model human mobility patterns and other urban processes, thereby offering valuable insights for enhancing infrastructure and services in the rapidly evolving urban environment (Jahromi et al., 2016).

The application of ABM in urban human mobility offers significant advantages for understanding the complexity of urban systems. Traditional modeling approaches often fall short when dealing with the unpredictable nature of human movement in cities (Zhang et al., 2018). In contrast, ABM provides the flexibility to incorporate diverse data sources, including mobile phone data and geographic information system (GIS) information, to simulate individual mobility decisions. This has proven particularly beneficial in the context of smart cities, where the need to anticipate and adapt to changing population dynamics is critical (El-Amine et al., 2017).

In the context of smart city development, ABM contributes to achieving smarter urban services and infrastructure management, particularly in areas such as transportation, resource allocation, and public safety (Hasan et al., 2012). By enabling city planners to simulate multiple scenarios, ABM helps to optimize traffic flow, improve resource distribution, and enhance public safety measures. The use of ABM for simulating urban dynamics not only allows for better anticipation of problems but also supports proactive decision-making to enhance the quality of urban life (Arora et al., 2023).

Early applications of ABM in traffic flow management have paved the way for innovative approaches to future urban planning. These models have provided a foundation for improving transportation efficiency and reducing congestion in smart cities. By simulating different traffic conditions and evaluating the impact of interventions, ABM has enabled urban planners to explore new ways to manage traffic more effectively, providing insights into potential improvements for urban mobility and infrastructure (Bastarianto et al., 2023; Cheshmehzangi & Pourroostaei Ardakani, 2021).

As a crucial component in the toolbox for smart city development, ABM's ability to model and predict complex interactions between individuals and the environment makes it particularly suitable for addressing the multifaceted challenges faced by modern cities (Weil et al., 2023). This chapter aims to explore the current progress, challenges, and future perspectives of ABM in the context of digital twin applications for urban human mobility.

12.2 CURRENT PROGRESS IN ABM AND HUMAN MOBILITY FOR SMART CITIES

ABM has become integral to the development and functionality of smart cities. The capability of ABM to simulate the behaviors and interactions of agents allows city planners and policymakers to better manage urban systems. This modeling approach supports a wide range of applications

DOI: 10.1201/9781003503262-15

from infrastructure management to the optimization of services in dynamic urban environments. The current progress of ABM in smart cities showcases its versatility and effectiveness in addressing key challenges in urban planning, transportation, resource management, public health, and technological advancements such as autonomous vehicles.

12.2.1 ABM in Smart City Transportation and Resource Management

The application of ABM in transportation has enabled significant strides in understanding and optimizing urban traffic flow and resource distribution. ABM's ability to simulate large-scale transportation networks provides policymakers with critical insights for forecasting urban travel behavior and crafting adaptive solutions (Kagho et al., 2020). The integration of network analysis within ABM frameworks has further enriched the analysis of congestion patterns and decision-making processes, offering strategies to alleviate traffic issues (Amézquita-López et al., 2021). Additionally, ABM's role in harnessing global positioning system (GPS) data has proven vital for revealing detailed mobility patterns and improving resource allocation (Jahromi et al., 2016). The inclusion of social behavior modeling within ABM frameworks has provided nuanced perspectives on infrastructure use and transportation policies (El-Amine et al., 2017). By fostering smart and adaptable mobility solutions, ABM enhances cities' abilities to respond to dynamic conditions and ensure efficient transportation systems (Del Vecchio et al., 2019; Wu et al., 2019).

12.2.2 ABM in Human Mobility Analysis for Urban Planning

ABM extends its utility beyond transportation into urban planning by simulating human mobility and spatial interactions. Incorporating temporal dynamics within ABM has allowed for more precise simulations of urban growth and the associated mobility demands (Wu et al., 2014). For instance, ABM has been applied to analyze city-wide subway systems, offering insights for optimizing transit management and enhancing urban mobility (Hasan et al., 2012). The versatility of ABM supports scenario testing that enables planners to anticipate and evaluate the outcomes of policy decisions (Crooks et al., 2021). Integrating geosocial media data into ABM models has enriched urban simulations, although challenges such as data accuracy remain (Wen & Li, 2020). Additionally, the use of GPS data in ABM has contributed to the development of realistic agent behaviors that reflect actual mobility patterns, aiding in strategic urban planning (Pang et al., 2018).

12.2.3 ABM and Epidemic Modeling for Public Health Impact Assessment

The intersection of ABM and public health has proven particularly valuable for modeling disease spread in urban contexts. By linking transport ABMs with epidemic models, researchers have been able to create simulations that capture individual infection probabilities and broader city-wide transmission patterns (Hackl & Dubernet, 2019). These models provide actionable insights into how public health policies impact urban mobility and inform crisis management strategies (Adnan et al., 2020). The integration of diverse transport modes within ABM has supported sustainable strategies for health and congestion management (Uthpala et al., 2023). ABM has also played a role in addressing health disparities, offering a platform for modeling equitable public health interventions (Basu et al., 2018). Furthermore, ABM's ability to identify high-risk areas supports targeted mitigation strategies and enhances public health response efforts (Richter et al., 2022).

12.2.4 ABM for Autonomous Vehicle Modeling

ABM has emerged as a critical tool for studying the integration of autonomous vehicles (AVs) within urban settings. Comprehensive ABM simulations have demonstrated the potential of AVs to improve urban mobility while addressing challenges related to agent behavior and data standardization

(Li et al., 2021). Recent progress in AV research has underscored the necessity for consistent methodologies and robust data for reliable simulations (Jing et al., 2020). ABM-based studies on shared autonomous micro-mobility have revealed operational challenges such as fleet management and the integration of real-time data (Sanchez et al., 2022). Additionally, ABM has been utilized to model the energy and transport impacts of AVs, contributing to sustainable urban development initiatives (Liu et al., 2025; Wu et al., 2020).

12.2.5 ABM Models for Smart Cities

The diversity of ABM models caters to a variety of urban challenges, ranging from basic to highly complex simulations. Platforms like SimMobility ST exemplify detailed calibration and validation, providing reliable models for multimodal transport and dynamic agent interactions (Azevedo et al., 2017). Agent-based policing models showcase the application of ABM in simulating real-world constraints effectively, with recommendations for integrating more comprehensive data to enhance accuracy (Wise & Cheng, 2016). Multi-agent frameworks designed for traffic signal regulation demonstrate ABM's adaptability and superior performance over traditional methods, with potential for broader system integration (Li et al., 2019). Data-driven strategies in ABM contribute to sustainable urban planning by balancing technological advances with community needs (Kaluarachchi, 2022). Innovative frameworks focusing on energy efficiency have shown promising results in real-time urban mobility simulations (Oshin et al., 2014). Lastly, ABM's role in shared autonomous mobility underscores its importance for optimizing fleet operations and real-time data use (Sanchez et al., 2022), while also highlighting improved traffic flow from smart vehicle penetration (Kamiński et al., 2020).

12.3 SOFTWARE USED IN ABM AND HUMAN MOBILITY

12.3.1 ABM Software for Realistic Human Behavior Simulation

ABM software stands out for its ability to simulate individual-level behaviors and detailed interactions, but it may lack the scalability of urban simulation platforms. *NetLogo* is an open-source tool with a user-friendly interface, making it ideal for beginners and advanced users alike; however, it is better suited for small to medium-sized simulations and may need extensions for larger projects. *Repast* is highly flexible and supports cross-domain research with complex agent interactions, but its extensive feature set can be difficult to master. *AnyLogic* excels in its multimethod approach, combining agent-based, system dynamics, and discrete event modeling, making it suitable for comprehensive urban analysis; however, its proprietary nature limits accessibility. *GAMA* is known for handling complex geospatial data and supports advanced ecological and urban modeling, though users must have significant geospatial expertise. *MASON* is efficient for large-scale agent simulations, favored in academia for its computational power but requiring extra customization for more advanced modeling. *Swarm*, one of the earliest ABM tools, supports diverse use cases from urban to ecological modeling but falls behind in supporting modern extensions and complex integrations.

12.3.2 Python-Based ABM Packages for Custom and Scalable Solutions

Python-based ABM packages offer unmatched flexibility and easy integration with data pipelines, but they may not provide the out-of-the-box capabilities found in standalone ABM software. *Mesa* is popular for its seamless integration with Python's data analysis and visualization libraries, enabling rapid prototyping of human mobility models; however, it struggles with high agent counts and large-scale datasets. *PyABM* offers basic ABM modeling features with minimal setup, ideal for small-scale projects, but lacks the advanced simulation tools needed for complex modeling. *Agents.jl*, while designed for high-performance simulations with Julia, integrates well with Python but requires advanced coding skills and has limited community resources. *DEAP* allows for evolutionary algorithm-based agent behaviors, supporting complex scenario building, though it has a steep learning curve, especially for nonprogrammers. *SimPy* is perfect for discrete event modeling in mobility simulations and can be combined with ABM, but it has basic visualization features that may need enhancement with other tools. *Mesa-Geo*, an extension of Mesa, adds geospatial components, making it suitable for spatially explicit simulations, though extensive customization is needed for larger projects. *POLISim* focuses on policy impact analysis related to population mobility, providing niche but powerful insights, although it is less widely adopted in mainstream urban studies.

12.3.3 Urban Simulation Platforms for Large-Scale Infrastructure Analysis

Urban simulation platforms excel in modeling large-scale city infrastructure and mobility, providing a broader perspective than ABM or Python-based tools, but they may lack the agent-level detail found in ABM software. *SUMO* is an open-source platform known for simulating large-scale city traffic and integrating Internet of Things (IoT) data for real-time analysis; however, it requires considerable configuration and setup. *MATSim* offers powerful tools for simulating origin-destination (OD)-based traffic and population movement, essential for assessing city-wide human mobility, though it demands significant data preprocessing and technical expertise. *UrbanSim* is highly valuable for land use and urban planning simulations and can integrate with platforms like *MATSim* to explore how urban changes affect mobility; its integration requires advanced configuration skills. *CityFlow* specializes in analyzing traffic flow and congestion patterns, suitable for quick traffic studies, but may not provide comprehensive support for multimodal transport analysis. *CityEngine* enables intricate three-dimensional urban visualization and interaction, ideal for scenario planning and stakeholder engagement, though its proprietary nature comes with high costs and limits customization.

12.4 CHALLENGES IN IMPLEMENTING ABM IN SMART CITIES

While ABM provides powerful capabilities for modeling and managing urban systems in smart cities, addressing challenges related to data collection, computational complexity, model verification, scalability, and social implications is essential for realizing its full potential.

12.4.1 Data Collection and Integration

One of the primary challenges in implementing ABM in smart cities is obtaining high-quality data and effectively integrating it. The reliability of data such as OD records and mobile device data is often compromised by privacy concerns and logistical difficulties (Jahromi et al., 2016). The complexity increases when integrating multisource data like traffic, demographic, and environmental data due to inconsistencies in format, temporal resolution, and coverage (Wu et al., 2014). The use of social media and mobile data further adds layers of complexity, particularly in ensuring data quality and addressing biases (Ronald et al., 2015; Zhou et al., 2018). This challenge underscores the need for advanced data management techniques to create robust ABM simulations.

12.4.2 Computational Complexity and Efficiency

ABM's computational demands present another major challenge, especially when modeling large-scale urban environments. The interaction among thousands of agents and the need for high granularity often lead to computational bottlenecks, making scalability a critical issue (Wu et al., 2014). Simulating diverse behaviors and complex interactions can be resource-intensive, posing difficulties in representing urban systems accurately (Nguyen et al., 2021). High-performance computing and parallel processing have been proposed to enhance the scalability and efficiency of ABM, but these solutions require significant computational resources and specialized expertise (Li et al., 2011, 2019).

12.4.3 Model Verification and Uncertainty

Due to the intricate nature of urban systems, verifying ABM models remains a significant challenge. Ensuring that models accurately reflect real-world dynamics requires careful calibration and validation using historical or observational data, which may be limited or unavailable (Hackl & Dubernet, 2019). The unpredictability of agent behavior and variability in input data contribute to inherent uncertainties within ABM, impacting the reliability of predictions (Fan et al., 2021; Massaro et al., 2019). This calls for strategies to quantify and manage uncertainty in ABM simulations, which is crucial for building trust in model outcomes (Al-Rahamneh et al., 2021; Breyer et al., 2020).

12.4.4 Scalability and Real-Time Data Integration

ABM's scalability, especially when incorporating real-time data streams like IoT sensor data, is another pressing challenge. Integrating real-time data requires substantial computational power and advanced algorithms capable of dynamically adjusting models to reflect changing urban conditions (Hasan et al., 2012; Khan et al., 2021). The complexity of processing and integrating large-scale, real-time data streams makes it difficult to maintain model responsiveness and accuracy (Hussain et al., 2024; Li et al., 2011). Addressing these issues is crucial for developing adaptive and real-time ABM applications that support effective decision-making (Tosi, 2017).

12.4.5 Social and Ethical Considerations

The use of ABM in smart cities also raises social and ethical concerns, particularly regarding data privacy and the representation of individuals within models. Detailed data collection can risk infringing on individual privacy, especially when tracking personal activities and behaviors (Kaluarachchi, 2022). Additionally, ensuring that ABM models do not reinforce existing biases or exclude vulnerable populations is a significant ethical consideration (Serok & Blumenfeld-Liebertha, 2015). Addressing these issues is critical for fostering public trust and ensuring that ABM applications are equitable and inclusive (Adnan et al., 2020; Bıyık et al., 2021; Zhou et al., 2018).

12.5 DISCUSSION AND FUTURE DIRECTIONS

12.5.1 Integrating Real-Time Data and IoT for Enhanced Responsiveness

The future of ABM in smart cities lies in its ability to integrate real-time data from IoT sensors and other dynamic sources. As cities become increasingly interconnected through smart devices, ABM can harness these data streams to deliver adaptive and responsive simulations, enabling city managers to react swiftly to evolving conditions (Chavhan et al., 2022). This integration will support more accurate modeling of urban dynamics, such as traffic flow and emergency response, ultimately improving the efficiency of urban management (Hussain et al., 2024; Tosi, 2017).

12.5.2 Advancing Computational Power for Greater Scalability

Improvements in computational power, including the use of high-performance computing and parallel processing, are expected to significantly boost the scalability and efficiency of ABM (Nguyen et al., 2021). These technological advancements will allow for more granular simulations across larger urban areas, providing a deeper understanding of complex urban systems (Oshin et al., 2014).

Cloud-based solutions can facilitate distributed simulations, enabling ABM models to operate on a broader scale while maintaining computational efficiency.

12.5.3 Integrating Machine Learning for Enhanced Predictive Capabilities

Combining ABM with machine learning and AI technologies offers a promising direction for enhancing the predictive and adaptive capabilities of urban simulations (Pang et al., 2020). Machine learning can help identify patterns and trends in historical data, which can be used to calibrate ABM models and improve their accuracy and reliability (Pang et al., 2020). Additionally, reinforcement learning can optimize agent behaviors within simulations, enabling the exploration of different strategies and policies in a more automated and intelligent manner (Del Vecchio et al., 2019; Khaidem et al., 2020).

12.5.4 Addressing Ethical and Social Challenges in ABM Implementation

As ABM becomes more embedded in urban management, addressing its ethical and social challenges is crucial. Future developments should prioritize data privacy and ensure that models do not perpetuate biases or exclude marginalized populations (Kaluarachchi, 2022). Establishing ethical guidelines for data collection and model implementation will be vital for building public trust and promoting fair, inclusive use of ABM applications (Groff et al., 2018; Zhou et al., 2018). Transparent and participatory modeling approaches can engage citizens in the planning process, fostering greater community involvement and acceptance (Yao & Wang, 2020).

12.5.5 Expanding Interdisciplinary Collaboration for Comprehensive Solutions

The future of ABM in smart cities also involves expanding interdisciplinary collaboration. Integrating insights from social sciences, environmental studies, and economics can provide a more comprehensive understanding of urban challenges and contribute to more holistic solutions (El-Amine et al., 2017; Yao & Wang, 2020). Collaboration among researchers, urban planners, and policymakers will be essential for developing ABM models that are not only scientifically robust but also practically relevant (Englund et al., 2021; Huang et al., 2022). This cross-disciplinary approach can help tackle interconnected urban issues, from climate change to social inequality, making ABM an ideal tool for complex problem-solving in modern cities.

12.6 CASE STUDY PRACTICE ON ABM FOR CRIME SIMULATION

12.6.1 Study Area and Objectives

This case study focuses on East Baton Rouge Parish, Louisiana, as the research area for simulating street robbery occurrences based on the paper publish in 2021 (Zhu & Wang, 2021). The objective is to utilize an ABM to understand the spatiotemporal distribution of street robberies, evaluate crime theories, and explore alternative policing strategies. By creating virtual scenarios, this research aims to provide insights into crime pattern formation and the effectiveness of different intervention measures without real-world disruptions.

12.6.2 Routine Activity Theory for ABM in Crime Simulation

The routine activity theory plays a central role in understanding how crime opportunities arise in this ABM simulation (Leukfeldt & Yar, 2016). According to this theory, both potential offenders and victims follow daily routines that lead them to specific locations at certain times, such as homes, workplaces, schools, and recreational areas. These daily routes form the foundation of an individual's activity space. Crime events are more likely when the activity spaces of offenders and victims overlap spatially and temporally, creating windows of opportunity for offenses to occur. By incorporating this theory, the model simulates realistic interactions that reflect the complex dynamics of urban environments, enhancing the predictive power of the simulation.

12.6.3 Methodology and Workflow

The main process of the ABM crime simulation follows these steps:

Initialization: The model initializes with spatial data representing East Baton Rouge, including the spatial grid, road networks, and demographic data. Agents are assigned daily routines with fixed (e.g., work, rest) and flexible activities (e.g., shopping).

Agent Movement and Interaction: Agents calculate potential path areas using network data, ensuring they reach scheduled activities on time. Movement paths minimize travel deviation and allow agents to interact with their environment.

Crime Opportunity Evaluation: Potential offenders scan their surroundings for suitable targets, assessing conditions like police presence and crowd size. If conditions are favorable, they attempt a robbery.

Outcome Determination: Robbery attempts consider the self-defense capability of the targeted agent and a probabilistic outcome. Success or failure influences future offender motivation.

Police Patrol Integration: Police agents patrol designated routes, deterring crimes within their awareness zone.

Data Analysis and Evaluation: Simulation results are analyzed using kernel density estimation maps to identify predicted hot spots, compared to actual crime data with metrics like hit rate (H), prediction accuracy index (PAI), and forecast accuracy index (FAI) to assess accuracy.

12.6.4 Study Data and Tutorials for Replication and Reproduction

The data for this study are provided in formats that are compatible with open-source Python GIS libraries, including GDAL, Shapely, and NetworkX. The primary dataset consists of a 500-meter resolution spatial grid covering the study area, which is essential for simulating agent movement and spatial indexing. Additionally, a road network layer representing road segments is used to guide agent travel paths.

Demographic data, sourced from census block groups, are used to create a realistic agent population with representative residential and employment distributions. The study also includes OD matrices to model commuting flow and ensure that agent travel patterns align with real-world scenarios. Randomly generated points on the road network are included to simulate potential targets and activity locations. These comprehensive datasets enable agents to simulate movement and interactions within a realistic urban environment, forming the foundation for effective crime simulations.

Details on step-by-step tutorials can be found in the main textbook using ArcGIS Pro (Wang & Liu, 2023) and the lab manual using KNIME (Liu & Wang, 2023). Major files related to the data are organized in the subfolder Data under the folder ABMSIM, published in the Harvard Dataverse and accessible at the following link: https://doi.org/10.7910/DVN/4CM7V4.

REFERENCES

Adnan, M., Outay, F., Ahmed, S., Brattich, E., Di Sabatino, S., & Janssens, D. (2020). Integrated Agent-Based Microsimulation Framework for Examining Impacts of Mobility-Oriented Policies. *Personal and Ubiquitous Computing*, *25*, 205–217. https://doi.org/10.1007/s00779-020-01363-w

Al-Rahamneh, A., Astrain, J. J., Villadangos, J., Klaina, H., Guembe, I. P., Lopez-Iturri, P., & Falcone, F. (2021). Enabling Customizable Services for Multimodal Smart Mobility with City-Platforms. *IEEE Access*, *9*, 41628–41646. https://doi.org/10.1109/access.2021.3065412

Amézquita-López, J., Valdés-Atencio, J., & Angulo-García, D. (2021). Understanding Traffic Congestion via Network Analysis, Agent Modeling, and the Trajectory of Urban Expansion: A Coastal City Case. *Infrastructures*, *6*(6), 85. https://doi.org/10.3390/infrastructures6060085

Arora, A., Jain, A., Yadav, D., Hassija, V., Chamola, V., & Sikdar, B. (2023). Next Generation of Multi-Agent Driven Smart City Applications and Research Paradigms. *IEEE Open Journal of the Communications Society*, *4*, 2104–2121. https://doi.org/10.1109/ojcoms.2023.3310528

Azevedo, C. L., Deshmukh, N. M., Marimuthu, B., Oh, S., Marczuk, K., Soh, H., Basak, K., Toledo, T., Peh, L.-S., & Ben-Akiva, M. E. (2017). SimMobility Short-Term: An Integrated Microscopic Mobility Simulator. *Transportation Research Record*, *2622*(1), 13–23. https://doi.org/10.3141/2622-02

Bastarianto, F. F., Hancock, T. O., Choudhury, C. F., & Manley, E. (2023). Agent-Based Models in Urban Transportation: Review, Challenges, and Opportunities. *European Transport Research Review*, *15*, 19. https://doi.org/10.1186/s12544-023-00590-5

Basu, R., Araldo, A., Akkinepally, A. P., Nahmias Biran, B. H., Basak, K., Seshadri, R., Deshmukh, N., Kumar, N., Azevedo, C. L., & Ben-Akiva, M. (2018). Automated Mobility-on-Demand vs. Mass Transit: A Multi-Modal Activity-Driven Agent-Based Simulation Approach. *Transportation Research Record*, *2672*(8), 608–618. https://doi.org/10.1177/0361198118758630

Bıyık, C., Abareshi, A., Paz, A., Ruiz, R. A., Battarra, R., Rogers, C. D., & Lizarraga, C. (2021). Smart Mobility Adoption: A Review of the Literature. *Journal of Open Innovation: Technology, Market and Complexity*, *7*(2), 146. https://doi.org/10.3390/joitmc7020146

Breyer, N., Rydergren, C., & Gundlegård, D. (2020). Comparative Analysis of Travel Patterns from Cellular Network Data and an Urban Travel Demand Model. *Journal of Advanced Transportation*, *2020*, 3267474. https://doi.org/10.1155/2020/3267474

Chavhan, S., Gupta, D., Gochhayat, S. P., N, C. B., Khanna, A., Shankar, K., & Rodrigues, J. J. (2022). Edge Computing AI-IoT Integrated Energy-Efficient Intelligent Transportation System for Smart Cities. *ACM Transactions on Internet Technology*, *22*(4), 1–18. https://doi.org/10.1145/3507906<

Cheshmehzangi, A., & Pourroostaei Ardakani, S. (2021). Urban Traffic Optimization Based on Modeling Analysis of Sector-Based Time Variable: The Case of Simulated Ningbo, China. *Frontiers in Sustainable Cities*, *3*, 629940. https://doi.org/10.3389/frsc.2021.629940

Crooks, A., Heppenstall, A., Malleson, N., & Manley, E. (2021). Agent-Based Modeling and the City: A Gallery of Applications. In: Shi, W., Goodchild, M.F., Batty, M., Kwan, M.P., Zhang, A. (eds), Urban Informatics. Springer, Singapore. https://doi.org/10.1007/978-981-15-8983-6_46

Del Vecchio, P., Secundo, G., Maruccia, Y., & Passiante, G. (2019). A System Dynamic Approach for the Smart Mobility of People: Implications in the Age of Big Data. *Technological Forecasting and Social Change*, *149*, 119771. https://doi.org/10.1016/j.techfore.2019.119771

El-Amine, S., Galland, S., & Koukam, A. (2017). Demand for Agent-Based Transportation Models & Social Behavioral Challenges. *Procedia Computer Science*, *113*, 210–216. https://doi.org/10.1016/j.procs.2017.08.351

Englund, C., Aksoy, E. E., Alonso-Fernandez, F., Cooney, M. D., Pashami, S., & Åstrand, B. (2021). AI Perspectives in Smart Cities and Communities to Enable Road Vehicle Automation and Smart Traffic Control. *Smart Cities*, *4*(2), 783–802. https://doi.org/10.3390/smartcities4020040

Fan, C., Jiang, X., & Mostafavi, A. (2021). Evaluating Crisis Perturbations on Urban Mobility Using Adaptive Reinforcement Learning. *Sustainable Cities and Society*, *75*, 103367. https://doi.org/10.1016/j.scs.2021.103367

Groff, E. R., Johnson, S. D., & Thornton, A. (2019). State of the Art in Agent-Based Modeling of Urban Crime: An Overview. *Journal of Quantitative Criminology*, *35*, 155–193. https://doi.org/10.1007/s10940-018-9376-y

Hackl, J., & Dubernet, T. (2019). Epidemic Spreading in Urban Areas Using Agent-Based Transportation Models. *Future Internet*, *11*(4), 92. https://doi.org/10.3390/fi11040092

Hasan, S., Schneider, C. M., Ukkusuri, S. V., & González, M. C. (2012). Spatiotemporal Patterns of Urban Human Mobility. *Journal of Statistical Physics*, *151*, 304–318. https://doi.org/10.1007/s10955-012-0645-0

Huang, J., Cui, Y., Zhang, L., Tong, W., Shi, Y., & Liu, Z. (2022). An Overview of Agent-Based Models for Transport Simulation and Analysis. *Journal of Advanced Transportation*, *2022*, 1252534. https://doi.org/10.1155/2022/1252534

Hussain, I., Elomri, A., Kerbache, L., & El Omri, A. (2024). Smart City Solutions: Comparative Analysis of Waste Management Models in IoT-Enabled Environments Using Multiagent Simulation. *Sustainable Cities and Society*, *103*, 105247. https://doi.org/10.1016/j.scs.2024.105247

Jahromi, K. K., Zignani, M., Gaito, S., & Rossi, G. P. (2016). Simulating Human Mobility Patterns in Urban Areas. *Simulation Modelling Practice and Theory*, *62*, 137–156. https://doi.org/10.1016/j.simpat.2015.12.002

Jing, P., Hu, H., Zhan, F., Chen, Y., & Shi, Y. (2020). Agent-Based Simulation of Autonomous Vehicles: A Systematic Literature Review. *IEEE Access*, *8*, 79089–79103. https://doi.org/10.1109/access.2020.2990295

Kagho, G. O., Balac, M., & Axhausen, K. W. (2020). Agent-Based Models in Transport Planning: Current State, Issues, and Expectations. *Procedia Computer Science*, *170*, 726–732. https://doi.org/10.1016/j.procs.2020.03.164

Kaluarachchi, Y. (2022). Implementing Data-Driven Smart City Applications for Future Cities. *Smart Cities*, *5*(2), 455–474. https://doi.org/10.3390/smartcities5020025

Kamiński, B., Kraiński, Ł., Mashatan, A., Prałat, P., & Szufel, P. (2020). Multiagent Routing Simulation with Partial Smart Vehicles Penetration. *Journal of Advanced Transportation*, *2020*, 3152020. https://doi.org/10.1155/2020/3152020

Khaidem, L., Luca, M., Yang, F., Anand, A., Lepri, B., & Dong, W. (2020). Optimizing Transportation Dynamics at a City-Scale Using a Reinforcement Learning Framework. *IEEE Access*, *8*, 171528–171541. https://doi.org/10.1109/access.2020.3024979

Khan, A., Aslam, S., Aurangzeb, K., Alhussein, M., & Javaid, N. (2021). Multiscale Modeling in Smart Cities: A Survey on Applications, Current Trends, and Challenges. *Sustainable Cities and Society*. https://doi.org/10.20944/preprints202106.0008.v1

Leukfeldt, E. R., & Yar, M. (2016). Applying Routine Activity Theory to Cybercrime: A Theoretical and Empirical Analysis. *Deviant Behavior*, *37*(3), 263–280.

Li, Z., Al Hassan, R., Shahidehpour, M., Bahramirad, S., & Khodaei, A. (2019). A Hierarchical Framework for Intelligent Traffic Management in Smart Cities. *IEEE Transactions on Smart Grid*, *10*(1), 691–701. https://doi.org/10.1109/tsg.2017.2750542

Li, Z., Chen, C., & Wang, K. (2011). Cloud Computing for Agent-Based Urban Transportation Systems. *IEEE Intelligent Systems*, *26*(1), 73–79. https://doi.org/10.1109/mis.2011.10

Li, J., Rombaut, E., & Vanhaverbeke, L. (2021). A Systematic Review of Agent-Based Models for Autonomous Vehicles in Urban Mobility and Logistics: Possibilities for Integrated Simulation Models. *Computers, Environment and Urban Systems*, *89*, 101686. https://doi.org/10.1016/j.compenvurbsys.2021.101686

Liu, C., Liu, L., Peng, Z., Wu, H., Wang, F., Jiao, H., & Wang, J. (2025). Planning public electric vehicle charging stations to balance efficiency and equality: A case study in Wuhan, China. *Sustainable Cities and Society*, *124*, 106314. https://doi.org/https://doi.org/10.1016/j.scs.2025.106314

Liu, L., & Wang, F. (2023). *Computational methods and GIS applications in social science-lab manual*. CRC Press.

Massaro, E., Kondor, D., & Ratti, C. (2019). Assessing the Interplay between Human Mobility and Mosquito Borne Diseases in Urban Environments. *Scientific Reports*, *9*, 16911. https://doi.org/10.1038/s41598-019-53127-z

Nguyen, J., Powers, S. T., Urquhart, N., Farrenkopf, T., & Guckert, M. (2021). An Overview of Agent-Based Traffic Simulators. *Transportation Research Interdisciplinary Perspectives*, *12*, 100486. https://doi.org/10.1016/j.trip.2021.100486

Oshin, T. O., Poslad, S., & Zhang, Z. (2014). Energy-Efficient Real-Time Human Mobility State Classification Using Smartphones. *IEEE Transactions on Computers*, *2014*, 1680–1693. https://doi.org/10.1109/TC.2014.2339846

Pang, Y., Tsubouchi, K., Yabe, T., & Sekimoto, Y. (2020). Intercity Simulation of Human Mobility at Rare Events via Reinforcement Learning. In *SIGSPATIAL '20: Proceedings of the 28th International Conference on Advances in Geographic Information Systems*, pp. 293–302. Association for Computing Machinery. https://doi.org/10.1145/3397536.3422244

Richter, M. A., Hagenmaier, M., Bandte, O., Parida, V., & Wincent, J. (2022). Smart Cities, Urban Mobility and Autonomous Vehicles: How Different Cities Needs Different Sustainable Investment Strategies. *Technological Forecasting & Social Change*, 184, 121857. https://doi.org/10.1016/j.techfore.2022.121857

Ronald, N., Thompson, R., & Winter, S. (2015). Simulating Demand-responsive Transportation: A Review of Agent-based Approaches. *Transport Reviews*, *35*(4), 404–421. https://doi.org/10.1080/01441647.2015.1017749

Sanchez, N. C., Martinez, I., Pastor, L. A., & Larson, K. (2022). On the Simulation of Shared Autonomous Micro-Mobility. *Communications in Transportation Research*, 2, 100065. https://doi.org/10.1016/j.commtr.2022.100065

Serok, N., & Blumenfeld-Lieberthal, E. (2015). A Simulation Model for Intra-Urban Movements. *PLoS One*, *10*(7), e0132576. https://doi.org/10.1371/journal.pone.0132576

Tosi, D. (2017). Cell Phone Big Data to Compute Mobility Scenarios for Future Smart Cities. *International Journal of Data Science and Analysis*, *4*, 265–284. https://doi.org/10.1007/s41060-017-0061-2

Uthpala, N., Hansika, N., Dissanayaka, S., Tennakoon, K., Dharmarathne, S., Vidanarachchi, R., Alawatugoda, J., & Herath, D. (2023). Analyzing Transportation Mode Interactions Using Agent-Based Models. *SN Applied Sciences*, *5*, 357. https://doi.org/10.1007/s42452-023-05609-z

Wang, F., & Liu, L. (2023). *Computational methods and GIS applications in social science.* CRC Press.

Weil, C., Bibri, S. E., Longchamp, R., Golay, F., & Alahi, A. (2023). Urban Digital Twin Challenges: A Systematic Review and Perspectives for Sustainable Smart Cities. *Sustainable Cities and Society*, *99*, 104862. https://doi.org/10.1016/j.scs.2023.104862

Wen, R., & Li, S. (2020). A Review of the Use of Geosocial Media Data in Agent-Based Models for Studying Urban Systems. *Big Earth Data*, *5*(1), 5–23. https://doi.org/10.1080/20964471.2020.1810492

Wise, S. C., & Cheng, T. (2016). How Officers Create Guardianship: An Agent-based Model of Policing. *Transactions in GIS*, *20*, 790–806. https://doi.org/10.1111/tgis.12173

Wu, W., Dong, B., Wang, Q. R., Kong, M., Yan, D., An, J., & Liu, Y. (2020). A Novel Mobility-Based Approach to Derive Urban-Scale Building Occupant Profiles and Analyze Impacts on Building Energy Consumption. *Applied Energy*, *278*, 115656. https://doi.org/10.1016/j.apenergy.2020.115656

Wu, H., Liu, L., Yu, Y., Peng, Z., Jiao, H., & Niu, Q. (2019). An Agent-Based Model Simulation of Human Mobility Based on Mobile Phone Data: How Commuting Relates to Congestion. *ISPRS International Journal of Geo-Information*, *8*(7), 313. https://doi.org/10.3390/ijgi8070313

Wu, L., Zhi, Y., Sui, Z., & Liu, Y. (2014). Intra-Urban Human Mobility and Activity Transition: Evidence from Social Media Check-In Data. *PLoS One*, *9*(5), e97010. https://doi.org/10.1371/journal.pone.0097010

Yanbo, P., Kota, T., Takahiro, Y., & Yoshihide, S. (2018). Replicating urban dynamics by generating human-like agents from smartphone GPS data. In *SIGSPATIAL '18: Proceedings of the 26th ACM SIGSPATIAL International Conference on Advances in Geographic Information Systems*, pp. 440–443. Association for Computing Machinery. https://doi.org/10.1145/3274895.3274935

Yao, F., & Wang, Y. (2020). Towards Resilient and Smart Cities: A Real-Time Urban Analytical and Geo-Visual System for Social Media Streaming Data. *Sustainable Cities and Society*, *63*, 102448. https://doi.org/10.1016/j.scs.2020.102448

Zhou, Y., Lau, B. P. L., Yuen, C., Tunçer, B., & Wilhelm, E. (2018). Understanding Urban Human Mobility through Crowdsensed Data. *IEEE Communications Magazine*, *56*(11), 52–59. https://doi.org/10.1109/mcom.2018.1700569

Zhu, H., & Wang, F. (2021). An Agent-Based Model for Simulating Urban Crime with Improved Daily Routines. *Computers, Environment and Urban Systems*, *89*, 101680.

Zhang, D., He, T., Zhang, F., & Xu, C. (2018). Urban-Scale Human Mobility Modeling With Multi-Source Urban Network Data. *IEEE/ACM Transactions on Networking*, *26*(2), 671–684. https://doi.org/10.1109/tnet.2018.2801598

13 Trajectory Privacy Protection Approaches

Jinmeng Rao and Song Gao

13.1 INTRODUCTION

The widespread adoption of location-based services (LBS) has resulted in the collection of extensive individual-level trajectory data. These data encompass spatiotemporal sequences of locations along with attributes, gathered from diverse sources such as mobile phones, wearable devices, global positioning system (GPS) units, and geotagged social media posts (Liu et al., 2015). The availability of this trajectory big data opens up new possibilities for researching human mobility patterns and interactions with the environment (Huang and Wong, 2015; Zhang et al., 2024), disaster response strategies (Huang and Xiao, 2015; Wang et al., 2016; Zhou et al., 2024), social economics (Huang et al., 2023; Fabbri et al., 2024), and public health concerns (Park and Kwan, 2017; Li et al., 2019; Han et al., 2024). However, it also presents significant challenges in safeguarding geoprivacy and addressing broader implications in behavioral, social, ethical, legal, and policy contexts (Keßler and McKenzie, 2018). In general, trajectory privacy refers to an individual's right to protect their trajectory identity and related sensitive personal locations from being disclosed (Kwan et al., 2004; Chow and Mokbel, 2011; Gao et al., 2019; McKenzie and Zhang, 2023). Ensuring the protection of trajectory privacy is crucial to maintaining public trust in the use of trajectory data and its potential applications. Without proper safeguards, there is a risk of compromising individuals' privacy, which could lead to adverse social consequences. In this chapter, we summarize several streams of trajectory privacy protection techniques, discuss the existing challenges in trajectory privacy protection nowadays, and provide a case study on how to address the privacy concerns in vehicle trajectory data with the help of geospatial artificial intelligence (GeoAI) (Gao et al., 2024) and differential privacy.

13.2 TRAJECTORY PRIVACY PROTECTION

Given the growing concerns about data breaches and the increasing public awareness of the need for trajectory privacy protection, various approaches have been developed to prevent the identification of users' trajectories. According to how the privacy concerns are addressed, these approaches can be broadly categorized into four main research areas: de-identification, geomasking, K-anonymity, and differential privacy.

13.2.1 De-identification

De-identification is a widely adopted technique aimed at safeguarding the privacy of trajectory data by removing or concealing user identities from the dataset, such as names, identifications (IDs), or other customized user identifiers. However, recent studies have indicated that de-identification alone may not be adequate to ensure individual privacy in trajectory data. Even when direct identifiers are stripped from the data, the spatial, temporal, and thematic attributes of the trajectories can still be utilized to re-identify individuals. These quasi-identifiers, which may include factors like gender, date of birth, frequent check-in locations and timestamps, and geotagged content, can still serve as strong markers that facilitate the relinking of trajectory data to specific individuals (Chow and Mokbel, 2011; Shao et al., 2023). Additionally, even when these quasi-identifiers are

DOI: 10.1201/9781003503262-16

anonymized, external data sources can be employed to re-identify users. For example, information derived from social media, public records, or online activities can be leveraged to infer the identities of individuals in datasets (Narayanan and Shmatikov, 2008; Shu et al., 2017). In summary, while de-identification is a commonly used method for trajectory privacy, it might not be sufficient to fully prevent privacy violations.

13.2.2 Geomasking

Geomasking is a technique that modifies the locations within trajectory data—by blurring, shifting, or masking—to protect privacy while maintaining the spatial characteristics of the data (Armstrong et al., 1999; Hampton et al., 2010; Gao et al., 2019; Swanlund et al., 2020; Houfaf-Khoufaf et al., 2024). Generally, a geomasking approach involves adding noise to locations or aggregating locations into geographic or administrative units, which helps obscure the original locations. However, a major limitation of this method is the difficulty in managing the trade-off between privacy protection and spatial or temporal resolution after aggregation. Research has shown that fine-grained aggregation offers only limited privacy protection, as significant information can still be extracted from aggregated data (De Montjoye et al., 2013; Xu et al., 2017; Gao et al., 2019). Coarser aggregation can mitigate this issue but at the expense of reduced spatial resolution and data utility (Gao et al., 2019; Arbia et al., 2023). Other geomasking strategies include random perturbation (Kwan et al., 2004; Seidl et al., 2016), which involves adding uniformly distributed random noise to trajectory locations; donut geomasking, which perturbs locations within a donut-shaped region (Hampton et al., 2010); and Gaussian geomasking (Armstrong et al., 1999), which adds Gaussian-distributed random noise. Various studies have explored the effectiveness of these geomasking techniques in balancing privacy and spatial analysis needs. For instance, Armstrong et al. (1999) assessed the privacy preservation and spatial analysis effectiveness of different geomasking methods. Kwan et al. (2004) similarly evaluated the spatial analysis effectiveness of three random perturbation geomasks in the context of lung cancer mortality. Seidl et al. (2016) also studied the privacy protection performance of grid masking and random perturbation applied to GPS trajectory data. Recent findings suggest that Gaussian geomasking is slightly more effective in protecting privacy compared to random perturbation and aggregation, likely due to its unbounded noise range (Gao et al., 2019; Rao et al., 2020, 2023). Overall, geomasking offers a promising approach to preserving privacy while retaining spatial information, but careful calibration is needed to balance privacy with data utility.

13.2.3 K-anonymity

K-anonymity can be utilized as a privacy protection strategy that frames the issue as a K-anonymization problem (Niu et al., 2014; Zhu et al., 2019; McKenzie and Zhang, 2023). The core idea of trajectory K-anonymity is to cluster and mix trajectories from different users so that individual trajectories cannot be distinguished. The K-anonymity criterion is met when each user's information cannot be distinguished from at least $k - 1$ other users within the dataset (Jain et al., 2016), making it an effective approach for protecting trajectory privacy (Gruteser and Grunwald, 2003; Palanisamy and Liu, 2011; Zhang et al., 2020; Yu et al., 2023; Rao et al., 2023). Techniques based on trajectory K-anonymity include spatial cloaking, which involves mixing trajectory points from K users within K-anonymized regions (Gruteser and Grunwald, 2003), and the mix-zones method, which anonymizes trajectory points within mix zones using pseudonyms, thereby breaking the link between trajectory segments passing through a mix zone (Palanisamy and Liu, 2011). Another approach involves generalization, where trajectory points are grouped into K-anonymized regions and then reassembled into K new trajectories by selecting points uniformly from each region and linking them together (Nergiz et al., 2008). Yu et al. (2023) proposed a dual-K mechanism (DKM), where multiple anonymizers are involved, along with a location selection mechanism and a dynamic pseudonym mechanism, to obfuscate the users' real locations and protect the users' trajectory privacy. Rao et al. (2023) proposed Conditional Adversarial Trajectory Synthesis (CATS), a GeoAI framework that learns to reconstruct trajectory topology and generate synthetic trajectories from K-anonymized human mobility matrices, better balancing privacy and utility. Extensions of K-anonymity, such as l-diversity (Machanavajjhala et al., 2007) and t-closeness (Li et al., 2007), have also been proposed to further improve the privacy guarantee.

13.2.4 Differential Privacy

Differential privacy (DP), a mathematically rigorous privacy definition introduced by Dwork (2006), has been widely applied in privacy-preserving data publication. By injecting noise, such as Laplace noise, into the query or analysis results related to a database, DP ensures that the presence or absence of any specific record cannot be inferred. DP has recently been adopted for trajectory privacy protection in various studies (Andrés et al., 2013; Li et al., 2017; Deldar and Abadi, 2018; Chen et al., 2019; Miranda-Pascual et al., 2023). For example, Andrés et al. (2013) introduced geo-indistinguishability, a DP-based method for trajectory data publication, which prevents the identification of users' real locations by adding controlled random noise, such as Planar Laplace noise. Similarly, Li et al. (2017) proposed a differentially private trajectory data publication algorithm that incorporates a bounded Laplace noise generation algorithm and a trajectory merging algorithm, and Deldar and Abadi (2018) introduced the concept of personalized-location differential privacy (PLDP) for trajectory databases and developed the PLDP-TD algorithm to provide nonuniform privacy guarantees. Chen et al. (2019) presented a differential privacy-based trajectory protection framework with real-path reporting in mobile crowdsensing, capable of effectively preserving trajectory privacy even against Bayesian inference attacks. Furthermore, Miranda-Pascual et al. (2023) systematically analyzed differential privacy adaptations to trajectory and the utility metrics, as well as discussed the challenges and limitations of applying differential privacy in the context of trajectory data. The application of DP in trajectory privacy protection highlights its potential to balance data utility with privacy, offering a promising direction for future research.

13.3 EXISTING CHALLENGES

Although the previously mentioned methods can be applied to protect trajectory privacy, several significant challenges persist.

13.3.1 Balancing Privacy and Utility

A primary objective of existing trajectory privacy protection techniques is to obscure or mix trajectory locations, thereby introducing greater uncertainty. However, managing the balance between effective privacy protection and maintaining data utility is challenging (Liu et al., 2018). This creates a "catch-22" situation (Rao et al., 2020), which has not been thoroughly explored or evaluated. For instance, while methods like random perturbation or geomasking can reduce the likelihood of user identification or the discovery of sensitive locations, they often do so at the expense of obscuring or losing important spatiotemporal data characteristics. Striking the right balance between privacy and utility is difficult. Many differential privacy methods allow for the adjustment of a privacy budget, denoted by ϵ, to control privacy loss with each differential change. A lower ϵ means more noise is added, offering better privacy protection, but determining the appropriate ϵ value is a complex task.

13.3.2 Issues of Data Sparsity

A common problem in existing trajectory datasets is their relatively sparse sampling rates or check-in frequencies, leading to data sparsity in both spatial and temporal dimensions. When trajectory privacy protection methods are applied, these issues can become even more pronounced (e.g., when trajectory points are aggregated into coarser administrative units or broader time windows). Sparse trajectory data may not be sufficient to support applications that require dense data, such as estimating travel times, profiling driving behaviors, or detecting lane changes (Andersen and Torp, 2017).

13.3.3 Lack of Data Representativeness

Many trajectory datasets are collected from a limited subset of the population due to privacy policies and device limitations, leading to concerns about data representativeness. For example, most available vehicle trajectory datasets consist of taxi trip data, which may not accurately reflect overall traffic patterns or capture the diverse driving behaviors of pedestrians or private car drivers who rarely share their trajectories. This can result in imbalances in trajectory coverage, types, and behaviors, hindering further applications.

13.4 VEHICLE TRAJECTORY PRIVACY PROTECTION: A CASE STUDY

This case study takes vehicle trajectory data as a representative example and illustrates how to tackle the challenges associated with vehicle trajectory privacy protection. We introduce the privacy-preserving Vehicle Trajectory Simulation and Visualization (VTSV) framework, which generates simulated vehicle trajectories with customized spatiotemporal resolutions and driving behaviors by leveraging GeoAI and differential privacy. Starting with a dataset of real-world sparsely sampled vehicle trajectories, differential privacy techniques are employed to safeguard trajectory privacy. Subsequently, driving routes between privacy-preserved origin-destination (OD) pairs in the dataset are planned along a specified road network. A spatially explicit deep reinforcement learning model is then used to generate dense vehicle trajectory data, including GPS coordinates and sensor data (such as speed and acceleration), with a customizable spatiotemporal resolution along the planned routes. Additionally, various parameters can be adjusted to simulate different driving behaviors (such as overspeeding or aggressive acceleration). The simulated trajectories are visualized as autonomous driving trips, enabling users to gain a more comprehensive understanding and analysis of driving behaviors.

13.4.1 Methods

Figure 13.1a illustrates the overall architecture of VTSV, and Figure 13.1b shows the visual interface of VTSV. It's important to note that VTSV also incorporates visualization and interaction capabilities for users. Below we introduce the method details of the main components, namely privacy protection, route planning, and driving simulation.

13.4.1.1 Privacy Protection

Vehicle trajectory data inherently contain location information, which raises significant privacy concerns as it can potentially lead to privacy breaches. For instance, the starting and ending points of a user's trips may inadvertently reveal sensitive locations, such as their home address. Moreover, by analyzing users' historical trajectories, it is possible to model their travel patterns and location preferences, increasing the risk of re-identifying users from their trajectories and linking them back to specific individuals (known as trajectory-user linking [TUL]) (Gao et al., 2017). These scenarios pose a threat to location privacy and user identification, which necessitates the implementation of effective privacy safeguards. The Privacy Protection Module addresses these concerns by applying differential privacy techniques to process the origin and destination data. In particular, we employ a mechanism based on controlled planar Laplacian noise, adhering to the ϵ-Geo-Indistinguishability standard (Andrés et al., 2013), a location privacy framework rooted in differential privacy principles. This approach involves adding random noise, distributed according to a planar Laplacian

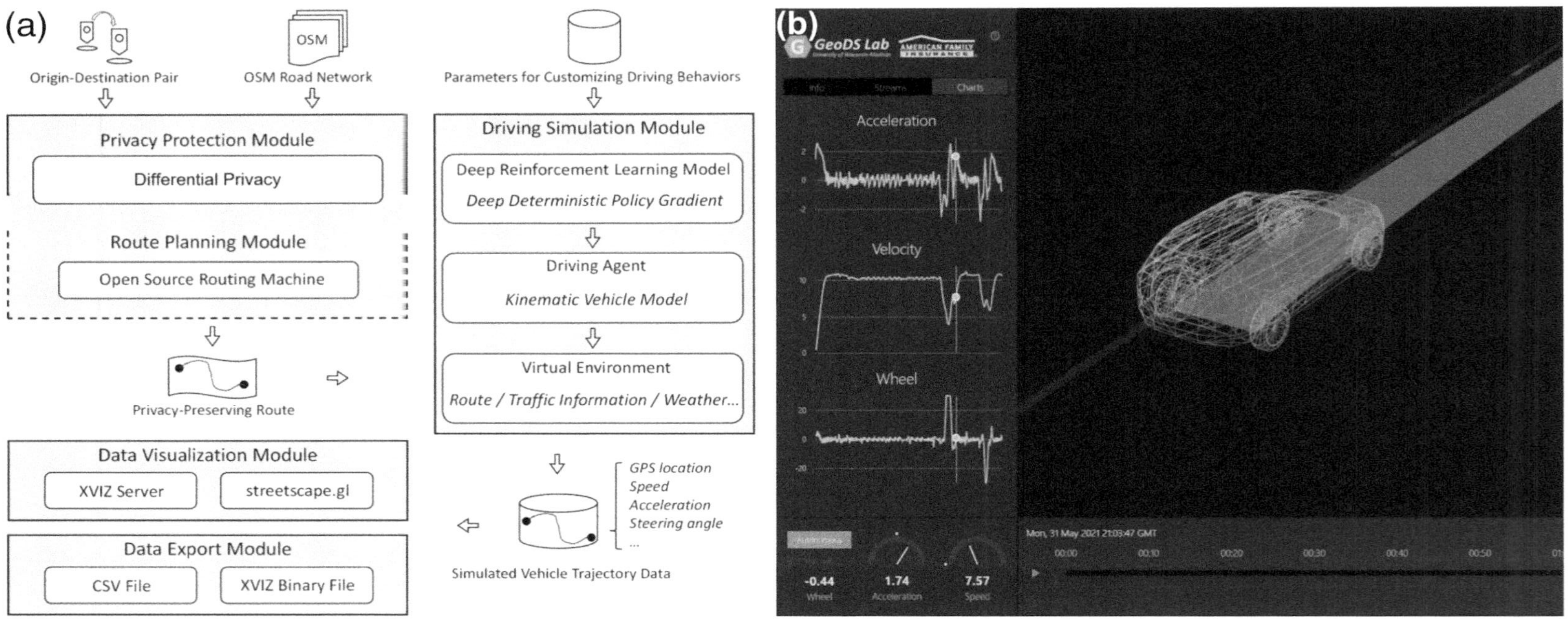

FIGURE 13.1 (a) The overall framework of VTSV and (b) the visual interface of VTSV. *Abbreviation:* OSM, OpenStreetMap.

distribution, to the latitude and longitude coordinates of each trajectory point in the dataset. In this context, ϵ represents the scaling factor, often referred to as the privacy budget in differential privacy. The Euclidean distance metric is used to calculate the distance between any two points. In VTSV, we have set ϵ to 100 and utilize the Euclidean distance metric for computing point-to-point distances.

13.4.1.2 Route Planning

Once the privacy-preserved origin and destination points are obtained from the processed dataset, the route planning module begins to calculate a synthetic route between these two locations. A driving route R is composed of a series of connected road segments S and intersections I. These intersections and the centerlines of the road segments are represented using GPS coordinates in the form of points (longitude, latitude) and polylines [(longitude1, latitude1), (longitude2, latitude2), ...]. The primary objective of the Route Planning Module is to determine a driving route between the user-selected origin and destination points. For this purpose, we utilize the Open Source Routing Machine (OSRM), an open-source route planning engine that relies on OpenStreetMap data. OSRM employs contraction hierarchies and the multilevel Dijkstra shortest path algorithm (Geisberger et al., 2008), which enables quick and adaptable routing, making it well-suited for web-based applications. By default, OSRM generates the shortest-path routes, but it also offers options for customizing the route planning process. For instance, the "steps" option allows for the return of detailed route steps for each road segment, while the "overview" option specifies the level of detail for the route (e.g., full or simplified). It is important to note that, since the origin and destination are processed via differential privacy and the planned synthetic route does not capture any real route information, trajectory privacy is well preserved.

13.4.1.3 Driving Simulation

We employ a Driving Simulation Module to generate high-spatiotemporal-resolution vehicle trajectory data along specified synthetic routes. This module operates by first training a deep reinforcement learning model to simulate a driving agent with customized driving behaviors, which is then deployed into a virtual environment constructed based on the provided routes. The driving agent interacts with the environment and simulates the driving process along the synthetic route planned from the privacy-preserved starting point to the destination, while its real-time GPS locations, acceleration, speed, and steering angle are recorded at a customized sample rate. The key concepts in this module include deep reinforcement learning, driving agent, driving behavior, and virtual environment.

- *Deep reinforcement learning (DRL)*: Reinforcement learning (RL) is a branch of machine learning that focuses on training an agent to take specific actions within an environment to maximize cumulative rewards. DRL integrates deep learning models with RL, enabling the agent to make optimal decisions without manual feature engineering. In our work, we utilize the Deep Deterministic Policy Gradient (DDPG), a model-free DRL algorithm designed for learning continuous actions (Lillicrap et al., 2015). DDPG uses an actor-critic approach, which involves training two distinct deep neural networks: an actor network that learns to estimate the optimal policy and a critic network that assesses the quality of the actor's actions. The actor network maps the input states to continuous actions that maximize the expected cumulative reward, while the critic network estimates the expected future reward based on the current state and action to evaluate the quality of the actor's decisions. These networks are updated iteratively: the actor's weights are adjusted to maximize the expected reward, and the critic's weights are tuned to minimize the discrepancy between the predicted and actual rewards. One of DDPG's significant advantages is its ability to learn from experience, allowing it to adapt to dynamic environments and make optimal decisions autonomously. Furthermore, the deep neural networks enable the algorithm to handle high-dimensional state spaces, making it applicable to various domains, such as robotic control and gaming. By applying DDPG on geospatial virtual environments with spatially explicit input, output, and optimization objective design, our framework follows spatially explicit AI (Janowicz et al., 2020; Mai et al., 2022) design and can simulate and refine the behavior of spatiotemporal vehicle trajectories, resulting in more accurate and realistic trajectory simulations.

- *Driving agent*: The driving agent is a simulated entity that represents a driver operating a vehicle—it perceives the environment, makes decisions, takes actions, updates its state, and makes subsequent observations. The decision-making process is supported by DDPG. To generate realistic vehicle trajectories, we utilize a simplified kinematic vehicle model to represent the vehicle. This model treats the vehicle as an object moving through space and captures its motion by tracking its position, velocity, and acceleration, which are updated based on the vehicle's current state and the actions taken by the driving agent. The kinematic model is computationally less intensive and simpler than more complex vehicle models, making it particularly suitable for our trajectory simulation needs. This model allows us to describe and control vehicle motion without needing to understand the vehicle's internal mechanisms. Key parameters used in the model are detailed in Table 13.1.
- *Driving behavior*: Driving behavior encompasses the habits and styles that characterize how a driver operates a vehicle, significantly influencing travel safety. Broadly, driving behavior can be categorized into aggressive behavior, normal behavior, and others. More specifically, driving behavior can be quantitatively assessed through various driving-related metrics. Table 13.2 outlines a set of indicators in the VTSV and how they can be adjusted to reflect different driving behaviors.
    - *Speed limit*: This indicator evaluates the driver's speed in comparison to the posted speed limit. Exceeding the speed limit is a clear sign of overspeeding, a typical aggressive driving behavior.
    - *Acceleration limit*: This indicator measures the driver's acceleration and deceleration against a predefined threshold. Exceeding this threshold suggests sudden acceleration or braking, which can be hazardous and increase the risk of accidents.
    - *Turning speed limit*: This indicator assesses the driver's speed during turns relative to a set limit. Exceeding this limit indicates aggressive turning, another form of aggressive driving behavior.
    - *Deviation from centerline*: This indicator tracks how much the driver strays from the road's centerline. Significant deviation from the centerline is indicative of risky lane changes, which can lead to collisions with other vehicles. By employing these indicators to quantify driving behavior, we can achieve a deeper understanding of the factors contributing to unsafe driving. This understanding is crucial for developing strategies aimed at mitigating these risks and enhancing overall transportation safety.

TABLE 13.1
Parameters in the Simplified Kinematic Vehicle Model

Parameter	Description
Length	The length of the vehicle
Position	The center of the vehicle
Orientation	The driving direction of the vehicle
Steering angle	The steering angle of the vehicle (for vehicle control)
Velocity	The current velocity of the vehicle

TABLE 13.2
Parameters for Customizing Driving Behaviors

Parameter	Example
Speed limit	Exceeding the speed limit is a clear sign of overspeeding
Acceleration limit	Exceeding this threshold suggests sudden acceleration or braking
Turning speed limit	Exceeding this limit indicates aggressive turning
Deviation to centerline	Significant deviation from the centerline is indicative of risky lane changes

- *Virtual environment*: To replicate realistic vehicle trajectories, a virtual environment is constructed based on the intended route and driving-related environmental factors, such as traffic information (e.g., road speed limits) and weather conditions. The primary objective of the virtual environment is to closely mimic real-world driving scenarios. This environment acts as the operational space for the driving agent, within which the agent interacts and generates movement trajectories. The driving agent's actions are shaped by the environment, which encompasses various real-world driving conditions, including speed limits and intersections. Specifically, six key observations are considered in our scenario: steering angle, acceleration, distance to the road centerline, speed difference between the speed limit and the current speed, the angle between the vehicle's orientation and the road centerline, and proximity to an intersection or stop sign. These six factors have proven effective in guiding the driving agent's actions to simulate realistic driving behavior. Throughout the simulation, the driving agent is rewarded or penalized based on how well it adheres to expected behavior, as assessed through these observations. We have designed five reward metrics: a speed reward (positive when the speed is near the limit, negative if it deviates too much), an acceleration reward (positive when acceleration is below the limit), a turning reward (negative for aggressive maneuvers), a centerline reward (negative for straying from the road centerline), and an included angle reward (negative when the angle between the vehicle's orientation and the road centerline is not zero). Each of these rewards is empirically weighted according to its significance in determining the overall reward value. By integrating a virtual environment, we can craft a realistic and dynamic driving scenario that allows for the simulation of a diverse array of driving conditions and behaviors. This approach provides us with a deeper understanding of driving patterns.

13.4.2 Research Area and Data

The research area chosen for this study is Dane County, Wisconsin, home to the state capital, Madison. To create a virtual environment, we acquired the major road centerline network (Figure 13.2) from the MadisonOpenData website. For simplicity, we standardized the speed limit across all road segments to 25 mph and did not consider traffic or weather conditions in our simulation. The experiment utilized the UberMedia Mobility Dataset (which was acquired by Near in 2021), containing data on geographic coordinates (latitude and longitude), timestamps (date, day of the week, time of day), and anonymized device IDs, representing large-scale, long-term individual trajectories. We focused on data collected between September 7, 2020, and November 29, 2020, with a temporal resolution of 1 hour. To ensure data quality, we filtered out trajectories from inactive users, defined as those with less than 20 hours of recorded movement per day and less than 80 days of recorded movement during the specified period. After preprocessing, the dataset comprised 90,395 hour-level trajectories, totaling 2,169,480 GPS records from 1,097 users. In our driving simulation case, these hour-level trajectories are sparse, making them ideal for evaluating the performance of our simulation framework.

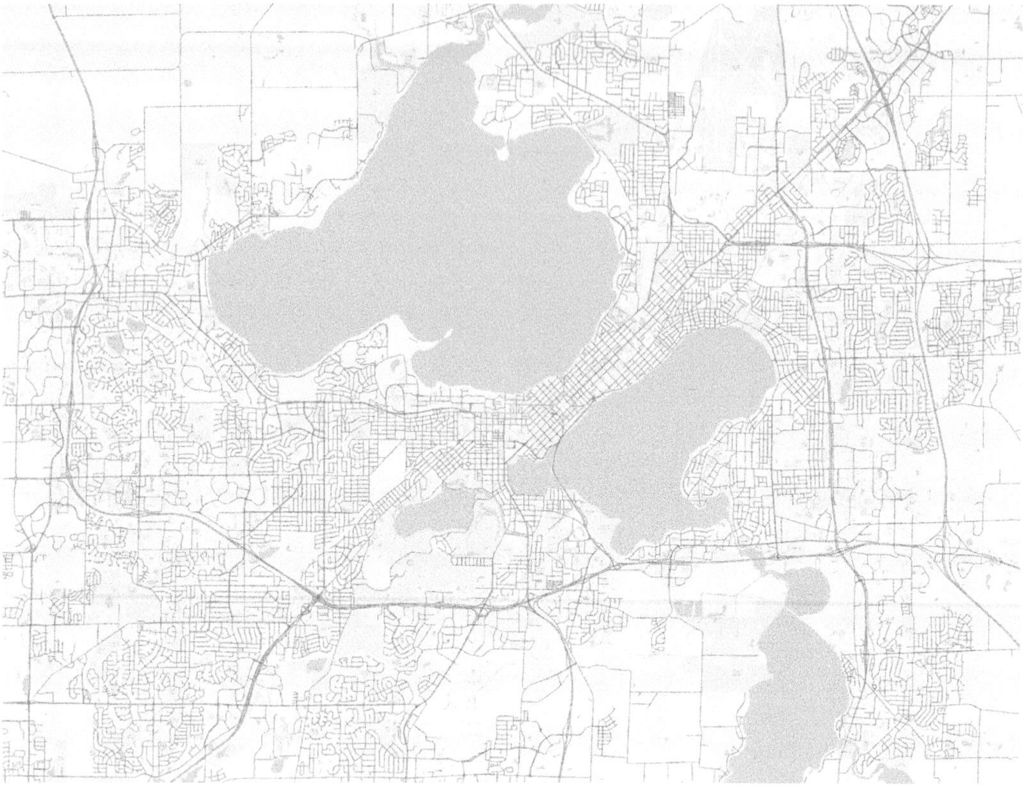

FIGURE 13.2 An overview of the major road network of downtown Madison, Wisconsin.

13.4.3 Driving Simulation Training

Our DDPG-based driving simulation model is developed using PyTorch, a versatile deep learning framework. The model undergoes training for 1,800 epochs on an NVIDIA Tesla K80 graphics processing unit (GPU). The training process and validation results, including the curves of the returned training reward and mean validation reward, are depicted in Figure 13.3. The model parameters adhere to the default settings of the original DDPG model, featuring a hidden layer size of 128, a batch size of 256, and a replay buffer size of 1×10^6. In terms of reward design, we prioritize speed with a reward factor of 5, as it is the most significant factor influencing driving events. Additionally, we assign a centerline reward factor of 2 to encourage the driving agent to maintain a straight path. The acceleration reward factor, turning reward factor, and included angle reward factor are each set to 0.5. These reward components are linearly combined with their respective factors to compute the final reward. To simulate the vehicle agent's behavior, we utilize a simplified kinematic model with a vehicle length of 4 meters and a maximum steering angle of 30 degrees. By adjusting these parameters and fine-tuning the reward factors, we can produce realistic vehicle trajectories that mimic real-world driving behavior.

13.4.4 Privacy-Preserving Origin-Destination Routing

We assess the effectiveness of the VTSV framework in preserving privacy during OD routing. Specifically, we investigate whether the framework, after applying its privacy protection module (such as differential privacy) to the raw dataset, can successfully prevent the identification of the original users from the processed data. If the framework proves effective in protecting

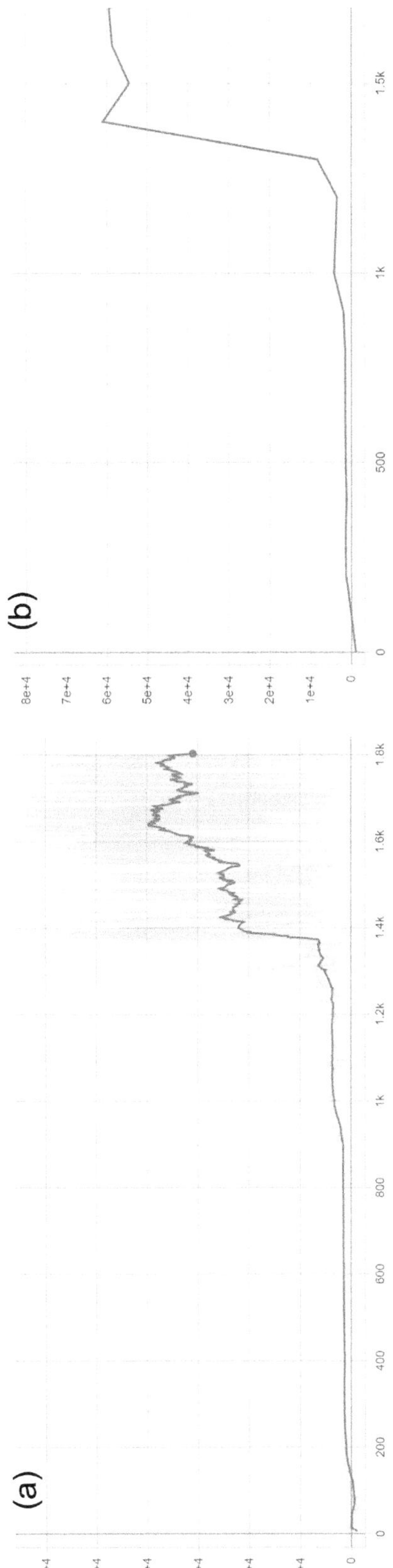

FIGURE 13.3 The curves of returned training reward (a) and the mean returned validation reward (b).

TABLE 13.3
The Trajectory-User Linking Measures on the Test Set of Raw Trajectory Dataset and Privacy-Preserved Trajectory Datasets

Dataset	Macro P	Macro R	Macro F1	Top-1 Acc	Top-5 Acc
Raw	0.242	0.150	0.167	0.524	0.864
RP	0.189	0.101	0.120	0.421	0.742
DP	**0.164**	**0.082**	**0.100**	**0.359**	**0.654**

Abbreviations: DP, datasets processed by differential privacy; Raw, raw dataset; RP, dataset processed by random perturbation.

user identities, it allows for the simulation of driving trajectories based on OD pairs in the processed dataset without risking privacy breaches. This evaluation is formulated as a TUL task, a well-known challenge in trajectory data publication and LBS, which aims to associate anonymous trajectories with their original users. To conduct this evaluation, we train a Multiple-Aspect tRajectory Classifier (MARC) (May Petry et al., 2020) using 80% of the raw trajectory dataset and then perform the TUL task on both the test sets of the raw and processed datasets, calculating the accuracy of the results. As shown in Table 13.3, the dataset processed with differential privacy achieves the lowest TUL scores, indicating superior privacy protection. Moreover, given that TUL is performed on users' complete hour-level trajectories, the routes generated through OD routing represent only short segments within a limited time frame, making it even less likely to compromise user identities.

13.4.5 Customized Driving Behavior Simulation

We assess the effectiveness of the VTSV framework in simulating customized driving behaviors. We adopted the thresholds outlined by Saiprasert et al. (2013) to differentiate between various driving behavior levels (e.g., normal/aggressive) and further categorized these behaviors into four distinct levels: normal (low), normal (medium), normal (high), and aggressive. Specifically, "normal (low)" indicates behavior that is normal with a low likelihood of becoming aggressive; "normal (medium)" signifies normal behavior with a moderate risk of aggression; "normal (high)" represents normal behavior with a high risk of aggression; and "aggressive" characterizes behavior that is already aggressive.

For the speed event, "normal (low)" is defined as the agent maintaining a speed strictly below the speed limit (e.g., 25 mph) throughout the driving process. "Normal (medium)" allows for the agent's speed to exceed the speed limit by up to 5 mph for no more than 10% of the driving duration. "Normal (high)" permits the speed to exceed the limit by up to 5 mph for up to 20% of the time. In contrast, "aggressive" behavior is when the speed exceeds the limit by more than 5 mph for more than 20% of the driving time or when it surpasses the limit by over 5 mph at any point.

For turning events, the classifications of normal (low), normal (medium), normal (high), and aggressive correspond to steering speeds of up to 10 degrees per second, 10–20 degrees per second, 20–30 degrees per second, and more than 30 degrees per second, respectively. Similarly, for acceleration/brake events, these categories are defined by the absolute value of acceleration: normal (low) corresponds to 0.1 G or less, normal (medium) falls between 0.1 and 0.2 G, normal (high) ranges from 0.2 to 0.3 G, and aggressive is defined by acceleration exceeding 0.3 G.

TABLE 13.4
The Average Scores of Customized Driving Behavior Simulation

Events	Normal (L)	Normal (M)	Normal (H)	Aggressive
Speed	100.0%	92.0%	82.4%	75.1%
Turning	100.0%	100.0%	100.0%	82.2%
Acceleration/brake	100.0%	100.0%	100.0%	95.6%

In our experiment, we sampled 1,000 OD pairs from the privacy-preserved dataset and generated routes for them using OSRM. In the driving simulation model, we then adjusted parameters to create specific driving behaviors (Table 13.2). These customized parameters influenced the observations made by the driving agent, such as the variance between its current speed and the road's speed limit or its deviation from the road's centerline. These parameters informed the agent's decision-making process, prompting actions that reflected the desired driving behaviors. For each route, we simulated multiple trajectories representing different driving behaviors and monitored them using a threshold-based method. The trajectories were evaluated based on how closely their behaviors aligned with the predefined behaviors. For instance, if the model was set to simulate a normal (low) speed behavior and consistently met the threshold criteria, it received a perfect score of 100.0%. However, any failure to meet the criteria resulted in a penalty proportional to the duration of non-compliance. For turning and acceleration/brake events, the difference in the number of events was used to calculate the scores.

The average scores from the 1,000 routes for the customized driving behavior simulations are summarized in Table 13.4. The results indicate that the simulation model performs well in replicating normal (low) and normal (medium) behaviors, although it occasionally struggles to meet threshold requirements for normal (high) or aggressive behaviors due to the increased complexity of controlling agents under these conditions.

Additionally, we provide a visual example showcasing three vehicle trajectories with different driving behaviors simulated by the VTSV framework (Figure 13.4). Figure 13.4a illustrates a trajectory with normal driving behavior [normal (low) speed within the road speed limit of 25 mph, approximately 11.18 m/s; normal (high) acceleration within 0.3 G, approximately 2.94 m/s^2]. Conversely, Figure 13.4b depicts a trajectory characterized by aggressive driving behavior (aggressive speed reaching up to 35 mph, approximately 15.65 m/s; aggressive acceleration up to 0.4 G, approximately 3.92 m/s^2). Figure 13.4c also shows a trajectory with similar aggressive driving behaviors as Figure 13.4b, but the difference is that it exhibits even more dangerous driving patterns (e.g., frequent swerving, leading to a zig-zag trace) that commonly appear in cases of drunk driving.

13.5 CONCLUSION

In this chapter, we reviewed various approaches to trajectory privacy protection, explored the current challenges in safeguarding trajectory data, and presented a case study demonstrating how GeoAI and differential privacy can be leveraged to address privacy concerns in vehicle trajectory data. We specifically introduced VTSV, a framework designed for privacy-preserving VTSV. This framework facilitates the planning of vehicle routes between privacy-preserved origin and destination points using differential privacy techniques and employs a DRL model to generate privacy-preserving vehicle trajectories with customizable spatiotemporal resolution and driving patterns. The findings indicate that VTSV effectively mitigates concerns and issues related to data privacy, sparsity, and underrepresentation in vehicle trajectory data, offering new insights into privacy-preserving trajectory generation and applications through the use of GeoAI methods.

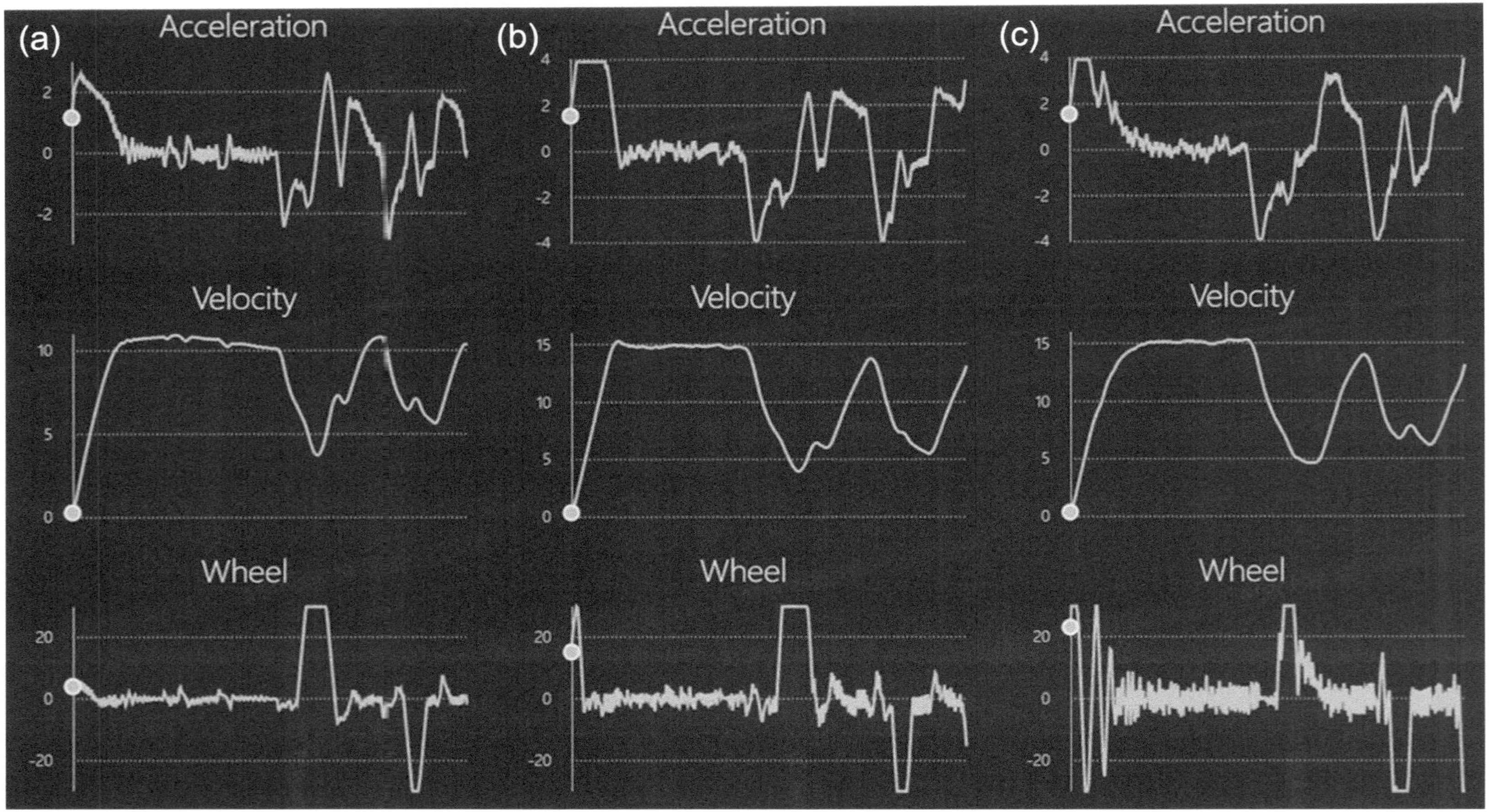

FIGURE 13.4 Customized driving behaviors. (a) Normal. (b and c) Aggressive. The units for acceleration, velocity, and wheel in the figure are m/s², m/s, and degree, respectively.

REFERENCES

Andersen, Ove, and Kristian Torp. "Sampling frequency effects on trajectory routes and road network travel time." In Proceedings of the 25th ACM SIGSPATIAL International Conference on Advances in Geographic Information Systems, pp. 1–10. 2017.

Andrés, Miguel E.; Nicolás E. Bordenabe; Konstantinos Chatzikokolakis, and Catuscia Palamidessi. "Geo-indistinguishability: Differential privacy for location-based systems." In Proceedings of the 2013 ACM SIGSAC conference on Computer & communications security, pp. 901–914. 2013.

Arbia, Giuseppe; Chiara Ghiringhelli, and Vincenzo Nardelli. "Effects of confidentiality-preserving geo-masking on the estimation of semivariogram and of the kriging variance." Geographical Analysis 55, no. 3 (2023): 466–481.

Armstrong, Marc P.; Gerard Rushton, and Dale L. Zimmerman. "Geographically masking health data to preserve confidentiality." Statistics in Medicine 18, no. 5 (1999): 497–525.

Chen, Xin; Xuangou Wu; Xiujun Wang; Wei Zhao, and Wei Xue. "Real-location reporting based differential privacy trajectory protection for mobile crowdsensing." In 2019 5th International Conference on Big Data Computing and Communications (BIGCOM), pp. 142–150. IEEE, 2019.

Chow, Chi-Yin, and Mohamed F. Mokbel. "Trajectory privacy in location-based services and data publication." ACM SIGKDD Explorations Newsletter 13, no. 1 (2011): 19–29.

Deldar, Fatemeh, and Mahdi Abadi. "PLDP-TD: Personalized-location differentially private data analysis on trajectory databases." Pervasive and Mobile Computing 49 (2018): 1–22.

De Montjoye, Yves-Alexandre; César A. Hidalgo; Michel Verleysen, and Vincent D. Blondel. "Unique in the crowd: The privacy bounds of human mobility." Scientific Reports 3, no. 1 (2013): 1–5.

Dwork, Cynthia. "Differential privacy." In International colloquium on automata, languages, and programming, pp. 1–12. Berlin, Heidelberg: Springer Berlin Heidelberg, 2006.

Fabbri, Giorgio; Salvatore Federico; Davide Fiaschi, and Fausto Gozzi. "Mobility decisions, economic dynamics and epidemic." Economic Theory 77, no. 1 (2024): 495–531.

Gao, Qiang; Fan Zhou; Kunpeng Zhang; Goce Trajcevski; Xucheng Luo, and Fengli Zhang. "Identifying human mobility via trajectory embeddings." IJCAI 17 (2017): 1689–1695.

Gao, Song; Yingjie Hu, and Wenwen Li. "Introduction to geospatial artificial intelligence (GeoAI)." In Handbook of geospatial artificial intelligence, pp. 3–16. CRC Press, 2024.

Gao, Song; Jinmeng Rao; Xinyi Liu; Yuhao Kang; Qunying Huang, and Joseph App. "Exploring the effectiveness of geomasking techniques for protecting the geoprivacy of Twitter users." Journal of Spatial Information Science 19 (2019): 105–129.

Geisberger, Robert; Peter Sanders; Dominik Schultes, and Daniel Delling. "Contraction hierarchies: Faster and simpler hierarchical routing in road networks." In Experimental Algorithms: 7th International Workshop, WEA 2008 Provincetown, MA, USA, May 30-June 1, 2008 Proceedings 7, pp. 319–333. Springer Berlin Heidelberg, 2008.

Gruteser, Marco, and Dirk Grunwald. "Anonymous usage of location-based services through spatial and temporal cloaking." In Proceedings of the 1st international conference on Mobile systems, applications and services, pp. 31–42. 2003.

Hampton, Kristen H.; Molly K. Fitch; William B. Allshouse; Irene A. Doherty; Dionne C. Gesink; Peter A. Leone; Marc L. Serre, and William C. Miller. "Mapping health data: Improved privacy protection with donut method geomasking." American Journal of Epidemiology 172, no. 9 (2010): 1062–1069.

Han, Xiaoyi; Yilan Xu; Yi Huang; Linlin Fan; Minhong Xu, and Song Gao. "Seeding Efficient Large-scale Public Health Interventions in Diverse Spatial-social Networks." Available at SSRN 4723268 (2024).

Houfaf-Khoufaf, Walid; Guillaume Touya, and Arnaud Le Guilcher. "Geographically masking addresses to study COVID-19 clusters." Cartography and Geographic Information Science 51, no. 2 (2024): 242–256.

Huang, Qunying, and David WS Wong. "Modeling and visualizing regular human mobility patterns with uncertainty: An example using Twitter data." Annals of the Association of American Geographers 105, no. 6 (2015): 1179–1197.

Huang, Qunying, and Yu Xiao. "Geographic situational awareness: Mining tweets for disaster preparedness, emergency response, impact, and recovery." ISPRS International Journal of Geo-Information 4, no. 3 (2015): 1549–1568.

Huang, Xiao; Xiaoqi Bao; Zhenlong Li; Shaozeng Zhang, and Bo Zhao. "Black businesses matter: A longitudinal study of black-owned restaurants in the COVID-19 pandemic using geospatial big data." Annals of the American Association of Geographers 113, no. 1 (2023): 189–205.

Jain, P.; M. Gyanchandani, and N. Khare. "Big data privacy: A technological perspective and review." Journal of Big Data 3 (2016): 25.

Janowicz, Krzysztof; Song Gao; Grant McKenzie; Yingjie Hu, and Budhendra Bhaduri. "GeoAI: Spatially explicit artificial intelligence techniques for geographic knowledge discovery and beyond." International Journal of Geographical Information Science 34, no. 4 (2020): 625–636.

Keßler, Carsten, and Grant McKenzie. "A geoprivacy manifesto." Transactions in GIS 22, no. 1 (2018): 3–19.

Kwan, Mei-Po; Irene Casas, and Ben Schmitz. "Protection of geoprivacy and accuracy of spatial information: How effective are geographical masks?" Cartographica: The International Journal for Geographic Information and Geovisualization 39, no. 2 (2004): 15–28.

Li, Meng; Liehuang Zhu; Zijian Zhang, and Rixin Xu. "Achieving differential privacy of trajectory data publishing in participatory sensing." Information Sciences 400 (2017): 1–13.

Li, Mingxiao; Song Gao; Feng Lu; Huan Tong, and Hengcai Zhang. "Dynamic estimation of individual exposure levels to air pollution using trajectories reconstructed from mobile phone data." International Journal of Environmental Research and Public Health 16, no. 22 (2019): 4522.

Li, Ninghui; Tiancheng Li, and Suresh Venkatasubramanian. "t-Closeness: Privacy beyond k-anonymity and l-diversity." In 2007 IEEE 23rd International Conference on Data Engineering, pp. 106–115. IEEE, 2007.

Lillicrap, T. P. "Continuous control with deep reinforcement learning." arXiv preprint arXiv:1509.02971 (2015).

Liu, X., Chen, H., and Andris, C. "trajGANs: Using generative adversarial networks for geo-privacy protection of trajectory data (vision paper)". Location Privacy and Security Workshop 2018 in conjunction with GIScience18 (2018): 1–7.

Liu, Yu; Xi Liu; Song Gao; Li Gong; Chaogui Kang; Ye Zhi; Guanghua Chi, and Li Shi. "Social sensing: A new approach to understanding our socioeconomic environments." Annals of the Association of American Geographers 105, no. 3 (2015): 512–530.

Machanavajjhala, Ashwin; Daniel Kifer; Johannes Gehrke, and Muthuramakrishnan Venkitasubramaniam. "*L*-diversity: Privacy beyond *k*-anonymity." ACM Transactions on Knowledge Discovery from Data 1, no. 1 (2007): 3–es.

Mai, Gengchen; Yingjie Hu; Song Gao; Ling Cai; Bruno Martins; Johannes Scholz; Jing Gao, and Krzysztof Janowicz. "Symbolic and subsymbolic GeoAI: Geospatial knowledge graphs and spatially explicit machine learning." Transactions in GIS 26, no. 8 (2022): 3118–3124.

May Petry, Lucas; Camila Leite Da Silva; Andrea Esuli; Chiara Renso, and Vania Bogorny. "MARC: A robust method for multiple-aspect trajectory classification via space, time, and semantic embeddings." International Journal of Geographical Information Science 34, no. 7 (2020): 1428–1450.

McKenzie, Grant, and Hongyu Zhang. "Platial k-Anonymity: Improving Location Anonymity Through Temporal Popularity Signatures." In 12th International Conference on Geographic Information Science (GIScience 2023). Schloss Dagstuhl-Leibniz-Zentrum für Informatik, 2023.

McKenzie, Grant; Hongyu Zhang, and Sébastien Gambs. "Privacy and ethics in GeoAI." In Handbook of geospatial artificial intelligence, pp. 388–405. CRC Press, 2023.

Miranda-Pascual, Àlex; Patricia Guerra-Balboa; Javier Parra-Arnau; Jordi Forné, and Thorsten Strufe. "SoK: Differentially private publication of trajectory data." In Proceedings on Privacy Enhancing Technologies, pp. 496–516. 2023.

Narayanan, Arvind, and Vitaly Shmatikov. "Robust de-anonymization of large sparse datasets." In 2008 IEEE Symposium on Security and Privacy (SP 2008), pp. 111–125. IEEE, 2008.

Nergiz, Mehmet Ercan; Maurizio Atzori, and Yucel Saygin. "Towards trajectory anonymization: a generalization-based approach." In Proceedings of the SIGSPATIAL ACM GIS 2008 International Workshop on Security and Privacy in GIS and LBS, pp. 52–61. 2008.

Niu, Ben; Qinghua Li; Xiaoyan Zhu; Guohong Cao, and Hui Li. "Achieving k-anonymity in privacy-aware location-based services." In IEEE INFOCOM 2014-IEEE Conference on Computer Communications, pp. 754–762. IEEE, 2014.

Palanisamy, Balaji, and Ling Liu. "Mobimix: Protecting location privacy with mix-zones over road networks." In 2011 IEEE 27th International Conference on Data Engineering, pp. 494–505. IEEE, 2011.

Park, Yoo Min, and Mei-Po Kwan. "Individual exposure estimates may be erroneous when spatiotemporal variability of air pollution and human mobility are ignored." Health & Place 43 (2017): 85–94.

Rao, Jinmeng; Song Gao; Yuhao Kang, and Qunying Huang. "LSTM-TrajGAN: A Deep Learning Approach to Trajectory Privacy Protection." In 11th International Conference on Geographic Information Science (GIScience 2021)-Part I. Schloss Dagstuhl-Leibniz-Zentrum für Informatik, 2020.

Rao, Jinmeng; Song Gao, and Sijia Zhu. "CATS: Conditional adversarial trajectory synthesis for privacy-preserving trajectory data publication using deep learning approaches." International Journal of Geographical Information Science 37, no. 12 (2023): 2538–2574.

Saiprasert, Chalermpol; Thunyasit Pholprasit, and Wasan Pattara-Atikom. "Detecting driving events using smartphone." In Proceedings of the 20th ITS World Congress, vol. 11, no. 11. 2013.

Seidl, Dara E.; Piotr Jankowski, and Ming-Hsiang Tsou. "Privacy and spatial pattern preservation in masked GPS trajectory data." International Journal of Geographical Information Science 30, no. 4 (2016): 785–800.

Shao, Jiangli; Yongqing Wang; Hao Gao; Boshen Shi; Huawei Shen, and Xueqi Cheng. "AsyLink: User identity linkage from text to geo-location via sparse labeled data." Neurocomputing 515 (2023): 174–184.

Shu, Kai; Suhang Wang; Jiliang Tang; Reza Zafarani, and Huan Liu. "User identity linkage across online social networks: A review." ACM SIGKDD Explorations Newsletter 18, no. 2 (2017): 5–17.

Swanlund, D.; N. Schuurman; P. Zandbergen, and M. Brussoni. "Street Masking: A network-based geographic mask for easily protecting geoprivacy." International Journal of Health Geographics 19 (2020): 1–11.

Wang, Zheye; Xinyue Ye, and Ming-Hsiang Tsou. "Spatial, temporal, and content analysis of Twitter for wildfire hazards." Natural Hazards 83 (2016): 523–540.

Xu, Fengli; Zhen Tu; Yong Li; Pengyu Zhang; Xiaoming Fu, and Depeng Jin. "Trajectory recovery from ash: User privacy is not preserved in aggregated mobility data." In Proceedings of the 26th International Conference on World Wide Web, pp. 1241–1250. 2017.

Yu, Wanshu; Haonan Shi, and Hongyun Xu. "A Trajectory K-Anonymity Model Based on Point Density and Partition." arXiv preprint arXiv:2307.16849 (2023).

Zhang, Guiming; Xi Gong, and Di Zhu. "Geographic proximity and homophily effects drive social interactions within VGI communities: An example of iNaturalist." International Journal of Digital Earth 17, no. 1 (2024): 2297948.

Zhang, Shaobo; Xinjun Mao; Kim-Kwang Raymond Choo; Tao Peng, and Guojun Wang. "A trajectory privacy-preserving scheme based on a dual-K mechanism for continuous location-based services." Information Sciences 527 (2020): 406–419.

Zhou, Ryan Zhenqi; Yingjie Hu; Lei Zou; Heng Cai, and Bing Zhou. "Understanding the disparate impacts of the 2021 Texas winter storm and power outages through mobile phone location data and nighttime light images." International Journal of Disaster Risk Reduction 103 (2024): 104339.

Zhu, Huaijie; Xiaochun Yang; Bin Wang; Leixia Wang, and Wang-Chien Lee. "Private trajectory data publication for trajectory classification." In Web Information Systems and Applications: 16th International Conference, WISA 2019, Qingdao, China, September 20–22, 2019, Proceedings 16, pp. 347–360. Springer International Publishing, 2019.

14 Deep Learning-Based Modeling of Human Trajectory Uncertainty in Urban Environments

Zhewei Liu, Yue Yu, Henyu Gu, Kai Yin, Xinghua Cheng, and Ruizhi Chen

14.1 INTRODUCTION: BACKGROUND

Accurate modeling of human trajectory movement uncertainty in urban indoor and outdoor environments is pivotal for enhancing geospatial data mining and urban planning applications. Traditional methods often fall short due to unrealistic assumptions and fixed error models that fail to capture the dynamic and complex nature of urban spaces. This chapter introduces innovative data-driven approaches that leverage deep learning techniques to overcome these limitations and improve uncertainty predictions for both indoor and outdoor trajectories.

Various tracking devices enable recording individuals' movement trajectories in indoor environments. Individuals' indoor movements are continuous and seamless both spatially and temporally, so representing the continuous trajectories with discrete sample points inevitably introduces errors and uncertainty, significantly affecting the reliability of related works such as indoor positioning, trajectory mining, and indoor map reconstruction (Gu et al. 2018; Zhou et al. 2018; Renaudin et al. 2019; Liu et al. 2020; Yu et al. 2021).

This chapter presents a forward-looking vision for the future of trajectory uncertainty modeling, emphasizing the transformative potential of data-driven approaches in this evolving field. By harnessing the power of advanced deep learning techniques, particularly long short-term memory (LSTM) and bidirectional LSTM (Bi-LSTM) architectures, we can overcome many of the limitations inherent in traditional methods. These neural network architectures offer a dynamic and adaptive way to model the uncertainty of human trajectories, providing a more comprehensive and accurate understanding of movement in complex environments. The approaches discussed in this chapter integrate a diverse array of features—such as stride length, heading angles, and time-variant gait characteristics—with global navigation satellite system (GNSS)-derived data. This integration allows for a more nuanced representation of movement uncertainty capable of adapting to the variability of both indoor and outdoor environments. These methods are designed not only to capture the intricate patterns of pedestrian movement but also to significantly enhance the accuracy and reliability of uncertainty predictions.

While our experiments demonstrate the superior performance of these deep learning models through comprehensive case studies, this chapter goes beyond empirical results. It sets the stage for future advancements in uncertainty modeling, highlighting the potential for these techniques to revolutionize the field. By providing reliable and precise uncertainty estimates, these models can support a wide range of smart city applications, from urban planning and navigation systems to enhanced spatial perception capabilities.

In envisioning the future of uncertainty modeling, this chapter advocates for continued innovation in applying machine learning frameworks to urban challenges. As cities become increasingly

DOI: 10.1201/9781003503262-17

complex and dynamic, the need for adaptive, data-driven models will only grow. The approaches outlined here offer a glimpse into a future where uncertainty modeling is not just more accurate, but also more responsive to the ever-changing nature of urban environments.

14.2 CRITICAL REVIEW ON EXISTING METHODS

The traditional methods for modeling trajectory uncertainty have been foundational in spatial analysis, particularly in outdoor environments. However, as the complexity of urban spaces increases, these methods show significant limitations. This section critically examines the existing approaches, highlighting their strengths and weaknesses, and identifying areas where they fall short in addressing modern urban challenges.

14.2.1 The Space–Time Prism and Its Adaptations

One of the most widely used tools for modeling the uncertainty of outdoor trajectories is the space–time prism. Introduced by Miller (1991) and Kwan (1998), the space–time prism defines the potential locations of a moving object within a three-dimensional spatiotemporal space based on two anchor points, the time available for travel, and the object's maximum speed. This model has been instrumental in various applications, including travel time uncertainty (Kuijpers and Othman 2009; Chen et al. 2013; Liao et al. 2014) and probability assessments of visiting specific locations within the prism (Winter and Yin 2010; Song and Miller 2014; Downs et al. 2014; Loraamm 2020).

While the space–time prism has proven useful in outdoor settings, it has limitations when extended to more complex scenarios, particularly in indoor environments. The model's assumptions, such as consistent maximum speed and uniform travel conditions, are often unrealistic in real-world urban spaces, where conditions are far more variable.

14.2.2 Challenges in Modeling Indoor Trajectory Uncertainty

The complexity of indoor environments poses a significant challenge to traditional uncertainty modeling techniques. Indoor spaces are characterized by irregular layouts, obstacles, and variable signal conditions, all of which contribute to increased uncertainty. Traditional methods, such as those proposed by Lu et al. (2016) and Li et al. (2018), attempted to adapt the space–time prism concept by generating elliptical uncertainty regions based on maximum speed and travel time. However, these adaptations still rely on overly simplistic assumptions that fail to capture the nuances of indoor movement.

Li et al. (2019) introduced a more advanced approach using neural networks to estimate location uncertainty based on received signal strength (RSS). While this method represents a step forward in leveraging machine learning, it is still constrained by the static assumptions of traditional models. Indoor environments are highly dynamic, with factors such as obstacles, human movement, and varying signal strengths leading to inconsistent measurement errors and fluctuating speeds—conditions that are not well-represented by existing models.

14.2.3 Limitations of Traditional Models in Dynamic Urban Environments

Traditional models, including the space–time prism and its derivatives, are limited by their reliance on static parameters and assumptions that do not account for the dynamic nature of urban environments. These models often employ fixed buffers, maximum speed constraints, and uniform error distributions, which are insufficient for capturing the variability inherent in real-world settings.

In outdoor environments, factors such as weather, traffic, and terrain can significantly influence movement patterns and trajectory uncertainty. The static nature of traditional models means they often fail to account for these dynamic factors, leading to inaccurate predictions.

The limitations are even more pronounced in indoor settings, where the environment is more complex and variable. The use of fixed assumptions, such as maximum speed and uniform error distribution, does not adequately reflect the variability in signal accuracy and movement patterns. As a result, traditional models often produce significant inaccuracies when applied to real-world scenarios.

14.2.4 The Need for Adaptive and Dynamic Models

The primary shortcoming of traditional methods is their lack of adaptability to the constantly changing nature of urban environments. Urban spaces are dynamic, with new infrastructure, changing movement patterns, and evolving environmental conditions continually altering the landscape. Traditional models are too rigid to accommodate these changes, limiting their effectiveness in dynamic or evolving urban contexts.

To address these limitations, there is a clear need for more advanced models that can dynamically adapt to the complexities of both indoor and outdoor environments. Future research should focus on developing methods that account for the myriad factors influencing movement patterns and uncertainty. This includes integrating real-time data, leveraging machine learning techniques, and incorporating adaptive algorithms that can adjust to changes in the environment.

14.3 PROBLEM STATEMENT AND METHODOLOGY

14.3.1 Problem Statement

This section addresses the core challenges in predicting pedestrian movement uncertainty across large urban areas. It includes the preparation and description of a real-world dataset for training purposes. A thorough and effective deep learning–based methodology is ultimately presented to resolve the identified challenges.

Definition 1: Space–Time Location Point (STP)

An STP is a two-dimensional geospatial point at a specific time, represented as $STP_i = (x,y,t)$, where (x, y) denotes the geospatial location and t the corresponding timestamp, provided by various positioning techniques.

Definition 2: Ground-Truth Trajectory (GT) Provided by SLAM Techniques

The ground-truth trajectory, represented as $GT = \{GT.STP_1, GT.STP_2 \ldots, GT.STP_n\}$, consists of a set of modeled STPs that indicate the pedestrian's reference locations. This trajectory is obtained using a combination of SLAM (simultaneous localization and mapping) systems and open street maps, providing centimeter-level accuracy for error comparison.

Definition 3: GNSS-Based Trajectory (HT) Provided by Hybrid Location Sources

In GNSS-denied areas, human motion detection is used to maintain trajectory continuity. The human motion–assisted GNSS trajectory $HT = \{HT.STP_1, HT.STP_2 \ldots HT.STP_n\}$ approximates an individual's path during temporary GNSS signal loss, where $HT.STP_n$ is the reconstructed trajectory point corresponding to $GT.STP_n$.

14.3.2 Feature Construction for Movement Uncertainty Prediction

The feature construction for movement uncertainty is as follows: (1) actual sampling interval between adjacent GNSS locations T_i and the current GNSS-reported location timestamp; (2) acquired gait length and heading vector between two adjacent GNSS-calculated locations; (3) coordinate differences provided by human motion detection, GNSS, and hybrid values along the x-axis and y-axis; (4) Euclidean distance between adjacent GNSS-based locations and gait/heading-based locations; (5) estimated GNSS-originated speed and location observations and gait/heading-originated observations; (6) virtual heading observations calculated from gait, heading information, and GNSS locations; and (7) average signal-to-noise ratio (SNR) of GNSS signals.

14.3.3 Proposed Model

This study employs a hybrid Bi-LSTM and multilayer perceptron (MLP) network for autonomous prediction of uncertainty errors using the constructed trajectory features. The LSTM-based network effectively predicts future data based on past data, making it suitable for continuous pedestrian trajectory prediction. However, LSTM units have limitations as they cannot use future data for prediction. To address this, a Bi-LSTM model is proposed, consisting of two LSTM layers with opposite output directions, enabling the use of both past and future information. The experiments compare the performance of Bi-LSTM and LSTM networks. The trajectory features described in the previous section are modeled as the input vector for the Bi-LSTM network, illustrated in Figure 14.1.

In the proposed Bi-LSTM model, the parameters of a single LSTM unit are as follows:

$$\begin{cases} f_t = \sigma \cdot \left(W_f \cdot h_{t-1} + U_f \cdot X_t + b_f\right) \\ i_t = \sigma \cdot \left(W_i \cdot h_{t-1} + U_i \cdot X_t + b_i\right) \\ a_t = tanh\left(W_a \cdot h_{t-1} + U_a \cdot X_t + b_a\right) \\ \tilde{C}_t = \tilde{C}_{t-1} \cdot f_t + i_t \cdot a_t \\ o_t = \sigma \cdot \left(W_o \cdot h_{t-1} + U_o \cdot X_t + b_o\right) \\ h_t = o_t \cdot tanh\left(C_t\right) \\ \widehat{y}_t = \sigma\left(Vh_t + c\right) \end{cases} \tag{14.1}$$

where i_t, f_t, and o_t represent the input, forget, and output units, respectively; x_t is the modeled input vector at each phase t according to the trajectory features; and h_t is the hidden state vector, which is the network's overall output. σ indicates the sigmoid function, and c_t is the memory state at time t, with the candidate vector mixed with the output vector.

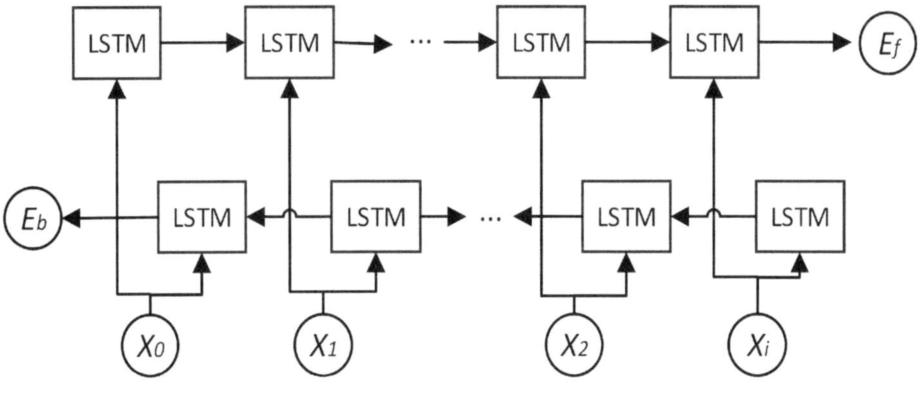

FIGURE 14.1 Structure of proposed Bi-LSTM network.

The output layer of the proposed Bi-LSTM units is used as the input vector for the MLP network:

$$\hat{E}_i = MLP(y_i) \tag{14.2}$$

Finally, the movement uncertainty error $\hat{E}_i$ is presented as the deviation between the acquired GNSS location and the ground-truth reference location $Dis(GT.\ STP_i,\ HT.\ STP_i)$.

14.4 EMPIRICAL VALIDATION

14.4.1 Data Collection

Our dataset was collected on the open campus of the Hong Kong Polytechnic University, where GNSS signals are disrupted by reflection and occlusion, leading to time-varying sampling errors and significant measurement inaccuracies, as shown in Figure 14.2. Ground-truth trajectories were obtained using a SLAM system with centimeter-level accuracy. The GNSS-based trajectory (HT) includes human movements and locations recorded by various smartphones. To reference the ground-truth trajectory, a SLAM system combining high-precision light detection and ranging (LiDAR), inertial sensors, and visual-based location data was employed

To evaluate our trained model's performance, 70% of the 7,954 GNSS points were used for model training, and the remaining 30% were used as the test dataset. The dataset statistics are shown in Table 14.1.

FIGURE 14.2 Experimental environments for data collection.

TABLE 14.1
Statistics of the Experiment Dataset

Number of Trajectories	Total Number of Recorded Points	Average Trajectory Length (m)	Maximum Speed (m/s)	Average Speed (m/s)	Average Sampling Interval (s)
15	7,954	981.7	55.6	1.7	1.1

The resulting dataset consists of over 15 trajectories and 7,954 spatiotemporal points (Table 14.1). The trajectories feature varying real-time sampling intervals due to complex environments, with an average sampling rate of 1.1 Hz using smartphones.

In the proposed Bi-LSTM framework, Adam was used as the optimizer due to its effectiveness with large training datasets. The input vector dimension was set to 14, and the hidden state's dimension to 80. Each training phase used a sequence of 20 continuous vectors as the input vector. Sensitivity analysis showed these parameters accurately represent the model's average performance.

14.4.2 Baseline Models and Evaluations

This study compares the proposed model with three baseline models:

- *UB*: A classic model creating elliptic uncertainty regions based on trajectories' starting/ending points, moving time, and maximum speed (Liu et al. 2022; Shi et al. 2022)
- *AUB*: Introduced the concept of "approximate upper bound distance" to limit the elliptic region size (Yu et al. 2023a)
- *BAEE*: Extends AUB using the Minkowski distance metric (Yu et al. 2023b)

For UB, the maximum speed of UB is set as 55.6 m/s with reference to Table 14.1, and the p value for BAEE is set as 1.5.

To quantitatively compare the performance of the existing models with our proposed Bi-LSTM approach, we used the following metrics: (1) coverage ratio—this metric indicates the proportion of ground-truth trajectories encompassed by the uncertain regions, assessing the regions' completeness; and (2) mean absolute error (MAE)—this metric measures the average error between the predicted trajectory uncertainty and the actual trajectory uncertainty. It evaluates the precision of the methods by comparing the ground-truth trajectory points with the corresponding GNSS-based trajectory points.

14.4.3 Performance Evaluation

The uncertain regions for our method and the baseline models were generated based on 15 experimental trajectories. In our model, these uncertain regions are constructed by creating a series of circles, each centered on a point from the GNSS-based trajectory with the radius representing the predicted uncertainty error. Figure 14.3 depicts the uncertain regions produced by our method on several sample trajectories.

We evaluated our method against baseline models for each trajectory. Figure 14.4 shows the uncertain regions generated by each method, highlighting variations in shape and coverage. These regions aim to capture the spatial area where ground-truth trajectories are likely to be found, making full coverage crucial. Table 14.2 presents the coverage ratios for each method, revealing that our method significantly outperforms baselines like BAEE and AUB. Our method achieves a coverage ratio exceeding 95% for most trajectories, with an average of 98.1%, compared to BAEE's 68.2% and AUB's 38.2%. While UB also performs well (97.9%), our method slightly surpasses it, demonstrating superior coverage of ground-truth trajectories.

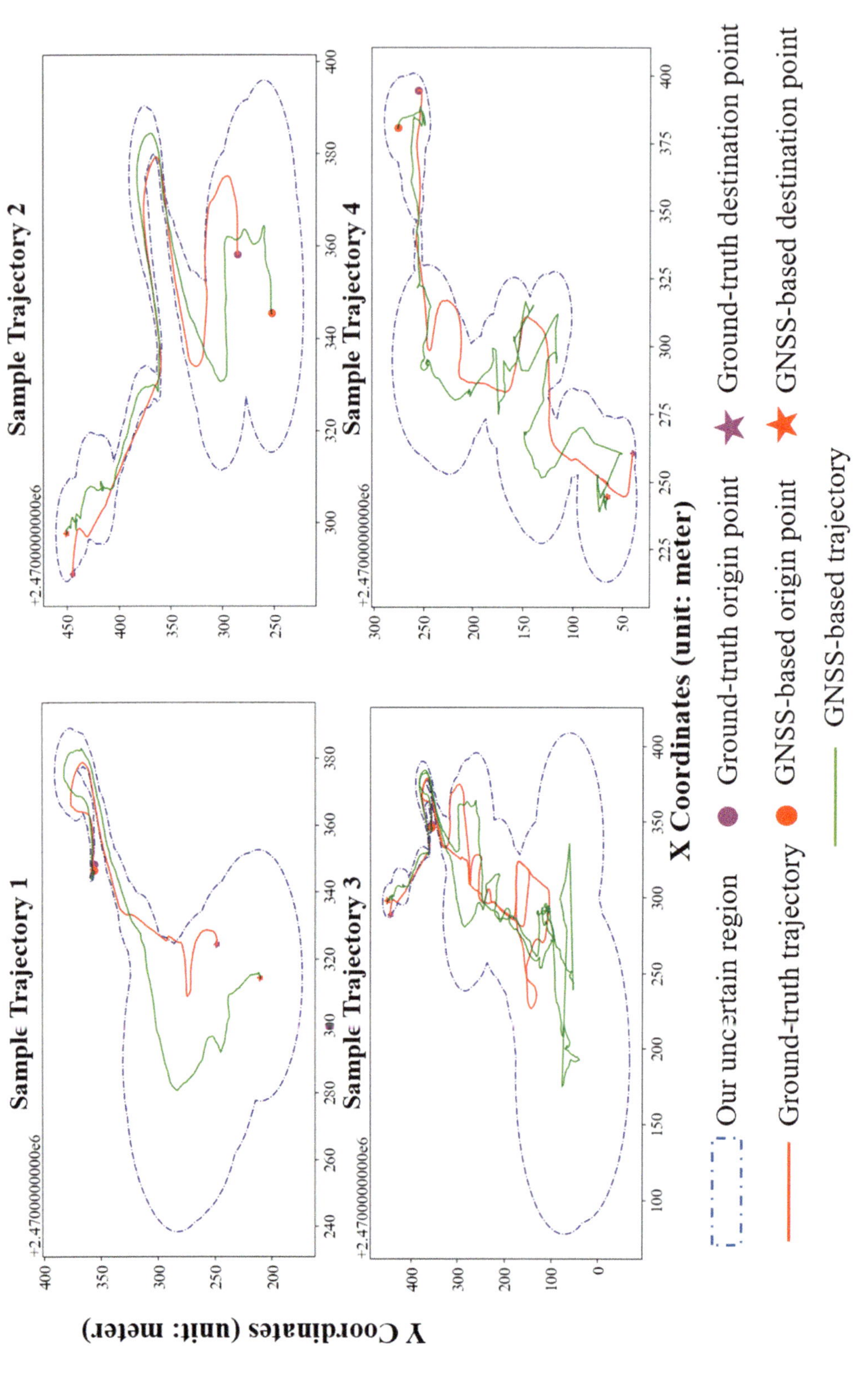

FIGURE 14.3 The uncertain regions generated by our methods on several sample trajectories.

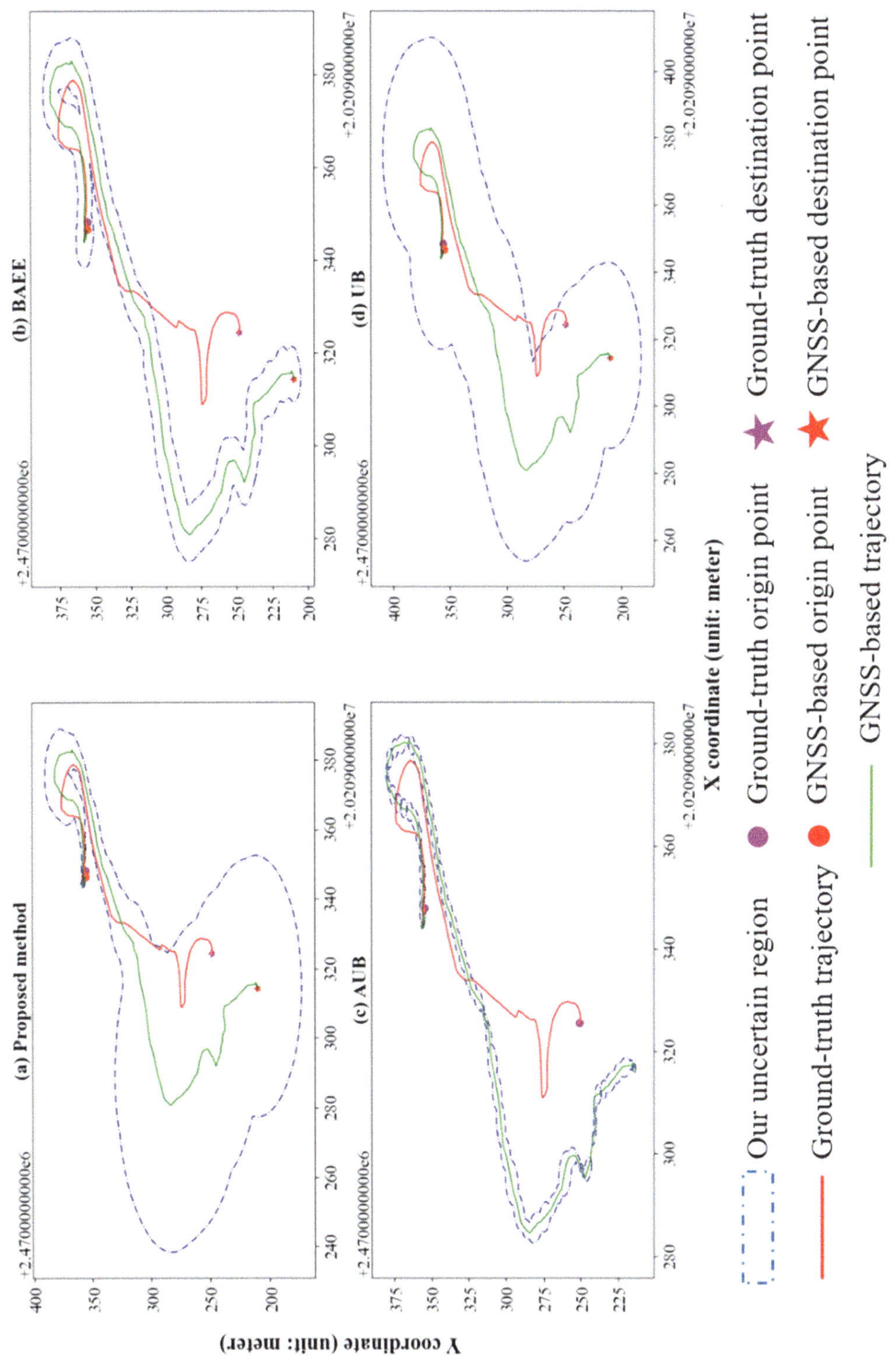

FIGURE 14.4 A snapshot of the uncertain regions generated by (a) our proposed method, (b) BAEE, (c) AUB, and (d) UB on one sample trajectory.

TABLE 14.2
Coverage Ratio of Respective Method on Each Trajectory

	Coverage Ratio			
	Proposed Method	BAEE	AUB	UB
1	94.7%	59.6%	70.2%	88.4%
2	97.5%	21.7%	23.8%	100%
3	100%	29.7%	35.2%	81.7%
4	83.8%	56.9%	61.0%	100%
5	96.8%	50.2%	15.2%	100%
6	100%	74.3%	36.0%	100%
7	97.9%	83.8%	64.9%	94.0%
8	100%	66.8%	51.5%	100%
9	97.5%	67.7%	52.3%	100%
10	100%	82.9%	59.5%	100%
11	98.2%	78.2%	23.6%	100%
12	100%	87.4%	46.7%	100%
13	99.0%	34.5%	48.7%	100%
14	99.5%	57.8%	32.8%	100%
15	99.3%	82.1%	46.9%	100%
Average	98.1%	68.2%	38.2%	97.9%

The coverage ratios of the methods are influenced by the size of the uncertain regions. Larger regions may cover more ground-truth trajectory points, increasing completeness but reducing precision in indicating the actual locations of the trajectories. An ideal uncertain region balances completeness and precision. Table 14.3 shows that while UB achieves a high coverage ratio due to larger uncertain regions, it is less precise, resulting in higher MAE. Our

TABLE 14.3
Mean Absolute Error (MAE) of Respective Method on Each Trajectory

	MAE (m)			
Trajectory	Proposed Method	BAEE	AUB	UB
1	0.95	13.36	14.13	37.30
2	2.63	54.41	54.41	21.62
3	1.69	50.92	51.78	15.52
4	1.34	7.10	7.56	44.20
5	1.58	31.95	32.39	29.23
6	1.31	7.01	9.39	43.15
7	2.53	12.05	13.41	47.09
8	2.32	12.05	10.98	41.90
9	2.15	7.57	8.73	47.96
10	2.10	9.63	10.61	44.80
11	1.23	4.10	5.70	45.83
12	1.83	7.95	8.86	44.97
13	1.86	10.04	11.28	43.62
14	1.98	8.16	9.63	47.22
15	1.73	8.25	9.66	45.68
Average	1.82	16.30	17.23	40.00

method, however, achieves a significantly lower MAE (1.82 on average) compared to the baselines, demonstrating better accuracy in predicting the uncertainty error between GNSS-based and ground-truth trajectories. This balance between completeness and precision is crucial. An overly large uncertain region may give a false sense of high coverage but lack precision, which is vital for applications like navigation and urban planning. Our model's ability to maintain high coverage while achieving low MAE shows it effectively balances these aspects, providing actionable insights.

Practically, our Bi-LSTM-based approach can enhance GNSS-based applications in urban environments. It can improve public transportation by optimizing routes and schedules and provide more reliable navigation directions. Its low MAE also suggests robustness across various urban contexts, making it adaptable to different urban landscapes, from dense cities to open areas.

14.5 ENHANCEMENTS AND FUTURE DIRECTIONS IN DEEP LEARNING-BASED TRAJECTORY UNCERTAINTY MODELING

The Bi-LSTM-based framework developed for human movement uncertainty modeling represents a significant step forward in understanding and predicting the complexities of crowdsourced trajectories. However, as urban environments become increasingly dynamic and mobile applications continue to evolve, several limitations of this framework must be addressed to ensure its applicability and robustness in real-world scenarios. These challenges highlight important avenues for future research and development.

14.5.1 Seamless Indoor–Outdoor Localization

One of the primary challenges in trajectory uncertainty modeling lies in seamlessly integrating indoor and outdoor localization. Human movement trajectories often span environments where GNSS signals vary significantly in strength or are entirely unavailable, particularly indoors. To overcome this, future research should focus on developing hybrid localization models that can seamlessly transition between indoor and outdoor environments. This will require a more sophisticated uncertainty prediction structure that can accommodate the unique characteristics of each environment. Techniques such as sensor fusion, where data from various sensors (e.g., inertial sensors, Wi-Fi, Bluetooth, and LiDAR) are combined, can play a crucial role in enhancing the accuracy and continuity of localization. Moreover, advancements in indoor positioning systems, such as the use of ultra-wideband (UWB) technology or vision-based localization, could further bridge the gap between indoor and outdoor trajectory tracking.

14.5.2 Enhanced Feature Incorporation and Data Enrichment

The precision of uncertainty predictions is heavily dependent on the richness of the input features provided to the deep learning model. Currently, the model's performance is constrained by the limitations of the available dataset, which may not fully capture the diversity of human movement patterns. Future research should explore the incorporation of additional trajectory-related features, such as environmental context (e.g., weather conditions, time of day, and urban density) and behavioral factors (e.g., walking speed, stop-and-go behavior). These features could be derived from various data sources, including Internet of Things (IoT) devices, social media activity, and urban sensor networks. Additionally, the development of synthetic data generation techniques could help augment the dataset, providing a more comprehensive representation of potential movement scenarios. This would not only improve the accuracy of the final uncertainty predictions but also enhance the model's ability to generalize to new and unseen environments.

14.5.3 Unsupervised and Semi-supervised Learning Techniques

The challenge of obtaining accurate ground-truth labels, particularly for indoor trajectories, limits the effectiveness of supervised learning approaches in uncertainty modeling. To address this, future research should investigate the application of unsupervised and semi-supervised learning techniques. These methods could enable the model to autonomously learn important features from unlabeled or partially labeled data, reducing the dependency on extensive labeled datasets. For instance, clustering techniques could be used to identify patterns in the data, which could then be used to train models without requiring explicit labels. Additionally, self-supervised learning, where the model is trained on pretext tasks that do not require manual labeling, could provide a way to learn useful representations from large amounts of unlabeled data.

14.5.4 Adapting State-of-the-Art Natural Language Processing Models for Trajectory Prediction

The remarkable success of transformer-based models in natural language processing (NLP) suggests new possibilities for improving trajectory uncertainty modeling. Both tasks involve sequence prediction, where the goal is to predict the next element in a sequence based on prior elements. By adapting models like GPT (Generative Pre-trained Transformer) or BERT (Bidirectional Encoder Representations from Transformers) to the domain of trajectory prediction, researchers could significantly enhance the model's ability to capture complex temporal dependencies and spatial patterns. This could involve developing trajectory-specific embeddings that represent the spatial and temporal characteristics of movement data, followed by fine-tuning these embeddings within transformer architectures. Additionally, integrating attention mechanisms, which allow the model to focus on relevant parts of the input sequence, could further improve the accuracy of uncertainty predictions by highlighting key features in the trajectory data.

14.5.5 Integration of Diverse Location Sources

Urban environments are characterized by a wide array of location data sources, including radiofrequency signals (e.g., Wi-Fi, Bluetooth), local magnetic fields, and visual landmarks. Future research should explore the integration of these diverse sources into the trajectory uncertainty modeling framework. By leveraging data from multiple sources, it becomes possible to compensate for the limitations of any single source, such as the unreliability of GNSS signals in dense urban canyons. This multisource approach could involve the development of sensor fusion algorithms that intelligently combine data from various inputs, taking into account their respective accuracies and uncertainties. Additionally, exploring the potential of emerging technologies, such as 5G networks and edge computing, could provide new opportunities for real-time trajectory tracking and uncertainty estimation in complex urban settings.

14.6 CONCLUSION

This chapter has explored the significant challenges and emerging opportunities in modeling human trajectory uncertainty in urban environments. While traditional methods, such as the space-time prism, have laid the groundwork for this field, their reliance on static assumptions and simplistic error models has proven inadequate for the dynamic and complex nature of modern urban spaces. Our work demonstrates that by leveraging advanced deep learning techniques, particularly the Bi-LSTM-based framework, we can significantly enhance the accuracy and reliability of uncertainty predictions.

Our experiments highlight the effectiveness of these deep learning models in capturing the intricate patterns of pedestrian movement. For instance, our Bi-LSTM-based approach achieved an impressive average MAE of 1.82 meters and a coverage ratio of 98.1% across various trajectories, significantly outperforming traditional models such as BAEE, AUB, and UB. These results underscore the potential of our approach to provide precise and reliable uncertainty estimates, which are crucial for applications in smart cities, such as urban planning, navigation, and public transportation optimization.

However, this chapter extends beyond the empirical findings, offering a vision for the future direction of uncertainty modeling. We advocate for continued innovation in this field, particularly in areas such as the seamless integration of indoor and outdoor localization, the incorporation of enriched and context-aware features, and the exploration of unsupervised and semi-supervised learning techniques. These advancements are essential for creating models that can dynamically adapt to the ever-changing conditions of urban environments.

Moreover, the potential to adapt state-of-the-art models from natural language processing (NLP), such as transformer-based architectures, to trajectory prediction presents a promising direction for future research. These models, which excel in sequence prediction tasks, could further enhance the adaptability and precision of uncertainty modeling, making them highly suitable for complex urban scenarios.

In conclusion, this chapter not only presents a robust framework for modeling human trajectory uncertainty but also charts a course for future advancements in the field. By embracing innovative, data-driven approaches and continuously refining our methodologies, we can significantly improve our ability to predict and manage uncertainty in urban environments, ultimately contributing to the development of smarter, more resilient cities. The results of our experiments serve as a testament to the effectiveness of these approaches, providing a strong foundation for future research and application in this critical area.

REFERENCES

Chen, B. Y., *et al.* 2013. Reliable space–time prisms under travel time uncertainty. *Annals of the Association of American Geographers*, 103(6), 1502–1521.

Downs, J. A., *et al.* 2014. Quantifying spatio-temporal interactions of animals using probabilistic space–time prisms. *Applied Geography*, 55, 1–8.

Gu, Y., *et al.* 2018. Trajectory estimation and crowdsourced radio map establishment from foot-mounted IMUs, Wi-Fi fingerprints, and GPS positions. *IEEE Sensors Journal*, 19(3), 1104–1113.

Kuijpers, B. and Othman, W. 2009. Modeling uncertainty of moving objects on road networks via space–time prisms. *International Journal of Geographical Information Science*, 23(9), 1095–1117.

Kwan, M. P. 1998. Space-time and integral measures of individual accessibility: A comparative analysis using a point-based framework. *Geographical Analysis*, 30(3), 191–216.

Li, H., *et al.* 2018. In search of indoor dense regions: An approach using indoor positioning data. *IEEE Transactions on Knowledge and Data Engineering*, 30(8), 1481–1495.

Li, Y., *et al.* 2019. Wireless fingerprinting uncertainty prediction based on machine learning. *Sensors*, 19(2), 324.

Liao, F., Rasouli, S. and Timmermans, H. 2014. Incorporating activity-travel time uncertainty and stochastic space–time prisms in multistate supernetworks for activity-travel scheduling. *International Journal of Geographical Information Science*, 28(5), 928–945.

Liu, T., *et al.* 2020. A simple positioning system for large-scale indoor patrol inspection using foot-mounted INS, QR code control points, and smartphone. *IEEE Sensors Journal*, 21(4), 4938–4948.

Liu, Z., Shi, W., Yu, Y., Chen, P. and Chen, B. Y. 2022. A LSTM-based approach for modelling the movement uncertainty of indoor trajectories with mobile sensing data. *International Journal of Applied Earth Observation and Geoinformation*, 108, 102758.

Loraamm, R. W. 2020. Incorporating behavior into animal movement modeling: A constrained agent-based model for estimating visit probabilities in space-time prisms. *International Journal of Geographical Information Science*, 34(8), 1607–1627.

Lu, H., *et al.* 2016. Finding frequently visited indoor POIs using symbolic indoor tracking data. In Proceedings of the 19th International Conference on Extending Database Technology (EDBT) (pp. 449–460). OpenProceedings.org. https://doi.org/10.5441/002/edbt.2016.41

Miller, H. J. 1991. Modelling accessibility using space-time prism concepts within geographical information systems. *International Journal of Geographical Information System*, 5(3), 287–301.

Renaudin, V., *et al.* 2019. Evaluating indoor positioning systems in a shopping mall: The lessons learned from the IPIN 2018 competition. *IEEE Access*, 7, 148594–148628.

Shi, W., Yu, Y., Liu, Z., Chen, R. and Chen, L. 2022. A deep-learning approach for modelling pedestrian movement uncertainty in large-scale indoor areas. *International Journal of Applied Earth Observation and Geoinformation*, 114, 103065.

Song, Y. and Miller, H. J. 2014. Simulating visit probability distributions within planar space-time prisms. *International Journal of Geographical Information Science*, 28(1), 104–125.

Winter, S. and Yin, Z.-C. 2010. Directed movements in probabilistic time geography. *International Journal of Geographical Information Science*, 24(9), 1349–1365.

Yu, Y., *et al.* 2021. Autonomous 3D indoor localization based on crowdsourced Wi-Fi Fingerprinting and MEMS sensors. *IEEE Sensors Journal*, 22(6), 5248–5259.

Yu, Y., Shi, W., Liu, Z., Tang, K., Chen, L. and Chen, R. 2023b. LSTM-MLP based uncertainty modelling approach for complex human indoor trajectory. *ISPRS Annals of the Photogrammetry, Remote Sensing and Spatial Information Sciences*, 10, 525–532.

Yu, Y., Yao, Y., Liu, Z., An, Z., Chen, B., Chen, L. and Chen, R. 2023a. A Bi-LSTM approach for modelling movement uncertainty of crowdsourced human trajectories under complex urban environments. *International Journal of Applied Earth Observation and Geoinformation*, 122, 103412.

Zhou, B., *et al.* 2018. A graph optimization-based indoor map construction method via crowdsourcing. *IEEE Access*, 6, 33692–33701.

15 Toward the Spatial Evolution Between Distribution Snapshots

A Network Perspective

Di Zhu and Zhongfu Ma

15.1 INTRODUCTION

15.1.1 Concept of Spatial Evolution

The evolution of geographical entities and systems has long been studied in diverse domains, including ecology, geomorphology, economy, epidemiology, and urban science, among others (Stoddart, 1966; Ostfeld et al., 2005; Batty, 2013; Boschma and Frenken, 2018). A large body of work has been devoted to representing and modeling the evolution as the continuous process of change that leads to a series of transformations, such as the continual variation of landscape and ecosystem (Tucker and Hancock, 2010; Moreno-Mateos et al., 2020), the development of industrial structures (Belussi and Gottardi, 2018), the spatial spreading of infectious diseases (Balcan et al., 2009), the dynamics of human activity (Ratti et al., 2006; Kang et al., 2010), and so on. The studies of these evolutionary phenomena provide crucial insight for understanding the physical and biological processes that collectively shape the natural landscape, informing strategies for sustainable economic growth, pandemic control, and urban planning.

As one of the most common ways to depict the progress of evolution, a temporal snapshot represents the state of affairs in a particular domain at a single moment in time, and one can think of snapshots as sampling dynamic phenomena at a sequence of temporal instants (Worboys 2005). Given the increasing volumes of spatiotemporal data, the evolution of geographical entities has been captured by various forms of snapshots, such as time-series remote sensing images, historical census data, temporal maps of disease cases, hourly mobile phone data, and many more. These snapshots represent the spatial distribution of land cover (Hibbard et al., 2010), socioeconomic characteristics (Anderson 2015), pandemic status (Kamel Boulos and Geraghty, 2020), and human activities (Louail et al., 2014). The evolution is then uncovered by comparing sequential snapshots one after another and tracking the changes in attributes over time.

Underlying the transformation between two snapshots, the changes in the distribution of geographic attributes are contributed by the process that involves spatial interactions (Fotheringham, 2001; Roy and Thill, 2003) among locations. These processes, such as the movement of population, virus, commodities, and information, led to the reorganization of mobile entities within the space, thereby transforming the distributions of geographic attributes (Ratti et al., 2010; Barthélemy 2011; Bartosz Hawelka et al., 2014 and Ratti 2014). We define such a process as spatial evolution, a detailed description of how the current snapshot is produced based on the previous snapshot. However, as discussed in Worboys (2005), the snapshot metaphor offers no mechanism for explicitly representing the time or occurrence of events. This limitation causes the information on spatial evolution to be completely lost when we sample the continuous evolution process as discrete snapshots.

DOI: 10.1201/9781003503262-18

Therefore, to understand the mechanisms driving the transformation between snapshots, it is necessary to find alternatives for modeling spatial evolution and inferring spatial evolution from changes between snapshots.

15.1.2 Modeling Spatial Evolution with Networks

The studies of spatial network (Barthélemy 2011) provide a solid foundation for extending network science to model spatial evolution, including the representation, analysis, and visualization of spatial evolution. Spatial networks have been widely used to characterize spatial interactions, namely spatial interaction networks. In these networks, the flows of passengers (Zhong et al., 2014), migration (Liu et al., 2014), currency, and trade among locations are represented by weighted edges among nodes that measure the strength of spatial interaction. Although flows in these spatial networks imply a dynamics of change, because it takes time for a flow to occur, the focus of most studies has been very much on space (Batty 2013), such as analyzing the attraction of places, the distance decay effect, and the spatial structure of a region. To represent spatial evolution, the hidden dynamic process in a spatial network should be explicitly described. Therefore, a spatial evolutionary network based on two spatial distribution snapshots can be introduced as an extension to the family of spatial networks. In this network, each edge represents the spatial evolutionary characteristics between two nodes (locations), weighted by the changes of spatial interaction volume in geographic attributes between two locations.

Modeling spatial evolution using the spatial evolutionary network allows us to take advantage of the analytical and visualization tools from network science. Classic network measures such as degree, closeness, and betweenness are commonly used to quantify the node centrality, which can reveal the sources and sinks in the evolution process (Kimbrell 2010). Network partition methods, such as community detection (Fortunato 2010) and hierarchical clustering (Yiu and Mamoulis, 2004), shed light on the spatial organization of the evolution process. To display spatial evolution in geographic space, network-based representation enables us to visualize the process via flow maps (Guo 2009), with various flow clustering methods that facilitate effective visual analysis. Last but not least, inferring the spatial evolutionary network from widely accessible snapshot data offers a more practical alternative for obtaining population-level movement data (Zhu et al., 2018) compared to the time-consuming workflow of collecting individual movement data and aggregating it across locations.

This chapter introduces the concept of spatial evolution, explores the potential of modeling the spatial evolution process using networks, and reviews methods for inferring spatial evolutionary networks. In Section 15.2, we discuss the current state-of-the-art methods for inferring spatial evolutionary networks and potential applications in this task. Then, in Section 15.3, we provide two examples to illustrate the modeling and inference of spatial evolution: one on national migration between population distribution snapshots and another on the spatial shift of the COVID-19 pandemic between snapshots of disease case distributions. Finally, in Section 15.5, we conclude this chapter by discussing future challenges and research directions in the study of spatial evolution.

15.2 RELATED METHODS

15.2.1 Network Optimization with Spatial Constraints

Assume that the time interval between two consecutive spatial distribution snapshots can ensure that only one spatial process exists to directly drive the observed temporal evolution of spatial distributions (Swihart and Slade, 1985). Zhu et al. (2018) proposed that the decision strategy for determining the underlying spatial interaction between spatial distribution snapshots, such as the migration flows underlying temporal population snapshots, is similar to the minimum-cost flow problem (Ahuja et al., 1993; Cook 1997; Holzhauser et al., 2016) in network optimization. The goal of this optimization is to find the minimum-cost flow configuration of a network with the requirement of conservation of flows at all nodes. The inferred spatial interaction pattern is a form of spatial evolution because it explicitly reflects the process that induces the change between spatial distribution snapshots. Each spatial unit has a potential bidirectional process with every other unit, and thus, a fully connected network can be constructed as the spatial evolutionary network. To determine a global evolution process from the many possible processes that match the observed temporal evolution of distributions, we can model the problem as a network optimization task. This task minimizes the total cost of flows that cause changes between spatial distribution snapshots, constrained by spatial configuration and the distance friction of human movement.

The network optimization for spatial evolution inference can be formalized as a linear program. Given two consecutive snapshots of population distributions, S_1 at time t_1 and S_2 at time t_2, we formalize S_1 and S_2 as

$$\begin{aligned} S_1 &= \left[s_1, s_2, \ldots, s_i, \ldots, s_n\right] \\ S_2 &= \left[s_1^*, s_2^*, \ldots, s_i^*, \ldots, s_n^*\right] \end{aligned} \tag{15.1}$$

where the study area consists of a set N of n spatial units. s_i and s_i^* are the observed geographic attribute (e.g., population) in unit $i \in \mathrm{N}$ at S_1 and S_2, respectively. The fully connected spatial evolutionary network among all units can be represented by an $n \times n$ matrix $F = \left\{f_{ij}\right\}$. f_{ij} is the flow from unit i to unit j $(i, j \in N)$ during the evolution from S_1 to S_2, without considering intra-unit interaction. Meanwhile, $n \times n$ matrix $C = \left\{c_{ij}\right\}$ is used to describe the movement costs among the spatial units. Derived from Newton's law of gravity, a gravity-based cost matrix is defined, where the unit cost of an interaction between unit i to unit j is $C_{ij} \propto \frac{d^\beta}{A_i A_j}$. A_i and A_j are the attractions of unit i and unit j, respectively, which are usually measured by population size. d^β is the distance friction function (Toole et al., 2012).

The cost matrix is assumed to be symmetric, and the entries of the main diagonal are all zero; that is, $\forall i, j \in N$, $c_{ij} = c_{ji}$, and $c_{ii} = 0$. Z is the variable representing the objective function. Then, the linear program can be constructed as follows:

$$\begin{aligned} &\text{minimize } C^T \times F \\ &\text{subject to } s_i - \sum_{j \in N} f_{ij} + \sum_{j \in N} f_{ji} = s_i^*, \forall i \in N \\ &f_{ij} \in Z^n, \forall i, j \in N. \end{aligned} \tag{15.2}$$

The solution for F can be derived by the simplex method, if the optimal basic solution does exist. Details including solving the linear program with simplex tableau and synthetic experiments of network optimization method can be found in Zhu et al. (2018). To the best of our knowledge, the above-mentioned network optimization method is one of the few exploratory works that can be directly applied in spatial evolution inference from a network perspective.

15.2.2 Markov Chain for Origin-Destination Matrix Inference

Spatial evolution can be described as a stochastic process because it aggregates the movement of individuals, inherently involving the uncertainty due to free will and arbitrariness in individuals' actions (Barbosa et al., 2018). This nature of uncertainty allows spatial evolution to be inferred by statistical methods. The Markov-chain Monte Carlo method has been widely used for inferring the origin-destination (O-D) matrix in networks of public transit route (Li 2009; Ji et al., 2015; Amblard et al., 2022). To demonstrate the connection between O-D matrix inference and spatial evolution inference, we present an example of O-D matrix inference for a bus transit route. In a bus transit route with N stops, the observed number of passengers boarding and aligning at stop i is denoted as y_i and z_i ($i = 1,2,3\ldots,N$), respectively. x_{ij} denotes the counts of passengers boarding at stop i and alighting at stop j. These observations can be considered as two snapshots of passenger distribution. The inferred O-D matrix $X = \{x_{ij}\}_{N\times N}$ depicts the expected volume of passengers traveling between any two traffic stops, capturing the transition from y_i to z_i. Therefore, this O-D matrix can be viewed as a representation of the spatial evolutionary network.

Rather than directly inferring the passenger flows that cause changes between distribution snapshots as a spatial evolution process, the Markov chain model characterizes the probability of a passenger traveling between any two stops, known as Markov transition probabilities. Consider the behaviors of a passenger boarding at one stop; there are only two options, aligning at the next stop or not. Therefore, let q_i be the probability that a passenger alights at stop i given that they are on board at stop $i-1$; then $1-q_i$ will be the probability that a passenger does not alight at stop i and keeps traveling. These two probabilities collectively define the Markov transition probabilities at one stop. Using the transition probabilities, the probabilities of passengers alighting at stop j given that they boarded at stop i, denoted as p_{ij}, can be calculated with the properties of Markov chains. For example, the probability of a passenger traveling from stop 1 to stop 3 can be calculated as follows:

$$\begin{aligned} p_{13} &= \Pr\{\text{alighting at stop } 3 \mid \text{boarding at stop } 1\} \\ &= \Pr\{\text{alighting at stop } 3, \text{not alighting at stop } 2 \mid \text{boarding at stop } 1\} \\ &= q_3\left(1-q_2\right). \end{aligned} \tag{15.3}$$

where $1-q_2$ is the probability of a passenger boarding at stop 1 and not aligning at stop 2. q_3 is the probability of a passenger boarding at stop 2 and aligning at stop 3. Given the probability p_{ij} and assuming all the passengers boarding at stop i share the same p_{ij}, the expected passenger flow $\hat{x}_{ij}$ that reflects the evolution process from i to j can be inferred as follows:

$$\hat{x}_{ij} = p_{ij}y_i\left(i = 1,\ldots,N-1; j = i+1,\ldots,N\right). \tag{15.4}$$

15.2.3 Deep Learning–Based Flow Generation Model

The aforementioned methods offer two perspectives of inferring spatial evolution only based on information from snapshots. The development of deep learning (DL) approaches enables researchers to uncover the nonlinear relationship between geographical features and human mobility, which also offers unprecedented opportunities to reveal how features, such as land use and the diversity of points of interest (POIs), influence the spatial evolution process. The DL-based flow generation models (Luca et al., 2021) have a great potential for spatial evolution inference These models aim to generate a network of realistic flows among locations given the geographical features and the distance among locations, without any historical information about the real flows. However, the commonly used flow generation models, such as the Deep Gravity model (Simini et al., 2021), cannot be directly applied to spatial evolution inference because they approach mobility patterns from a static view and the generated flow pattern cannot explicitly reflect the dynamics of spatial evolution processes.

Emerging studies (Liu et al., 2020; Rong et al., 2023; Shi et al., 2024) that generate flows by modeling supply and demand characteristics with DL approaches may shed light on inferring spatial evolution. These flow generation models assume that flows are determined by the supply characteristics of the origin geographical unit and the demand characteristics of the destination geographical unit. In terms of spatial evolution, the process can be determined by the initial state at the origin and the final state at the destination. This similarity allows us to leverage the strategy of capturing supply and demand characteristics separately from geographical features used in these flow generation models. For example, Liu et al. (2020) employ two separate graph attention networks (GATs) to encode the geographical features into two different embedding spaces. Two GATs have the same structure, but the attention mechanism in GAT will assign different weights to different features based on the origin or destination roles, thus modeling supply and demand characteristics. This paralleling structure can be adopted in spatial evolution inference by extracting the characteristics of the initial and final state from two snapshots using two GATs, respectively.

15.3 EXAMPLES

In Section 15.2, we reviewed potential methods for spatial evolution inference. Among these methods, network optimization has already demonstrated its feasibility in different applications. Here we provide two examples of how network optimization can be used to infer spatial evolution and how these inferred processes facilitate understanding the change between snapshots. Both examples follow a network optimization–based framework named inferring interactions from distribution snapshots (IIDS) (Zhu et al., 2018). Section 15.3.1 showcases the inference of inferring interurban flows of migrants as the spatial evolution underlying the transformation between population distributions during the Chinese Spring Festival. Section 15.3.2 illustrates how spatial evolutionary networks, inferred from the snapshots of COVID-19 confirmed case distribution, can help reveal the spatial shift of the pandemic risks.

15.3.1 Inferring Interurban Human Migration

15.3.1.1 Construction of Population Distribution Snapshots

A positioning dataset, covering mainland China, is collected from Tencent, a major Chinese Internet company. This dataset contains the daily number of active users in each 0.01° × 0.01° geographic cell collected from related Tencent Mobile Apps for 28 consecutive days (January 16 to February 12, 2016). More than 38.4 billion user records are obtained in total, representing an average of 1.37 billion records per day, which is roughly equal to China's population. Two subdatasets are extracted from the original dataset and generate two snapshots (Figure 15.1) of the population as the input data for the IIDS method. The first snapshot is the distribution of the average number of daily active users from January 18 to January 22, which can be considered as a reflection of the population distribution during a normal working week (Figure 15.1a). The second snapshot is the distribution of the average number of daily active users from February 7 to February 11 (February 7, 2016, is New Year's Eve), which reflects the population distribution when people are enjoying the Spring Festival (Figure 15.1b).

15.3.1.2 Analysis of Inferred Spatial Evolution

The inferred spatial evolution process using a gravity model–based cost matrix with the power law distance friction parameter $\beta = 1.0$ is illustrated in Figure 15.2. The inferred number of aggregated flows is 364, with more than 335 million individual migrations in total. As expected, four urban agglomerations in China can be easily found (Figure 15.2), that is, the Yangtze River Delta urban agglomeration in East China (Shanghai, Suzhou, and Hangzhou as core cities), the Beijing–Tianjin–Hebei area in North China, the Pearl River Delta urban agglomeration in South China (Guangzhou and Shenzhen as core cities), and the Chengdu–Chongqing urban agglomeration in Southwest China. Meanwhile, most provincial capital cities act as the sources of migration to lower-level cities within the province. With the geographic constraint of the gravity model–based cost matrix, the inferred long-distance movements are smaller than the short-distance movements, while some long-distance flows between large cities, such as the most prominent flow from Shenzhen to Chongqing, can also be identified.

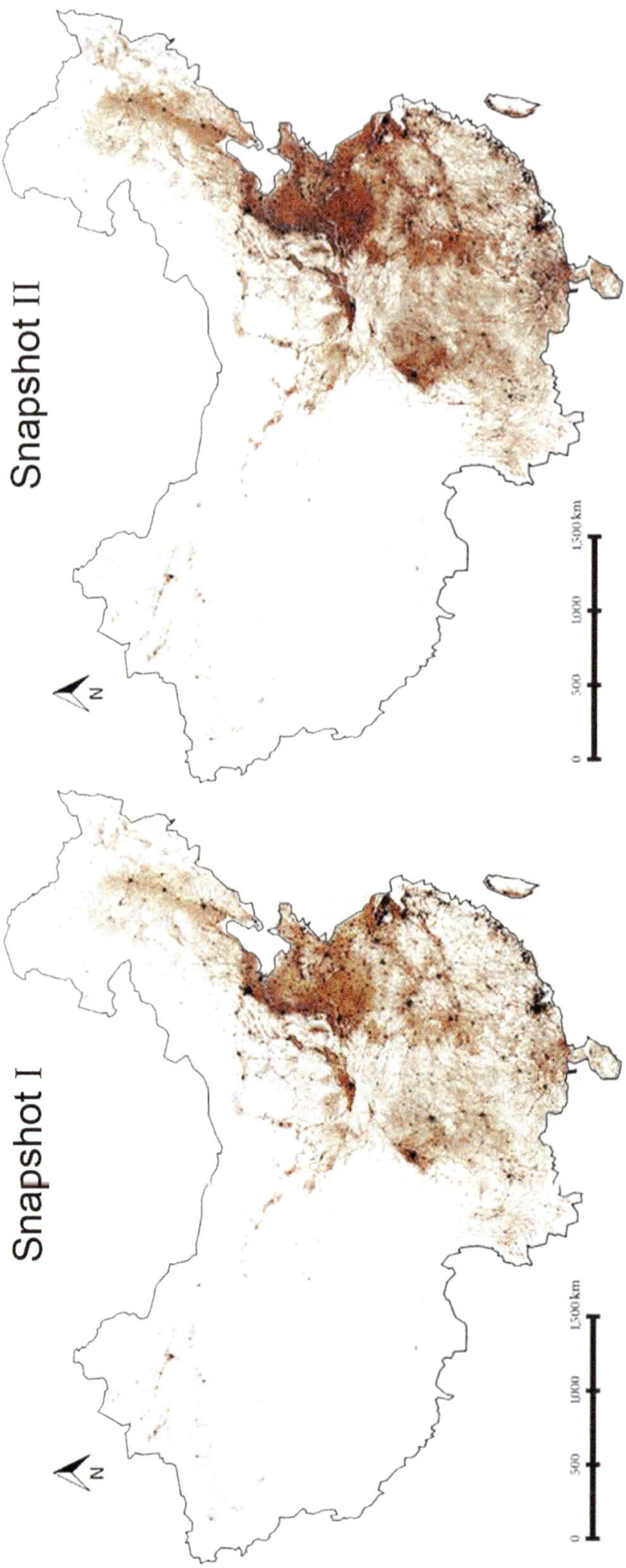

FIGURE 15.1 (a) Snapshot I, the spatial distribution of daily average active users before the Spring Festival (January 18, 2016–January 22, 2016). (b) Snapshot II, the spatial distribution of daily average active users during the Spring Festival (February 7, 2016–February 11, 2016).

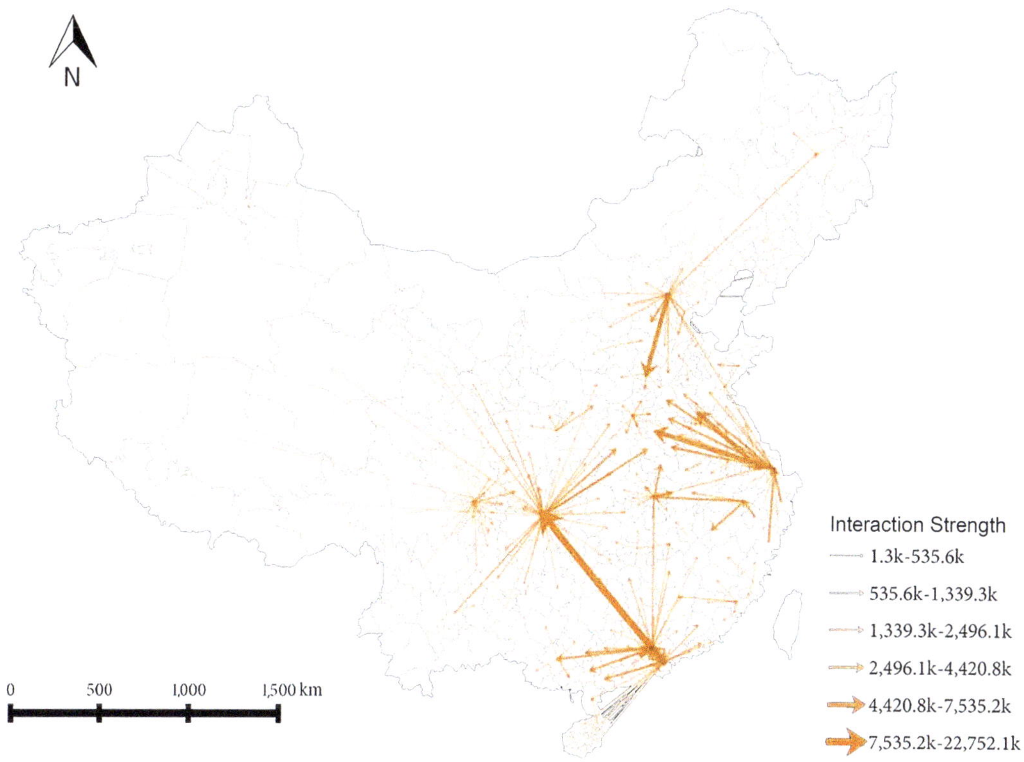

FIGURE 15.2 The inferred spatial evolutionary network with the gravity cost matrix $cij \propto d\beta/(Ai\ Aj)$ and $\beta = 1.0$.

Different inferred spatial evolution with several other cost matrices is compared in Figure 15.3. By changing the distance–decay parameter β in the gravity model–based cost matrix, the influence of the spatial displacement on the occurrence of the evolution process. A higher β represents a stronger blocking effect, while a lower β denotes less rejection for long-distance processes. Figures 15.3a and 15.3b illustrate the inferred results with β = 0.8 and 0.4, respectively. Compared to Figure 15.2, long-distance flows are more likely to be inferred when decreasing the value of β in Figure 15.3a, even though most long-distance flows have relatively low weights. Additionally, β = 0.4 generates a greater diffusional process in which more long-distance flows are inferred (Figure 15.3b), and provincial capital cities tend to have larger affiliated hinterland areas. Figure 15.3c shows the results of a cost matrix that considers only the inverse proportional relationship between the flow cost and the attraction of cities. For this result, the spatial evolution process occurs among big cities and forms a diamond-like pattern. It is difficult to identify the central role of the provincial capital cities because the flows from lower-level cities are rearranged to achieve the optimal total cost. Figure 15.3d shows the result of a cost matrix that considers only the distance decay effect, which makes the inferred result unreasonable compared to Figure 15.2. Figure 15.3e shows the results of an identical cost matrix, where the cost between any two cities is fixed as one. The result is simply one of the optimal solutions and exhibits no apparent regular pattern.

15.3.1.3 Validation of Inferred Spatial Evolution

Because this study does not have access to the actual national-scale human migration data during the 2016 Chinese Spring Festival, a dataset collected by Baidu (a major Chinese web services company) was adopted as a substitute for the real migrant flows to evaluate our inferred spatial evolution

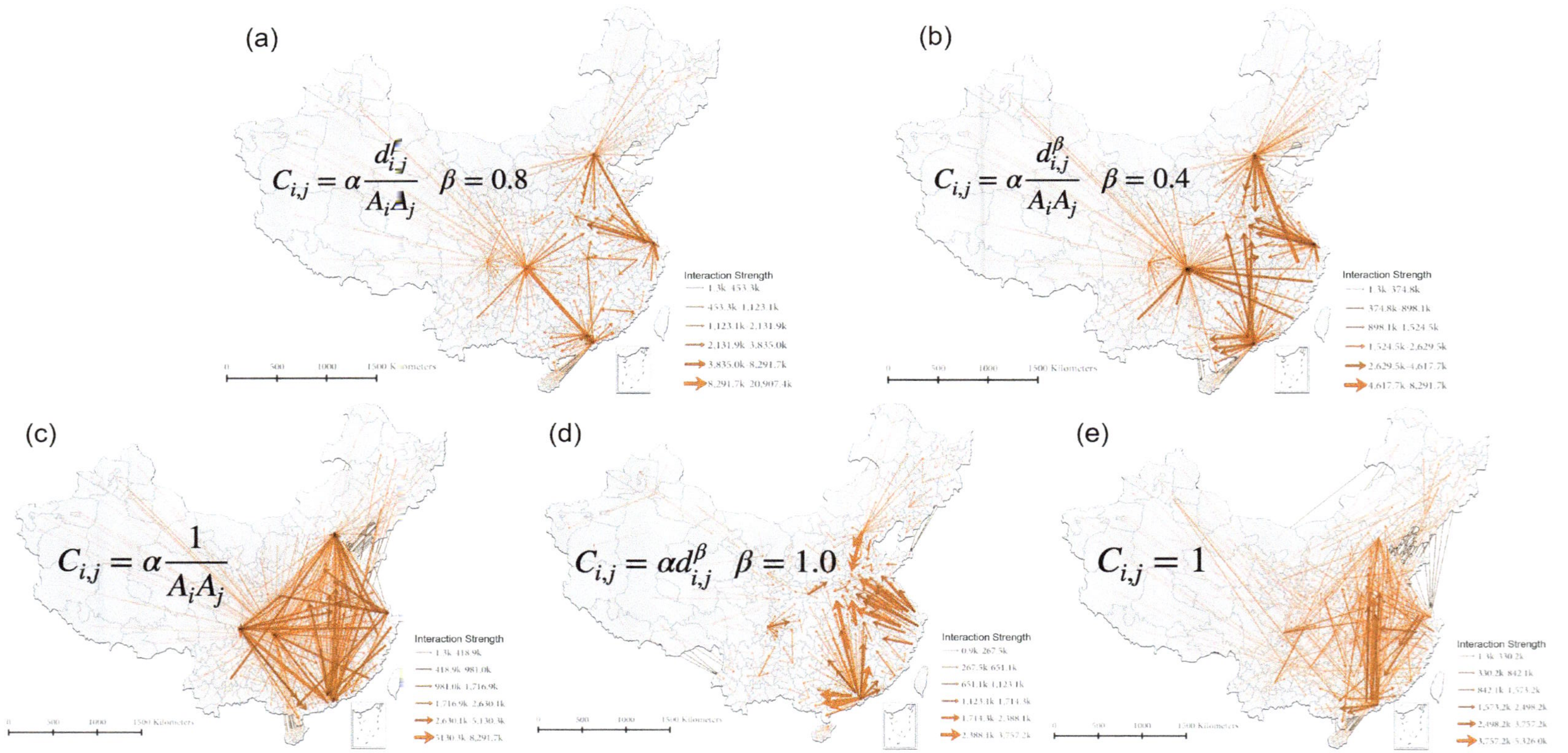

FIGURE 15.3 The inferred spatial evolutionary network using IIDS with the different cost matrix. (a) Inferred result with a gravity cost matrix and $\beta = 0.8$. (b) Inferred result with a gravity cost matrix and $\beta = 0.4$ (c) Inferred result with an inverse attraction–based cost matrix. (d) Inferred result with a distance decay–based cost matrix. (e) Inferred result with identical costs.

process by IIDS. This dataset includes the city-level population migration numbers (quarterly) from the fourth quarter of 2015 (2015Q4) to the first quarter of 2016 (2016Q1). The real numbers of migrations are scaled down to another dimension for data confidentiality. The scale of Baidu's migration data is thus different from that of the inferred flows. The inferred spatial evolution process related to Beijing, Chengdu, and Shanghai is selected and searched for their corresponding migration numbers in the Baidu dataset. The correlation is investigated between the inferred evolution process and the real migration. The chosen cities in this evaluation experiment are relatively large, and their migration patterns are considered more stable compared to lower-level cities.

The correlation between the inferred evolution process (during the 2016 Chinese Spring Festival) and Baidu's quarterly migration dataset (from 2015Q4 to 2016Q1) is evaluated for three major cities, Beijing, Chengdu, and Shanghai, using ordinary least squares (OLS) regression. For Beijing, the Pearson correlation coefficient r is 0.758 when fitting the migration numbers, and the Spearman's correlation coefficient for ranked data ρ is 0.923 when fitting the city ranks of migration numbers. Shanghai exhibits a similar result to Beijing, with a lower $r = 0.733$ and a higher $\rho = 0.844$ in. For Chengdu, the correlation coefficients are approximately 0.9 in both situations. The correlation analysis indicates that, for the representative cities, the inferred evolution process is positively associated with the real process from the numerical perspective ($r > 0.75$); the correlation is even stronger from the rank perspective ($\rho \approx 0.9$ in) when reducing the impacts of extreme values on the calculation. Although the studied time intervals are not exactly the same between the case study and the Baidu dataset, the correlation analysis results are satisfying enough to demonstrate the practicability of network optimization under real scenarios of inferring large-scale spatial evolution.

15.3.2 Revealing the Spatial Shifting Pattern of the COVID-19 Pandemic

15.3.2.1 Sequential Snapshots of Pandemic Distribution

This example explores the spatial shift of the COVID-19 pandemic between the snapshots of pandemic case distribution using the network optimization method (Zhu et al., 2021). Spatial shift refers to the transfer of the pandemic across space, representing the spatial evolution process as the pandemic progresses. Three datasets are collected in this study, including the COVID-19 case data, Twitter movement data, and state-level population data. The COVID-19 data were collected from the *New York Times*, based on reports from state and local health agencies. The reported data of cumulative counts of confirmed coronavirus cases can be used to draw epidemic snapshot maps at the state or county level over time. The study examines seven snapshots and six periods along the timeline from January 31, 2020, to August 09, 2020. Figure 15.4a shows the timeline of six pandemic phases that correspond to key moments in the pandemic in the United States. The snapshots of six pandemic phases are shown in Figure 15.4b, presenting the spatial distribution of the total confirmed cases. The number of cases between each of the two snapshots is rescaled to ensure that the total number of confirmed cases (CCs) remains unchanged.

15.3.2.2 Spatial Evolution of Pandemic at Global Scale

Based on six snapshots, the spatial evolution between each two snapshots is inferred using the IIDS method. In Figure 15.5, the inferred spatial evolution reflects a stable spatial evolution process of pandemic centers throughout the six phases, represented as spatial evolutionary networks. As the total CCs consistently increase, major population centers including California (CA), New York (NY), Texas (TX), and Illinois (IL) remain as the local concentration centers in the network. Long-range and strong evolution processes can be observed between states such as NY and CA (P1, P2, P5, P6), NY and IL (P1, P3, P4), CA and TX (P1, P3, P4, P6), and TX and NY (P4, P5). These dominant states exhibit in-shifts from distant states and also shift out to nearby states with fewer COVID-19 cases across phases. Seeing these evolution processes is helpful in seeing how states with larger economies and populations tend to have stronger evolution processes in the system.

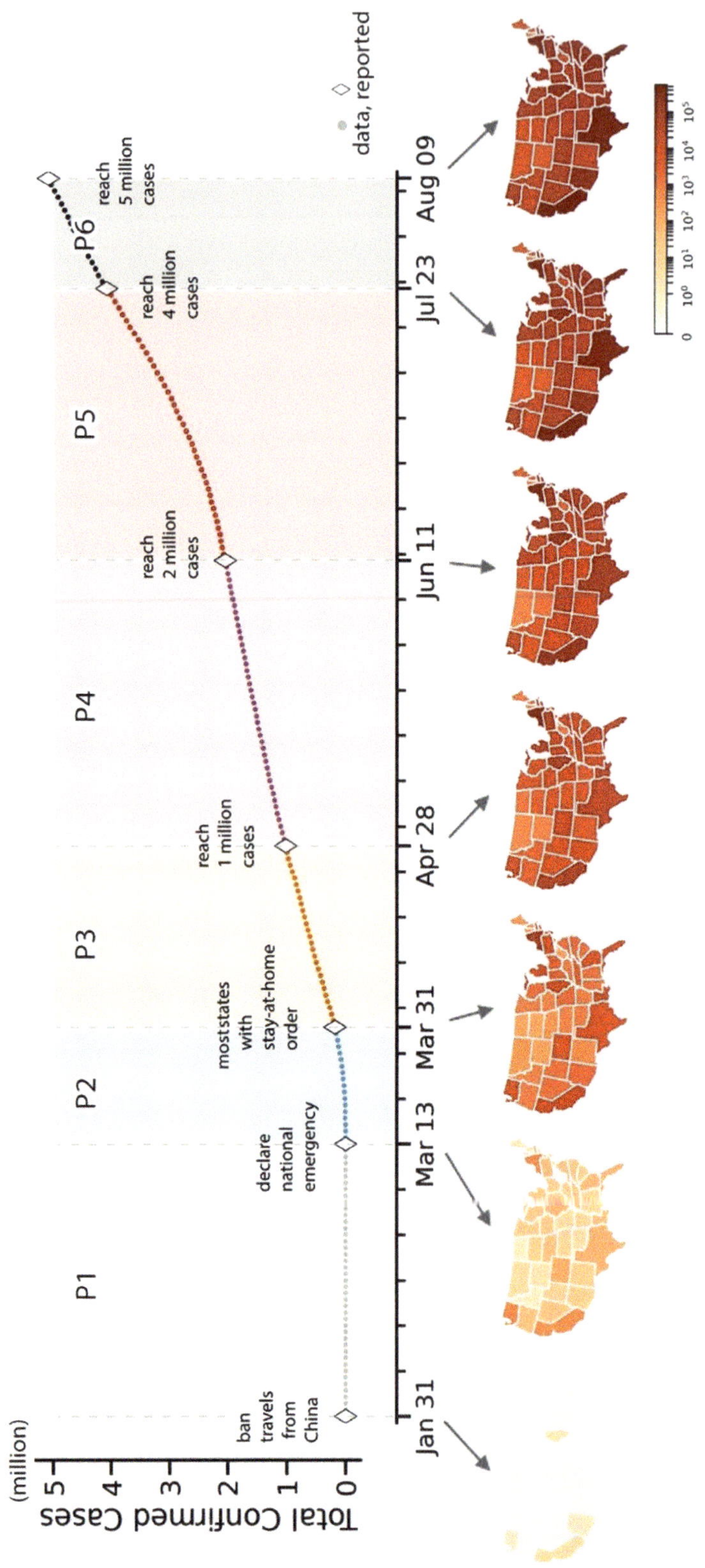

FIGURE 15.4 Six phases of the COVID- 9 pandemic and their corresponding inferred spatial evolution process in the United States. (a) Timeline of the total confirmed cases (CCs) in the United States from January 31 to August 9, 2020, which is divided into six phases. (b) Seven epidemic snapshots of CCs at the state level. (*Source:* Adapted from Zhu et al. 2021.)

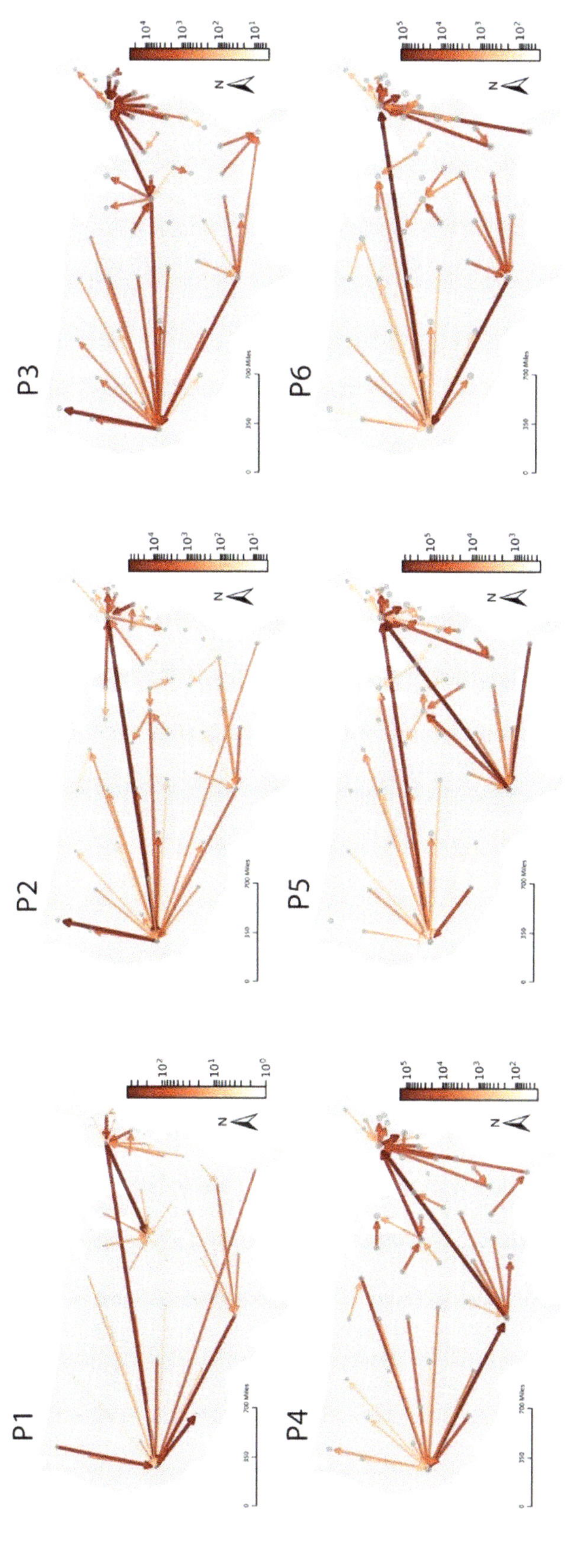

FIGURE 15.5 Flow maps of the inferred spatial evolution for the six phases. (*Source:* Adapted from Zhu et al. 2021.)

Specifically, CA, NY, IL, and TX exchange major processes as well as the movement of cases into surrounding states. Meanwhile, some states with nonnegligible outbreaks during P4 and P5, such as New Jersey (NJ), Massachusetts (MA), and Georgia (GA), exhibit a pattern of first shifting out and then receiving in-shifts. We can also see how GA and Florida (FL) emerge in the flow maps of P4, P5, and P6 with strong out-shifts, implying potential outbreaks in later phases.

15.3.2.3 Spatial Evolution of Pandemic at State and Local Scales

Apart from looking at national-scale statistical metrics, the approaches developed here can also shed light on state and local dynamics. Doing so may give insight into how the local pandemic centers are moving or how well state reaction control measures are working or provide a better understanding of where potential outbreaks in other high-risk regions may occur.

Take as an example the spatial evolution from P2 to P5 in NY. The flow maps of local spatial evolution around NY are illustrated in Figure 15.6a, where the coloring of arrows is the same as that of Figure 15.5. Figure 15.6b is a heat map of the shift matrix clarifying where the in-shifts to NY are coming from and where the out-shifts from NY are going to in each phase. Overall, NY transitions from being a “black hole” to a “volcano” in its relationships with other states. During P2, NY started to show the potential for becoming the hub of the pandemic in the Northeast United States, with out-shifts to CA, MA, and Virginia (VA), and in-shifts from NJ, Connecticut (CT), and Michigan (MI). In-shifts and out-shifts are roughly balanced in intensity, and most shifts are within the Northeast except for the out-shift to CA. In P3, we see a local “black hole” effect in the sense that there are far more in-shifts than out-shifts from almost all nearby states, including CT, Delaware (DE), Maryland (MD), MA, New Hampshire (NH), NJ, North Carolina (NC), Ohio (OH), Pennsylvania (PA), and VA, while out-shifts only occur for Maine (ME) and Vermont (VT). In P4, NY started again to show out-shifts to nearby states, especially NJ. Strong in-shifts were coming from farther states such as FL, GA, IL, NC, TX, and VA. This is a sign of NY becoming more influential regarding spatial shifts and having a bigger impact on nearby states via a spatial spillover effect (Chen et al. 2017). In P5, we can see an “active volcano” effect when NY was receiving strong in-shifts from distant states, including CA, GA, TX, and NC, and out-shifting to its nearby states in Northeast America.

The shifting nature of the pandemic at the state level can also be analyzed using a temporal scatter plot (Figure 15.6c). Each point denotes the intensity of daily in-shift and daily out-shift of a state for each phase. We may use arrows to connect all points of a state over time to check how the local evolution process is changing across phases. As shown in Figure 15.6c, NY saw a significant drop in out-shifts and a rise of in-shifts between P2 and P3, which indicates a local concentration within the state. After P3, NY gradually returned to a situation marked by increasing out-shifts corresponding to local spillover from NY to nearby states. In sum, state-scale spatial evolution is useful for examining local and state dynamics.

15.4 DISCUSSION

The methods reviewed in Section 15.2 offer insights into inferring spatial evolution from a network perspective. However, inferring the realistic processes between snapshots, such as bidirectional migrant flows or the actual number of disease case transfers, remains a challenging task that current methods cannot effectively address. The inferred spatial evolution process based on network optimization can only reflect the net flux that directly causes the reconfiguration of spatial distribution snapshots (Zhu et al., 2018). While the Markov chain method can estimate the most likely process from observed snapshots, it constrains the direction of spatial evolution between locations to the traffic transit route. DL approaches have been widely used for generating bidirectional spatial flows, but few explicitly treat the flow as the process between spatial distribution snapshots. Recent developments in integrating causal inference with spatial process (Akbari et al., 2023) offer opportunities to improve DL-based flow generation models for spatial evolution inference. For example, Zeng

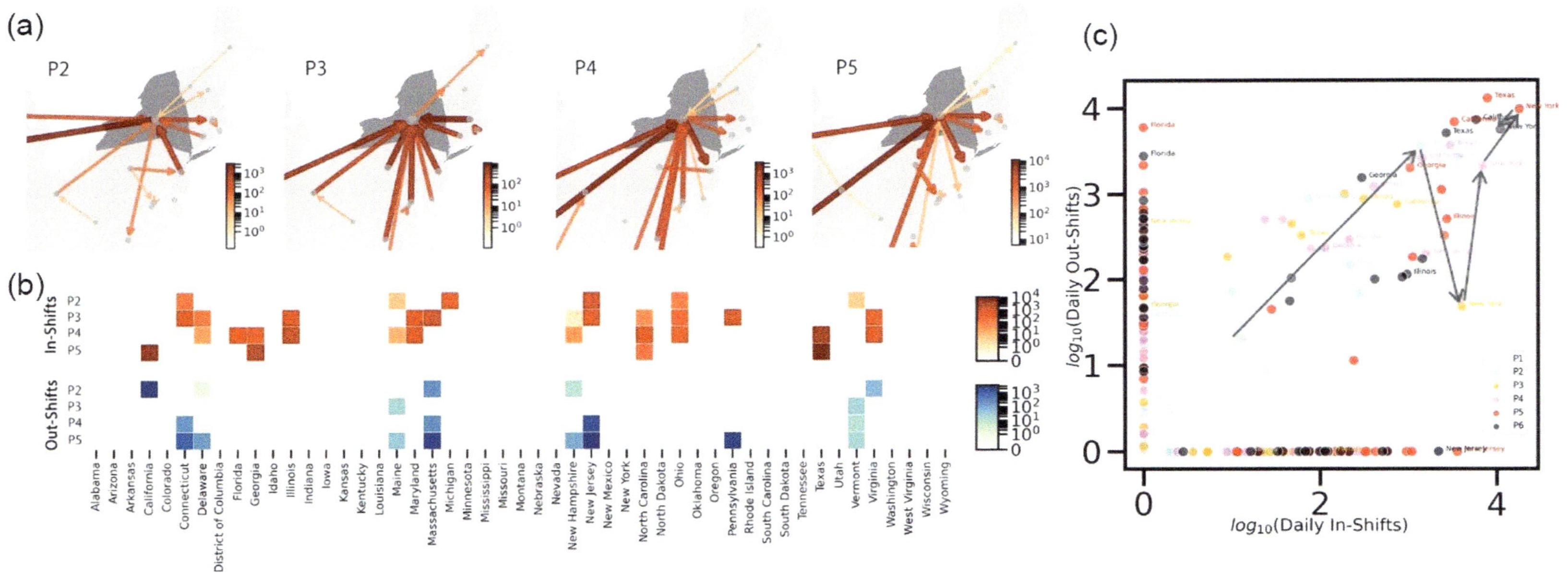

FIGURE 15.6 Local spatial evolution process indicates the changing situation of a state. (a) Flow maps of local spatial evolution around New York (NY) in P2, P3, P4, and P5. (b) Heat maps of the shift matrix, with NY as the origin or destination in the four phases. (c) Temporal scatter plot of the daily in-shift and out-shift intensities. Arrows are used to connect all data points of NY in a temporal sequential order. (*Source*: Adapted from Zhu et al. 2021.)

et al. (2022) predict future spatial flows by inferring the causality between historical flow changes and geographic feature changes. Similarly, the integration of causal inference and flow generation models might offer a new venue for spatial evolution inference by disentangling the causality between the generated flow and the change in spatial distribution snapshots.

Modeling spatial evolution between snapshots can offer knowledge that extends beyond migration and epidemics. Advancements in big geodata with high spatiotemporal resolution allow for the observation of significant transformations in geographical entities over short time intervals, offering an unprecedented opportunity to study spatial evolution at a finer scale. Future studies of spatial evolution may be conducted for contextualized explanations of human activity changes and mobility patterns during social events or emergencies. For example, to understand the festival's impacts on travel behavior, the spatial evolution process can be uncovered between snapshots of human activity distribution before and during Christmas. Moreover, during emergencies, such as wildfires and earthquakes, spatial evolution underlying the hourly snapshots of mobile phone distribution may explain the detail of the evacuation process, which provides critical insights for the collective response of human populations (Bagrow et al., 2011).

15.5 SUMMARY

In this chapter, we introduced the concept of spatial evolution and the ways to approach it based on the spatial evolutionary network. By reviewing existing methods, including network optimization, Markov chain models, and DL-based flow generation, we demonstrated how spatial evolution can be inferred from sequential snapshots of spatial distributions. The examples provided, focusing on national migration during the Chinese Spring Festival and the spatial shift of the COVID-19 pandemic in the United States, illustrate the inference of spatial evolution in different practical contexts. These inferred processes, represented by spatial evolutionary networks, provide a detailed description of the underlying changes in population and pandemic distributions, offering valuable insights for human mobility analysis and urban planning. Challenges and opportunities are discussed regarding the inference models and future visions of spatial evolution.

REFERENCES

Ahuja, Ravindra K., Thomas L. Magnanti, and James B. Orlin. "Network Flows: Theory, Algorithms, and Applications." (1993): Chapter 11, 402–453. New Jersey, NJ: Prentice Hall.

Akbari, Kamal; Stephan Winter, and Martin Tomko. "Spatial causality: A systematic review on spatial causal inference." *Geographical Analysis*, 55, no. 1 (2023): 56–89.

Amblard, Victor; Amir Dib; Noëlie Cherrier; and Guillaume Barthe. "A Bayesian Markov model for station-level origin-destination matrix reconstruction." In Joint European Conference on Machine Learning and Knowledge Discovery in Databases, pp. 538–553. Cham: Springer Nature Switzerland, 2022.

Anderson, Margo J. *The American census: A social history.* Yale University Press, 2015.

Bagrow, James P; Dashun Wang, and Albert-Laszlo Barabasi. "Collective response of human populations to large-scale emergencies." *PloS One*, 6, no. 3 (2011): e17680.

Balcan, Duygu; Vittoria Colizza; Bruno Gonçalves; Hao Hu; José J. Ramasco, and Alessandro Vespignani. "Multiscale mobility networks and the spatial spreading of infectious diseases." *Proceedings of the National Academy of Sciences*, 106, no. 51 (2009): 21484–21489.

Barbosa, Hugo; Marc Barthelemy; Gourab Ghoshal; Charlotte R James; Maxime Lenormand; Thomas Louail; Ronaldo Menezes; Jos´e J Ramasco; Filippo Simini, and Marcello Tomasini. "Human mobility: Models and applications." *Physics Reports*, 734 (2018): 1–74.

Barthélemy, Marc. "Spatial networks." *Physics Reports*, 499, no. 1–3 (2011): 1–101.

Batty, Michael. *The new science of cities.* MIT press, 2013.

Belussi, Fiorenza, and Giorgio Gottardi. *Evolutionary patterns of local industrial systems.* Routledge, 2018.

Boschma, Ron, and Koen Frenken. "Evolutionary economic geography." *The new Oxford handbook of economic geography*, 2018, 213–229.

Chen, Xiaoyu; Shuai Shao; Zhihua Tian; Zhen Xie, and Peng Yin. "Impacts of air pollution and its spatial spillover effect on public health based on China's big data sample." *Journal of Cleaner Production*, 142 (2017): 915–925.

Cook, W J. "Minimum-cost flow problems." In Combinatorial Optimization, (1997): chapter 4, 91–126. Hoboken, NJ: John Wiley & Sons,

Fortunato, Santo. "Community detection in graphs." *Physics Reports*, 486, no. 3–5 (2010): 75–174.

Fotheringham, Alexander S. "Spatial interaction models." *International Encyclopedia of the Social and Behavioral Sciences*, 2001.

Guo, Diansheng. "Flow mapping and multivariate visualization of large spatial interaction data." *IEEE Transactions on Visualization and Computer Graphics*, 15, no. 6 (2009): 1041–1048.

Hawelka, Bartosz, Izabela Sitko, Euro Beinat, Stanislav Sobolevsky, Pavlos Kazakopoulos, and Carlo Ratti. "Geo-Located Twitter as Proxy for Global Mobility Patterns." *Cartography and Geographic Information Science*, 41, no. 4 (2014): 260–271. doi:10.1080/15230406.2014.890072.

Hibbard, Kathy; Anthony Janetos; Detlef P van Vuuren; Julia Pongratz; Steven K Rose; Richard Betts; Martin Herold, and Johannes J Feddema. "Research priorities in land use and land-cover change for the Earth system and integrated assessment modelling." *Int. J. Climatol*, 30, no. 13 (2010): 2118–2128.

Holzhauser, Michael; Sven O Krumke; and Clemens Thielen. "Budget-constrained minimum cost flows." *Journal of Combinatorial Optimization*, 31, no. 4 (2016): 1720–1745.

Ji, Yuxiong; Qiyuan You; Shengchuan Jiang, and Hongjun Michael Zhang. "Statistical inference on transit route-level origin–destination flows using automatic passenger counter data." *Journal of Advanced Transportation*, 49, no. 6 (2015): 724–737.

Kamel Boulos, Maged N., and Estella M. Geraghty. "Geographical tracking and mapping of coronavirus disease COVID-19/severe acute respiratory syndrome coronavirus 2 (SARS-CoV-2) epidemic and associated events around the world: How 21st century GIS technologies are supporting the global fight against outbreaks and epidemics." *International Journal of Health Geographics*, 19, no. 1 (2020): 1–12.

Kang, Chaogui; Song Gao; Xing Lin; Yu Xiao; Yihong Yuan; Yu Liu, and Xiujun Ma. "Analyzing and geo-visualizing individual human mobility patterns using mobile call records." In *2010 18th International Conference on Geoinformatics*, 1–7. IEEE, 2010.

Kimbrell, Tristan. "Canalization and adaptation in a landscape of sources and sinks." *Evolutionary Ecology*, 24 (2010): 891–909.

Li, Baibing. "Markov models for Bayesian analysis about transit route origin–destination matrices." *Transportation Research Part B: Methodological*, 43, no. 3 (2009): 301–310.

Liu, Yu; Zhengwei Sui; Chaogui Kang, and Yong Gao. "Uncovering patterns of inter-urban trip and spatial interaction from social media check-in data." *PloS One*, 9, no. 1 (2014): e86026.

Liu, Zhicheng; Fabio Miranda; Weiting Xiong; Junyan Yang; Qiao Wang, and Claudio Silva. "Learning geo-contextual embeddings for commuting flow prediction." In *Proceedings of the AAAI Conference on Artificial Intelligence*, 34, no. 01 (2020): 808–816.

Louail, Thomas; Maxime Lenormand; Oliva G. Cantu Ros; Miguel Picornell; Ricardo Herranz; Enrique Frias-Martinez; José J. Ramasco, and Marc Barthelemy. "From mobile phone data to the spatial structure of cities." *Scientific Reports*, 4, no. 1 (2014): 5276.

Luca, Massimiliano; Gianni Barlacchi; Bruno Lepri, and Luca Pappalardo. "A survey on deep learning for human mobility." *ACM Computing Surveys (CSUR)*, 55, no. 1 (2021): 1–44.

Moreno-Mateos, David; Antton Alberdi; Elly Morriën; Wim H. van der Putten; Asun Rodríguez-Uña, and Daniel Montoya. "The long-term restoration of ecosystem complexity." *Nature Ecology & Evolution*, 4, no. 5 (2020): 676–685.

Ostfeld, Richard S.; Gregory E. Glass, and Felicia Keesing. "Spatial epidemiology: An emerging (or re-emerging) discipline." *Trends in Ecology & Evolution*, 20, no. 6 (2005): 328–336.

Ratti, Carlo; Dennis Frenchman; Riccardo Maria Pulselli, and Sarah Williams. "Mobile landscapes: Using location data from cell phones for urban analysis." *Environment and Planning B: Planning and Design*, 33, no. 5 (2006): 727–748.

Ratti, Carlo; Stanislav Sobolevsky; Francesco Calabrese; Clio Andris; Jonathan Reades; Mauro Martino; Rob Claxton, and Steven H. Strogatz. "Redrawing the map of Great Britain from a network of human interactions." *PloS One*, 5, no. 12 (December 2010): 1–6.

Rong, Can; Jie Feng, and Jingtao Ding. "GODDAG: Generating origin-destination flow for new cities via domain adversarial training." *IEEE Transactions on Knowledge and Data Engineering*, 35, no. 10 (2023): 10048–10057.

Roy, John R, and Jean-Claude Thill. "Spatial interaction modelling." *Papers in Regional Science*, 83 (2003): 339–361.

Shi, Qingli; Li Zhuo; Haiyan Tao, and Junying Yang. "A fusion model of temporal graph attention network and machine learning for inferring commuting flow from human activity intensity dynamics." *International Journal of Applied Earth Observation and Geoinformation*, 126 (2024): 103610.

Simini, Filippo; Gianni Barlacchi; Massimilano Luca, and Luca Pappalardo. "A deep gravity model for mobility flows generation." *Nature Communications*, 12, no. 1 (2021): 6576.

Stoddart, David R. "Darwin's impact on geography." *Annals of the Association of American Geographers*, 56, no. 4 (1966): 683–698.

Swihart, Robert K, and Norman A Slade. "Testing for independence of observations in animal movements." *Ecology*, 66, no. 4 (1985): 1176–1184.

Toole, Jameson L.; Michael Ulm; Marta C. González, and Dietmar Bauer. "Inferring land use from mobile phone activity." In *Proceedings of the ACM SIGKDD international workshop on urban computing*, pp. 1–8. 2012.

Tucker, Gregory E, and Gregory R Hancock. "Modelling landscape evolution." *Earth Surface Processes and Landforms*, 35, no. 1 (2010): 28–50.

Worboys, Michael. "Event-oriented approaches to geographic phenomena." *International Journal of Geographical Information Science*, 19, no. 1 (2005): 1–28.

Yiu, Man Lung, and Nikos Mamoulis. "Clustering objects on a spatial network." In *Proceedings of the 2004 ACM SIGMOD international conference on Management of data*, 443–454. 2004.

Zeng, Jinwei; Guozhen Zhang; Can Rong; Jingtao Ding; Jian Yuan, and Yong Li. "Causal learning empowered OD prediction for urban planning." In *Proceedings of the 31st ACM International Conference on Information & Knowledge Management*, 2455–2464. 2022.

Zhong, Chen; Stefan Müller Arisona; Xianfeng Huang; Michael Batty, and Gerhard Schmitt. "Detecting the dynamics of urban structure through spatial network analysis." *International Journal of Geographical Information Science*, 28, no. 11 (2014): 2178–2199.

Zhu, Di; Zhou Huang; Li Shi; Lun Wu, and Yu Liu. "Inferring spatial interaction patterns from sequential snapshots of spatial distributions." *International Journal of Geographical Information Science*, 32, no. 4 (2018): 783–805.

Zhu, Di; Xinyue Ye, and Steven Manson. "Revealing the spatial shifting pattern of COVID-19 pandemic in the United States." *Scientific Reports*, 11, no. 1 (2021): 8396.

Section IV

Advanced Topics and Future Directions

16 Human Mobility and Public Health

Tao Hu, Yaxiong Shao, and Changzhen Wang

16.1 INTRODUCTION

Human mobility has predominantly been used to denote residential movements by individuals and households (Kwan and Schwanen 2016). However, it extends beyond the mere act of moving from one place to another. Mobility also encompasses the frequency, distance, and reasons behind these movements. These movements—whether short term or long term, local or global—are shaped by various factors, including travel purposes, social and economic structures, employment opportunities, housing affordability, education, and environmental conditions. They can have profound implications for access to resources, exposure to risks, and overall well-being.

The application of mobility in public health has garnered increasing attention, particularly in the wake of the COVID-19 pandemic. During the pandemic, tracking human mobility patterns was a crucial tool for understanding the spread of infectious diseases and implementing effective containment actions. However, the application of mobility in public health is not limited to infectious diseases. It plays a vital role in various subdisciplines of public health, such as health accessibility, healthcare utilization, environmental exposure, and social determinants of health. For instance, mobility data can be used to assess the accessibility of healthcare facilities, identify underserved populations, and optimize the allocation of health resources. It can also provide insights into how the environmental exposures, such as air pollution or extreme weather events, affect human health (Chen, Wang, and Crooks 2024; McConnell et al. 2024). Moreover, understanding mobility patterns of various socioeconomic groups across different geographic regions can help in designing interventions to address health inequities (Yu et al. 2024).

The primary objective of this chapter is to review and analyze existing studies that leverage human mobility data to address a wide range of public health challenges. By examining the methodologies, applications, and outcomes of these studies, this chapter aims to highlight the potential of mobility data in enhancing our understanding of public health issues and advancing health interventions. This chapter is organized into five sections. Following the introduction, the second section discusses several metrics that incorporate human mobility into the measurement of access to healthcare. The third section offers a comprehensive review of using human mobility in studying infectious diseases, particularly in the epidemiological modeling process. The fourth section addresses the limitations of current research and explores future opportunities for advancing the use of mobility data in public health. The last section concludes the chapter by summarizing the key insights and implications discussed.

16.2 ACCESS TO HEALTHCARE

16.2.1 Background

Access to healthcare services is a critical determinant of health outcomes and an important measure of achieving health equity. Understanding the factors that influence access to care is essential for designing effective public health interventions to reduce health disparities and improve overall health outcomes.

Access to healthcare can be classified into different categories, such as virtual access and physical access. Virtual access pertains to telehealth usage, whereas physical access refers to in-person

DOI: 10.1201/9781003503262-20

visits. Further, access can be classified into four groups based on two dichotomous dimensions: potential versus realized and spatial versus aspatial. This results in four categories: potential spatial access, realized spatial access, potential aspatial access, and realized aspatial access (Khan, 1992; Wang and Liu 2023).

Accurately measuring access to healthcare is a crucial step toward achieving health equity. While most studies have focused on measuring potential spatial access, relatively few have measured realized spatial accessibility. This limitation is primarily due to restricted access to mobility and health-related data, as well as a lack of clarity regarding the Health Insurance Portability and Accountability Act (HIPAA) and similar privacy regulations. These challenges have hindered researchers' ability to fully analyze how individuals utilize healthcare services in spatial contexts.

In recent years, a growing number of researchers have leveraged human mobility data to develop various metrics for assessing and quantifying healthcare accessibility. These mobility data, often derived from mobile phone records, transportation surveys, and global positioning system (GPS) tracking, show the movement frequency, pattern, and travel time of individuals. They can provide more accurate assessments of access to healthcare, allowing planners and policymakers to design interventions that better reflect the realities of healthcare accessibility.

16.2.2 Mobility-Enhanced Accessibility Measurement

16.2.2.1 Actual Utilization of Health Services

Traditional methods for evaluating healthcare utilization primarily depend on statistical surveys and administrative records. These methods often require considerable time and resources to carry out large-scale studies, and data quality can be challenging to secure. However, the increasing availability of mobile device location data provides new opportunities to better understand how people utilize health services. For example, Jing et al. (2023) combined the human mobility data with Centers for Disease Control and Prevention (CDC) mental health data to develop the mental health service visit-to-need ratio index R_i, which can be calculated using Equation 16.1.

$$R_i = \frac{V_i}{N_i} * 1{,}000 \tag{16.1}$$

where:

- V_i is the visits at the census tract i extracted from human mobility data (e.g., SafeGraph Visit Pattern dataset).
- N_i is the number of "mental health not good" at the census tract i extracted from CDC PLACES data.

16.2.2.2 Mobility-Enhanced Geospatial Accessibility Measures

The Two-Step Floating Catchment Area (2SFCA) model is widely recognized in measuring spatial accessibility to healthcare and other resources. Its simplicity and practicality have made it a popular choice for identifying geographic disparities in access. However, the model has notable limitations, such as the assumption of a consistent catchment area and the oversimplification or omission of the distance decay effect associated with travel behavior in seeking health services. These limitations have motivated the development of several improved versions, such as the Variable Two-Step Floating Catchment Area (V2SFCA) (Luo and Whippo 2012), which incorporates variable catchment sizes, the Three-Step Floating Catchment Area (3SFCA) (Wan, Zou, and Sternberg 2012), which adds an additional step to account for competition among providers, and the Supply-Demand Adjusted Two-Step Floating Catchment Area (SDA-2SFCA) (Shao and Luo 2022), which integrates supply-demand adjustments into accessibility calculations.

The 2SFCA model can be summarized as follows:

Step 1. For each supply location *j*, supply-to-demand ratio R_j is calculated by searching all demand locations (*k*) that are within a threshold (travel distance d_0 or travel time t_0) from supply location *j*. The formula is summarized in Equation 16.2.

$$R_j = \frac{S_j}{\sum_{k \in \{t_{kj} \le t_0\}} D_k} \tag{16.2}$$

where:

- S_j is the supply capacity at location *j*.
- D_k is the demand at demand location *k*.
- t_{kj} is the travel time between location *k* and *j*.
- t_0 is the threshold travel time that defines the catchment area.

Step 2. For each demand location *i*, the potential spatial accessibility A_i^P is calculated by searching all supply locations *j* that are within the travel time threshold t_0 from location *i* and summing up the supply-to-demand ratio R_j in Equation 16.3.

$$A_i^P = \sum_{j \in \{t_{ij} \le t_0\}} R_j = \sum_{j \in \{t_{ij} \le t_0\}} \frac{S_j}{\sum_{k \in \{t_{kj} \le t_0\}} D_k} \tag{16.3}$$

where:

- A_i^P represents the potential spatial accessibility score at location *i*.

However, the 2SFCA model has notable limitations, such as the assumption of a consistent catchment area and the oversimplification or omission of the distance decay effect associated with travel behavior in seeking health services. These limitations have motivated the development of several improved versions. One popular model is the Enhanced Two-Step Floating Catchment Area (E2SFCA) model. By incorporating a distance decay function *f(t)* in the 2SFCA model in Equation 16.3, the E2SFCA model can be expressed in Equation 16.4.

$$A_i^P = \sum_{j \in \{t_{ij} \le t_0\}} \frac{S_j f(t_{ij})}{\sum_{k \in \{t_{kj} \le t_0\}} f(t_{kj}) D_k} \tag{16.4}$$

The distance decay function reflects how the interaction between supply and demand diminishes as distance increases. Previous studies often used an empirical value. By analyzing human mobility data, such as information on where people live (demand) and where they receive healthcare services (supply), more precise distance decay functions can be derived, reflecting actual travel behaviors. For instance, SafeGraph data, which provide detailed insights into population movement patterns, can be integrated with the E2SFCA model to improve the accuracy of spatial accessibility measures and better understand healthcare utilization dynamics.

First, estimate hospital visits for the entire population. Compute adjusted population visits (AV_i) for census block group *i* based on the SafeGraph sample data in Equation 16.5.

$$AV_i = RV_i * \frac{P_i}{ND_i} \tag{16.5}$$

where:

- RV_i is the sampled raw visit count in census block group i.
- P_i is the population within census block group i.
- ND_i is the number of devices sampled in census block group i.

Next, the 2SFCA model can be applied to the adjusted population visits. Because the human mobility data already have the relationship between the demand and supply, there is no need to search in the nearby areas. The steps can be summarized as follows:

Step 1. For each supply location j, sum all adjusted visits from demand locations k visiting supply location j, and compute the supply-to-demand ratio R_j. This step can be summarized in Equation 16.6:

$$R_j = \frac{S_j}{\sum_{k \in K_j} AV_k} \tag{16.6}$$

where:

- S_j is the supply capacity at location j.
- K_j is the set of demand locations that have patients visiting hospital j.
- AV_k is the number of adjusted visits from location k visiting hospital j.

Step 2. For each demand location i, identify all supply locations that patients from location i have visited, and sum up the supply-to-demand ratio R_j (Equation 16.7) of these hospitals:

$$A_i^R = \sum_{j \in J_i} R_j = \sum_{j \in J_i} \frac{S_j}{\sum_{k \in K_j} AV_k} \tag{16.7}$$

where:

- A_i^R is the realized accessibility index at demand location i.
- J_i is the set of hospitals visited by patients from location i.

In this realized spatial accessibility evaluation model, the incorporation of human mobility data enhances the traditional potential spatial accessibility model, enabling a more comprehensive measurement of spatial access across the entire population from a new perspective. This innovative approach offers planners and policymakers valuable insights, facilitating more informed and effective decision-making.

16.2.3 Applications

Human mobility data provide more spatially and temporally detailed behaviors on how care seekers access health services. The use of human mobility data in healthcare access studies presents a significant number of new opportunities to improve the measurement of healthcare access. Integrating mobility datasets in healthcare access measurement shows significant promise from a new perspective without concerns associated with HIPAA regulations.

16.2.3.1 Health Services Utilization Patterns

It is undeniable that mental health is equally as important as physical illness, and it can affect anyone, regardless of age, race, income, or education attainment. Human mobility data have proven valuable in mental health research. For example, Jing et al. (2023) utilized SafeGraph data to explore the relationship between immigrant concentration and mental health service utilization in the United States, as studies have shown immigrants generally have lower rates of mental health services utilization. The authors used SafeGraph data to track mental health service visits and employed a visit-to-ratio index to estimate average mental health service utilization.

In addition to mental health service utilization, mobility data have also demonstrated the potential in improving mental well-being. Jing et al. (2024) used SafeGraph mobility data to investigate the neighborhood-level depression prevalence in the United States. Their study found that visits to drinking places were positively associated with neighborhood-level depression, while visits to fitness centers were negatively associated. These findings suggest that reducing alcohol consumption and increasing physical exercise could help alleviate neighborhood-level depression.

16.2.3.2 Healthcare Accessibility

Access to healthcare has become a critical issue, especially in the aftermath of the COVID-19 outbreak, which has placed unprecedented attention on the need for equitable healthcare resources distribution. The rational allocation of healthcare resources is a key factor in the success of any healthcare system. Shao (2023) emphasized the significance of integrating human mobility data into spatial access evaluation models. In their study, a human mobility dataset comprising more than 45 million devices was combined with the U.S. census population to estimate the healthcare facility visitations across the entire U.S. population. Using this estimation, the authors calculated a mobility data–enhanced geographical accessibility metric to quantify healthcare access in the United States. This study offers policymakers a novel approach to gaining a comprehensive understanding of realized spatial accessibility on a large scale.

Mobility data have also been used in other studies to model access to healthcare. Du et al. (2024) incorporated over 8.8 million taxi trajectory data records to investigate the realized spatial accessibility to hospital in Beijing, China. Their findings revealed that the potential spatial accessibility index tends to underestimate access in suburban areas compared to the realized access index. Similarly, Su et al. (2022) used the taxi trajectories to measure access to care for acute myocardial infarction patients in Beijing, highlighting the importance of incorporating human mobility data to enhance healthcare access evaluations.

16.3 SPREAD OF INFECTIOUS DISEASES

16.3.1 Background

Understanding the dynamics of infectious disease transmission is crucial for public health planning, intervention, and prevention. The spread of infectious diseases is influenced by a complex interplay of biological, environmental, and social factors. Among these, human mobility plays a pivotal role in determining how and where diseases propagate. As people move, they can carry pathogens with them, introducing diseases to new areas and facilitating the spread within and between populations.

The integration of mobility data into Susceptible-Infectious-Recovered (SIR) (Hethcote, 1989, 2000) and SIR-related models (Ndaïrou et al. 2020) has proven to be a powerful approach in the study of infectious diseases. By capturing real-world movement patterns, these models can provide more accurate predictions and insights into how diseases spread across different regions. The application of such models in various disease studies, from malaria in Kenya to the global spread of COVID-19, demonstrates their versatility and importance in public health planning and intervention. As mobility data continue to become more accessible and detailed, their integration into epidemiological models will likely play an increasingly vital role in managing and mitigating the spread of infectious diseases.

16.3.2 Mobility-Enhanced Epidemiological Model

The SIR model is a foundational epidemiological model used to describe the spread of infectious diseases within a population. In its basic form, the SIR model assumes a closed population with no births, deaths, or migration, and it does not account for spatial heterogeneity or human mobility. However, integrating mobility data into the SIR model can significantly enhance its ability to accurately simulate disease dynamics across different regions or subpopulations.

The basic SIR model divides the population into three compartments:

- $S(t)$: The number of susceptible individuals at time t.
- $I(t)$: The number of infectious individuals at time t.
- $R(t)$: The number of recovered (and immune) individuals at time t.

The dynamics of these compartments are governed by the following differential equations:

$$\frac{dS(t)}{dt} = -\beta \frac{S(t)I(t)}{N} \tag{16.8}$$

$$\frac{dI(t)}{dt} = \beta \frac{S(t)I(t)}{N} - \gamma I(t) \tag{16.9}$$

$$\frac{dR(t)}{dt} = \gamma I(t) \tag{16.10}$$

where:

- β is the transmission rate, which determines how often a susceptible-infectious contact results in a new infection.
- γ is the recovery rate, which is the rate at which infectious individuals recover and move to the recovered compartment.
- $N = S(t) + I(t) + R(t)$ is the total population size.

When integrating mobility data, the population is typically divided into multiple subpopulations or geographic regions, each with its own SIR dynamics. Mobility data allow us to model the movement of individuals between these subpopulations, thereby capturing the spatial spread of the disease.

Let's consider a metapopulation model with N distinct regions, each having its own SIR compartments: $S_i(t)$, $I_i(t)$, and $R_i(t)$ for region i. The extended SIR model with mobility can be represented as follows.

$$\frac{dS_i(t)}{dt} = -\beta_i \frac{S_i(t)I_i(t)}{N_i} + \sum_{j=1}^{n} m_{ji} S_j(t) - \sum_{j=1}^{n} m_{ij} S_i(t) \tag{16.11}$$

$$\frac{dI_i(t)}{dt} = \beta_i \frac{S_i(t)I_i(t)}{N_i} - \gamma_i I_i(t) + \sum_{j=1}^{n} m_{ji} I_j(t) - \sum_{j=1}^{n} m_{ij} I_i(t) \tag{16.12}$$

$$\frac{dR_i(t)}{dt} = \gamma_i I_i(t) + \sum_{j=1}^{n} m_{ji} R_j(t) - \sum_{j=1}^{n} m_{ij} R_i(t) \tag{16.13}$$

where:

- m_{ji} is the mobility rate from region i to region j, which can be derived from mobility data (e.g., mobile phone data, transportation networks).
- $N_i = S_i(t) + I_i(t) + R_i(t) + \Sigma_{j=1;j\neq i}^{n} m_{ji}(t) - \Sigma_{j=1;j\neq i}^{n} m_{ij}(t)$ is the total population in region i at time t.

The terms $\Sigma_{j=1}^{n} m_{ji} S_j(t)$ and $\Sigma_{j=1}^{n} m_{ij} S_i(t)$ represent the influx and outflux of susceptible individuals into and out of region i at time t, respectively. Similar terms apply to the infectious and recovered compartments.

In this mobility-extended SIR model, the dynamics of disease in each region are influenced not only by the local transmission and recovery rates (β_i and γ_i) but also by the movement of individuals between regions. Mobility data thus allow the model to capture the spatial diffusion of the disease, which is particularly important for understanding the spread of diseases in highly connected regions.

16.3.3 Applications

In the study of infectious diseases, mobility data play a crucial role in understanding and predicting the spatial spread of pathogens. Various forms of mobility data, including mobile phone data, airline transportation data, commuting data, and migration data, are commonly used to quantify human movements between regions or countries. The integration of these mobility datasets into SIR models and their extended versions has been instrumental in modeling the transmission dynamics of various infectious diseases, including H1N1 influenza (Hufnagel et al., 2004; Colizza et al., 2006; Balcan et al., 2009), COVID-19 (Chinazzi et al., 2020; Kraemer et al., 2020; Pullano et al., 2020), malaria (Wesolowski et al., 2012; Pindolia et al., 2013), and dengue fever (Wesolowski et al., 2015).

16.3.3.1 Dengue Fever

Human movement has a significant impact on the spatial distribution of the virus that causes dengue, an illness spread by mosquitoes. By concentrating on dengue disease in Pakistan, Tizzoni et al. (2014) illustrated the importance of including mobility data into epidemiological models. In this work, human mobility patterns were modeled using mobile phone data from millions of users, and the results were then included into a dengue transmission model. It proved that human mobility patterns can be utilized to forecast dengue transmission areas, offering vital information for focused public health initiatives.

Other studies have also used mobility data to model dengue transmission. For example, Wesolowski et al. (2015) employed mobile phone data to analyze the spatial dynamics of dengue in Thailand. Their findings showed that mobility patterns could explain much of the spatial heterogeneity in dengue incidence across the country. Similarly, Pindolia et al. (2013) used a combination of mobile phone and transportation data to predict malaria spread in East Africa, highlighting the importance of both local and long-distance human movements in the disease's transmission dynamics.

16.3.3.2 Influenza and Severe Acute Respiratory Syndrome

In addition to mobile phone data, airline transportation data have been widely used to model the spread of diseases on a global scale. In order to simulate the spread of the H1N1 influenza virus, Balcan et al. (2009) incorporated global airline transportation data into a metapopulation SIR model. The dataset included detailed information on the number of passengers, the frequency of flights, and the specific routes followed when commuting between various airports. By integrating the airline data, the researchers were able to simulate the global spread of H1N1 with a high degree of accuracy. The study found that international air travel was a critical factor in the rapid dissemination of the virus, highlighting the importance of mobility data in pandemic modeling. The findings further underscored the need for monitoring and potentially regulating air travel during the early stages of a pandemic to slow down the global spread of infectious diseases.

Colizza et al. (2006) used a similar approach to model the global spread of influenza using airline transportation data. Their model, which integrated real-time air traffic data, was able to predict the timing and location of influenza outbreaks with remarkable precision. Another study by Hufnagel et al. (2004) used a network-based model incorporating airline data to understand the global spread of the severe acute respiratory syndrome (SARS) coronavirus. These studies collectively demonstrate that airline transportation data are critical for modeling the international spread of airborne infectious diseases.

16.3.3.3 COVID-19 Pandemic

The COVID-19 pandemic further emphasized the need for mobility data in understanding and managing disease spread. For example, Kraemer et al. (2020) used mobile phone data from China to analyze the impact of human mobility on the early spread of COVID-19. Their findings indicated that mobility reductions, driven by lockdowns and travel restrictions, were crucial in controlling the outbreak in its initial stages. Similarly, Pullano et al. (2020) combined mobile phone and transportation data to model the spread of COVID-19 in France, revealing that mobility patterns could be used to predict the course of the epidemic and the effectiveness of intervention measures.

Different from previous studies, Chinazzi et al. (2020) used a comprehensive mobility dataset that included international air travel, ground transportation data, and local mobility patterns generated from cell phone data. This integrated dataset was used in a Susceptible-Exposed-Infectious-Recovered (SEIR) model to assess the impact of various travel restrictions and mobility changes on the spread of COVID-19. The study showed that early travel limitations considerably slowed the global spread of the virus but were unable to totally contain it. This finding emphasized the intricate relationship between global mobility and infectious disease dynamics, as well as the importance of having complete mobility data to inform public health actions. Thus, the integration of multiple sources of mobility data enabled a more nuanced understanding of how different types of human movement contribute to the spread of COVID-19, providing valuable lessons for managing future pandemics.

16.4 DISCUSSION

16.4.1 Limitations of Mobility Data in Public Health

Mobility data offer invaluable insights into human behavior and its implications for public health, but it has several significant limitations. First, it lacks socioeconomic and demographic information of individuals (Brazil, Chakalov, and Ko 2024). Currently, most human mobility datasets, such as those derived from public transportation, smartphone, or migration data, primarily provide aggregate population counts without detailed socioeconomic or demographic information. Further, existing mobility datasets are typically sampled rather than comprehensive (Hu et al. 2021). For instance, data collected through smartphones or specific apps represent only a subset of the population, potentially excluding groups such as older adults, low-income individuals, or those without access to technology. Moreover, some smartphone-based datasets that track visits to health facilities do not explicitly clarify the purpose of these visits (Yu et al. 2024). For example, a trip to a health facility for a family visit may be incorrectly interpreted as a service-related visit.

Furthermore, the integration of data from multiple sources—necessary for comprehensive analysis—poses additional challenges. Different datasets may not align due to discrepancies in data collection methods, timescales, or formats, requiring extensive data harmonization efforts that are not only resource-intensive but also prone to introducing errors. Lastly, the dynamic nature of human mobility and behavior, especially in response to public health crises, can quickly render collected data outdated, necessitating constant updates and real-time data processing capabilities, which are not always feasible for many public health organizations.

16.4.2 Human Mobility and Health Disparities

Human mobility and health disparities are closely linked with each other. On the one hand, human mobility can reveal the spatial behavior of patients in utilizing medical services, such as cancer care (Lin et al. 2015; Wang, Wang, and Onega 2021) and detect the mismatch between health care delivery and demand (Koylu et al. 2018). Disparity in mobility could be a contributor to the increased exposure risk to infectious disease (Coleman et al. 2022). On the other hand, the inequitable distribution of healthcare facilities and different types of disease and health problems may lead to different travel behaviors, patterns, and frequencies of medical visits. For example, patients with nonprivate insurance and those residing in low-density oncologist areas were less likely to receive adjuvant chemotherapy (Lin et al. 2015). Rare cancer patients traveled longer than common cancer patients to receive surgical care, especially those living in rural areas (Onega et al. 2023). People with advanced diseases such as late-stage cancer patients may travel more frequently than those with general diseases.

Both human mobility and health disparities are shaped by social and economic characteristics. A study in Florida found that high-income, younger, privately insured, and White patients tended to travel longer for hospitalization, and so did the rural patients, although the former patients had better mobility and more options while rural patients suffered from poorer access to healthcare (Jia, Wang, and Xierali 2019). Another study in China demonstrated that female, senior, and marginal patients traveled shorter distances than their counterparts, and more disparities were found in low-tier hospital utilization (Yan et al. 2022). In addition, the choices of travel destinations and transportation modes, as well as where people live, work, study, and entertain, may further affect the use of healthcare services and somehow exacerbate the existing health disparities among different population groups and across different geographic areas. A recent study in the United States found that areas with a higher concentration of Latin American immigrants were associated with lower mental health visits and visit-to-need ratio, particularly in the West region. However, areas with higher concentrations of Asian and European immigrants were associated with higher visit-to-need ratios but not with mental health visits (Jing et al. 2023). Another study in China found that cross-city healthcare utilization declined with a longer distance at a varying degree for different tiers of hospitals (Yan et al. 2022).

The interaction between human mobility and health disparities is disrupted by unexpected events, such as road closures caused by extreme weather or the outbreak of infectious diseases (e.g., the COVID-19 pandemic and H1N1 influenza). Wang et al. (2021) analyzed the spatial and temporal patterns of healthcare visits during the COVID-19 pandemic in North Carolina and found that the census block groups with more seniors, minorities, and low-income individuals had lower access to healthcare before the pandemic, but they were the most at risk for reduced healthcare access during the pandemic. Hu et al. (2022) examined racial-ethnic disparities in the United States across a spatiotemporal scale and found that racial differences in human mobility varied dramatically at the beginning of the COVID-19 pandemic but remained stable thereafter, while racial differences in COVID-19 infections and deaths broadly decreased over time.

These interesting findings on human mobility and health disparities benefit from advanced geospatial methods and the increasingly available mobility data from multiple sources, such as mobile phone–based visitation data from SafeGraph (Wang, Wang, and Onega 2021; Hu et al., 2022; Jing et al. 2023) and patient flow data from healthcare utilization records or claims data (Lin et al. 2015; Koylu et al. 2018; Jia, Wang, and Xierali 2019; Yan et al. 2022; Onega et al. 2023). While they provide new insights into developing more effective policies to reduce health disparities and updating existing transportation design and planning to reduce mobility barriers, more in-depth analysis of granular mobility data, such as individual-level data for medical visits, is needed. Given that most research considers physical visits to healthcare, the rise of telehealth in reducing travel burden and even health disparities is worth investigating as well. In addition, the survey-based mobility data and perspectives from sociologists and healthcare providers may complement existing findings and advance our understanding of the interaction between human mobility and health disparities.

16.4.3 Potential Research Directions in Mobility and Health

As human mobility data become increasingly accessible to the research community, the development of new metrics is essential for enhancing the measurement and understanding of travel behavior to health services. A recent study by Yu (2024) exemplifies this effort by leveraging smartphone data to propose an innovative metric for estimating actual travel time. This metric was then compared to the traditional proximity metric, which only accounts for driving time to the nearest health facility. The findings revealed dramatic disparities between the two metrics, highlighting significant regional differences across the United States and variations among demographic groups. These results suggest that traditional travel time calculations substantially underestimate the challenges faced by certain populations in reaching health services. This study emphasizes the importance of reevaluating conventional methodologies and adopting more nuanced approaches that incorporate real-world travel behaviors to better inform health equity policies and interventions.

Artificial intelligence (AI) has been advancing rapidly in recent years, particularly with the emergence of foundation models like Google TimesFM and the Population Dynamics Foundation Model. These cutting-edge techniques empower researchers to develop robust and generalizable models for analyzing human behavior in health access. For example, Park et al. (2023) developed AI agents capable of interactively simulating human behavior. These AI agents could be adopted to model healthcare-seeking behavior, providing valuable insights into healthcare access measurement. In addition to offering insights into mobility behaviors, these models provide immense potential for predicting spatiotemporal patterns of disease spread and health service utilization. By integrating large-scale data and advanced machine learning algorithms, foundation models can provide actionable insights that improve public health planning, enhance resource allocation, and support the development of targeted interventions.

However, these AI models face a significant limitation in their interpretability. This challenge makes it difficult to explain the results to health professionals and policymakers, hindering the broader adoption of AI models in public health decision-making. To address this issue, eXplainable AI (XAI) models have emerged as a promising solution, such as Shapley and LIME (Local Interpretable Model-agnostic Explanations) (Lundberg, 2017). XAI techniques aim to enhance the interpretability of AI models by providing clear insights into the importance and contribution of factors to the outcomes. For instance, XAI methods can identify which mobility patterns most influence health service utilization or disease spread, enabling researchers to share findings in a way that is understandable and actionable for nontechnical stakeholders. By bridging the gap between advanced AI techniques and practical applications, XAI models have the potential to significantly boost the usability and trust of AI-driven insights in public health and policymaking.

16.5 CONCLUSION

Human mobility plays a critical role in shaping public health outcomes, influencing everything from geographic access to healthcare facilities to the spread of infectious diseases. As populations become more mobile and global interconnectedness increases, the ability to understand and model mobility patterns becomes essential for developing effective public health strategies. This chapter has explored the multifaceted relationship between human mobility and public health, focusing on two major areas: geographic access to healthcare and the integration of mobility data into epidemiological models. By leveraging mobility data, public health strategies can be better informed, leading to more equitable healthcare access and more effective management of infectious diseases. Moving forward, addressing the challenges of data integration, privacy, and equity will be essential to maximizing the benefits of mobility-driven health interventions.

REFERENCES

Balcan, Duygu, Vittoria Colizza, Bruno Gonçalves, Hao Hu, José J. Ramasco, and Alessandro Vespignani. 2009. "Multiscale Mobility Networks and the Spatial Spreading of Infectious Diseases." *Proceedings of the National Academy of Sciences* 106 (51): 21484–89.

Brazil, Noli, Bozhidar T. Chakalov, and Michelle Ko. 2024. "The Health Implications of Neighborhood Networks Based on Daily Mobility in US Cities." *Social Science & Medicine* 354: 117058.

Chen, Qingqing, Boyu Wang, and Andrew Crooks. 2024. "Community Resilience to Wildfires: A Network Analysis Approach by Utilizing Human Mobility Data." *Computers, Environment and Urban Systems* 110: 102110.

Chinazzi, Matteo, Jessica T. Davis, Marco Ajelli, Corrado Gioannini, Maria Litvinova, Stefano Merler, Ana Pastore y Piontti, et al. 2020. "The Effect of Travel Restrictions on the Spread of the 2019 Novel Coronavirus (COVID-19) Outbreak." *Science* 368 (6489): 395–400.

Coleman, Natalie, Xinyu Gao, Jared DeLeon, and Ali Mostafavi. 2022. "Human Activity and Mobility Data Reveal Disparities in Exposure Risk Reduction Indicators among Socially Vulnerable Populations during COVID-19 for Five US Metropolitan Cities." *Scientific Reports* 12 (1): 15814.

Colizza, Vittoria, Alain Barrat, Marc Barthélemy, and Alessandro Vespignani. 2006. "The Role of the Airline Transportation Network in the Prediction and Predictability of Global Epidemics." *Proceedings of the National Academy of Sciences* 103 (7): 2015–20.

Du, Fangye, Jiaoe Wang, Liang Mao, and Yan Liu. 2024. "Spatial Equity in Healthcare Access: An Opportunity-Utilization Perspective." *Cities* 155: 105424.

Hethcote, Herbert W. 1989. "Three Basic Epidemiological Models." In Levin, S.A., Hallam, T.G., Gross, L.J. (eds) *Applied Mathematical Ecology*, 119–44. Berlin, Heidelberg: Springer.

Hethcote, Herbert W. 2000. "The Mathematics of Infectious Diseases." *SIAM Review* 42 (4): 599–653.

Hu, Tao, Siqin Wang, Bing She, Mengxi Zhang, Xiao Huang, Yunhe Cui, Jacob Khuri et al. 2021. "Human Mobility Data in the COVID-19 Pandemic: Characteristics, Applications, and Challenges." *International Journal of Digital Earth* 14 (9): 1126–47.

Hufnagel, Lars, Dirk Brockmann, and Theo Geisel. 2004. "Forecast and Control of Epidemics in a Globalized World." *Proceedings of the National Academy of Sciences* 101 (42): 15124–29.

Jia, Peng, Fahui Wang, and Imam M. Xierali. 2019. "Differential Effects of Distance Decay on Hospital Inpatient Visits among Subpopulations in Florida, USA." *Environmental Monitoring and Assessment* 191(Suppl 2): 381.

Jing, Fengrui, Zhenlong Li, Shan Qiao, Huan Ning, Suhong Zhou, and Xiaoming Li. 2023. "Association between Immigrant Concentration and Mental Health Service Utilization in the United States over Time: A Geospatial Big Data Analysis." *Health & Place* 83: 103055.

Jing, Fengrui, Zhenlong Li, Shan Qiao, Huan Ning, M. Naser Lessani, and Xiaoming Li. 2024. "From Neighborhood Contexts to Human Behaviors: Cellphone-Based Place Visitation Data Contribute to Estimating Neighborhood-Level Depression Prevalence in the United States." *Cities* 148: 104905.

Khan, Abdullah A. 1992. "An Integrated Approach to Measuring Potential Spatial Access to Health Care Services." *Socio-Economic Planning Sciences* 26 (4): 275–87.

Koylu, Caglar, Selman Delil, Diansheng Guo, and Rahmi Nurhan Celik. 2018. "Analysis of Big Patient Mobility Data for Identifying Medical Regions, Spatio-Temporal Characteristics and Care Demands of Patients on the Move." *International Journal of Health Geographics* 17: 1–17.

Kraemer, Moritz U. G., Chia-Hung Yang, Bernardo Gutierrez, Chieh-Hsi Wu, Brennan Klein, David M. Pigott, Open COVID-19 Data Working Group, et al. 2020. "The Effect of Human Mobility and Control Measures on the COVID-19 Epidemic in China." *Science* 368 (6490): 493–97.

Kwan, Mei-Po, and Tim Schwanen. 2016. "Geographies of Mobility." *Annals of the American Association of Geographers* 106 (2): 243–56.

Lin, Chun Chieh, Suanna S. Bruinooge, M. Kelsey Kirkwood, Christine Olsen, Ahmedin Jemal, Dean Bajorin, Sharon H. Giordano, et al. 2015. "Association between Geographic Access to Cancer Care, Insurance, and Receipt of Chemotherapy: Geographic Distribution of Oncologists and Travel Distance." *Journal of Clinical Oncology* 33 (28): 3177–85.

Lundberg, Scott. 2017. "A Unified Approach to Interpreting Model Predictions." *arXiv preprint arXiv:1705.07874*.
Luo, Wei, and Tara Whippo. 2012. "Variable Catchment Sizes for the Two-Step Floating Catchment Area (2SFCA) Method." *Health & Place* 18 (4): 789–95.
McConnell, Kathryn, Elizabeth Fussell, Jack DeWaard, Stephan Whitaker, Katherine J. Curtis, Lise St. Denis, Jennifer Balch, and Kobie Price. 2024. "Rare and Highly Destructive Wildfires Drive Human Migration in the US." *Nature Communications* 15 (1): 6631.
Ndaïrou, Faïçal, Iván Area, Juan J. Nieto, and Delfim F. M. Torres. 2020. "Mathematical Modeling of COVID-19 Transmission Dynamics with a Case Study of Wuhan." *Chaos, Solitons & Fractals* 135: 109846.
Onega, Tracy, Jennifer Alford-Teaster, Chris Leggett, Andrew Loehrer, Julie E. Weiss, Erika L. Moen, Catherine C. Pollack, and Fahui Wang. 2023. "The Interaction of Rurality and Rare Cancers for Travel Time to Cancer Care." *The Journal of Rural Health* 39 (2): 426–33.
Park, Joon Sung, O'Brien Joseph, Carrie Jun Cai, Meredith Ringel Morris, Percy Liang, and Michael S. Bernstein. 2023. "Generative Agents: Interactive Simulacra of Human Behavior." In *Proceedings of the 36th Annual ACM Symposium on User Interface Software and Technology*, 1–22.
Pindolia, Deepa K., Andres J. Garcia, Zhuojie Huang, David L. Smith, Victor A. Alegana, Abdisalan M. Noor, Robert W. Snow, and Andrew J. Tatem. 2013. "The Demographics of Human and Malaria Movement and Migration Patterns in East Africa." *Malaria Journal* 12: 1–12.
Pullano, Giulia, Eugenio Valdano, Nicola Scarpa, Stefania Rubrichi, and Vittoria Colizza. 2020. "Population Mobility Reductions during COVID-19 Epidemic in France under Lockdown." *MedRxiv* 29: 2020.
Shao, Yaxiong. 2023. *Modeling Spatial Access to Healthcare*. Graduate Research Theses & Dissertations, Northern Illinois University. https://huskiecommons.lib.niu.edu/allgraduate-thesesdissertations/7848.
Shao, Yaxiong, and Wei Luo. 2022. "Supply-Demand Adjusted Two-Steps Floating Catchment Area (SDA-2SFCA) Model for Measuring Spatial Access to Health Care." *Social Science & Medicine* 296: 114727.
Su, Yuwei, Zhengying Liu, Jie Chang, Qiuju Deng, Yuyang Zhang, Jing Liu, and Ying Long. 2022. "Measuring Accessibility to Healthcare Using Taxi Trajectories Data: A Case Study of Acute Myocardial Infarction Cases in Beijing." *International Journal of Health Policy and Management* 12: 6653.
Tizzoni, Michele, Paolo Bajardi, Adeline Decuyper, Guillaume Kon Kam King, Christian M. Schneider, Vincent Blondel, Zbigniew Smoreda, Marta C. González, and Vittoria Colizza. 2014. "On the Use of Human Mobility Proxies for Modeling Epidemics." *PLOS Computational Biology* 10 (7): e1003716.
Wan, Neng, Bin Zou, and Troy Sternberg. 2012. "A Three-Step Floating Catchment Area Method for Analyzing Spatial Access to Health Services." *International Journal of Geographical Information Science* 26 (6): 1073–89.
Wang, Changzhen, Fahui Wang, and Tracy Onega. 2021. "Spatial Behavior of Cancer Care Utilization in Distance Decay in the Northeast Region of the US." *Travel Behaviour and Society* 24: 291–302.
Wang, Fahui, and Lingbo Liu. 2023. *Computational Methods and GIS Applications in Social Science*. CRC Press.
Wang, Jueyu, Noreen McDonald, Abigail L. Cochran, Lindsay Oluyede, Mary Wolfe, and Lauren Prunkl. 2021. "Health Care Visits during the COVID-19 Pandemic: A Spatial and Temporal Analysis of Mobile Device Data." *Health & Place* 72: 102679.
Wesolowski, Amy, Nathan Eagle, Andrew J. Tatem, David L. Smith, Abdisalan M. Noor, Robert W. Snow, and Caroline O. Buckee. 2012. "Quantifying the Impact of Human Mobility on Malaria." *Science* 338 (6104): 267–70.
Wesolowski, Amy, Taimur Qureshi, Maciej F. Boni, Pål Roe Sundsøy, Michael A. Johansson, Syed Basit Rasheed, Kenth Engø-Monsen, and Caroline O. Buckee. 2015. "Impact of Human Mobility on the Emergence of Dengue Epidemics in Pakistan." *Proceedings of the National Academy of Sciences* 112 (38): 11887–92.
Yan, Xiang, Lu Shan, Shenjing He, and Jiekui Zhang. 2022. "Cross-City Patient Mobility and Healthcare Equity and Efficiency: Evidence from Hefei, China." *Travel Behaviour and Society* 28: 1–12.
Yu, Lixiaona, Tao Hu, Taiping Liu, and Yunyu Xiao. 2024. "Using Smartphone User Mobility to Unveil Actual Travel Time to Healthcare: An Example of Mental Health Facilities." *Health & Place* 90: 103375.

17 The Challenges and Future Research for Human Mobilities in Nature Disasters

Junkang Xu and Chao Fan

17.1 INTRODUCTION

Natural disasters have been causing widespread social harm, including casualties and economic losses. The United Nations Office for Disaster Risk Reduction pointed out that natural disasters resulted in 1.2 million deaths and affected the lives of 2.9 billion people between 2000 and 2012 (UNISDR, 2013). The effectiveness of people's preparation and responses largely determines the disaster impact. Human mobility has emerged as a critical tool for understanding how populations respond to such crises and how disasters alter people's behaviors. The development and widespread adoption of various positioning technologies and social media platforms have provided advanced technical support to collect and analyze human mobility data, providing capabilities of observing anomalous behaviors during disasters and assessing societal resilience in the aftermath, thereby informing strategies for targeted interventions and the safe, efficient mobility of people. However, highly aggregated data may not be able to reflect the details of mobility and their associated sociodemographic attributes, which may hinder the understanding and predictability of human behaviors. Hence, it is critical to look into individual-level human mobility data in the context of natural disasters. However, effective representation and examination of mobility at the individual level remain underdeveloped due to data limitations and ethical problems. The subsequent sections explore the societal challenges of individual mobility in disaster management and outline potential future research directions to address these gaps.

17.2 STATE-OF-THE-ART RESEARCH ON HUMAN MOBILITY IN DISASTERS

17.2.1 The Application of Human Mobility: During and After Disasters

Mobility patterns undergo significant disruptions during disasters, with routine movements largely altered by evacuations, infrastructure damage, and risk perception. Understanding these disruptions is crucial for assessing their effects on individuals' daily activities, access to essential services, and overall societal resilience. Real-time insights into how disasters affect population movements can be obtained from global positioning system (GPS) data of mobile devices, geolocated social media posts, and other digital trace platforms. For example, GPS data have been instrumental in analyzing evacuation patterns during events like Hurricane Sandy, revealing how people respond to evacuation orders and the effectiveness of these orders (Wang & Taylor, 2014). The large-scale, high-resolution location data collected from anonymous mobile phone users help to describe human mobility patterns in coastal counties in the United States and the resulting flood risks, indicating that mobility behavior plays an extremely important role in expanding the flood exposure of different sociodemographic groups (Li et al., 2024). In addition to GPS data, geolocated social media has emerged as a powerful tool for tracking both movement and sentiment during crises. Studies following events such as the 2011 Tohoku earthquake in Japan have shown that Twitter data can be used to map evacuation behaviors and public responses in real time, providing a crucial layer of

DOI: 10.1201/9781003503262-21

situational awareness (Sakaki et al., 2010; Acar & Muraki, 2011). An analysis of social media videos from China helped reveal the occurrence and processes of natural and built environment failures during Typhoon Lekima, providing insights into the full scope of complex events triggered by the typhoon (Guo et al., 2024).

Transportation network data have been integrated with mobility data to assess the functionality of critical infrastructure during disasters. This integration provides new insights regarding how disruptions in transportation systems exacerbate or mitigate the impact on human mobility, offering a comprehensive picture of disaster impacts. In particular, during Hurricane Harvey, mobility data combined with road network information helped identify which areas were most affected by flooding and how this influenced evacuation routes and access to emergency services (Balakrishnan et al., 2020; Gori et al., 2020). The integration of such diverse data sources, including satellite imagery and sensor networks, also enhances the granularity of mobility studies, allowing for detailed assessments of how different communities are impacted based on their location and infrastructure resilience (Goodchild & Glennon, 2010).

After a disaster, the return to normal mobility patterns is often slow and uneven, reflecting challenges in recovery efforts such as the restoration of infrastructure, reestablishment of public services, and the psychological impact on affected populations (Cutter et al., 2003; Bonanno et al., 2010). These disruptions may have long-term effects on community resilience, as evidenced by the protracted recovery periods following events like Hurricane Katrina (Comfort et al., 2010). Mobility behaviors provide a vital lens through which this resilience can be assessed and characterized, with resilience referring to a population's ability to absorb, recover, and adapt following a disruptive event. By analyzing mobility data captured through GPS from mobile devices, geolocated social media posts, and transportation network data, community resilience is quantifiable by observing how quickly and effectively mobility patterns return to normal after a disruption. For instance, Lu et al. (2012) demonstrated that population displacement patterns following the Haiti earthquake could measure resilience by tracking how quickly people returned to their original locations. Similarly, studies on Hurricane Sandy revealed that mobility data could identify areas with delayed recovery, indicating lower resilience (Wang & Taylor, 2014).

Social media data have proven useful in capturing real-time responses and adaptations during crises. For example, Sakaki et al. (2010) showed how Twitter data could be used to detect early signs of disaster impact and human responses. Combining GPS and social media data helps construct a comprehensive picture of resilience, highlighting both the physical return to normalcy and the emotional and social adaptations that occur in the wake of a disaster. These insights are crucial for policymaking and disaster preparedness and responses. Furthermore, integrating mobility data sources with emerging technologies like artificial intelligence (AI)-driven predictive models enables a holistic view of both the quantitative shifts in movement and the qualitative experiences of affected populations, which not only aids in immediate disaster response but also informs long-term recovery and resilience planning, ensuring that interventions are more effectively targeted and equitable (Lu et al., 2012; Song et al., 2014).

17.2.2 High Interconnectivity of Human Mobility: With Social and Infrastructure Systems

Human mobility patterns, when analyzed in combination with social and infrastructure systems, provide valuable insights into the possible damages during disasters and postdisaster recovery experiences in different communities. When coupled with social systems, mobility data can reveal how individuals of different social strata interact and access resources during and after a disaster. For instance, mobility behaviors indicate how people from different socioeconomic backgrounds frequently show up in locations, such as restaurants or community centers, reflecting their social networks and access to resources (Eagle et al., 2009), which highlights disparities in how various groups experience and recover from a disaster.

Examining mobility alongside infrastructure systems helps to understand how the built environment influences movement and accessibility during a crisis. For example, individuals living in remote areas or near major highways may experience vastly different mobility disruptions compared to those in urban centers. People residing in areas with well-developed infrastructure, such as near highways or public transit, often have more robust mobility options, enabling quicker evacuation or access to emergency services(Arrighi et al., 2019; Murray-Tuite & Wolshon, 2013). Conversely, those in isolated or less developed areas may face significant mobility challenges, increasing their exposure and sensitivity to disasters, while decreasing their adaptability and capacity to respond effectively. Research has shown that mobility patterns reflect the resilience of infrastructure networks, as seen in studies of urban areas where the range and flexibility of activities are directly tied to the quality of local infrastructure (González et al., 2008). Integrating mobility data with analyses of social and infrastructure systems helps develop a new understanding of the hot spots of disaster impacts and the varied adaptive strategies employed by different populations in response to disasters, informing strategies to enhance resilience and disaster preparedness.

17.3 SOCIETAL CHALLENGES OF INDIVIDUAL MOBILITY FOR DISASTER MANAGEMENT

Understanding individual mobility during disasters presents significant challenges, particularly due to the diversity of movement patterns and the anonymous nature of the data typically collected, such as GPS traces or social media posts. While these datasets allow one to observe general mobility trends, the lack of individual information severely limits the ability to understand the specific characteristics and needs of different populations. For instance, without knowing an individual's socioeconomic background, age, health status, or physical abilities, it becomes nearly impossible to fully grasp how different groups are affected by a disaster (Baker, 2011). This gap in data prevents disaster management strategies from being tailored effectively to the needs of vulnerable populations, such as the elderly, disabled, or those with chronic health conditions, who may require more specialized assistance during evacuations (Cutter et al., 2003; Rufat et al., 2015).

Mobility disruptions during disasters can exacerbate these challenges by affecting individuals' abilities to evacuate, access emergency services, and maintain social and economic activities. Transportation infrastructure is often at the frontier in the face of disasters, where road closures, damaged bridges, and disrupted public transit systems significantly hinder mobility activities (Litman, 2006). For populations reliant on public transportation or those with limited mobility, these disruptions can be particularly devastating, trapping them in hazardous areas or preventing them from reaching essential services (Tri et al., 2020). Without detailed data linking mobility patterns to specific individuals, it becomes difficult to assess which groups are most at risk and how best to support them during such events.

Environmental factors further complicate the understanding of mobility behaviors during disasters. Geographic features, such as mountains, rivers, and coastlines, may add an extra burden to mobility and limit the effectiveness of transportation networks (Kates et al., 2006; Wang & Taylor, 2014). Additionally, environmental conditions such as extreme weather can have varying impacts on different segments of the population (Gu et al., 2024). For instance, extreme heat may disproportionately affect older adults, while younger individuals might be less impacted (Sampson et al., 2013). However, without individual-level data, the differential impact of such environmental factors remains largely unknown, making it challenging to predict or mitigate risks effectively (Mitchell, 2003; González et al., 2008).

The diversity of individual mobility presents substantial challenges for disaster management. Wealthier individuals often have access to personal vehicles, enabling rapid and flexible mobility modes and routes, while lower-income populations might rely on public transportation, which is frequently disrupted during disasters (Tierney, 2006; Baker, 2011). Cultural and linguistic barriers can also impact how different communities respond to emergency orders and access information, potentially leading to delayed or incomplete response actions (Lindell & Perry, 2003; Wang et al., 2021). Vulnerable populations, such as the elderly, disabled, or those with chronic health conditions, may face significant challenges in mobility, requiring tailored evacuation strategies that include assistance and special transportation (Cutter et al., 2003; Rufat et al., 2015).

In summary, the interactions among mobility behaviors, individual characteristics, and environmental contexts during disasters underscore the need for more granular data collection that respects privacy while enabling a nuanced understanding. Understanding these complexities will enable disaster management efforts to be equitable, minimizing harm across all segments of society and ensuring that emergency services can be deployed efficiently (Aldrich, 2012; Esnard & Sapat, 2014).

17.4 FUTURE RESEARCH DIRECTIONS

The future of disaster management research increasingly depends on the understanding of individual mobility behaviors, particularly through methods that respect privacy while yielding actionable insights. A significant challenge in current mobility data collection is the anonymity of the datasets, which, while protecting privacy, limits our ability to link mobility patterns to the personal characteristics that influence them. This gap makes it difficult to fully understand how specific individuals or groups are impacted by disasters and, consequently, to design effective interventions. Privacy-preserving techniques such as differential privacy and federated learning offer a promising solution, enabling the analysis of large-scale mobility data without compromising individual identities (Dwork & Roth, 2014; Kairouz et al., 2021). These techniques allow researchers to draw meaningful conclusions about behavioral patterns across different demographic groups, thereby facilitating more targeted and equitable disaster responses.

In addition to privacy-preserving analytics, coupling individual mobility data with broader social theories presents a critical research direction. By integrating data from mobility studies with sociological, psychological, and economic theories, researchers can gain deeper insights into the decision-making processes of individuals during disasters. For example, Granovetter's theory of social networks can help explain how social ties influence decisions such as whether to evacuate or shelter in place during emergencies (Granovetter, 1973). Similarly, Simon's concept of bounded rationality provides a framework for understanding how people make decisions under the stress and uncertainty typical of disaster scenarios (Simon, 1955). This interdisciplinary approach can uncover the underlying social, cultural, and economic factors that shape individual behaviors, leading to more accurate models of human response and more effective disaster management strategies.

Simulating individual mobility in the context of environmental extremes represents another vital research direction. Integrating mobility data with social theories and environmental variables can lead to the creation of sophisticated models that predict how people might behave under various disaster scenarios, such as hurricanes, floods, or wildfires (Helbing & Molnár, 1995; Thiele et al., 2014). These models are crucial for disaster preparedness, allowing authorities to visualize potential evacuation routes, identify vulnerable populations, and design more resilient infrastructures. For example, simulations could be used to anticipate how extreme heat might disproportionately affect older adults or those with chronic illnesses, enabling more effective mitigation strategies (Collins, 2005; Batty, 2008). However, the anonymization of individual data poses a challenge in accurately linking these simulations to real-world behaviors, making it essential to improve the granularity of data collection while maintaining privacy.

Understanding the role of the environment in mobility behaviors is crucial, particularly in how different environmental factors impact various segments of the population. For instance, extreme weather conditions, such as heat waves or flooding, may cause more damage to vulnerable groups, such as the elderly or economically disadvantaged, compared to others (González et al., 2008; Cutter & Finch, 2008). The lack of individual-level data makes it difficult to assess these differential impacts accurately, underscoring the need for future research to focus on developing more granular and context-aware data collection methods. These advances will enable a better understanding of how environmental factors influence mobility during disasters, leading to more effective and tailored disaster management strategies.

The integration of privacy-preserving analytics with social theories and advanced environmental simulations offers a groundbreaking approach to understanding and managing human mobility in disaster scenarios. These methods enhance predictive accuracy by anonymizing individual data while still linking mobility patterns to personal and environmental factors, allowing for a more nuanced understanding of social dynamics and decision-making processes during crises (Cutter & Finch, 2008; Aldrich, 2012). These approaches ensure equity in disaster response by tailoring interventions to the needs of different population segments, particularly those most vulnerable. Ultimately, this holistic framework contributes to sustainable resilience-building, strengthening communities over time and preparing for future challenges in a changing climate.

REFERENCES

Acar, A. & Muraki, Y. (2011). "Twitter for crisis communication: Lessons learned from Japan's tsunami disaster." International Journal of Web Based Communities, 7(3), 392–402.

Aldrich, D. P. (2012). Building resilience: Social capital in post-disaster recovery. University of Chicago Press.

Arrighi, C.; Pregnolato, M.; Dawson, R. J., & Castelli, F. (2019). Preparedness against mobility disruption by floods. Science of The Total Environment, 654, 1010–1022.

Baker, E. J. (2011). "Household preparedness for the aftermath of hurricanes in Florida." Applied Geography, 31(1), 46–52.

Balakrishnan, S.; Zhang, Z.; Machemehl, R., & Murphy, M. R. (2020). Mapping resilience of Houston freeway network during Hurricane Harvey using extreme travel time metrics. International Journal of Disaster Risk Reduction, 47, 101565.

Batty, M. (2008). "The size, scale, and shape of cities." Science, 319(5864), 769–771.

Bonanno, G. A.; Brewin, C. R.; Kaniasty, K., & Greca, A. M. L. (2010). "Weighing the costs of disaster: Consequences, risks, and resilience in individuals, families, and communities." Psychological Science in the Public Interest, 11(1), 1–49.

Collins, T. W. (2005). "Households, forests, and fire hazard vulnerability in the American West: A case study of a California community." Environmental Hazards, 6(1), 23–37.

Comfort, L. K.; Ko, K., & Zagorecki, A. (2010). "Coordination in rapidly evolving disaster response systems: The role of information." American Behavioral Scientist, 48(3), 295–313.

Cutter, S. L. (2003). "The Vulnerability of Science and the Science of Vulnerability." Annals of the Association of American Geographers, 93(1), 1–12.

Cutter, S. L.; Boruff, B. J., & Shirley, W. L. (2012). Social vulnerability to environmental Hazards. In Hazards vulnerability and environmental justice (pp. 143–160). Routledge, 2012. 143–160.

Cutter, S. L. & Finch, C. (2008). "Temporal and spatial changes in social vulnerability to natural hazards." Proceedings of the National Academy of Sciences, 105(7), 2301–2306.

Dwork, C. & Roth, A. (2014). "The algorithmic foundations of differential privacy." Foundations and Trends in Theoretical Computer Science, 9(3–4), 211–407.

Eagle, N.; Macy, M., & Claxton, R. (2009). "Network diversity and economic development." Science, 328(5981), 1029–1031.

Esnard, A.-M. & Sapat, A. (2014). Displaced by disaster: Recovery and resilience in a globalizing world. Routledge.

González, M. C.; Hidalgo, C. A., & Barabási, A.-L. (2008). "Understanding individual human mobility patterns." Nature, 453(7196), 779–782.

Goodchild, M. F. & Glennon, J. A. (2010). "Crowdsourcing geographic information for disaster response: A research frontier." International Journal of Digital Earth, 3(3), 231–241.

Gori, A.; Gidaris, I.; Elliott, J. R.; Padgett, J.; Loughran, K.; Bedient, P.; Panakkal, P., & Juan, A. (2020). "Accessibility and recovery assessment of Houston's roadway network due to fluvial flooding during Hurricane Harvey." Natural Hazards Review, 21(2), 4020005.

Granovetter, M. S. (1973). "The strength of weak ties." American Journal of Sociology, 78(6), 1360–1380.

Gu, X.; Chen, P., & Fan, C. (2024). "Socio-demographic inequalities in the impacts of extreme temperatures on population mobility11This research received no specific grant from any funding agency, commercial or not-for-profit sectors." Journal of Transport Geography, 114, 103755.

Guo, J.; Du, Y., & Fan, C. (2024). Inferring cascading effects of natural and built environments failures from social media videos in disasters. International Journal of Disaster Risk Reduction, 104, 104378.

Helbing, D. & Molnár, P. (1995). "Social force model for pedestrian dynamics." Physical Review E, 51(5), 4282–4286.

Kairouz, P.; McMahan, H. B.; Avent, B.; Bellet, A.; Bennis, M.; Bhagoji, A. N., … & Zhao, S. (2021). Advances and open problems in federated learning. Foundations and trends® in Machine Learning, 14(1–2): 1–210.

Kates, R. W.; Colten, C. E.; Laska, S., & Leatherman, S. P. (2006). "Reconstruction of New Orleans after Hurricane Katrina: A research perspective." Proceedings of the National Academy of Sciences, 103(40), 14653–14660.

Li, B.; Fan, C.; Chien, Y.-H.; Hsu, C.-W., & Mostafavi, A. (2024). Mobility behaviors shift disparity in flood exposure in U.S. population groups. International Journal of Disaster Risk Reduction, 108, 104545.

Lindell, M. K. & Perry, R. W. (2003). Communicating environmental risk in multiethnic communities. Sage publications.

Litman, T. (2006). "Lessons from Katrina and Rita: What major disasters can teach transportation planners." Journal of Transportation Engineering, 132(1), 11–18.

Lu, X.; Bengtsson, L. & Holme, P. (2012). "Predictability of population displacement after the 2010 Haiti earthquake." Proceedings of the National Academy of Sciences, 109(29), 11576–11581.

Mitchell, J. K. (2003). "European river floods in a changing world." Risk Analysis, 23(3), 567–574.

Murray-Tuite, P. & Wolshon, B. (2013). Evacuation transportation modeling: An overview of research, development, and practice. Transportation Research Part C: Emerging Technologies, 27, 25–45.

Rufat, S.; Tate, E.; Burton, C. G., & Maroof, A. S. (2015). Social vulnerability to floods: Review of case studies and implications for measurement. International Journal of Disaster Risk Reduction, 14, 470–486.

Sakaki, T.; Okazaki, M., & Matsuo, Y. (2010). "Earthquake shakes Twitter users: Real-time event detection by social sensors." In Proceedings of the 19th International Conference on World Wide Web (pp. 851–860).

Sampson, N. R.; Gronlund, C. J.; Buxton, M. A.; Catalano, L.; White-Newsome, J. L.; Conlon, K. C.; O'Neill, M. S.; McCormick, S., & Parker, E. A. (2013). "Staying cool in a changing climate: Reaching vulnerable populations during heat events." Global Environmental Change, 23(2), 475–484.

Simon, H. A. (1955). "A behavioral model of rational choice." The Quarterly Journal of Economics, 69(1), 99–118.

Song, C.; Qu, Z.; Blumm, N., & Barabási, A.-L. (2014). "Limits of predictability in human mobility." Science, 327(5968), 1018–1021.

Thiele, J. C.; Kurth, W., & Grimm, V. (2014). "Facilitating parameter estimation and sensitivity analysis of agent-based models: A cookbook using NetLogo and R." Journal of Artificial Societies and Social Simulation, 17(3), 11.

Tierney, K. (2006). Social inequality, hazards, and disasters. In On risk and disaster: Lessons from Hurricane Katrina (pp. 109–128). University of Pennsylvania Press.

Tri, N. K.; Wright, S. J.; Betts, J. M., & Vu, H. L. (2020). Studying the Impact of Public Transport on Disaster Evacuation. 2020 IEEE 23rd International Conference on Intelligent Transportation Systems (ITSC), 1–6.

UNISDR (The United Nations Office for Disaster Risk Reduction) (2013) Disasters Impacts 2000–2012. Available: http://www.preventionweb.net/files/31737_20130312disaster20002012copy.pdf. Accessed: 22 October 2014.

Wang, Q. & Taylor, J. E. (2014). "Quantifying human mobility perturbation and resilience in Hurricane Sandy." PLOS ONE, 9(11), e112608.

Wang, Y.; Kyriakidis, M., & Dang, V. N. (2021). Incorporating human factors in emergency evacuation – An overview of behavioral factors and models. International Journal of Disaster Risk Reduction, 60, 102254.

18 From Vulnerability to Resilience

Rethinking Urban Transportation in a Changing Climate

Yiyi He

18.1 INTRODUCTION

Anthropogenic climate change is changing the frequency, intensity, and duration of extreme weather events such as hurricanes, floods, heatwaves, and wildfires (National Academies of Sciences, Engineering, and Medicine et al. 2016; Seneviratne et al. 2021; Clarke et al. 2022). These events pose severe risks to critical infrastructure systems, mainly urban transportation networks. Flooding from rising sea levels and increased rainfall can damage roads, bridges, and railways. Prolonged heatwaves can cause pavements to buckle and rails to warp, leading to significant service disruptions. For example, Hurricane Sandy in 2012 caused the unprecedented shutdown of the New York City subway system, crippling one of the world's largest transit networks. Similarly, in 2021, a major winter storm in Texas brought record-low temperatures and icy conditions, resulting in widespread road closures, flight cancellations, and public transit disruptions. These incidents highlight the vulnerability of transportation infrastructure to climate extremes and underscore the need for proactive strategies to enhance resilience.

This chapter investigates climate change's diverse and far-reaching impacts on urban transportation networks, focusing on how key climate extremes disrupt their functionality. From physical damage to infrastructure and operational breakdowns to rising maintenance demands, the challenges posed by climate change necessitate a fundamental reevaluation of the resilience and adaptability of transportation systems. The chapter emphasizes the intricate interconnectedness of transportation networks, highlighting how these connections, while essential for efficiency, can amplify vulnerabilities and trigger cascading failures during disruptions. Finally, the chapter underscores the critical role of a systems approach in developing comprehensive strategies to address vulnerabilities and foster resilience in transportation networks.

18.2 RETHINKING THE INTER- AND INTRACONNECTIVITY OF TRANSPORTATION NETWORKS

Modern transportation networks are becoming increasingly interconnected, creating unprecedented complexity in their structure, operation, and management (He, Lindbergh, et al. 2021). A helpful framework to conceptualize this complexity is through the lens of an old concept: the *rhizome*. A rhizome is a plant-root system characterized as "a prostrate or subterranean root-like stem emitting roots and usually producing leaves at its apex" (Kövesi 2016). Beyond botany, the concept of the rhizome has been adopted in philosophy by postmodern theorists Gilles Deleuze and Félix Guattari. They propose the rhizome as a structure of interconnected, nonlinear, and nonhierarchical fibers with no fixed origin or endpoint. Unlike the tree structure, where all growth emerges from a single

DOI: 10.1201/9781003503262-22

trunk, the rhizome grows and spreads organically, with its nodes connected so that any point can be accessed independently of the others.

The concept of the rhizome offers a compelling lens through which to understand the structure and organization of modern transportation networks. Physically, transportation networks and rhizomes share several similarities. Some connections in transportation systems are visible—like roads, railways, or flight routes—while others, such as air traffic control systems or freight logistics, operate less conspicuously. From a high-level perspective, transportation systems often appear as independent and distinct. However, closer observation reveals that these networks are deeply interconnected, forming a singular, interdependent organism. Different transportation networks are interconnected, allowing goods, passengers, and data to transfer seamlessly across modes. These interconnections mirror the rhizome's hidden root structure, which links its seemingly independent stems into one cohesive system.

While the rhizome's interconnectivity fosters organic resilience, transportation networks are often more vulnerable to disruptions. Deleuze and Guattari describe the principle of asignifying rupture in rhizomes: even when a rhizome's stem is broken or damaged, the organism continues to grow and spread, reestablishing itself over time. Transportation networks, however, are far less adaptable to disruptions. A localized disruption, such as a railway breakdown, can ripple through interconnected systems, triggering further delays in freight delivery, passenger commutes, or airport operations. Unlike the rhizome, where redundancy and adaptability are inherent, transportation networks often lack sufficient redundancy due to cost constraints. Additionally, while rhizomes are self-sustaining and can regenerate independently, transportation systems rely on various coordinated resources and infrastructures. For example, air transportation networks depend on seamlessly integrating ground operations, fuel supply, air traffic control, and weather monitoring systems. A disruption in any of these subsystems—such as a power outage affecting air traffic control—can significantly impact the entire network (Little 2004; Schäfer et al. 2018). Similarly, in freight transportation, maritime ports such as the Port of Los Angeles depend on interconnected rail and trucking systems for the onward movement of goods. The entire network's efficiency is affected if a critical component, such as a rail line, is compromised.

The rhizome metaphor reveals both the strengths and limitations of modern transportation networks. While their interconnectivity and multimodality mirror the organic characteristics of rhizomes, transportation systems remain constrained by human design choices and resource dependencies. The redundancy inherent in natural rhizomatic systems is rarely present in human-constructed networks, where planners prioritize efficiency over adaptability. Unlike the rhizome, transportation networks are vulnerable to external disruptions, namely climate extremes' impacts. The challenge is building resilience in transportation systems in a changing climate. Addressing this requires rethinking traditional network design models and incorporating adaptive principles to ensure these interconnected systems can adapt and thrive amid evolving climate conditions.

18.3 CLIMATE EXTREMES: KEY INFLUENCERS

Extreme climate events have a profound impact on transportation networks, causing physical damage and operational disruptions. This section examines three critical hazards—floods, heatwaves, and earthquakes—and their specific implications for transportation systems.

18.3.1 Floods

Many regions worldwide, particularly urbanized areas, have witnessed increased flood intensity and frequency under a changing climate (Westra et al. 2014; Najibi and Devineni 2018; Breinl et al. 2021). Flood hazards can put a strain on transportation networks both short and long-term, causing direct physical damage to infrastructure and indirect impacts through a cascade of adverse effects that can severely impact human mobility, supply chain, and connecting infrastructure networks, leading to system-level failures (Carrera et al. 2015; Rebally et al. 2021). Floodwater inundates roads, bridges, railways, and tunnels, rendering them impassable and leading to widespread travel delays and congestion. Public transit systems, such as buses and subways, are particularly vulnerable, often necessitating service suspensions and rerouting (He, Thies, et al. 2021). The disruption affects daily commuters and impedes emergency services and supply chains as flooded routes impede the movement of ambulances, fire trucks, and other critical services, exacerbating the broader societal impact (Coles et al. 2017). In many developing nations, the repair and reconstruction efforts strain municipal budgets. At the same time, the loss of productivity due to disrupted commutes and delivery delays affects businesses and the local economy (Jha et al. 2011). Additionally, frequent and severe flooding events can deter investment in affected areas, leading to long-term economic consequences.

18.3.2 Heatwaves

Often known as the "invisible killer," heatwaves impact urban transportation systems physically and operationally. From a physical perspective, prolonged periods of extremely high temperatures can adversely affect road pavements, increasing rutting, cracking, and even melting in surface materials (Mulholland and Feyen 2021). These types of deformation reduce the lifespan of road transportation networks and increase the risk of vehicle damage and accidents. Similarly, rail tracks are susceptible to thermal expansion, resulting in buckling, derailments, and service delays or cancellations (Arkell and Darch 2006; Dobney et al. 2010; McEvoy, Ahmed, and Mullett 2012). Traditionally, the selection of materials and the design of transportation infrastructure have been, unfortunately, based on historical climatic data, which assumes relatively stable temperature ranges. However, with the anticipated increase in the frequency and intensity of extreme heat events due to climate change (Russo et al. 2019), these infrastructure networks will increasingly operate beyond their intended design parameters. From an operational standpoint, excessive heat strains many transportation networks, especially public transit systems. Higher temperatures increase the likelihood of equipment failures in buses, trains, and other transit vehicles. Components such as engines, electrical systems, and cooling mechanisms are particularly vulnerable to overheating, leading to more frequent breakdowns and the need for intensified maintenance efforts. This compromises service reliability and raises operational costs for transit agencies.

Moreover, the heightened demand for air conditioning during extreme heat events puts additional pressure on the electrical grid. This surge in energy consumption can overwhelm power infrastructure, increasing the risk of blackouts. In turn, power outages can severely disrupt transportation systems, resulting in the malfunction of critical elements such as traffic signals, subway systems, and electric-powered trains. These disruptions not only cause delays and inconvenience for commuters but also create safety risks, particularly in densely populated urban areas. Finally, heatwaves pose health risks to commuters, particularly those using non-air-conditioned transport modes or waiting at exposed transit stops (Fraser and Chester 2017). This can lead to heat-related illnesses, necessitating increased medical response efforts. The economic impact of heat waves on urban transportation is also significant. Increased maintenance and repair costs, combined with the potential for reduced public transit ridership due to discomfort and health concerns, can strain municipal budgets. Furthermore, delays and disruptions can affect productivity, as workers face longer commutes and businesses deal with supply chain interruptions.

18.3.3 Earthquakes

Earthquakes pose a significant threat to urban transportation systems, often causing extensive disruptions and infrastructure damage. The intense shaking from an earthquake can lead to the collapse of bridges, overpasses, and elevated roads, making key transportation routes impassable (Aghababaei, Costello, and Ranjitkar 2020; Yonson et al. 2020). Underground structures such as tunnels and subways are also at risk, as seismic activity can result in cave-ins, flooding from ruptured water mains, and electrical failures. In the immediate aftermath, transportation networks often experience severe traffic congestion as emergency response teams and evacuees navigate damaged and debris-laden roads. Additionally, rail systems can be derailed, and airports may be temporarily shut down for inspections and repairs (Wang et al. 2021; Gu et al. 2024). Public transit systems, including buses, light rail, commuter trains, and subways, are particularly vulnerable, with tracks and stations potentially suffering structural damage that necessitates prolonged service interruptions (Giuliano and Golob 1998).

The challenges are further compounded by disruptions in communication and coordination across different transportation modes, which can hinder evacuation and emergency response efforts. The disruptions and failures in multiple networks can impede the delivery of essential services and goods, exacerbating the crisis by limiting access to medical care, food supplies, and shelter. The economic consequences are equally profound: repairing damaged infrastructure, lost productivity, and disrupted supply chains can impose substantial costs on the urban economy (Aksoy et al. 2024). Businesses dependent on just-in-time delivery systems may face severe operational challenges, leading to financial losses or permanent closures. Beyond these tangible effects, the psychological toll on residents, coupled with diminished mobility, can undermine community resilience and recovery efforts in the aftermath of an earthquake.

18.4 FROM SINGLE HAZARD TO COMPOUND HAZARD

Historically, hazard risk management in the transportation sector has been reactive. The field's evolution has resulted from catastrophes, calamities, heightened risks, and newly identified threats that affect the population, economic stability, and infrastructure. This reactive approach works well with independent hazards (hazards are independent and caused by different triggers). If that is the case, the risk can be calculated by adding the average annual losses for various hazards. In such cases, transportation infrastructure damage from an earthquake, such as collapsed bridges or rail disruptions, can be considered separately from flooding events, such as washed-out roadways or inundated subway systems. Existing multihazard risk assessment tools commonly operate under this assumption, simplifying modeling by treating these hazards independently.

However, upon closer examination, the relationships between hazards and their impacts on transportation networks are frequently more intricate. For instance, ground shaking during an earthquake may destabilize cliffs and steep slopes, triggering landslides or rockfalls that block major transportation routes or damage rail lines. When compounded by extreme precipitation, landslides can exacerbate flooding by altering natural water flow patterns, further complicating recovery efforts and crippling transportation systems. These cascading impacts demonstrate the inadequacy of isolating hazards when assessing risks to interconnected infrastructure systems.

There is an urgent need to understand the spatial and temporal convergence of multiple hazards, their interactions with transportation systems, and the mechanisms by which one hazard induces another. The aggregate impact of such compound events can devastate infrastructure, communities, and economies. With increasing complexity and interconnectivity within transportation networks, compounded by greater climatic variability and uncertainty, the social and economic impacts of multihazard events are expected to escalate. Transportation networks, as critical lifelines, face heightened vulnerabilities in this context.

Despite growing recognition of the need to model compound hazards and conduct multihazard risk assessments, significant challenges remain. First, the characteristics of different hazards vary widely, necessitating distinct modeling approaches that are not easily integrated. For transportation networks, this might involve reconciling earthquake-induced structural failures with flood-induced road washouts in a single framework. Second, the underlying processes connecting one hazard to another—such as landslide-induced flooding following an earthquake—are often difficult to disentangle and model effectively. Third, the impacts of these processes on transportation systems are highly variable; for example, road networks may suffer prolonged closures, while rail systems may face operational derailments. Methods to describe and quantify vulnerability within transportation infrastructure differ significantly between hazard types. Finally, although various risk quantification measures exist, they require substantial adaptation to enable meaningful integration of risks across multiple hazards and their associated impacts on transportation networks.

18.5 BUILDING RESILIENCE IN TRANSPORTATION NETWORKS: A SYSTEMS PERSPECTIVE

Transportation networks are complex systems comprising physical infrastructure, users, and governance frameworks. These elements interact dynamically and are deeply intertwined with economic, social, and environmental systems. Climate extremes—such as hurricanes, floods, and heatwaves—can trigger cascading failures, disrupting mobility, financial stability, and emergency response efforts. In the face of these challenges, systems theory and systems thinking provide valuable frameworks that capture the interconnected nature of transportation networks and their interactions with broader environments. This is critical for understanding how vulnerabilities emerge and how to build resilience in these systems.

General Systems Theory, introduced by Ludwig von Bertalanffy, emphasizes that systems, whether biological, social, or mechanical, consist of interconnected elements that interact dynamically with their environments. In 1968, Bertalanffy wrote that a system is "a complex of interacting elements and that they are open to and interact with their environments" (Laszlo and Krippner 1998). This perspective is especially pertinent to transportation networks, which function as open systems that continuously exchange energy, information, and material with their surroundings. Transportation systems are not static; they exhibit emergent properties and evolve in response to external pressures such as climate extremes.

Building on these theoretical foundations, systems thinking, introduced by Jay W. Forrester and further developed in interdisciplinary contexts, enables researchers and practitioners to analyze how transportation networks interconnect with broader social, economic, and environmental systems. According to Forrester, systems thinking is an approach to analysis that zeroes in on how the different parts of a system interrelate and how systems work within the context of other, more extensive systems (Gharajedaghi 2011). This holistic approach is indispensable for addressing climate-related challenges in the transportation sector. Applying systems thinking to transportation resilience begins with recognizing the interconnectedness of hazards, vulnerabilities, and capacities within the system. For instance, a systems analysis of flood-prone urban areas might involve mapping the interactions between rainfall patterns, drainage infrastructure, land use, and population density. Such an analysis would reveal not only direct vulnerabilities (e.g., low-lying roads susceptible to inundation) but also indirect impacts (e.g., disruptions to emergency response routes).

Climate resilience in transportation networks also benefits from integrating systems thinking into disaster management strategies. As highlighted in integrated disaster management, resilience planning must consider the interactions among the natural environment, human behavior, and the built environment. For example, developing a comprehensive evacuation plan during a hurricane requires identifying at-risk transportation routes, coordinating with emergency services, assessing community vulnerabilities, and considering environmental factors such as flooding patterns. This interdisciplinary approach aligns with von Bertalanffy's assertion that systems thinking involves a fundamental shift in scientific and practical methodologies—a move from reductionist to holistic problem-solving paradigms. In this context, transportation resilience is not simply about fortifying infrastructure but also about fostering adaptive capacities within the system.

Ultimately, building resilience in transportation networks in the face of climate extremes requires reimagining planning and management practices. By leveraging the principles of systems theory and the analytical tools of systems thinking, policymakers, engineers, and planners can develop strategies that account for the complex interplay of factors influencing transportation resilience. These strategies include designing redundant and flexible infrastructure, enhancing real-time communication systems, and fostering collaboration across sectors and jurisdictions. Transportation networks, as open systems, must adapt to the evolving realities of climate change. Systems thinking provides a lens through which these networks can be understood and strengthened, ensuring their ability to withstand and recover from the impacts of climate extremes while continuing to support societal and economic functions.

18.6 CONCLUDING REMARKS

Recognizing and addressing the inter- and intraconnectivity within and between transportation systems is paramount for accurately assessing climate risks and planning effective resilience strategies. Oversimplified models that overlook these intricate interdependencies risk underestimating the impacts of climate hazards and producing inadequate measures to safeguard transportation systems. As transportation networks grow increasingly complex and interconnected and uncertainties surrounding future climatic conditions intensify, the field of transportation planning and disaster risk management faces a critical juncture.

This chapter underscores the necessity of adopting new paradigms and methodologies tailored to transportation systems under climate stress. Emerging geospatial tools, network metrics, and integrated spatial modeling offer promising pathways to assess vulnerabilities and predict disruptions across multimodal transportation networks. These approaches enable planners to simulate cascading effects, identify critical vulnerabilities, and incorporate uncertainty into decision-making processes, ensuring a more robust foundation for risk management.

Planning must evolve beyond traditional, siloed approaches and embrace holistic, systems-oriented strategies to build resilience in transportation systems. This includes fostering interdisciplinary collaboration, leveraging technological advancements, and addressing the unique complexities of transportation systems as lifelines for urban mobility. Adaptive design principles must guide the development of redundant and flexible transportation infrastructure, supported by real-time monitoring and enhanced coordination across transit modes. These measures are essential for withstanding and recovering from the compounding effects of climate extremes.

Mahatma Gandhi noted, "The future depends on what you do today." This sentiment is especially relevant to building resilient transportation systems that support urban communities and economies in an era of climate uncertainty. Proactive, scientifically informed planning is essential to safeguarding transportation networks from immediate threats and ensuring their sustainability as critical arteries of mobility and economic activity.

REFERENCES

Aghababaei, Mohammad, Seosamh B Costello, and Prakash Ranjitkar. 2020. "Transportation impact assessment following a potential Alpine fault earthquake in New Zealand." *Transportation Research Part D: Transport and Environment* 87:102511.

Aksoy, Cevat Giray, Maxim Chupilkin, Zsoka Koczan, and Alexander Plekhanov. 2024. "Unearthing the impact of earthquakes: A review of economic and social consequences." *Journal of Policy Analysis and Management*, 1–22.

Arkell, Brian P, and GJC Darch. 2006. "Impact of climate change on London's transport network." In *Proceedings of The Institution of Civil Engineers- Municipal Engineer*, 159:231–237. 4. Thomas Telford Ltd.

Breinl, Korbinian, David Lun, Hannes Müller-Thomy, and Günter Blöschl. 2021. "Understanding the relationship between rainfall and flood probabilities through combined intensity-duration-frequency analysis." *Journal of Hydrology* 602:126759.

Carrera, Lorenzo, Gabriele Standardi, Francesco Bosello, and Jaroslav Mysiak. 2015. "Assessing direct and indirect economic impacts of a flood event through the integration of spatial and computable general equilibrium modelling." *Environmental Modelling & Software* 63:109–122.

Clarke, Ben, Friederike Otto, Rupert Stuart-Smith, and Luke Harrington. 2022. "Extreme weather impacts of climate change: An attribution perspective." *Environmental Research: Climate*, 1(1):012001.

Coles, Daniel, Dapeng Yu, Robert L Wilby, Daniel Green, and Zara Herring. 2017. "Beyond 'flood hotspots': Modelling emergency service accessibility during flooding in York, UK." *Journal of Hydrology* 546:419–436.

Dobney, Kay, CJ Baker, Lee Chapman, and AD Quinn. 2010. "The future cost to the United Kingdom's railway network of heat-related delays and buckles caused by the predicted increase in high summer temperatures owing to climate change." *Proceedings of the Institution of Mechanical Engineers, Part F: Journal of Rail and Rapid Transit*, 224(1):25–34.

Fraser, Andrew M, and Mikhail V Chester. 2017. "Transit system design and vulnerability of riders to heat." *Journal of Transport & Health* 4:216–225.

Gharajedaghi, Jamshid. 2011. *Systems thinking: Managing chaos and complexity: A platform for designing business architecture.* Elsevier.

Giuliano, Genevieve, and Jacqueline Golob. 1998. "Impacts of the Northridge earthquake on transit and highway use." *Journal of Transportation and Statistics*, 1(2): 1–20.

Gu, Yue, Mirjam Wiedemann, Robert Freestone, Henrik Rothe, and Nicholas Stevens. 2024. "The impacts of shock events on airport management and operations: A systematic literature review." *Transportation Research Interdisciplinary Perspectives* 27:101182.

He, Yiyi, Sarah Lindbergh, Yang Ju, Marta Gonzalez, and John Radke. 2021. "Towards resilient critical infrastructures: Understanding the impact of coastal flooding on the fuel transportation network in the san francisco bay." *ISPRS International Journal of Geo-Information*, 10(9):573.

He, Yiyi, Stephan Thies, Paolo Avner, and Jun Rentschler. 2021. "Flood impacts on urban transit and accessibility—A case study of Kinshasa." *Transportation Research Part D: Transport and Environment* 96:102889.

Jha, Abhas, Jessica Lamond, Robin Bloch, Namrata Bhattacharya, Ana Lopez, Nikolaos Papachristodoulou, Alan Bird, David Proverbs, John Davies, and Robert Barker. 2011. "Five feet high and rising: cities and flooding in the 21st century." *World Bank Policy Research Working Paper,* no. 5648.

Kövesi, Simon. 2016. "John Clare & Deleuze and Guattari's Rhizome." In *Ecology and the Literature of the British Left*, 75–88. Routledge.

Laszlo, Alexander, and Stanley Krippner. 1998. "Systems theories: Their origins, foundations, and development." In *Advances in psychology*, 126:47–74. Elsevier.

Little, Richard. 2004. "A socio-technical systems approach to understanding and enhancing the reliability of interdependent infrastructure systems." *International Journal of Emergency Management*, 2(1–2):98–110.

McEvoy, Darryn, Iftekhar Ahmed, and Jane Mullett. 2012. "The impact of the 2009 heat wave on Melbourne's critical infrastructure." *Local Environment*, 17(8):783–796.

Mulholland, Eamonn, and Luc Feyen. 2021. "Increased risk of extreme heat to European roads and railways with global warming." *Climate Risk Management* 34:100365.

Najibi, Nasser, and Naresh Devineni. 2018. "Recent trends in the frequency and duration of global floods." *Earth System Dynamics*, 9(2):757–783.

National Academies of Sciences, Engineering, and Medicine; Division on Earth and Life Studies; Board on Atmospheric Sciences and Climate; Committee on Extreme Weather Events and Climate Change Attribution. 2016. *Attribution of extreme weather events in the context of climate change*. National Academies Press. https://doi.org/10.17226/21852

Rebally, Aditya, Caterina Valeo, Jianxun He, and Saeid Saidi. 2021. "Flood impact assessments on transportation networks: A review of methods and associated temporal and spatial scales." *Frontiers in Sustainable Cities* 3:732181.

Russo, Simone, Jana Sillmann, Sebastian Sippel, Monika J Barcikowska, Claudia Ghisetti, Marek Smid, and Brian O'Neill. 2019. "Half a degree and rapid socioeconomic development matter for heatwave risk." *Nature Communications*, 10(1):136.

Schäfer, Benjamin, Dirk Witthaut, Marc Timme, and Vito Latora. 2018. "Dynamically induced cascading failures in power grids." *Nature Communications*, 9(1):1975.

Seneviratne, Sonia I, Xuebin Zhang, Muhammad Adnan, Wafae Badi, Claudine Dereczynski, A Di Luca, Subimal Ghosh, Iskhaq Iskandar, James Kossin, and Sophie Lewis, et al. 2021. "*Weather and climate extreme events in a changing climate*." In Masson-Delmotte, V. P., Zhai, A., Pirani, S. L. and Connors, C. (eds.) *Climate Change 2021: The Physical Science Basis: Working Group I contribution to the Sixth Assessment Report of the Intergovernmental Panel on Climate Change.* Cambridge University Press.

Wang, Qianzhi, Kai Liu, Ming Wang, and Elco E Koks. 2021. "A river flood and earthquake risk assessment of railway assets along the belt and road." *International Journal of Disaster Risk Science* 12:553–567.

Westra, Seth, Hayley J Fowler, Jason P Evans, Lisa V Alexander, Peter Berg, Fiona Johnson, Elizabeth J Kendon, Geert Lenderink, and NM Roberts. 2014. "Future changes to the intensity and frequency of short-duration extreme rainfall." *Reviews of Geophysics*, 52(3):522–555.

Yonson, Rio, Ilan Noy, Vivienne C Ivory, and Chris Bowie. 2020. "Earthquake-induced transportation disruption and economic performance: The experience of Christchurch, New Zealand." *Journal of Transport Geography* 88:102823.

19 Human Mobility in the Postpandemic World

Yihong Yuan, Pingping Wang, Grant McKenzie, Mengqi Sun, Yang Xu, and Priyanka Verma

19.1 INTRODUCTION

The COVID-19 pandemic has undeniably reshaped urban life across the globe, particularly in how people move within and between urban spaces (de Palma, Vosough, and Liao 2022; Kopecka-Piech and Łódzki 2022; Loa et al. 2021). The disruption caused by the pandemic provided an unprecedented challenge for policymakers to revisit and innovate transportation and planning strategies. These changes affect everything from daily commutes to international travel (Huang et al. 2022a; Ren et al. 2022). As cities begin to recover from the impacts of the pandemic, it is crucial to examine these transformations to understand the future of urban mobility (Qiang and McKenzie 2024).

Cities are complex systems made by various processes and elements (Batty 2005; Yuan and Raubal 2016). The behavior of individuals, who serve as the basic elements in an urban system, can be broadly abstracted into the spatial and temporal characteristics of that system (Ahas et al. 2010; Yuan, Raubal, and Liu 2012). Therefore, individual mobility influences economic dynamics, social interactions, and access to services and opportunities (Li and Liu 2019; Xu et al. 2021). The importance of studying urban mobility in a postpandemic world cannot be overstated. The changes witnessed during the pandemic have highlighted the vulnerabilities and strengths of existing transport systems (Loa et al. 2021; Qiang and McKenzie 2024). For example, social distancing measures led to a drastic decrease in the use of public transport, whereas the fear of contagion drove an increase in alternative modes of transport such as biking and walking (De Vos 2020). These shifts have prompted cities to rethink their urban spaces and reevaluate existing infrastructure.

Moreover, the pandemic has underscored the inequalities in urban mobility that affect different demographic groups in various ways (Chow et al. 2020; Huang et al. 2022b). Low-income communities often faced the dual challenge of greater exposure to the virus due to dependency on public transport and greater economic vulnerability due to job losses in sectors relying more on physical presence, such as the service industry. Understanding these disparities is crucial for updating policies that promote more equitable and sustainable mobility.

This chapter aims to provide a comprehensive overview of the immediate and long-term effects of the pandemic, the socioeconomic inequalities it has intensified or lessened, and the changes in mobility patterns across different sectors. By understanding these changes, urban planners, policymakers, and researchers can better plan for resilient and equitable urban transport systems. We will highlight not only the challenges faced but also the opportunities that lie ahead for reimagining urban mobility in the postpandemic era. As we move forward from the pandemic, it becomes increasingly clear that the decisions made today will shape the urban mobility landscape of tomorrow.

The rest of this chapter is organized as follows. Section 19.2 examines the immediate impacts of the pandemic through a literature review and a case study. The case study focuses on the geographic heterogeneity of urban mobility changes during COVID-19. Section 19.3 dives into changes within specific sectors such as public transportation, tourism, retail, and commuting. Section 19.4 discusses the policy implications and future outlook of urban mobility in a world that continues to adapt to the realities after COVID-19. We conclude this work in Section 19.5.

DOI: 10.1201/9781003503262-23

19.2 THE IMMEDIATE IMPACTS OF COVID-19 ON URBAN MOBILITY

As mentioned in Section 19.1, the COVID-19 pandemic brought about immediate and significant changes in urban mobility patterns worldwide (Gao et al. 2020; Xiong et al. 2020). With the enforcement of lockdowns, stay-at-home mandates, and travel restrictions, there was a dramatic reduction in movement within cities. People drove less, and public transport usage significantly declined (Bouzaghrane et al. 2024). For instance, the Netherlands experienced a dramatic reduction in travel during the early lockdown period (van der Drift, Wismans, and Olde Kalter 2022).

The pandemic inevitably triggered changes in activity patterns and schedule habits. Contrary to the expectations that flexible work arrangements and telecommuting would lead to more irregular schedules, people developed stronger schedule habits (Edsberg Møllgaard, Lehmann, and Alessandretti 2022; Huang et al. 2020; Islind, Óskarsdóttir, and Steingrímsdóttir 2020). This suggests a complex relationship between the pandemic and individual behavior. A longitudinal study from the United States revealed that while certain mobility indicators, like the number of weekly trips, returned to prepandemic levels, others such as at-home dwell times were still quite different from prepandemic times, which suggests a permanent behavioral shift (Nikiforiadis et al. 2022).

To illustrate these changes through a case study, we analyzed the 10 most populated Metropolitan Statistical Areas (MSAs) in the United States during the summer months (June to August) of 2019 and 2020. The mobility data were sourced from SafeGraph's Social Distancing Metrics V2.1 (SafeGraph 2021). The Mobility Index (MI) measures the proportion of devices not stationary at home relative to the total number of observed devices. This is a helpful indicator for understanding general mobility's activeness among the population.

In the summer of 2020, all 10 MSAs saw significant reductions in MI compared to 2019. For instance, in the Washington MSA, the mean MI decreased by 0.08. This decline reflects compliance with movement restrictions, stay-at-home mandates, travel restrictions, and business closures aimed at limiting virus spread. Notably, in early August 2020, the Miami MSA experienced a significant drop in MI from 0.71 on July 31 to 0.58 on August 2. This significant decrease reveals that a considerable number of residents decreased their trips outside the home during that time. This decline could be attributed to increased awareness and caution following the summer surges in COVID-19 cases. Additionally, localized health advisories, travel restrictions, and lockdown measures likely urged people to stay home.

Examining the geographic distribution of MI offers further insights into how mobility behaviors adjusted to COVID-19 restrictions. As shown in Figures 19.1 and 19.2, in the New York MSA, coastal areas were more active before COVID-19, with higher MI values. During the pandemic, the pattern became more scattered, indicating a broader distribution of mobility across the region. This dispersion can be attributed to strict lockdown measures and individual health concerns that pushed people to avoid crowded city centers and the shift to remote work.

In the Los Angeles MSA, peripheral census block groups (CBGs) were more active in 2019. During COVID-19, coastal areas became more active. This shift likely reflects people's preference for outdoor spaces and less densely populated areas.

On the contrary, before COVID-19, mobility was relatively scattered across the midwestern and southern MSAs, such as Chicago, Dallas, Houston, Philadelphia, Atlanta, and Boston. However, during the pandemic, mobility patterns shifted to focus on the peripheral areas. This change suggests a migration to suburban or rural areas to work remotely or minimize health risks while still participating in economic activities.

In the Washington MSA, high MI value was concentrated in the central region before COVID-19 but shifted westward during the pandemic. This move suggests a preference for less densely populated neighborhoods while maintaining proximity to economic activities. Similarly, in the Miami MSA, mobility was concentrated near coastal areas in 2019. However, during COVID-19, these areas saw reduced MI values, while inland regions showed increased mobility.

These geographic shifts highlight how mobility behaviors adjusted to COVID-19 restrictions, revealing regional disparities and new trends in mobility patterns. Overall, these patterns reveal a

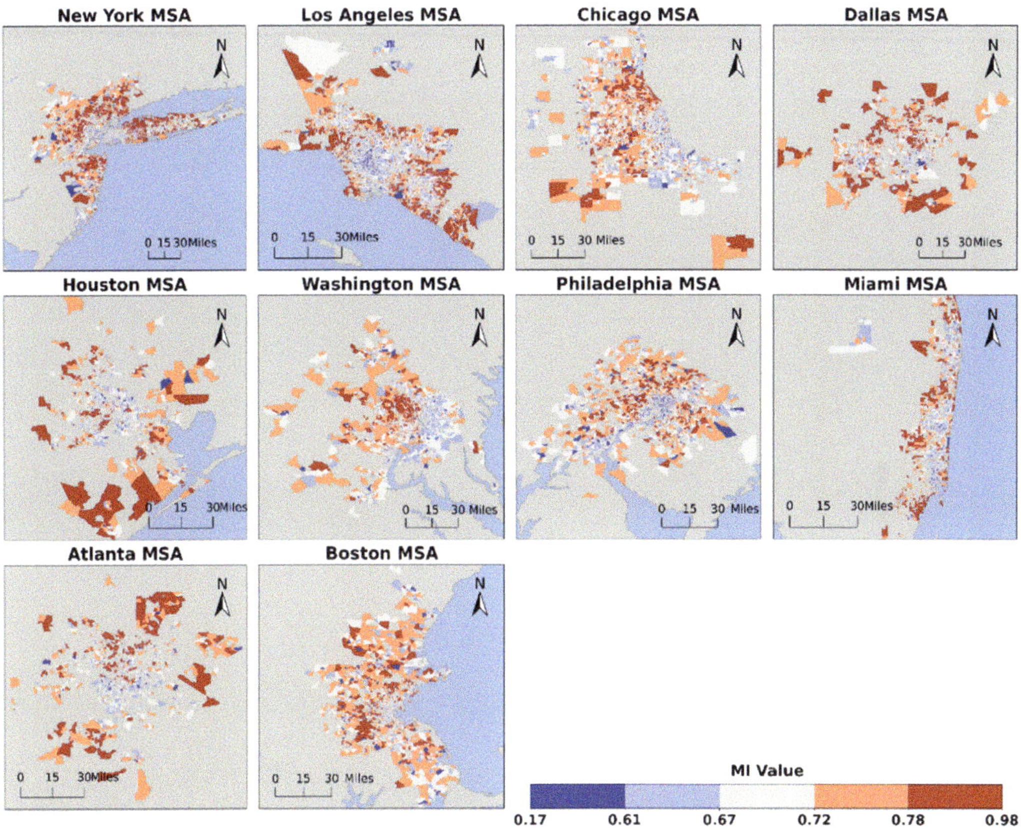

FIGURE 19.1 Distribution of Mobility Index (MI) in summer of 2019 across the 10 Metropolitan Statistical Areas (MSAs).

clear shift in mobility behavior during the pandemic, reflecting not only reductions in movement but also changes in how people adjusted their activity habits during this time.

Meanwhile, the pandemic highlighted existing disparities in transportation use (Sy et al. 2021; Wang et al. 2022) and residential mobility (Iglesias et al. 2019), where movement between neighborhoods varied based on socioeconomic contexts (Modai-Snir and Plaut 2019). The pandemic's impact on urban mobility was uneven across socioeconomic groups (Hong et al. 2021), with lower-income and less-educated individuals continuing to travel more due to their inability to work remotely (Brough, Freedman, and Phillips 2021). They faced higher viral exposure due to economic necessity and less favorable home conditions for prolonged isolation. Counties with a higher level of social poverty also showed a greater increase in COVID-19 cases and deaths, underscoring the disproportionate impact of the pandemic on disadvantaged communities (Garnier et al. 2021; Lucchini et al. 2023; Ossimetha et al. 2021). Understanding the diverse impact of the pandemic on different groups is crucial for developing equitable transportation policies and infrastructure that support all demographic groups, especially the most vulnerable. Future urban planning must prioritize resilience and accessibility to ensure that all communities can navigate both everyday life and crises effectively.

19.3 SECTOR-SPECIFIC CHANGES IN URBAN MOBILITY

In this section, we review the pandemic's impact on various sectors, such as public transportation, tourism, retail and service, and so on. These sector-specific changes offer insights into how urban mobility and economic activities have been reshaped by the pandemic, which provides a foundation for future planning and policy development.

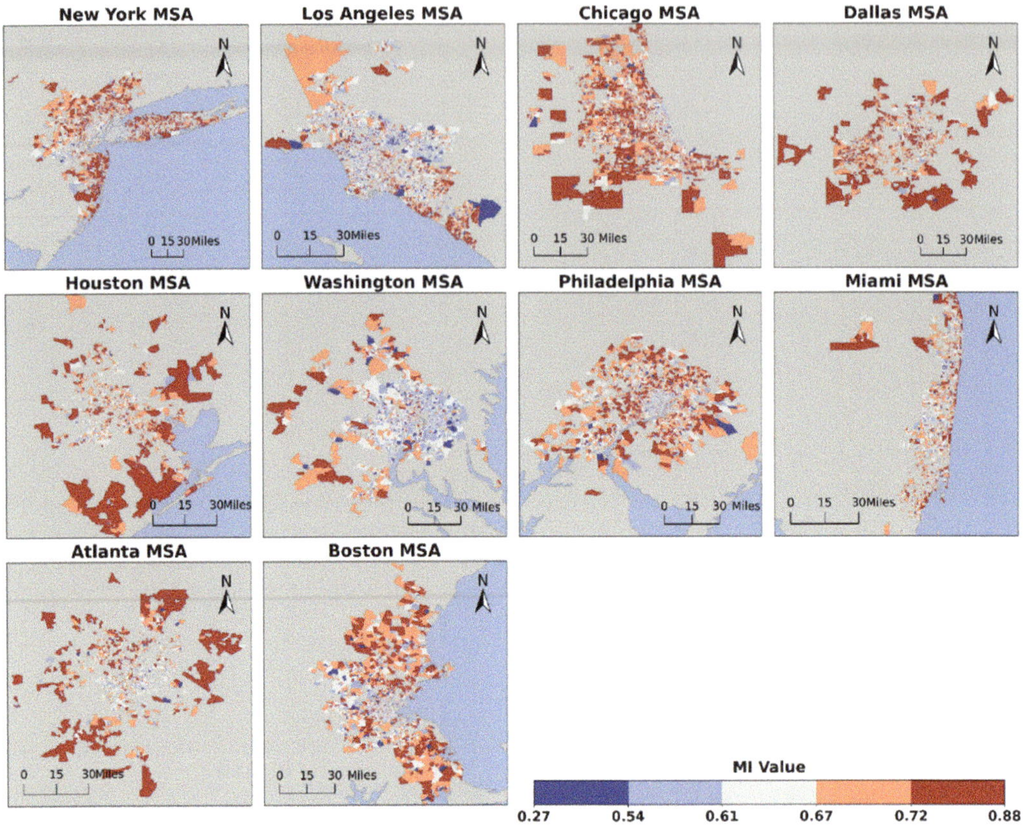

FIGURE 19.2 Distribution of Mobility Index (MI) in summer of 2020 across the 10 Metropolitan Statistical Areas (MSAs).

19.3.1 Public Transportation

The pandemic's long-term impacts on commuting behavior were notable, with a shift from public transport to individual transport modes such as personal cars and motorcycles (Badr et al. 2020). This shift was primarily driven by the perceived risk associated with different transportation modes during the pandemic (Dingil and Esztergár-Kiss 2021). In particular, the COVID-19 pandemic greatly impacted global urban transportation systems, particularly public transportation.

Ridership dropped across various countries and regions due to health concerns, lockdown measures, and other related factors (Hu and Chen 2021; Lin et al. 2023; Qi et al. 2023; Vickerman 2021). For instance, during the pandemic peak, tube and bus trips in the United Kingdom dropped by 94% and 83%, respectively (Qi et al. 2023). In the United States, Chicago was hit hard, with 95% of transit stations experiencing significant impacts and an average ridership decline of 72.4% (Hu and Chen 2021). Lin et al. (2023) analyzed public transit usage in Jeju, South Korea, using a 2-year smart card dataset, and they observed clear distinctions in hot spots of public transit usage at a station level during different pandemic stages (Figure 19.3). The authors found clear spatial heterogeneity in terms of public transit ridership changes. The ridership in popular areas, such as tourist sites, declined more substantially when the pandemic situation was severe, as people were likely to take caution and avoid possibly crowded sites.

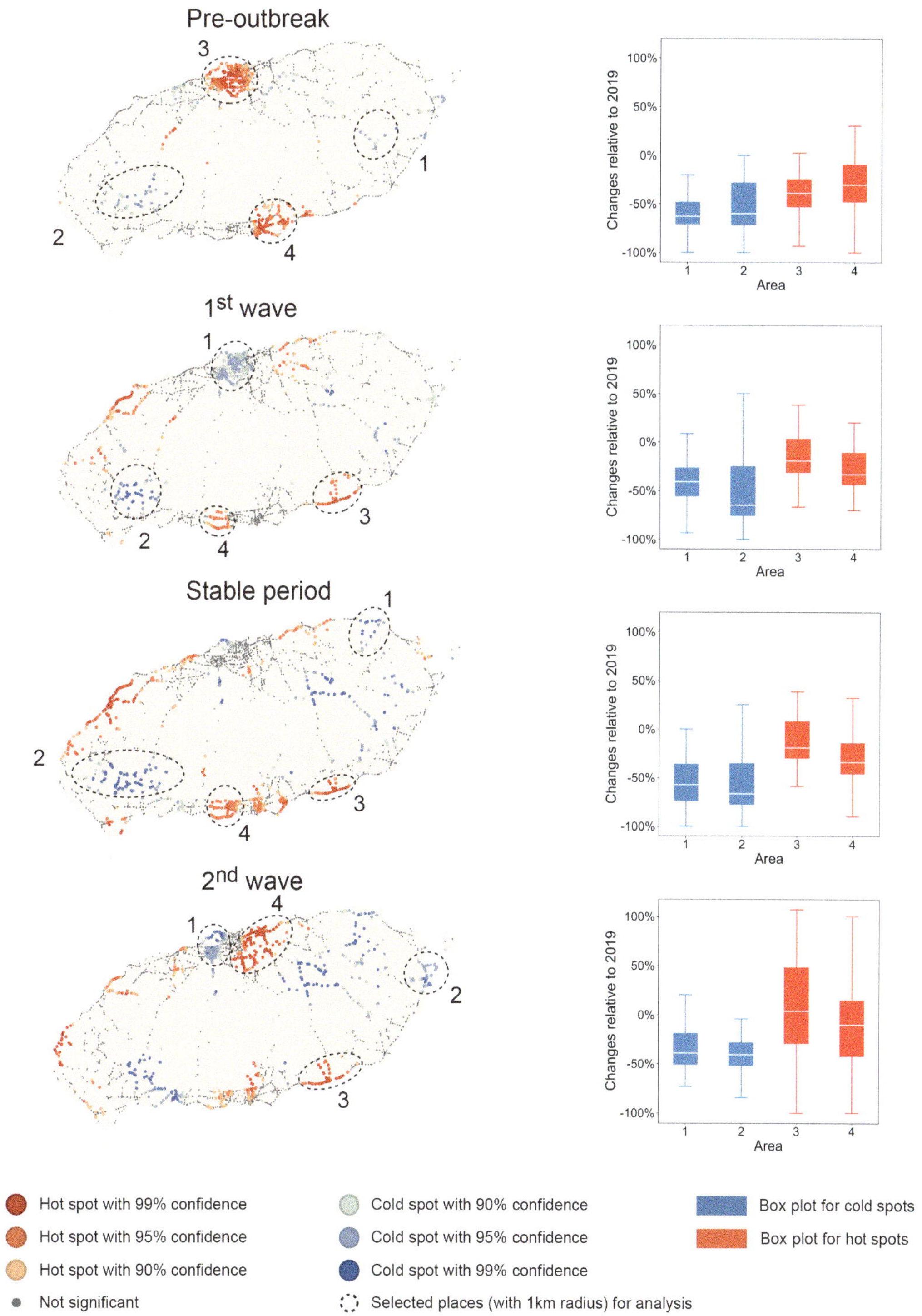

FIGURE 19.3 Observed hot spots of the changing public transit ridership at station level during the four pandemic stages. (This graphic was reproduced from Figure 7 in Lin et al. [2023] with written permission from the authors and publisher.)

As the pandemic eased, there was a general rebound in the use of public transportation. However, recent statistics from the United Kingdom indicated that public transport usage has not yet fully returned to its prepandemic level (Department for Transport 2024). This partial recovery can be attributed to several factors. First, the perception of risk continues to influence individuals' willingness to use public transit (Dong et al. 2021; Lin et al. 2024). The fear of being infected by the virus in crowded spaces like buses and trains deterred many people from using these public services. Second, the rise of telecommuting and remote working has had a profound effect on commuting patterns. With more people working from home, the demand for daily commutes has decreased, leading to a reduction in peak-hour traffic and, consequently, a decreased need for public transportation services (Vickerman 2021). Many companies have adopted hybrid work modes, which means fewer employees are traveling daily, impacting the overall demand for public transit.

In the postpandemic era, there has been a noticeable shift toward alternative modes of transportation. In many cities, the use of active transportation methods such as walking and cycling has increased, and these transportation methods are considered safer and more sustainable (Xu et al. 2023; Li et al. 2021). To support this shift, some cities have redefined road spaces, enhanced infrastructure, invested in facilities to accommodate bicycles and scooters, and encouraged the adoption of these modes of transport (Aloi et al. 2020; McKinsey 2020). This shift toward active transportation is not only a response to the pandemic but also a step toward achieving sustainability goals.

To sum up, the COVID-19 pandemic has reshaped public transportation. Although there has been a partial return to prepandemic usage levels, the long-term effects of increased health risk awareness, the rise in remote working, and a shift toward more active and sustainable transportation modes remain significant. These changes, although presenting challenges, offer cities a unique opportunity to transition toward more sustainable transportation systems.

19.3.2 Tourism and Travel

The tourism industry was one of the sectors hit the hardest by the COVID-19 pandemic, with travel restrictions and health risks significantly reducing tourist flows, leading to significant economic losses (González-Torres et al. 2021; Ren et al. 2022; Yang et al. 2020). During the pandemic, nearly 96% of the global population experienced some form of international travel restrictions, which resulted in a far-reaching impact on global tourism (Gössling et al. 2020; UNWTO 2020).

In response to the pandemic, there was a notable change in tourist flows and destination preferences. Historically popular and densely populated tourist hot spots experienced a decline in visitor numbers as travelers sought safer, less crowded alternatives. This led to a surge in interest toward remote, less populated destinations, which were considered to be safer and more controlled (Zenker and Kock 2020).

With the relaxation of travel restrictions, the phenomenon of "revenge travel" emerged, reflecting a surge of enthusiasm and urgency to resume travel and make up for lost time (Abdullah 2021). This trend is characterized by a rebound in demand for travel to fulfill postponed travel plans. Despite the rebound in demand, fear of traveling continues to linger among certain populations due to ongoing concerns about health and safety (Zenker et al. 2021). In addition, "untact" tourism has gained attention, particularly in South Korea, where minimizing physical contact has become a top priority in reducing health risks (Bae and Chang 2021). This trend has led to the rise of self-driving tours and camping, allowing tourists to enjoy their vacations while maintaining social distancing.

The accelerated digitalization of tourism has been another significant development. Virtual reality (VR) and augmented reality (AR) technologies have provided innovative ways for individuals to experience destinations from the comfort of their homes. This trend has not only provided a safe alternative for travel during the pandemic but also made tourism more accessible to those who are unable to travel due to physical or financial constraints (Talwar et al. 2023). Furthermore, the rise of the "workation" trend—where individuals merge work and leisure—has also shaped new travel patterns. This trend has been facilitated by the rise of remote work, allowing individuals to work from anywhere with a stable Internet connection (Bassyiouny and Wilkesmann 2023).

Overall, the postpandemic era has brought a range of new trends and changes to the tourism and travel sector, driven by changing behaviors, technological advancements, and evolving health and safety priorities. These developments will likely define the future trajectory of global tourism.

19.3.3 Retail and Services

Retail and service businesses in urban areas saw a significant decline in foot traffic as lockdowns and health concerns kept people at home (Rajpurohit 2021; Shah 2021). For instance, the foot traffic near stores in metropolitan areas remains 10% to 20% below prepandemic levels (Mischke et al. 2023). However, the number of active stores was almost the same as pre-COVID levels, with most store closures happening among independent stores, particularly in smaller municipalities. Independent stores accounted for over 90% of the overall decline in active stores and more than 70% of the drop in total foot traffic volume (Wang 2023).

The drop in foot traffic led to a rethinking of store accessibility. Retailers began to focus more on integrating new digital technologies with traditional retailing (Kotb 2020). For example, some added social media sales to their services (Cruz-Cárdenas et al. 2021). In the Puget Sound area, online shopping is a substitution for in-store shopping for groceries, while it complements in-store shopping for food and other items (Diaz-Gutierrez, Mohammadi-Mavi, and Ranjbari 2023). The number of businesses that operate both online and physical locations, known as click-and-mortar businesses, also grew during this period (Lashgari and Shahab 2022).

In addition, businesses had to innovate to survive the new normal. Many turned to delivery and curbside pickup models to maintain sales. For example, some added home delivery services to their previous services (Cruz-Cárdenas et al. 2021). Many businesses offered free contactless home delivery services and click-and-collect options (Grimmer 2022). Some restaurants expanded their delivery services, partnering with platforms like Uber Eats and Deliveroo (Ecker and Strüver 2022, Goodman 2023).

Additionally, some services went digital. Health and wellness services, like fitness classes and therapy sessions, moved online, offering virtual appointments (Silva-Jose et al. 2022; Wu et al. 2023). Retailers also invested in technology like augmented reality to allow customers to "try on" clothes or see how furniture would look in their homes without visiting a store (ICE Cobotics 2024; ThinkSys Inc. 2023). These changes in urban mobility for the retail and service sectors may have lasting effects, pushing businesses to continue integrating flexible, customer-centric service delivery models.

19.3.4 Work and Commuting

The shift to remote work during the pandemic has also been highlighted in studies across various disciplines like geography, transportation, urban planning, and economics. Studies show that remote work has reduced the need for commuting and contributed to a decrease in urban traffic congestion and emissions (Beck and Hensher 2020; Florida, Rodríguez-Pose, and Storper 2023; Lennon 2023). These studies also highlighted the potential long-term changes in how cities might be planned with a focus on resilience and sustainability.

However, this transition also emphasized the digital divide that may affect productivity and accessibility (Fairlie 2020). The digital divide describes the gap between demographics and regions that have access to modern information and communication technology and those that do not, and this issue became increasingly significant with the increase in teleworking. This divide affects not only access to technology but also the capability to engage effectively in a remote environment. A broader analysis by Robinson et al. (2020) highlighted how different levels of access to high-speed Internet and digital literacy skills can impact employee productivity and job opportunities. Without strategic interventions, the digital divide may cause bigger socioeconomic inequalities, particularly in less urbanized areas.

19.4 POLICY IMPLICATIONS AND FUTURE OUTLOOK

As discussed in Sections 19.1–19.3, the COVID-19 pandemic prompted a reevaluation of urban mobility and transport policies within cities. Here we review the policy implications and discuss the outlook of human mobility, current strategies adopted by cities, and recommendations for urban planners in the postpandemic world.

19.4.1 Strategies Implemented by Cities

As the pandemic accelerated existing trends like remote work and food delivery, it fundamentally altered urban life (Edsall 2022; Lai et al. 2020; Sharifi and Lee 2024). A key focus for city officials and transportation planners has been adapting to changing mobility patterns. Interestingly, many of these mobility strategies align with broader climate change goals. Here, we synthesize some of the broader strategies being implemented.

- *Urban Space Redesign*: Cities have responded by implementing various urban redesign initiatives to create more resilient, equitable, and healthy environments. To accommodate evolving mobility demands, urban centers have expanded bike lanes, implemented bike-sharing programs, and created pedestrian-friendly zones. Some cities have even closed streets to traffic, either seasonally or permanently. The concept of "pop-up" infrastructure has emerged as a strategy for rapid adaptation to changing conditions (Moran 2022).

  Green spaces have also gained prominence as essential components of urban well-being (Noszczyk et al. 2022). Cities are investing in parks, green roofs, and urban forests to improve air quality, mitigate climate change, and provide spaces for community engagement. These spaces are being designed with accessibility and inclusivity in mind, making them destinations for residents from all backgrounds (Jian, Luo, and Chan 2020).
- *Changes to Public Transport*: Vulnerabilities in public transportation systems were exposed during the pandemic, forcing transport officials to make substantial changes. Responses came in the form of adjusting schedules, managing capacity, and reconfiguring routes. Initially, diminished ridership led to reduced service frequency. As restrictions eased, cities increased service levels (UITP 2024) while grappling with the challenge of ensuring essential workers' mobility. On-demand transit and schedule prioritization for essential works (often from lower income communities) have emerged as potential solutions (Ashour et al. 2024).

  To address safety concerns, capacity limitations were implemented through measures such as passenger limits and seat-blocking (ITF 2023). While these restrictions are easing, capacity management remains crucial. Furthermore, altered travel patterns, including increased commuting to suburban areas and the rise of e-commerce, prompted route adjustments. Cities are adapting their networks to better serve these evolving transportation needs (APTA 2021).
- *Technology Integration*: The pandemic also accelerated the adoption of technology in urban transportation. Smart city initiatives that leverage sensor data and analytics are growing in interest (Megahed and Abdel-Kader 2022). These systems offer insights into mobility patterns, congestion, and public health concerns. Contactless payment systems, while massively beneficial for tracking and efficiency, have also raised concerns about accessibility for lower-income individuals (Golub et al. 2022).

- *Environmental Implications and Strategies*: It is worth mentioning that all these monumental shifts in mobility patterns led to a noticeable decrease in air pollution and greenhouse gas emissions. Cities around the world reported cleaner air and improved air quality indices (Slezakova and Pereira 2021). While emissions have rebounded (ESA 2021), cities are working to sustain these gains. Strategies to enhance air quality include promoting remote work, expanding public transit, and investing in active transportation infrastructure. A shift toward electric vehicles, both personal and shared (e.g., buses), is also underway. These initiatives address both climate change and public health concerns. Governments are implementing long-term environmental policies. These include carbon pricing (Kennedy, Ibrahim, and Hoornweg 2014), prioritizing low-carbon development and green spaces (Nieuwenhuijsen 2021), investing in clean(er) technologies (Hoang, Nguyen, and Others 2021), and collaborating across agencies, cities, and governments to address global climate change challenges.

19.4.2 Outlook and Recommendations

While cities have made significant strides in adapting to these changes, the path toward resilient, sustainable, and equitable urban mobility is long. To build upon the progress made, cities must prioritize long-term strategies that foster resilience, inclusivity, and environmental sustainability. Investing in sustainable infrastructure, creating adaptable urban environments, and ensuring equitable access to transportation are important steps toward achieving these goals. To prioritize public health, cities must continue to improve hygiene measures and capacity management and invest in real-time data collection and analytics. To enhance urban mobility resilience and promote equitable transportation solutions in the postpandemic world, we suggest that city officials, transportation planners, and policymakers focus their efforts on a few key areas, including but not limited to the following areas:

- *Invest in Sustainable and Resilient Infrastructure*: Cities should prioritize infrastructure that supports sustainable and adaptable mobility. This includes expanding public transportation options like electric buses and trains, as well as investing in bike-sharing systems. Additionally, continuing to implement smart technology to collect and analyze transportation data is essential for optimizing traffic flow, public transit schedules, and resource allocation.
- *Build Adaptive and Flexible Urban Systems*: To withstand future disruptions, cities must design flexible transportation systems. This involves creating multimodal spaces that can accommodate various transportation modes, such as bikes, e-scooters, cars, and buses. Infrastructure should be adaptable to rapidly changing mobility patterns, ensuring the city can respond effectively to future crises.
- *Prioritize Equity and Inclusion in Transportation*: Equitable access to transportation for all residents, including those in underserved communities, is essential. Those in marginalized communities were most impacted by the most recent global pandemic, so every effort should be made to engage with these communities when investing in accessible mobility solutions. Cities must address the disproportionate impact of transportation challenges on marginalized communities through targeted investments and inclusive policymaking. Meaningful community engagement is essential for developing transportation solutions that meet the diverse needs of the population.

Realizing these ambitions requires a proactive and adaptive approach, and continuous monitoring and evaluation of transportation policies and programs are essential for identifying emerging challenges and seizing new opportunities. A data-driven and evidence-based approach, coupled with active public engagement, will be instrumental in shaping the future of urban mobility.

19.5 CONCLUSION

In conclusion, the COVID-19 pandemic has caused lasting changes in urban mobility and revealed the vulnerabilities within our transport systems. This chapter comprehensively examined these shifts, their immediate impacts, enduring changes, and the broader implications for urban planning and policy. As we reflect on the past few years, it is evident that the pandemic did not merely disrupt normality in our lives; it also pushed us into a period of rapid adaptation and reevaluation of our urban spaces and mobility frameworks.

The initial shock of the pandemic resulted in a dramatic reduction in movement, as observed in Section 19.2. This reduction was not just a temporary response but has led to lasting changes in how we view and utilize transportation. Public transportation, for instance, has seen fluctuating ridership with a persistent concern about health risks. Meanwhile, the surge in alternative modes of transport like walking and biking underscores a shift toward more sustainable and health-conscious urban mobility. These shifts, while challenging, offer cities an unprecedented opportunity to reimagine and redesign urban spaces for greater sustainability, resilience, and equity.

Moreover, the pandemic has exposed the socioeconomic disparities in urban mobility, which demands a more focused approach to ensure that transportation systems are inclusive and equitable. The alleviated vulnerability of certain demographic groups during the pandemic has highlighted the need for policies that do not merely return to the prepandemic status quo but strive to rectify these inequalities. As cities plan going forward, integrating technology and innovative transport solutions will be key in developing responsive and adaptive urban transport systems that cater to the needs of all citizens.

Looking ahead, the resilience of urban mobility systems will heavily depend on the lessons drawn from this pandemic. Policymakers, urban planners, and researchers must collaborate to harness the insights gained from these difficult times to foster transportation systems that are not only efficient but also robust and inclusive. The path forward should aim not just to recover but to transform urban mobility in ways that better prepare cities for any future challenges.

This reevaluation and transformation of urban mobility are not merely a response to a crisis but proactive steps toward future-proofing our cities. As we navigate out of the pandemic era, the decisions we make today will undoubtedly shape the landscape of urban mobility for generations to come, ideally toward more equitable, sustainable, and resilient urban environments.

REFERENCES

Abdullah, M. N. A. 2021. "Revenge tourism: Trend or impact post-pandemic Covid-19?" In *Promoting Creative Tourism: Current Issues in Tourism Research*. Routledge.

Ahas, R., A. Aasa, S. Silm, and M. Tiru. 2010. "Daily rhythms of suburban commuters' movements in the Tallinn metropolitan area: Case study with mobile positioning data." *Transportation Research Part C-Emerging Technologies* 18 (1):45–54. https://doi.org/10.1016/j.trc.2009.04.011.

Aloi, A., B. Alonso, J. Benavente, R. Cordera, E. Echániz, F. González, C. Ladisa, Lezama-Romanelli, R., López-Parra, Á., Mazzei, V., Perrucci, L., Prieto-Quintana, D., Prieto-Quintana, D., Rodríguez, A., and Sañudo, R. 2020. "Effects of the COVID-19 lockdown on urban mobility: Empirical evidence from the city of santander (Spain)." *Sustainability* 12 (9):3870. https://doi.org/10.3390/su12093870.

APTA. 2021. "On the Horizon: Planning for Post-Pandemic Travel." https://www.apta.com/research-technical-resources/research-reports/on-the-horizon-planning-for-post-pandemic-travel/

Ashour, L. A., Q. Shen, A. Moudon, M. Cai, Y. Wang, and M. Brown. 2024. "Post-pandemic transit commute: Lessons from focus group discussions on the experience of essential workers during COVID-19." *Journal of Transport Geography* 116:103832.

Badr, H. S., H. Du, M. Marshall, E. Dong, M. M. Squire, and L. M. Gardner. 2020. "Association between mobility patterns and COVID-19 transmission in the USA: A mathematical modelling study." *The Lancet Infectious Diseases* 20 (11):1247–1254.

Bae, S. Y., and P.-J. Chang. 2021. "The effect of coronavirus disease-19 (COVID-19) risk perception on behavioural intention towards 'Untact' tourism in south korea during the first wave of the pandemic (March 2020)." *Current Issues in Tourism* 24 (7):1017–1035. https://doi.org/10.1080/13683500.2020.1798895.

Bassyiouny, M., and M. Wilkesmann. 2023. "Going on workation – Is tourism research ready to take off? Exploring an emerging phenomenon of hybrid tourism." *Tourism Management Perspectives* 46:101096. https://doi.org/10.1016/j.tmp.2023.101096.

Batty, M. 2005. *Cities and Complexity: Understanding Cities with Cellular Automata, Agent-Based Models, and Fractals.* Cambridge, Mass: MIT Press.

Beck, M. J., and D. A. Hensher. 2020. "Insights into the impact of COVID-19 on household travel and activities in Australia – The early days under restrictions." *Transport Policy* 96:76–93. https://doi.org/10.1016/j.tranpol.2020.07.001

Bouzaghrane, M. A., H. Obeid, M. González, and J. Walker. 2024. "Human mobility reshaped? Deciphering the impacts of the Covid-19 pandemic on activity patterns, spatial habits, and schedule habits." *EPJ Data Science* 13 (1):24.

Brough, R., M. Freedman, and D. C. Phillips. 2021. "Understanding socioeconomic disparities in travel behavior during the COVID-19 pandemic." *Journal of Regional Science* 61 (4):753–774.

Chow, T. E., Y. Choi, M. Yang, D. Mills, and R. Yue. 2020. "Geographic pattern of human mobility and COVID-19 before and after Hubei lockdown." *Annals of GIS* 27 (2):127–138. https://doi.org/10.1080/19475683.2020.1841828.

Cruz-Cárdenas, J., E. Zabelina, J. Guadalupe-Lanas, A. Palacio-Fierro, and C. Ramos-Galarza. 2021. "COVID-19, consumer behavior, technology, and society: A literature review and bibliometric analysis." *Technological Forecasting and Social Change* 173:121179.

de Palma, A., S. Vosough, and F. Liao. 2022. "An overview of effects of COVID-19 on mobility and lifestyle: 18 months since the outbreak." *Transportation Research Part A: Policy and Practice* 159:372–397. https://doi.org/10.1016/j.tra.2022.03.024.

De Vos, J. 2020. "The effect of COVID-19 and subsequent social distancing on travel behavior." *Transportation Research Interdisciplinary Perspectives* 5:100121. https://doi.org/10.1016/j.trip.2020.100121.

Department for Transport. 2024. "Domestic transport usage by mode." GOV.UK. https://www.gov.uk/government/statistics/transport-use-during-the-coronavirus-covid-19-pandemic/domestic-transport-usage-by-mode.

Diaz-Gutierrez, J. M., H. Mohammadi-Mavi, and A. Ranjbari. 2023. "COVID-19 impacts on online and in-store shopping behaviors: Why they happened and whether they will last post pandemic." *Transportation Research Record* 2678 (12):580–598. https://doi.org/10.1177/03611981231155169.

Dingil, A. E., and D. Esztergár-Kiss. 2021. "The influence of the Covid-19 pandemic on mobility patterns: The first Wave's results." *Transportation Letters* 13 (5–6):434–446.

Dong, H., S. Ma, N. Jia, and J. Tian. 2021. "Understanding public transport satisfaction in post COVID-19 pandemic." *Transport Policy* 101:81–88. https://doi.org/10.1016/j.tranpol.2020.12.004.

Ecker, Y., and A. Strüver. 2022. "Towards alternative platform futures in post-pandemic cities? A case study on platformization and changing socio-spatial relations in on-demand food delivery." *Digital Geography and Society* 3:100032.

Edsall, T. 2022. "Opinion | How a 'Golden Era for Large Cities' Might Be Turning Into an 'Urban Doom Loop'" (Published 2022) — nytimes.com.

Edsberg Møllgaard, P., S. Lehmann, and L. Alessandretti. 2022. "Understanding components of mobility during the COVID-19 pandemic." *Philosophical Transactions of the Royal Society A* 380 (2214):20210118.

ESA. 2021. "Air pollution returning to pre-COVID levels." https://www.esa.int/Applications/Observing_the_Earth/Copernicus/Sentinel-5P/Air_pollution_returning_to_pre-COVID_levels

Fairlie, R. 2020. "The impact of COVID-19 on small business owners: Evidence from the first three months after widespread social-distancing restrictions." *Journal of Economics & Management Strategy* 29 (4):727–740. https://doi.org/10.1111/jems.12400.

Florida, R., A. Rodríguez-Pose, and M. Storper 2023. "Critical commentary: Cities in a post-COVID world." *Urban Studies* 60 (8):1509–1531. https://doi.org/10.1177/00420980211018072.

Gao, S., J. Rao, Y. Kang, Y. Liang, and J. Kruse. 2020. "Mapping county-level mobility pattern changes in the United States in response to COVID-19." *SIGSpatial Special* 12 (1):16–26.

Garnier, R., J. R. Benetka, J. Kraemer, and S. Bansal. 2021. "Socioeconomic disparities in social distancing during the COVID-19 pandemic in the United States: Observational study." *Journal of Medical Internet Research* 23 (1):e24591.

Gössling, S., D. Scott, and C. Michael Hall. 2020. "Pandemics, tourism and global change: A rapid assessment of COVID-19." *Journal of Sustainable Tourism* 29 (1):1–20. https://doi.org/10.1080/09669582.2020.1758708.

Golub, A., A. Brown, C. Brakewood, J. MacArthur, S. Lee, and A. Ziedan. 2022. "Equity and exclusion issues in cashless fare payment systems for public transportation." *Transportation Research Interdisciplinary Perspectives* 15:100628.

González-Torres, T., J.-L. Rodríguez-Sánchez, and E. Pelechano-Barahona. 2021. "Managing relationships in the tourism supply chain to overcome epidemic outbreaks: The case of COVID-19 and the hospitality industry in spain." *International Journal of Hospitality Management* 92:102733. https://doi.org/10.1016/j.ijhm.2020.102733.

Goodman, D. 2023. "Between Physical Space and Digital Space: COVID-19, Platform Capitalism, and Changing Patterns of Food Provisioning." In *Transforming Agriculture and Foodways*, 71–87. Bristol University Press.

Grimmer, L. 2022. "Lessons from the COVID19 pandemic: The case of retail and consumer service firms." *Journal of Retailing and Consumer Services* 68:103012.

Hoang, A. T., and X. P. Nguyen, and Others. 2021. "Integrating renewable sources into energy system for smart city as a sagacious strategy towards clean and sustainable process." *Journal of Cleaner Production* 305:127161.

Hong, B., B. J. Bonczak, A. Gupta, L. E. Thorpe, and C. E. Kontokosta. 2021. "Exposure density and neighborhood disparities in COVID-19 infection risk." *Proceedings of the National Academy of Sciences* 118 (13):e2021258118.

Hu, S., and P. Chen. 2021. "Who left riding transit? Examining socioeconomic disparities in the impact of COVID-19 on ridership." *Transportation Research Part D: Transport and Environment* 90:102654. https://doi.org/10.1016/j.trd.2020.102654.

Huang, J., H. Wang, M. Fan, A. Zhuo, Y. Sun, and Y. Li. 2020. "Understanding the impact of the COVID-19 pandemic on transportation-related behaviors with human mobility data." Proceedings of the 26th ACM SIGKDD International Conference on Knowledge Discovery & Data Mining. pp. 3443–3450. https://doi.org/10.1145/3394486.3412856.

Huang, X., J. Y. Lu, S. Gao, S. C. Wang, Z. W. Liu, and H. X. Wei. 2022a. "Staying at home is a privilege: Evidence from fine-grained Mobile phone location data in the United States during the COVID-19 pandemic." *Journal of Planning Literature* 37 (1):116–116.

Huang, X., J. Y. Lu, S. Gao, S. C. Wang, Z. W. Liu, and H. X. Wei 2022b. "Staying at home is a privilege: Evidence from fine-grained Mobile phone location data in the United States during the COVID-19 pandemic." *Annals of the American Association of Geographers* 112 (1):286–305. doi: 10.1080/24694452.2021.1904819.

ICE Cobotics. 2024. "The Future of Retail: What Retailers Are Doing to Stay Viable." accessed July 14, 2024. https://icecobotics.com/blog/the-future-of-retail-what-retailers-are-doing-to-stay-viable.

Iglesias, V., F. Giraldez, I. Tiznado-Aitken, and J. C. Muñoz. 2019. "How uneven is the urban mobility playing field? Inequalities among socioeconomic groups in Santiago De Chile." *Transportation Research Record* 2673 (11):59–70.

Islind, A. S., M. Óskarsdóttir, and H. Steingrímsdóttir. 2020. "Changes in mobility patterns in Europe during the COVID-19 pandemic: Novel insights using open source data." *arXiv preprint arXiv:2008.10505.*

ITF. 2023. "Shaping Post-Covid Mobility in Cities: Summary and Conclusions." https://www.itf-oecd.org/sites/default/files/docs/shaping-post-covid-mobility-cities.pdf

Jian, I. Y., J. Luo, and E. H. W. Chan. 2020. "Spatial justice in public open space planning: Accessibility and inclusivity." *Habitat International* 97:102122.

Kennedy, C. A., N. Ibrahim, and D. Hoornweg. 2014. "Low-carbon infrastructure strategies for cities." *Nature Climate Change* 4 (5):343–346.

Kopecka-Piech, K., and B. Łódzki. 2022. "The Covid-19 Pandemic as a Challenge for media and Communication Studies." In *Routledge Research in Cultural and media Studies*, Abingdon, Oxon; New York, NY: Routledge.

Kotb, I. 2020. "Smart retailing in COVID-19 world: Insights from Egypt." *European Journal of Marketing and Economics* 3 (2):132–156.

Lai, K. Y., C. Webster, S. Kumari, and C. Sarkar. 2020. "The nature of cities and the Covid-19 pandemic." *Current Opinion in Environmental Sustainability* 46:27–31.

Lashgari, Y. S., and S. Shahab. 2022. "The impact of the COVID-19 pandemic on retail in city centres." *Sustainability* 14 (18):11463.

Lennon, M. 2023. "Planning and the post-pandemic City." *Planning Theory & Practice* 24 (1):140–143. doi: 10.1080/14649357.2021.1960733.

Li, D., and J. M. Liu. 2019. "Uncovering the relationship between point-of-interests-related human mobility and socioeconomic status." *Telematics and Informatics* 39:49–63. doi: 10.1016/j.tele.2019.01.001.

Li, H., Y. Zhang, M. Zhu, and G. Ren. 2021. "Impacts of COVID-19 on the Usage of public bicycle share in london." *Transportation Research Part A: Policy and Practice* 150:140–155. https://doi.org/10.1016/j.tra.2021.06.010.

Lin, Y., Y. Xu, Z. Zhao, S. Park, S. Su, and M. Ren. 2023. "Understanding changing public transit travel patterns of urban visitors during COVID-19: A multi-stage study." *Travel Behaviour and Society* 32:100587. doi: 10.1016/j.tbs.2023.100587.

Lin, Y., Y. Xu, Z. Zhao, W. Tu, S. Park, and Q. Li. 2024. "Assessing effects of pandemic-related policies on individual public transit travel patterns: A Bayesian online changepoint detection based framework." *Transportation Research Part A: Policy and Practice* 181:104003. https://doi.org/10.1016/j.tra.2024.104003.

Loa, P., S. Hossain, S. M. Mashrur, Y. Liu, K. Wang, F. Ong, and K. N. Habib. 2021. "Exploring the impacts of the COVID-19 pandemic on modality profiles for non-mandatory trips in the Greater Toronto Area." *Transport Policy* 110:71–85. https://doi.org/10.1016/j.tranpol.2021.05.028.

Lucchini, L., O. Langle-Chimal, L. Candeago, L. Melito, A. Chunet, A. Montfort, B. Lepri, N. Lozano-Gracia, and S. P. Fraiberger. 2023. "Socioeconomic disparities in mobility behavior during the COVID-19 pandemic in developing countries." *arXiv preprint arXiv:2305.06888.*

McKinsey. 2020. "From no mobility to future mobility: Where COVID-19 has accelerated change | McKinsey." McKinsey & Company. https://www.mckinsey.com/industries/automotive-and-assembly/our-insights/from-no-mobility-to-future-mobility-where-covid-19-has-accelerated-change.

Megahed, N. A., and R. F. Abdel-Kader. 2022. "Smart Cities after COVID-19: Building a conceptual framework through a multidisciplinary perspective." *Scientific African* 17:e01374.

Mischke, J., R. Luby, B. Vickery, J. Woetzel, O. White, A. Sanghvi, J. Rhee, A. Fu, R. Palter, and A. Dua. 2023. Empty spaces and hybrid places: The pandemic's lasting impact on real estate. McKinsey Global Institute. https://www.mckinsey.com/mgi/our-research/empty-spaces-and-hybrid-places

Modai-Snir, T., and P. Plaut. 2019. "The analysis of residential sorting trends: Measuring disparities in socio-spatial mobility." *Urban Studies* 56 (2):288–300.

Moran, M. E. 2022. "Treating COVID with bike lanes: Design, spatial, and network analysis of 'pop-up' bike lanes in Paris." Findings. https://doi.org/10.32866/001c.33765.

Nieuwenhuijsen, M. J. 2021. "New urban models for more sustainable, liveable and healthier cities post covid19; Reducing air pollution, noise and heat island effects and increasing green space and physical activity." *Environment International* 157:106850.

Nikiforiadis, A., L. Mitropoulos, P. Kopelias, S. Basbas, N. Stamatiadis, and S. Kroustali. 2022. "Exploring mobility pattern changes between before, during and after COVID-19 lockdown periods for young adults." *Cities* 125:103662.

Noszczyk, T., J. Gorzelany, A. Kukulska-Kozieł, and J. Hernik. 2022. "The impact of the COVID-19 pandemic on the importance of urban green spaces to the public." *Land Use Policy* 113:105925.

Ossimetha, A., A. Ossimetha, C. M. Kosar, and M. Rahman. 2021. "Socioeconomic disparities in community mobility reduction and COVID-19 growth." *Mayo Clinic Proceedings* 96(1):78–85.

Qi, Y., J. Liu, T. Tao, and Q. Zhao. 2023. "Impacts of COVID-19 on public transit ridership." *International Journal of Transportation Science and Technology* 12 (1):34–45. https://doi.org/10.1016/j.ijtst.2021.11.003.

Qiang, D., and G. McKenzie. 2024. "Navigating the post-pandemic urban landscape: Disparities in transportation recovery & regional insights from New York City." *Computers, Environment and Urban Systems* 110:102111. https://doi.org/10.1016/j.compenvurbsys.2024.102111.

Rajpurohit, A. 2021. "What Happened to Shopping Center Foot Traffic in Pandemic World? The Role of Socio-Demographics and Transport Modes." Concordia University.

Ren, M., S. Park, Y. Xu, X. Huang, L. Zou, M. S. Wong, and S.-Y. Koh. 2022. "Impact of the COVID-19 pandemic on travel behavior: A case study of domestic inbound travelers in Jeju, Korea." *Tourism Management* 92:104533. https://doi.org/10.1016/j.tourman.2022.104533.

Robinson, L., J. Schulz, A. Khilnani, H. Ono, S. R. Cotten, N. McClain, L. Levine, W. Chen, G. Huang, A. A. Casilli, P. Tubaro, M. Dodel, A. Quan-Haase, M. L. Ruiu, M. Ragnedda, D. Aikat, and N. Tolentino. 2020. "Digital inequalities in time of pandemic: COVID-19 exposure risk profiles and new forms of vulnerability." *First Monday* 25 (7). doi: 10.5210/fm.v25i7.10845.

SafeGraph, I. 2021. SafeGraph's Social Distancing Metrics V2.1 edited by Inc. SafeGraph.

Shah, S. I. M. A. 2021. "Decrease of in-store traffic due to the coronavirus pandemic and turning to e-commerce as a solution." Technische Hochschule Ingolstadt.

Sharifi, A., and C. Y. Lee. 2024. "The appeal of cities may not wane due to the COVID-19 pandemic and remote working." *Npj Urban Sustainability* 4 (1):11.

Silva-Jose, C., T. S. Nagpal, J. Coterón, R. Barakat, and M. F. Mottola. 2022. "The 'new normal' includes online prenatal exercise: Exploring pregnant women's experiences during the pandemic and the role of virtual group fitness on maternal mental health." *BMC Pregnancy and Childbirth* 22 (1):251.

Slezakova, K., and M. C. Pereira. 2021. "2020 COVID-19 lockdown and the impacts on air quality with emphasis on urban, suburban and rural zones." *Scientific Reports* 11 (1):21336.

Sy, K. T. L., M. E. Martinez, B. Rader, and L. F. White. 2021. "Socioeconomic disparities in subway use and COVID-19 outcomes in New York City." *American Journal of Epidemiology* 190 (7):1234–1242.

Talwar, S., P. Kaur, R. Nunkoo, and A. Dhir. 2023. "Digitalization and sustainability: Virtual reality tourism in a post pandemic world." *Journal of Sustainable Tourism* 31 (11):2564–2591. https://doi.org/10.1080/09669582.2022.2029870

ThinkSys Inc. 2023. "Embracing the Future with Immersive Retail." accessed July 14, 2024. https://thinksys.com/retail/immersive-retail/.

UNWTO. 2020. "COVID-19 response: 96% of global destinations impose travel restrictions, UNWTO reports | UN tourism." https://www.unwto.org/news/covid-19-response-travel-restrictions.

UITP. 2024. "COVID-19 pandemic: Resuming public transport services post-lockdown." https://cms.uitp.org/wp/wp-content/uploads/2020/07/Knowledge-Brief-Covid19.pdf

van der Drift, S., L. Wismans, and M.-J. Olde Kalter. 2022. "Changing mobility patterns in the Netherlands during COVID-19 outbreak." *Journal of Location Based Services* 16 (1):1–24.

Vickerman, R. 2021. "Will Covid-19 put the public back in public transport? A UK perspective." *Transport Policy* 103:95–102. https://doi.org/10.1016/j.tranpol.2021.01.005.

Wang, B. 2023. "From Consumer Cities to Consumer Towns? The Post-Pandemic Geography of Retail in the US." *MIT Center for Real Estate Research Paper* (23/21).

Wang, J., N. Kaza, N. C. McDonald, and K. Khanal. 2022. "Socio-economic disparities in activity-travel behavior adaptation during the COVID-19 pandemic in North Carolina." *Transport Policy* 125:70–78.

Wu, J., R. E. Bolton, C. Anwar, B. G. Bokhour, A. Khanna, R. S. Mullur, S. L. Taylor, and J. Hyde. 2023. "Modifying whole health services for successful telehealth delivery: Lessons from veterans health Administration's rapid transition during the COVID-19 pandemic." *Journal of Integrative and Complementary Medicine* 29 (12):792–804.

Xiong, C., S. Hu, M. Yang, H. Younes, W. Luo, S. Ghader, and L. Zhang. 2020. "Mobile device location data reveal human mobility response to state-level stay-at-home orders during the COVID-19 pandemic in the USA." *Journal of the Royal Society Interface* 17 (173):20200344.

Xu, N., Q. Nie, J. Liu, and S. Jones. 2023. "Post-pandemic shared mobility and active travel in alabama: A machine learning analysis of COVID-19 survey data." *Travel Behaviour and Society* 32, 100584. https://doi.org/10.1016/j.tbs.2023.100584.

Xu, Y., J. Y. Xue, S. Park, and Y. Yue. 2021. "Towards a multidimensional view of tourist mobility patterns in cities: A mobile phone data perspective." *Computers Environment and Urban Systems* 86:101593. https://doi.org/10.1016/j.compenvurbsys.2020.101593.

Yang, Y., H. Zhang, and X. Chen. 2020. "Coronavirus pandemic and tourism: Dynamic stochastic general equilibrium modeling of infectious disease outbreak." *Annals of Tourism Research* 83:102913. https://doi.org/10.1016/j.annals.2020.102913.

Yuan, Y., and M. Raubal 2016. "Analyzing the distribution of human activity space from mobile phone usage: An individual and urban-oriented study." *International Journal of Geographical Information Science* 30 (8):1594–1621. doi: 10.1080/13658816.2016.1143555.

Yuan, Y., M. Raubal, and Y. Liu. 2012. "Correlating mobile phone usage and travel behavior - a case study of Harbin, China." *Computers, Environment and Urban Systems* 36 (2):118–130.

Zenker, S., E. Braun, and S. Gyimóthy. 2021. "Too afraid to travel? Development of a pandemic (COVID-19) anxiety travel scale (PATS)." *Tourism Management* 84:104286. https://doi.org/10.1016/j.tourman.2021.104286.

Zenker, S., and F. Kock. 2020. "The coronavirus pandemic – A critical discussion of a tourism research agenda." *Tourism Management* 81:104164. https://doi.org/10.1016/j.tourman.2020.104164.

20 The Influence of the COVID-19 Pandemic on Human Daily Mobility and Time Allocation

Hui Shi, Konstadinos G. Goulias, and Yuyang Wu

20.1 INTRODUCTION

COVID-19 is the third major coronavirus outbreak of the 21st century, with a documented mortality rate significantly higher than that of its predecessors (Joshee, Vatti, and Chang 2022). From December 2019 to July 2021, the global impact of COVID-19 led to over 4 million deaths and affected approximately 200 million people (WHO 2023). In response, governments implemented various preventive measures to slow down the virus's spread, including closing entertainment venues, imposing travel restrictions, and isolating individuals diagnosed with coronavirus (Chen et al. 2023). Additionally, many companies adopted remote work policies to maintain operations (Huang, Loo, and Axhausen 2023). In this context, COVID-19 caused an unprecedented global disruption, significantly impacting individuals' time allocation, travel behavior, and purchasing habits (Aaditya and Rahul 2023; Batur et al. 2022).

For example, Restrepo and Zeballos observed that in response to the COVID-19 pandemic, many companies introduced teleworking policies to safeguard employee health (Restrepo and Zeballos 2022). This shift led to notable changes in daily routines, particularly for telecommuters, who experienced dramatic reductions in trip duration and frequency and visits to central business districts (Huang, Loo, and Axhausen 2023). According to a study by Shi et al. (Shi, Su, and Goulias 2023), during the pandemic, people adopted a new time allocation pattern, spending much more time at home. dos Santos Quaresma et al. (2021) further pointed out that COVID-19 affected time spent on specific activities, such as exercise and sleep. The reduction in outdoor activities may have been driven by fear of infection and lockdown measures, which also likely contributed to the surge in online shopping during the outbreak (Javadinasr et al. 2022; Shi and Goulias 2024a).

Although extensive research has examined the impact of COVID-19 on daily mobility and time allocation (Arroyo, Mars, and Ruiz 2021; Batur et al. 2022; Chen et al. 2021; Haghani et al. 2023), few studies have focused on its long-term effects, likely due to limited data availability. However, exploring these long-term impacts is crucial, as they may be linked to economics (Agrawal and Shybalkina 2023), health (Shi, Su, and Goulias 2023), the environment (Shaer et al. 2021), technologies (Campisi et al. 2023), and other key facets (Adam, Sayagh, and Buhler 2023; Haghani et al. 2023). While the widespread acceptance of telecommuting was mainly driven by the pandemic, many companies have found it effective and plan to continue even after the immediate threat of COVID-19 has passed (McLean 2020; Mohammadi et al. 2022; Salon et al. 2022). This shift could have lasting effects on travel behavior, influencing trip frequency, destinations, activity duration, modes of transport, and more.

20.2 CONSTRUCTING LABELED MOTIFS

An individual's spatial mobility structure can be represented through the construction of mobility motifs. These motifs capture the sequential locations an individual visits, and illustrating their movement patterns across different destinations i can be shown as $L(i) = \{Location_1, Location_2, \ldots, Location_N\}$,

DOI: 10.1201/9781003503262-24

where N represents the total number of separate visits to locations. A mobility graph $M(i) = \{Node_i, Link_i\}$ can be constructed. The node set $Node(i) = \{Node_1, Node_2, \ldots, Node_n\}$ represents all unique locations visited by an individual, where n indicates the total number of distinct locations. The link set $Link(u) = \{\ell(x, y \mid x, y), x, y \in Node(i) \wedge i \neq j\}$ encompasses all directed trips represented by $\ell(x, y)$, which denotes the trip from node x to node y. Each individual's daily trips can be represented by a travel motif, illustrated by a mobility graph where each node represents a unique destination visited, and each edge signifies the direction of travel between specific pairs of nodes. While this construction method has been widely utilized in prior research (Joy and Vijaya 2018; Schneider et al. 2013; Su, McBride, and Goulias 2020), it does not consider the types of locations visited. For instance, we designate 1 for the respondent's home, 2 for their workplace, and 3 for the shopping mall. According to existing methodologies, if respondent A travels from home to work in the morning and then returns home in the afternoon, while respondent B goes shopping and then returns home, their motifs would appear identical despite their differing schedules. In response to this limitation, this chapter proposes a new approach that develops motifs incorporating location labels to reflect the specific places visited.

20.3 TIME ALLOCATION INDICATOR

While labeled motifs can provide valuable insights into individuals' daily mobility, they do not account for the specifics of activities undertaken. To address this, an indicator of time allocation, referred to as fragmentation, is computed. Fragmentation of activities and travel is defined as a sequence of both long and short activities and trips throughout a person's day, reflecting how individuals allocate their time. This measure captures potential increases in the frequency of switching between activities (Couclelis 2004; McBride, Davis, and Goulias 2019; Shi et al. 2022; Su, McBride, and Goulias 2021).

A sequence comprises discrete time intervals, specifically 1 minute in this chapter, each associated with a distinct "state" (this chapter employs states such as H_Work, H_Leisure, NH_Education, and other activities). An individual can switch from one "state" to another, with each minute assigned to one of the 38 specified categories. Consequently, the total length of a daily activity sequence is 1,440 minutes (equivalent to 24 hours). Unlike general statistical descriptions typically used in previous research (Cheng, Lavieri, and Astroza 2022; Stone et al. 2018), sequences account for activity types and durations, as well as transitions between activities, allowing for a comprehensive analysis of all activities performed throughout a single day as an integrated schedule.

In this context, fragmentation can be quantified using an indicator termed complexity. This indicator integrates entropy, as defined in Equation 20.1, with transitions between distinct states within a sequence (x) outlined in Equation 20.2. This approach aligns with the methodologies described by McBride et al. and Xiao et al. (McBride, Davis, and Goulias 2019; Xiao et al. 2020).

$$h(x) = h(\pi_1, \pi_2, \ldots, \pi_s) = -\sum_{i=1}^{s} \pi_i \log(\pi_i) \tag{20.1}$$

$$C(x) = \sqrt{\frac{(l_d(x) - 1)h(x)}{(l(x) - 1)h_{max}}} \tag{20.2}$$

Here $h(x)$ denotes the entropy function of the sequence x, indicating the diversity of states within the sequence. n represents the total number of different states, while π_i signifies the proportion of occurrences of the i_{th} state in the sequence. To consider the number of transitions among states,

Complexity $C(x)$ combines entropy and the number of transitions $l_d(x)-1$ in the sequence x. This is normalized by the h_{max}, which occurs when individuals allocate equal amounts of time to each state, and by the sequence length $l(x)$, which is equal to 1,440.

20.4 HIERARCHICAL CLUSTERING

In line with the methodologies of McBride, Davis, and Goulias (2019) and Ravulaparthy et al. (2012), this chapter employs the agglomerative nesting clustering approach to detect individuals' daily patterns. This bottom-up technique begins with each observation acting as its own cluster. Clusters are then merged based on their similarity. Specifically, the "distance" between two sequences is defined as the number of operations required to make them identical, serving as a measure of similarity. The average silhouette coefficient (silhouette) and the within-cluster sum of squares are commonly used for determining the optimal number of clusters (Vagni and Cornwell 2018).

Figure 20.1 presents a stacked bar chart illustrating the daily time allocation patterns for a group of individuals. As indicated in the graph legend, there are 38 distinct activity states. The x-axis is the time of day in minutes, beginning at 4:00 a.m. and concluding at 3:59 a.m. the following day. The y-axis denotes the relative frequency of individuals participating in one of the 38 activities. The example depicted in Figure 20.1 shows a typical commuting pattern, where the majority of individuals commute to work in the morning, remain at their workplaces throughout the day (with a few taking breaks during lunchtime), and return home after 4 p.m. Subsequently, they spend the evening at home. Additionally, this group allocates a moderate amount of time to travel and engages in activities both outside the home and at home after work.

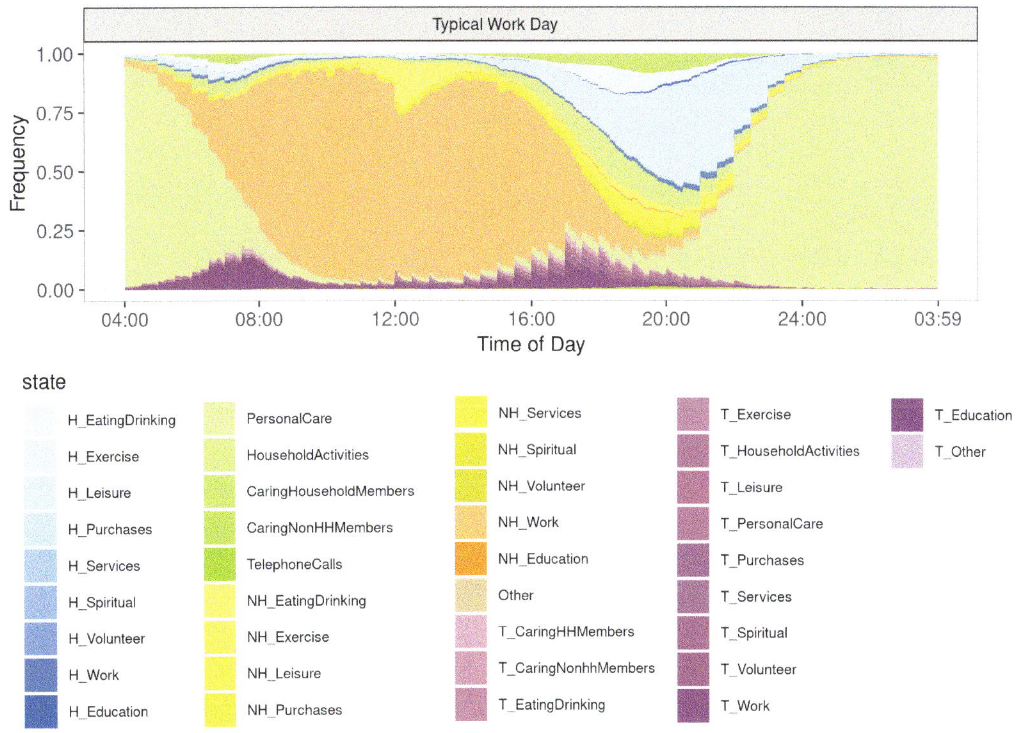

FIGURE 20.1 An example of daily pattern of minute-by-minute activity–travel sequences.

20.5 CASE STUDY IN THE UNITED STATES

20.5.1 COVID-19 and Daily Motifs

Table 20.1 illustrates the evolution of major motifs, displaying only those with unweighted frequencies exceeding 30 and weighted proportions greater than 1%. Previous analyses highlight that the motif reflecting a full day spent at home accounts for the largest percentage (CBS Minnesota 2023; theHRD 2020). Between 2019 and 2022, there was a significant rise in the percentage of individuals remaining at home, as expected. This increase peaked in 2020 before declining in 2021 and 2022. However, the proportion did not return to pre-COVID levels, possibly due to individuals adapting to home-based lifestyles (e.g., online shopping). This indicates that the pandemic may have long-lasting effects on individuals' daily routines (Aaditya and Rahul 2023). It is therefore unsurprising that the motif labeled as "Home with loop trips" (where both the origin and destination are home) continued to rise even after vaccinations were implemented. Throughout the pandemic, the participation of "Home-Workplace-Home" trips increased, correlating with declines in more complex commuting patterns, such as "Home-Workplace-Restaurant-Workplace-Home" and "Home-Workplace-Home-Workplace-Home." The comparison of the overall proportions of motifs involving workplaces from 2019 to 2022 reveals that the COVID year (2020 in the United States) had the lowest percentage, aligning with findings from previous studies (Shi, Su, and Goulias 2023; Shi & Goulias, 2024b).

Overall, people tended to adhere to simple motifs throughout these 4 years. Several popular motifs, including "Home-OtherHome-Home," "Home-Mall-Home," and "Home-Grocery-Home," experienced growth during the pandemic and maintained relatively high levels even in 2022. This trend may be attributed to individuals simplifying their schedules during the outbreak and choosing to stick to their updated daily routines.

20.5.2 COVID-19 and Time Allocation

According to the complexity (a time allocation indicator), fragmentation was highest before COVID-19, with a score of 0.065 in 2019. This was followed by 2021 and 2022, both recording a complexity of 0.062. As expected, the lowest level occurred during the outbreak in 2020, at 0.061.

TABLE 20.1
Weighted Proportions in Daily Percentage of Major Motifs from 2019 to 2022

Motifs	2019	2020	2021	2022
Home	12.92	22.62	20.21	18.08
Home-Workplace-Home	10.37	10.62	11.19	10.45
Home with loop trips	3.79	4.48	4.03	5.08
Home-Restaurant-Home	3.04	2.50	2.94	2.74
Home-OtherHome-Home	2.62	3.40	3.03	3.15
Home-Mall-Home	2.55	3.14	2.83	2.95
Home-Grocery-Home	2.37	2.60	3.05	3.11
Home-School-Home	1.73	<1.00	1.38	1.64
Home-Outdoors-Home	1.62	5.21	3.89	3.16
Home-Other-Home	1.55	2.03	1.75	1.69
Home-Worship-Home	1.22	<1.00	<1.00	<1.00
Home-Workplace-Restaurant-Workplace-Home	1.19	<1.00	<1.00	<1.00
Home-Workplace-Home-Workplace-Home	1.02	<1.00	1.08	1.01

Note: Unweighted frequencies are larger than 30 and weighted proportions are larger than 1%.

These findings suggest that the coronavirus significantly impacted residents' time fragmentation, and this effect may be long-lasting, as the complexity has not returned to its prepandemic level.

Figure 20.2 illustrates the average activity durations for the entire population, while Figure 20.3 provides the same information specifically for participants—those from the American Time Use Survey (ATUS) sample who engaged in a particular activity at least once in a day. Figure 20.2b is a zoomed-in version of Figure 20.2a, focusing solely on the duration of activities within an hour. Notably, the bar heights in Figure 20.2 are shorter than those in Figure 20.3. Overall, from 2019

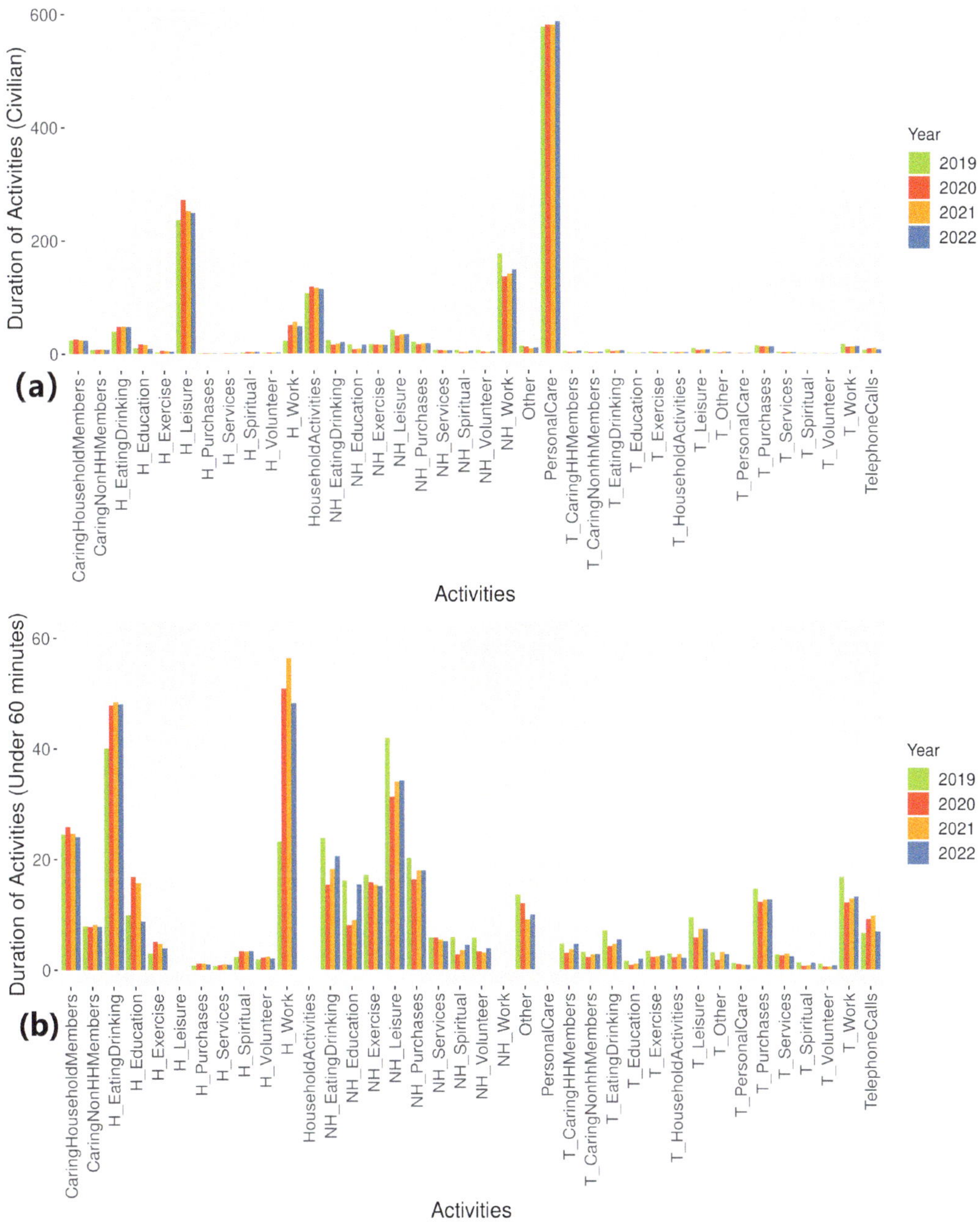

FIGURE 20.2 Average duration per day for civilian population.

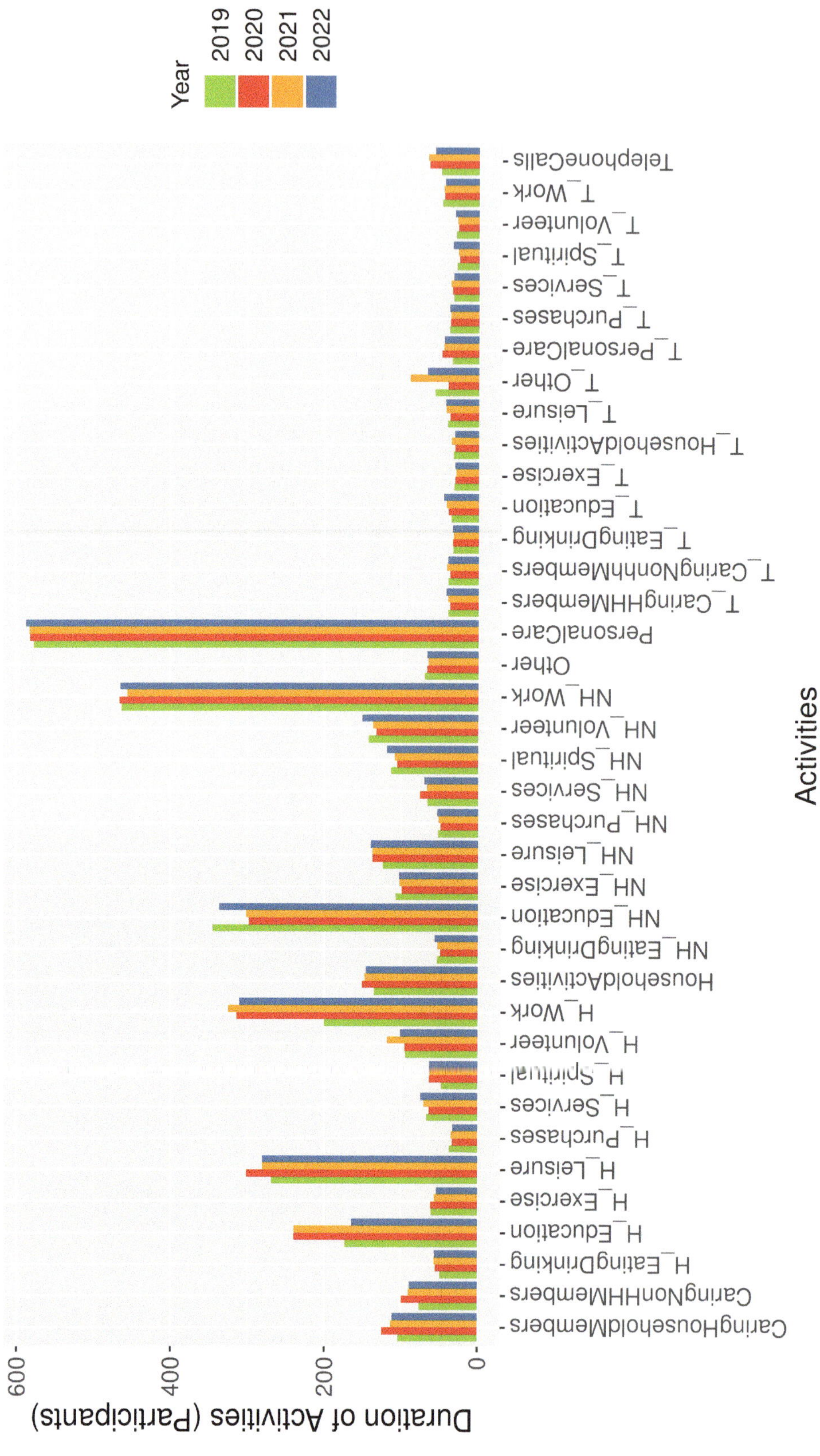

FIGURE 20.3 Average activity duration per day for participants.

to 2020, the duration of home activities (designated with "H_," such as H_Purchases, H_Leisure, and H_Work) increased, followed by a slight decline. In contrast, the duration of out-of-home activities, beginning with "NH_" or "T_," decreased from 2019 to 2020, then climbed again. However, it is worth noting that the duration of NH_Work for participants in Figure 20.3 did not follow the same trend; instead, it slightly increased during the COVID-19 period. This is understandable, as work hours could not be shortened for individuals required to leave their homes for work, including plumbers, doctors, and police officers. During COVID-19, additional issues may have increased the workload for them. However, on average, fewer individuals went to work during the pandemic, leading to a significant decrease in the NH_Work duration depicted in Figure 20.2. This decline resulted from the pandemic's impact on daily commuting patterns and work arrangements.

When considering the number of episodes (as illustrated in Figures 20.4 and 20.5, with Figure 20.4b zooming in on the average frequency of activities with fewer than two), it is evident that individuals engaged in a higher frequency of home activities, particularly related to eating, drinking, and leisure. In contrast, while the frequencies of outdoor activities initially declined due to COVID-19 and later increased alongside activity duration, American outdoor activity frequencies did not go back to prepandemic levels. This decline was particularly noticeable for activities such as dining outside the home, outdoor recreation, and in-person work, along with related travel. These trends suggest that the virus has long-lasting effects on daily schedules and activity patterns.

20.5.3 COVID-19 and Daily Activity–Travel Patterns

This section examines the changes in time allocation patterns between 2019 and 2022 using hierarchical clustering (see Figures 20.6–20.9). The y-axis represents the percentage of individuals participating in each activity type as indicated in the legend, while the x-axis denotes the time of day in minutes. The numbers in parentheses reflect the weighted proportion of each daily pattern. The ATUS data for 2019, 2020, and 2022 reveal four distinct daily patterns, whereas the 2021 data show five distinct patterns. Although the daily patterns differ from year to year, all years include mixed days, home discretionary days, and typical work days. Notably, home work days were not detected prior to the pandemic, while discretionary days appeared only in 2019 and 2021.

The mixed day pattern exhibits the highest weighted percentage across the four datasets. Individuals following this pattern engaged in a variety of at-home and out-of-home activities throughout the day, typically allocating time for leisure and exercise at home and traveling for errands and shopping. The proportion of mixed days decreased from 34.65% in 2019 to 29.42% in 2020, followed by a dramatic increase to 43.59% in 2022. Additionally, the yellow belt, reflecting outdoor activities, was thinner during the pandemic year compared to 2019, indicating a reduction in outdoor engagement during that period.

The typical work day pattern includes a significant number of individuals following a standard commuter schedule. Those adhering to this routine generally commuted to work in the morning, took a lunch break, continued their work in the afternoon, and engaged in activities such as shopping, picking up others, and dining after work. The proportion of this cluster declined from 33.24% in 2019 to 22.7% in 2020. Despite this notable drop, essential professionals, such as firefighters, doctors, and plumbers, continued to leave their homes for work. Although there was a subsequent increase in this pattern, the percentage in 2022 remained lower than in 2019, suggesting that COVID-19 has long-term implications for some individuals' workplaces.

Respondents classified under the home discretionary day pattern allocated a substantial amount of time to home leisure activities, such as watching television, playing games, and exercising. In contrast to the mixed day pattern, individuals in this category spent the majority of their time at home, with only occasional outings probably for errands and shopping. As anticipated, the proportion of this pattern initially increased from 24.14% in 2019 to 38.79% in 2020, before declining to 21.44% in 2022.

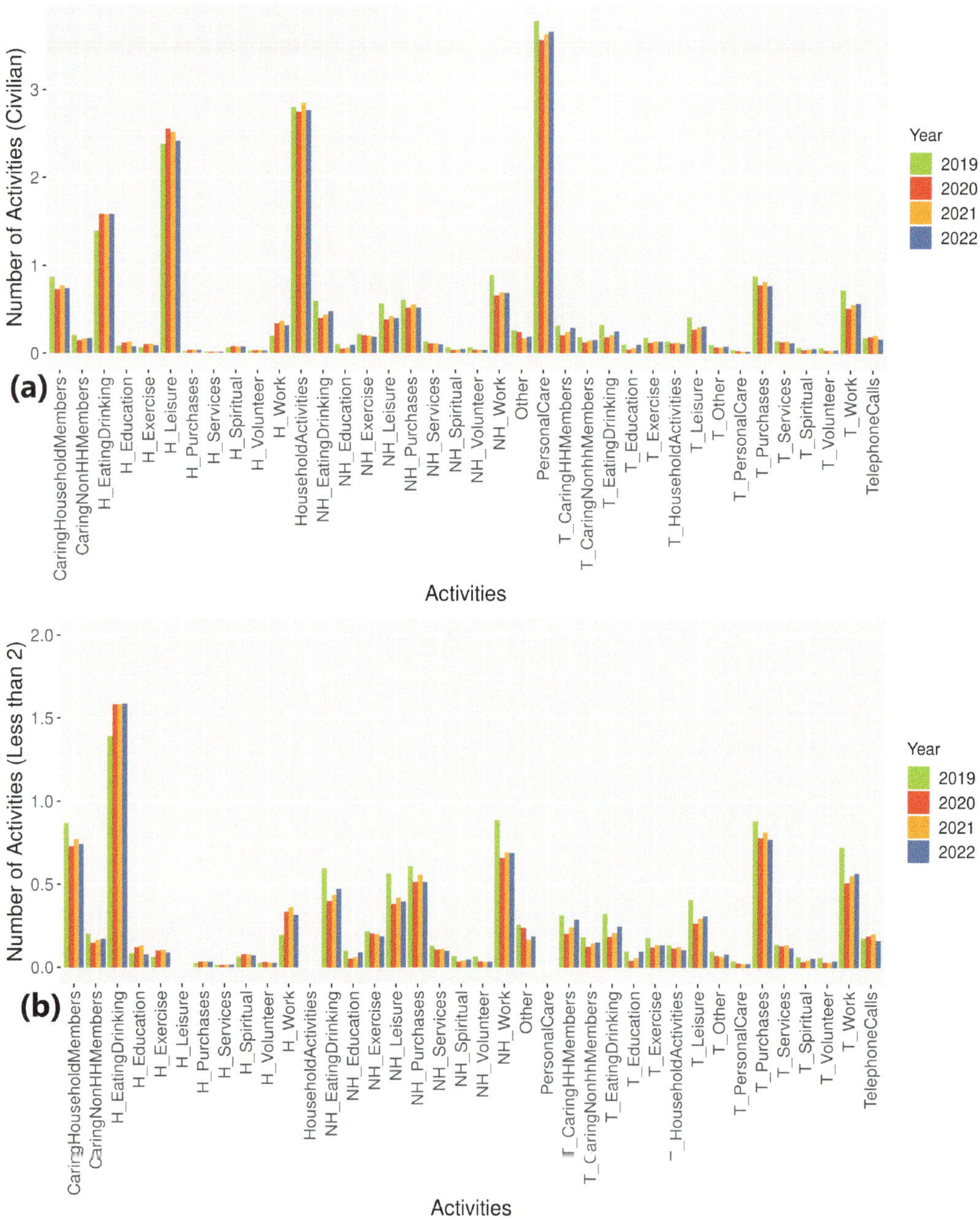

FIGURE 20.4 Average number of activities per day for civilian population.

Along with the three common patterns previously mentioned, the discretionary day pattern emerged in both 2019 (7.97%) and 2021 (8.69%). Respondents in this cluster dedicated a significant amount of time to travel and outdoor activities. While it may seem surprising that no discretionary days were recorded in 2022, examining the percentages of mixed days and home discretionary days reveals that individuals classified as discretionary days in 2021 could have transitioned to either mixed days or home discretionary days in 2022.

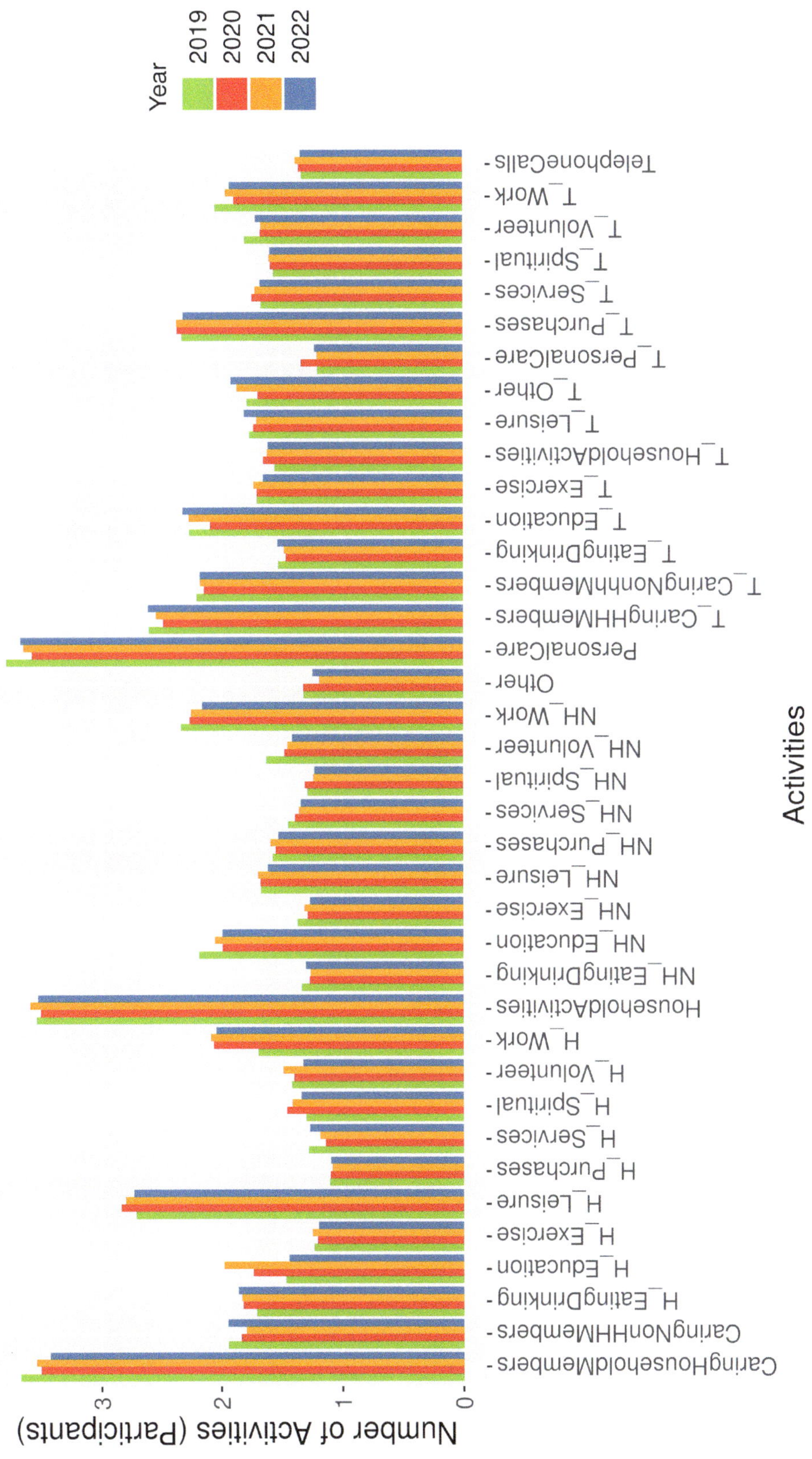

FIGURE 20.5 Average number of activities per day for participants.

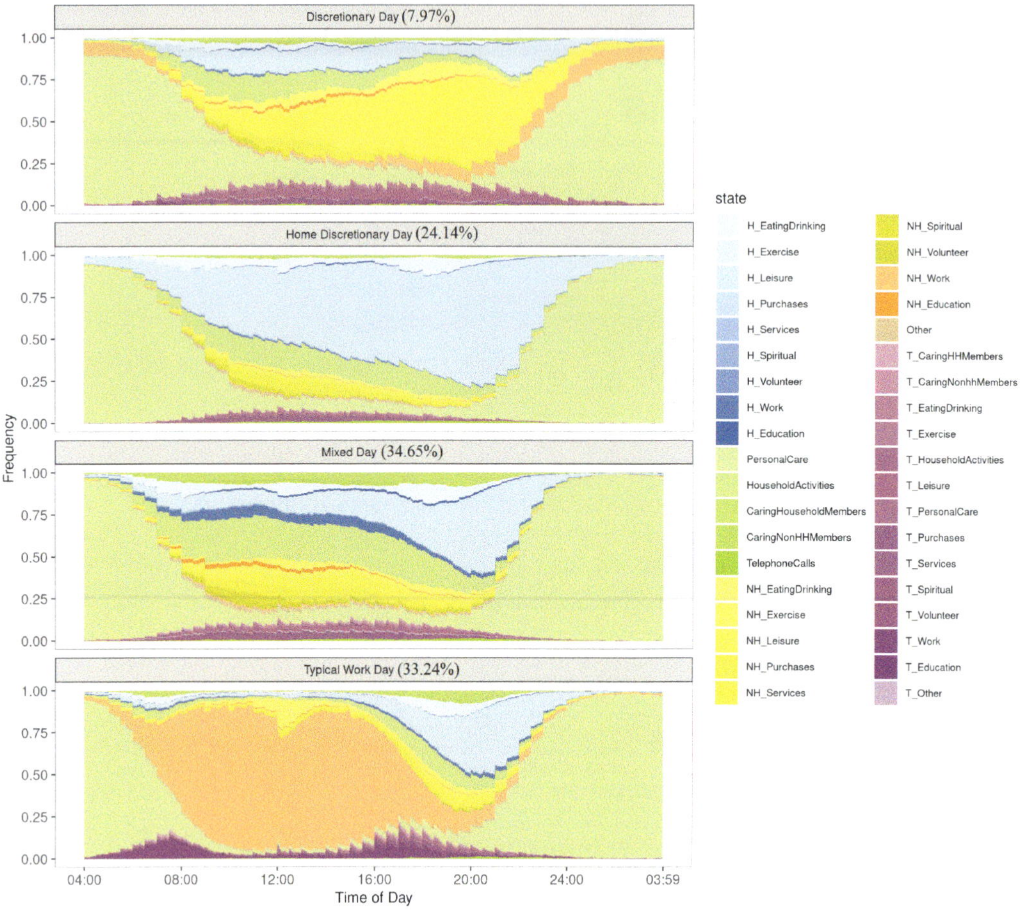

FIGURE 20.6 Daily patterns of activity–travel sequences in 2019.

The most notable change is the emergence of the home work day, which accounted for 9.01% in 2020, 10.14% in 2021, and 8.10% in 2022. This shift was driven by numerous firms implementing work-from-home policies in 2020 to maintain social distance and reduce infections (Parker, Horowitz, and Minkin 2022; Yan et al. 2021), a trend that persisted into 2022. The shape of the home work day is also significant because it closely resembles the typical work day, featuring the typical morning and afternoon peaks alongside a lunch break for employees.

20.6 CONCLUSION

As expected, travel declined during the COVID-19 outbreak but has subsequently increased alongside vaccination rates. However, most trips have not yet returned to prepandemic levels. This may be attributed to people getting used to home-based lifestyles, including online purchasing, suggesting that the COVID-19 pandemic could have lasting impacts on daily routines (Mohammadi et al. 2022; Rahman Fatmi, Orvin, and Thirkell 2022; Zhu et al. 2023). Furthermore, the labeled motifs in our analysis indicate that the stay-at-home all day pattern has the highest percentage, with individuals tending to maintain motifs that involve fewer nodes.

In terms of time allocation, schedule fragmentation was highest before COVID-19, followed by 2021 and 2022. It was lowest during the most severe year of the pandemic, as anticipated. This

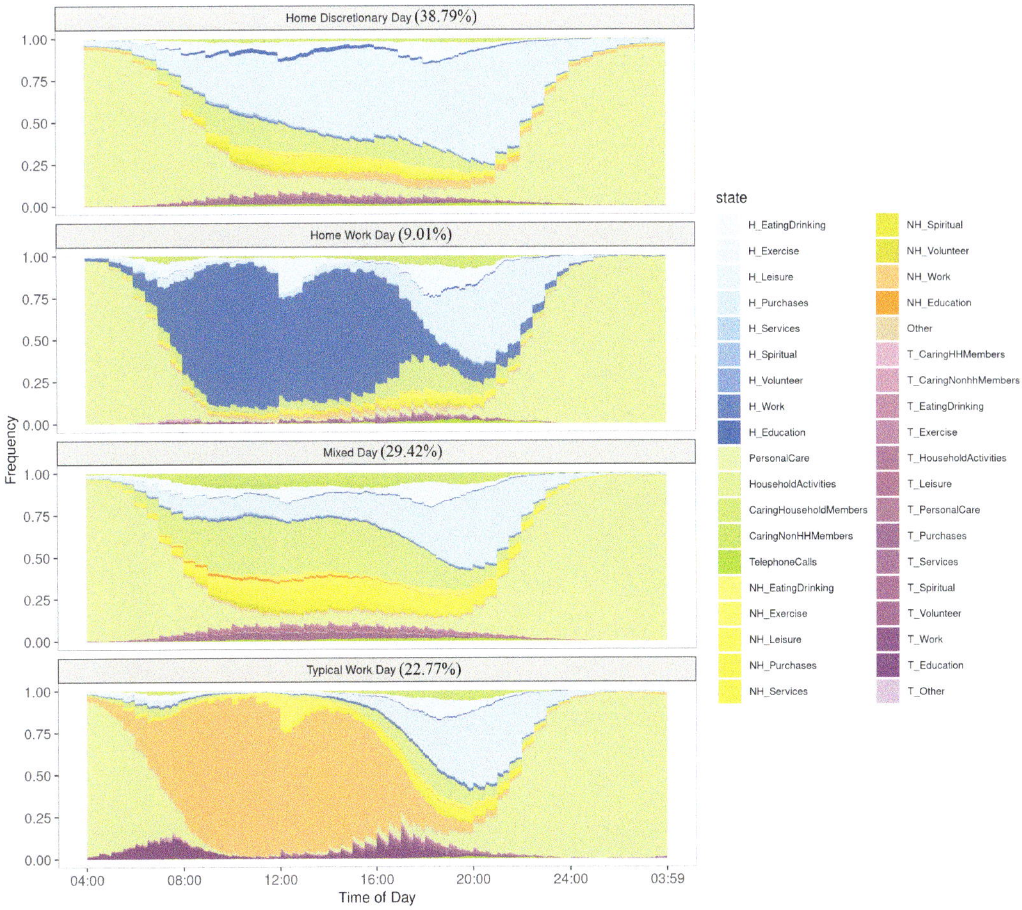

FIGURE 20.7 Daily patterns of activity–travel sequences in 2020.

suggests that the pandemic significantly affected people's activities and travel patterns, with potentially long-lasting effects, as Complexity has not yet returned to pre-outbreak levels. Contributing to this trend is the duration of domestic activities, which increased from 2019 to 2020, followed by a slight drop. This change may be attributed to some people not needing to commute during the pandemic (Mohammadi et al. 2022; Shi, Su, and Goulias 2023), allowing more time for eating and leisure activities at home. Conversely, the duration of outdoor activities decreased between 2019 and 2020 and then rose.

The frequency of outdoor activities in the United States remained lower in 2022 than before the pandemic. This is primarily due to individuals becoming accustomed to performing various activities at home, leading them to abandon their pre-COVID routines. Additionally, during the pandemic, the duration of NH_Work for participants increased slightly. This is reasonable as work hours could not be reduced for essential workers, such as doctors, firefighters, and police officers, who had to leave their homes. However, fewer people, on average, went to work during the pandemic, resulting in a substantial decrease in NH_Work duration (civilian).

In terms of daily patterns, the ATUS data in 2019, 2020, and 2022 reveal four different daily patterns, while the data in 2021 exhibit five distinct patterns. Although the daily patterns vary each year, they consistently include mixed days, home discretionary days, and typical work days. Moreover, home work days, which were not observed prior to the pandemic due to a limited

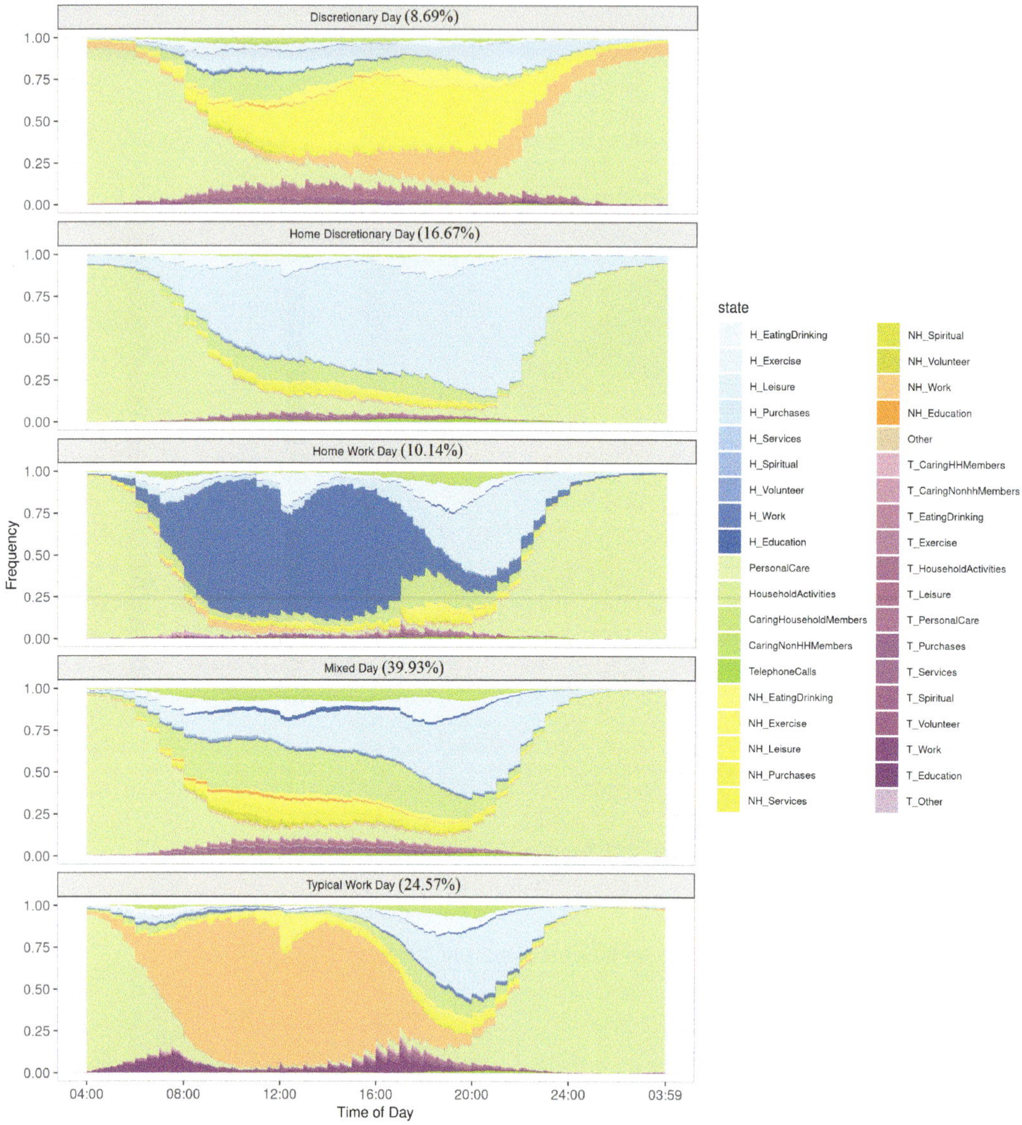

FIGURE 20.8 Daily patterns of activity–travel sequences in 2021.

number of individuals working from home, emerged during the pandemic and persisted into 2022. This trend can largely be attributed to the ongoing telecommuting adopted by many businesses, which offers various benefits such as reducing traffic congestion and energy consumption, improving family care, and enhancing job–family balance (Huang, Loo, and Axhausen 2023; Salon et al. 2022). Individuals in this group are typically young employees with long commutes, high incomes, and high education levels (Mohammadi et al. 2022). Furthermore, as noted by Shi, Su, and Goulias (2023), a correlation exists between daily time allocation and subjective well-being. They suggest that ATUS respondents during the COVID year experienced heightened negative emotions and that hybrid indoor–outdoor schedules positively impact mental health.

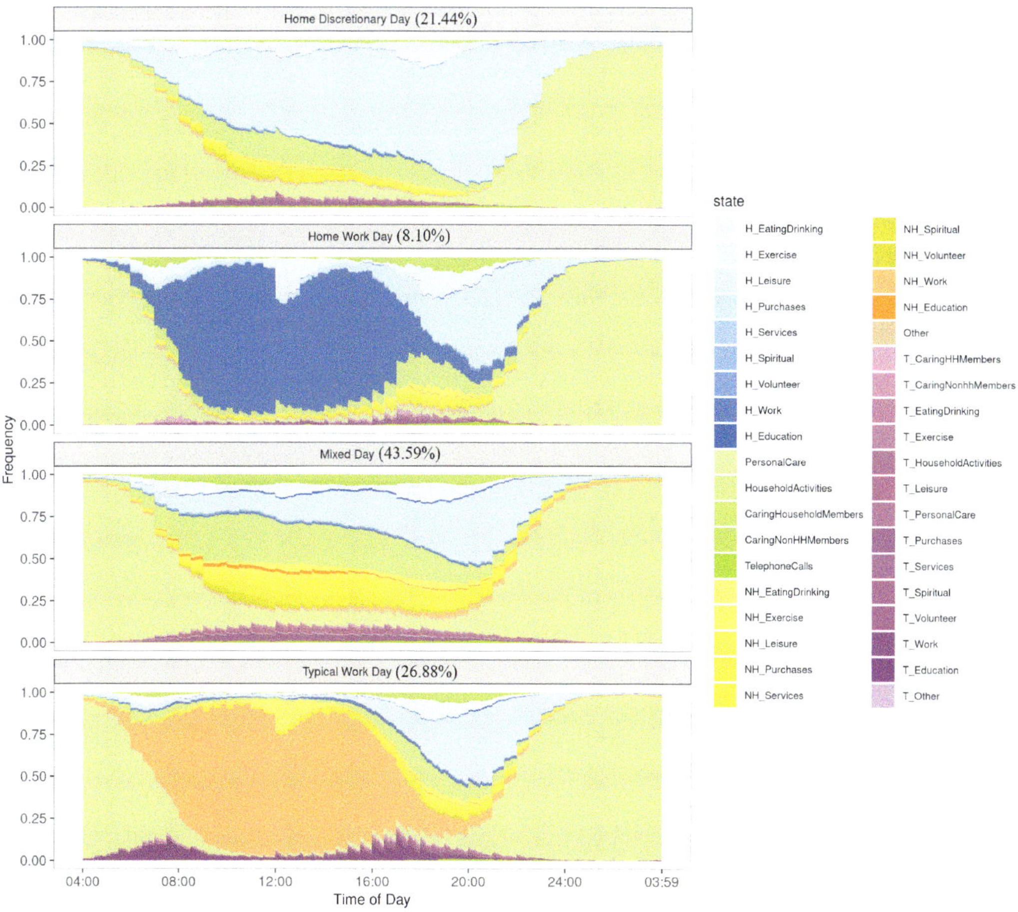

FIGURE 20.9 Daily patterns of activity–travel sequences in 2022.

ACKNOWLEDGMENTS

The results (including the table and figures) are derived from the authors' previous publications (Copyright Elsevier) (Shi and Goulias 2024a, 2024b).

REFERENCES

Aaditya, B. H., and T. M. Rahul. 2023. "Long-Term Impacts of COVID-19 Pandemic on Travel Behaviour." *Travel Behaviour and Society* 30: 262–70. https://doi.org/10.1016/j.tbs.2022.10.005

Adam, Matthieu, David Sayagh, and Thomas Buhler. 2023. "Changes in Cycling Practices in France during the Covid-19 Pandemic. An Illusory Reduction in Inequalities." *Journal of Transport & Health* 32: 101655. https://doi.org/10.1016/j.jth.2023.101655

Agrawal, David R., and Iuliia Shybalkina. 2023. "Online Shopping Can Redistribute Local Tax Revenue from Urban to Rural America." *Journal of Public Economics* 219: 104818. https://doi.org/10.1016/j.jpubeco.2022.104818

Arroyo, Rosa, Lidón Mars, and Tomás Ruiz. 2021. "Activity Participation and Wellbeing During the Covid-19 Lockdown in Spain." *International Journal of Urban Sciences* 25 (3): 386–415. https://doi.org/10.1080/12265934.2021.1925144

Batur, Irfan, Abbie C. Dirks, Ram M. Pendyala, Chandra R. Bhat, Steven E. Polzin, Cynthia Chen, and Ram M. Pendyala. 2022. "Analysis of Changes in Time use and Activity Participation in Response to the COVID-19 Pandemic in the United States: Implications for Well-Being." *Transportation Research Record* 2678(12): 512–526. https://doi.org/10.1177/03611981231165020

Campisi, Tiziana, Antonio Russo, Socrates Basbas, Efstathtios Bouhouras, and Giovanni Tesoriere. 2023. "A Literature Review of the Main Factors Influencing the E-Commerce and Last-Mile Delivery Projects during COVID-19 Pandemic." *Transportation Research Procedia*, AIIT 3rd International Conference on Transport Infrastructure and Systems (TIS ROMA 2022), 15th-16th September 2022, Rome, Italy, 69:552–59. https://doi.org/10.1016/j.trpro.2023.02.207.

"CBS Minnesota." April 25, 2023. https://www.cbsnews.com/minnesota/news/survey-shows-nearly-two-thirds-of-americans-stay-home-more-now-than-before-pandemic/

Chen, Liwei, Jian Li, Tong Xia, Timothy A. Matthews, Tung-Sung Tseng, Lu Shi, Donglan Zhang, et al. 2021. "Changes of Exercise, Screen Time, Fast Food Consumption, Alcohol, and Cigarette Smoking During the COVID-19 Pandemic Among Adults in the United States." *Nutrients* 13 (10): 3359. https://doi.org/10.3390/nu13103359

Chen, Lujie, Taiyu Li, Fu Jia, and Tobias Schoenherr. 2023. "The Impact of Governmental COVID-19 Measures on Manufacturers' Stock Market Valuations: The Role of Labor Intensity and Operational Slack." *Journal of Operations Management* 69 (3): 404–25. https://doi.org/10.1002/joom.1207

Cheng, Yu-Tong, Patricia Sauri Lavieri, and Sebastian Astroza. 2022. "How Did the COVID-19 Pandemic Impact the Location and Duration of Work Activities? A Latent Class Time-Use Study." *Findings*, 35621. https://doi.org/10.32866/001c.35621

Couclelis, Helen. 2004. "Pizza Over the Internet: E-Commerce, the Fragmentation of Activity and the Tyranny of the Region." *Entrepreneurship & Regional Development* 16 (1): 41–54. https://doi.org/10.1080/0898562042000205027

Haghani, Milad, Rico Merkert, Ali Behnood, Chris De Gruyter, Khashayar Kazemzadeh, Hadi Ghaderi, Zahra Shahhoseini, et al. 2023. "How COVID-19 Transformed the Landscape of Transportation Research: An Integrative Scoping Review and Roadmap for Future Research." *Transportation Letters* 16 (1): 43–88. https://doi.org/10.1080/19427867.2022.2160294

Huang, Zhiran, Becky P. Y. Loo, and Kay W. Axhausen. 2023. "Travel Behaviour Changes Under Work-from-Home (WFH) Arrangements During COVID-19." *Travel Behaviour and Society* 30: 202–11. https://doi.org/10.1016/j.tbs.2022.09.006

Javadinasr, Mohammadjavad, Tassio Maggasy, Motahare Mohammadi, Kouros Mohammadain, Ehsan Rahimi, Deborah Salon, Matthew W. Conway, Ram Pendyala, and Sybil Derrible. 2022. "The Long-Term Effects of COVID-19 on Travel Behavior in the United States: A Panel Study on Work from Home, Mode Choice, Online Shopping, and Air Travel." *Transportation Research Part F: Traffic Psychology and Behaviour* 90: 466–84. https://doi.org/10.1016/j.trf.2022.09.019

Joshee, Shreeya, Nikhil Vatti, and Christopher Chang. 2022. "Long-Term Effects of COVID-19." *Mayo Clinic Proceedings* 97 (3): 579–99. https://doi.org/10.1016/j.mayocp.2021.12.017

Joy, J. Jenniffer, and Samundeeswari Vijaya. 2018. "Analysing Work Tour MOTIFS from GPS Trajectory Data." *I-Manager's Journal on Communication Engineering and Systems* 7 (2): 31–36. https://doi.org/10.26634/jcs.7.2.14442

McBride, Elizabeth Callahan, Adam Wilkinson Davis, and Konstadinos G. Goulias. 2019. "Fragmentation in Daily Schedule of Activities Using Activity Sequences." *Transportation Research Record* 2673 (4): 844–54. https://doi.org/10.1177/0361198119837501

McLean, Rob. 2020. "CNN Business." These Companies Plan to Make Working from Home the New Normal. https://www.cnn.com/2020/05/22/tech/work-from-home-companies/index.html

Mohammadi, Motahare (Yalda), Ehsan Rahimi, Amir Davatgari, Mohammadjavad Javadinasr, Abolfazl (Kouros) Mohammadian, Matthew Wigginton Bhagat-Conway, Deborah Salon, Sybil Derrible, Ram M. Pendyala, and Sara Khoeini. 2022. "Examining the Persistence of Telecommuting after the COVID-19 Pandemic." *Transportation Letters* 15: 608–21. https://doi.org/10.1080/19427867.2022.2077582

Parker, Kim, Juliana Menasce Horowitz, and Rachel Minkin. 2022. "COVID-19 Pandemic Continues to Reshape Work in America." https://policycommons.net/artifacts/2256820/covid-19-pandemic-continues-to-reshape-work-in-america/3015476/

Quaresma, Marcus Vl Dos Santos, Camila Guazzelli Marques, Ana Carolina Oumatu Magalhães, and Ronaldo Vagner Thomatieli dos Santos. 2021. "Emotional Eating, Binge Eating, Physical Inactivity, and Vespertine Chronotype Are Negative Predictors of Dietary Practices During COVID-19 Social Isolation: A Cross-Sectional Study." *Nutrition* 90: 111223. https://doi.org/10.1016/j.nut.2021.111223

Rahman Fatmi, Mahmudur, Muntahith Mehadil Orvin, and Corrie Elizabeth Thirkell. 2022. "The Future of Telecommuting Post COVID-19 Pandemic." *Transportation Research Interdisciplinary Perspectives* 16: 100685. https://doi.org/10.1016/j.trip.2022.100685

Ravulaparthy, Srinath, Pamela Dalal, Yali Chen, and Konstadinos G. Goulias. 2012. "Exploratory Analysis of Spatial Hierarchical Clustering in Los Angeles County, California: Relationship of Opportunity-Based Accessibility, Reported Land Values, and Resident Characteristics." *Transportation Research Record* 2307 (1): 132–40. https://doi.org/10.3141/2307-14

Restrepo, Brandon J., and Eliana Zeballos. 2022. "Work from Home and Daily Time Allocations: Evidence from the Coronavirus Pandemic." *Review of Economics of the Household* 20 (3): 735–58. https://doi.org/10.1007/s11150-022-09614-w

Salon, Deborah, Laura Mirtich, Matthew Wigginton Bhagat-Conway, Adam Costello, Ehsan Rahimi, Abolfazl (Kouros) Mohammadian, Rishabh Singh Chauhan, Sybil Derrible, Denise da Silva Baker, and Ram M. Pendyala. 2022. "The COVID-19 Pandemic and the Future of Telecommuting in the United States." *Transportation Research Part D: Transport and Environment* 112: 103473. https://doi.org/10.1016/j.trd.2022.103473

Schneider, Christian M., Vitaly Belik, Thomas Couronné, Zbigniew Smoreda, and Marta C. González. 2013. "Unravelling Daily Human Mobility Motifs." *Journal of The Royal Society Interface* 10 (84): 20130246. https://doi.org/10.1098/rsif.2013.0246

Shaer, Amin, Meysam Rezaei, Behnam Moghani Rahimi, and Fatemeh Shaer. 2021. "Examining the Associations between Perceived Built Environment and Active Travel, before and after the COVID-19 Outbreak in Shiraz City, Iran." *Cities* 115: 103255. https://doi.org/10.1016/j.cities.2021.103255

Shi, Hui, and Konstadinos G. Goulias. 2024a. "Long-Term Effects of COVID-19 on Time Allocation, Travel Behavior, and Shopping Habits in the United States." *Journal of Transport & Health* 34: 101730. https://doi.org/10.1016/j.jth.2023.101730

Shi, Hui, and Konstadinos G. Goulias. 2024b. "Year-to-Year Time Allocation and Spatial Structure of Americans' Daily Schedules from 2019 to 2022 and a Detailed Analysis of the Stay-at-Home All-Day Patterns." *Transportation Research Part A: Policy and Practice*, 104190. https://doi.org/10.1016/j.tra.2024.104190

Shi, Hui, Rongxiang Su, and Konstadinos G. Goulias. 2023. "Exploring the Impact of COVID-19 Pandemic on Americans Time Use Related Subjective Wellbeing." *Wellbeing, Space and Society* 4: 100148. https://doi.org/10.1016/j.wss.2023.100148

Shi, Hui, Rongxiang Su, Jingyi Xiao, and Konstadinos G. Goulias. 2022. "Spatiotemporal Analysis of Activity-Travel Fragmentation Based on Spatial Clustering and Sequence Analysis." *Journal of Transport Geography* 102: 103382. https://doi.org/10.1016/j.jtrangeo.2022.103382

Stone, Arthur A., Stefan Schneider, Alan Krueger, Joseph E. Schwartz, and Angus Deaton. 2018. "Experiential Wellbeing Data from the American Time Use Survey: Comparisons With Other Methods and Analytic Illustrations With Age and Income." *Social Indicators Research* 136 (1): 359–78. https://doi.org/10.1007/s11205-016-1532-x

Su, Rongxiang, Elizabeth C. McBride, and Konstadinos G. Goulias. 2021. "Unveiling Daily Activity Pattern Differences between Telecommuters and Commuters Using Human Mobility Motifs and Sequence Analysis." *Transportation Research Part A: Policy and Practice* 147: 106–32. https://doi.org/10.1016/j.tra.2021.03.002

Su, Rongxiang, Elizabeth Callahan McBride, and Konstadinos G. Goulias. 2020. "Pattern Recognition of Daily Activity Patterns Using Human Mobility Motifs and Sequence Analysis." *Transportation Research Part C: Emerging Technologies* 120: 102796. https://doi.org/10.1016/j.trc.2020.102796

"theHRD." June 14, 2020. *theHRDIRECTOR* (blog). https://www.thehrdirector.com/business-news/work_life_balance/data-reveals-60-of-people-want-to-stay-at-home-after-covid-19/

Vagni, Giacomo, and Benjamin Cornwell. 2018. "Patterns of Everyday Activities Across Social Contexts." *Proceedings of the National Academy of Sciences* 115 (24): 6183–88. https://doi.org/10.1073/pnas.1718020115

"WHO." 2023. https://covid19.who.int/

Xiao, Jingyi, Rongxiang Su, Elizabeth C. McBride, and Konstadinos G. Goulias. 2020. "Exploring the Correlations between Spatiotemporal Daily Activity-Travel Patterns and Stated Interest and Perception of Risk With Self-Driving Cars." *AGILE: GIScience Series* 1: 1–15. https://doi.org/10.5194/agile-giss-1-22-2020

Yan, Youpei, Amyn A. Malik, Jude Bayham, Eli P. Fenichel, Chandra Couzens, and Saad B. Omer. 2021. "Measuring Voluntary and Policy-Induced Social Distancing Behavior During the COVID-19 Pandemic." *Proceedings of the National Academy of Sciences* 118 (16): e2008814118. https://doi.org/10.1073/pnas.2008814118

Zhu, Pengyu, Deborah Salon, Abolfazl Mohammadian, and Yuqing Guo. 2023. "Looking Forward: The Long-Term Implications of COVID-19 for Transportation." *Transportation Research Part D: Transport and Environment* 123: 103910. https://doi.org/10.1016/j.trd.2023.103910

21 Sustainability Implications of Digital Twins for Urban Mobility

Mahdi Aghaabbasi, Soheil Sabri, and Xinyue Ye

21.1 INTRODUCTION

Urban mobility systems face growing pressure from rapid urbanization, congestion, and environmental concerns. Maintaining and operating transportation infrastructure imposes significant economic burdens. In the United States, the American Society of Civil Engineers (2021) reports a $786 billion backlog in road and bridge needs. Globally, the G20 Global Infrastructure Outlook projects a $15 trillion gap between required and actual infrastructure investment by 2040 (Global Infrastructure Outlook 2024). Transportation issues also impact health and mortality, with the World Health Organization (2023) reporting 1.19 million annual road traffic deaths. Additionally, the American Public Health Association (2010) estimates U.S. health costs from transportation-related air pollution at $50–80 billion annually.

Digital twins (DTs), used across disciplines like manufacturing, supply chains, healthcare, and smart cities (Sabri, Alexandridis, and Lee 2024), have emerged as a promising technology to address urban mobility challenges by optimizing and managing transportation networks (Feng, Lv, and Lv 2023). Additionally, research explores DT applications in urban contexts, including infrastructure development (Alexandridis et al. 2024) and disaster management (Sabri et al. 2023; Ye et al. 2023). Studies also highlight the potential and limitations of DTs in achieving the United Nations' Sustainable Development Goals (SDGs) (Tzachor et al. 2022).

DTs create virtual replicas of urban systems by integrating real-time data, simulations, and visualization tools to represent, analyze, and predict city behavior. These digital models offer a comprehensive view of urban infrastructure, processes, and interactions, supporting more informed decision-making and planning.

DTs have been widely adopted in transportation to enhance efficiency, optimize traffic management, and improve overall mobility (Wang 2022; Kušić, Schumann, and Ivanjko 2023). Recent studies demonstrate that DTs contribute to environmental sustainability by reducing emissions (Xu, Jiang, and Zheng 2022; Faiad et al. 2023), improving traffic flow (Kušić, Schumann, and Ivanjko 2023), and promoting sustainable transport modes (Kajba, Jereb, and Cvahte Ojsteršek 2023). They also enhance economic sustainability through more efficient infrastructure management (Ammar, Nassereddine, and Dadi 2022; Tu et al. 2022) and advance social sustainability by improving transportation safety and accessibility (Adibfar and Costin 2022; Saba, Ullah, and Tariq 2024). While many DTs incorporate elements of complexity science and public engagement, Caldarelli et al. (2023) argue that further integration can enhance their potential as adaptable, efficient, resilient, and sustainable urban solutions.

The implementation of DTs in urban mobility faces several obstacles. Significant concerns about data privacy and security exist (Banerjee et al. 2023; Astarita et al. 2024), while technological barriers and digital divide issues hinder widespread adoption (Liu et al. 2022; Jafari et al. 2023). Additionally, integrating DTs with other smart city initiatives presents complex challenges and opportunities (Adibfar and Costin 2022).

 DOI: 10.1201/9781003503262-25

While the value of DTs for urban mobility is well-established across various scientific and professional venues (Dembski et al. 2020; Sabri, Winter, and Rajabifard 2023), comprehensive studies are still needed to synthesize their diverse applications and assess their long-term impacts on sustainability.

The priority of sustainability dimensions—environmental, economic, and social—varies based on urban contexts and goals. Environmental concerns, particularly reducing greenhouse gas emissions and improving air quality, often take precedence due to urgent climate change needs. Economic considerations, like infrastructure cost optimization, follow closely as cities face limited budgets. Sometimes, improvements in environmental and economic areas address social dimensions, crucial for equitable urban development, rather than serving as the primary focus.

This chapter explores the sustainability implications of DTs for urban mobility, examining their potential benefits and challenges across environmental, economic, and social dimensions. By synthesizing current research and emerging practices, we aim to offer insights for urban planners, policymakers, and researchers pursuing sustainable urban transportation solutions.

21.2 THEORETICAL FOUNDATIONS OF SUSTAINABILITY

Sustainability is a complex notion encompassing various aspects of the environment, economy, and society (Brønn and Brønn 2018). Maintaining a balance between these aspects is crucial for ensuring fair resource allocation, optimal economic efficiency, and minimal environmental impact.

Transportation is essential for societal functioning but significantly impacts the environment. Sustainable transportation aims to meet current needs while minimizing environmental harm for future generations (Rodrigue 2024). Transforming transportation networks into sustainable systems is a long-term process due to the longevity of infrastructure and the complexity of urban systems. Although rapid change is often desired, urban transportation transformations typically occur incrementally due to budget constraints, political processes, and the need for operational continuity during changes.

Moreover, the global inclination toward automobiles (Carruthers 2003), driven by convenience and affordability, poses challenges in promoting other transportation modes. However, various methods exist to achieve sustainable transportation, including transport *demand* management and enhancement of transport *supply*.

The *demand* strategy involves tactics like full-cost pricing, which considers external factors such as pollution, and parking limits, which can reduce car trips by 10%–30% in urban areas (Litman 2025). Promoting methods to reduce unnecessary trips, such as telework and e-commerce, has shown promise; for instance, during the COVID-19 pandemic, widespread telework reduced peak-hour trips in some cities (Obeid et al. 2024). Traffic bans, like car-free days, can reduce air pollutants by up to 40% during implementation (Nieuwenhuijsen and Khreis 2016). These strategies aim to discourage car usage and promote sustainable alternatives.

The supply emphasizes enhancing transportation options through energy-efficient vehicles and less carbon-intensive fuels. It advocates for compact urban areas to reduce travel distances and support efficient transport modes while promoting context-specific solutions for different situations. Additionally, it highlights integrating micromobility options like walking, cycling, and e-scooters for short trips and connections to public transit.

Sustainability increasingly emphasizes decarbonizing transportation as a key approach, focusing on reducing carbon emissions in three areas: primary emissions from cars, secondary emissions from fuel and electricity generation, and emissions from related activities like employee commuting. Several methods exist to achieve this, such as using lower-carbon-content materials, promoting energy-efficient modes in infrastructure construction, and transitioning to electric vehicles and alternative fuels. Additionally, optimizing vehicle use and implementing pricing mechanisms that account for carbon emissions can encourage a shift toward sustainable transportation.

These challenges relate to applying DTs in urban transportation. DTs can simulate the impact of strategies like full-cost pricing, parking restrictions, and traffic prohibitions. A digital replica of an urban transport network can assess how road pricing affects traffic movement and shifts in transportation modes. Moreover, DTs play a crucial role in creating, testing, and optimizing these solutions virtually before real-world implementation. There's a global trend toward adopting electric vehicles and alternative fuels, and DTs can facilitate this transition by simulating electric vehicle charging impacts on the power grid, modeling fuel type efficiency, and optimizing traffic flow to reduce congestion and emissions.

21.3 THE ROLE OF DTs IN SUSTAINABLE URBAN MOBILITY

21.3.1 Technological Overview of Urban Digital Twins

DTs for urban transport and mobility leverage various fundamental technologies (Figure 21.1). These technologies work together to create a robust platform for simulating, assessing, and optimizing transportation networks.

21.3.1.1 Digital Modeling Layer

This layer serves as the fundamental digital representation of physical transportation infrastructure and its associated assets. It encompasses various essential modeling techniques, including physical or mechanical modeling, which deepens our understanding of the behavior and characteristics of transportation assets like bridges and tunnels. This approach enables simulation of their structural integrity, load capacity, and responses to different stresses and strains.

Another important technique is using geographic information systems (GIS) to integrate, map, and analyze spatial data (Sabri and Rajabifard 2019). This includes static elements like road networks and transportation infrastructure locations, as well as dynamic aspects such as traffic flow patterns and accessibility analysis (Reisi et al. 2020; Sun et al. 2023). While static elements contribute to the physical description, dynamic aspects represent the system's behavior over time. The DT gains insights into the geographical context of urban mobility.

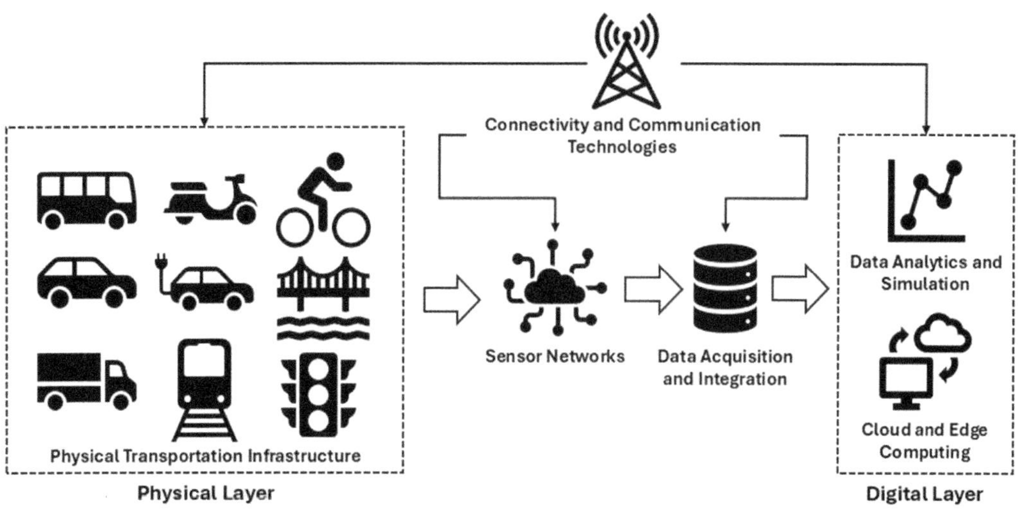

FIGURE 21.1 Components and interactions in an urban digital twin.

21.3.1.2 Real-Time Data Acquisition and Integration

Real-time data feeds are essential for a digital replica to accurately mirror the current transportation system condition. This involves deploying sensor networks throughout the transportation network, which consist of various embedded sensors, including:

- Traffic cameras for visual monitoring
- Global positioning system (GPS) devices in vehicles for location tracking
- Environmental sensors for air quality and weather conditions
- Inductive loop detectors for vehicle counting and classification
- Radar and light detection and ranging (LiDAR) sensors for speed detection and object recognition
- Bluetooth and Wi-Fi sensors for travel time estimation
- Weight-in-motion sensors for heavy vehicle monitoring
- Acoustic sensors for noise level measurement

These diverse sensor types provide a comprehensive view of the transportation network's state and performance, continuously collecting data on traffic flow, vehicle position, air quality, and other relevant aspects. We then fuse and integrate this data. Integrating data from various sources, each with unique formats, accuracy levels, and update frequencies, poses a significant challenge. Robust data fusion methods are necessary to ensure the DT accurately and reliably represents the real-world system.

21.3.1.3 Data Analytics and Simulation

The gathered data are processed and analyzed to understand the transportation system's behavior, forecast future conditions, and identify optimization opportunities. There is a growing trend toward using machine learning (ML) algorithms to evaluate extensive transportation data, identify patterns, and predict traffic congestion, optimize routes, and detect potential disruptions. Using the digital modeling layer and real-time data, simulation models can examine various scenarios, including changes in traffic patterns, infrastructure enhancements, and new mobility services. This promotes a deeper understanding of how the transportation system may respond to different interventions, facilitating data-driven decision-making.

21.3.1.4 Cloud and Edge Computing

Cloud computing systems provide the processing power and storage capacity needed to manage the vast data volumes generated by a DT. The edge computing paradigm involves processing closer to the data source, reducing transmission delays and enabling immediate responses. This is especially crucial for time-critical applications like autonomous vehicle navigation.

21.3.1.5 Connectivity and Communication Technologies

Efficient connectivity between the physical and digital elements of a DT is essential for instant data sharing and control. Internet of Things (IoT) technologies play a vital role in connecting sensors, vehicles, and infrastructure, enabling data exchange with the DT platform. Additionally, vehicle-to-everything (V2X) communication technologies, including vehicle-to-vehicle (V2V) and vehicle-to-infrastructure (V2I), are critical for sharing real-time data on vehicle position, speed, and road conditions, significantly enhancing traffic control and safety.

21.3.2 Environmental Sustainability Implications

Environmental sustainability can be enhanced through DTs, particularly when integrated with artificial intelligence (AI) and ML (Kamal et al. 2024; Xiao et al. 2024). For instance, Kamal et al. (2024) propose a DT-based adaptive traffic signal control system using the multiagent deep deterministic policy gradient algorithm to optimize signals for reduced fuel consumption and carbon dioxide (CO_2) emissions. By simulating various traffic scenarios with a DT of an urban network and using real-world traffic data from Amman, Jordan, during peak hours, their DT-based deep reinforcement learning approach significantly reduced CO_2 emissions by 11.78% and fuel consumption by 4.57% in a 1-hour rush period. Additionally, they achieved a 42% reduction in average travel time, clearing traffic in 4,600 seconds compared to 4,800 seconds in the real system.

Xiao et al. (2024) highlight the role of DTs in developing sustainable route planning for smart city public transport systems. They present an approach using multiobjective optimization techniques and the modified bat algorithm (MBA) to minimize operating costs and emissions. Their study shows that MBA consistently achieves lower costs and emissions than other algorithms. For example, in one scenario, MBA resulted in a total cost of $100,000, which was 5%–10% lower than alternative methods. It also reduced emissions, with one route showing 500 kg compared to 520 kg for other algorithms, a 4% improvement. The study explains that the DT model simulates real-time transport system aspects, allowing for scenario exploration and adaptable route plans, leading to significant economic and environmental benefits.

Liao et al. (2022) focus on DT technology to enhance highway on-ramp merging, resulting in reduced fuel consumption and emissions. They demonstrate that simulating a cooperative merging system using vehicle-to-cloud communication can improve merging maneuvers, thereby lowering fuel use and pollution.

DTs enhance the adaptability of urban mobility systems through real-time data integration, simulation, and optimization with AI and ML. They gather real-time data from various sources, including sensors, infrastructure, and user devices, creating a dynamic virtual representation that reflects actual traffic conditions, infrastructure status, and user behavior. This representation provides a comprehensive understanding of the urban mobility system's current state (Chen, Sun, and Lin 2024; Kumarasamy et al. 2024).

21.3.3 Economic Sustainability Implications

DTs significantly enhance the cost-effectiveness of urban transport systems. Kušić, Schumann, and Ivanjko (2023) highlight DTs' role in real-time traffic management on motorways, asserting that they can revolutionize traffic simulations through real-time analysis and control optimization. This capability leads to more efficient traffic management, reducing congestion and improving flow, resulting in significant long-term cost savings. Similarly, Chang et al. (2024) note the growing interest in DTs for managing transportation infrastructure health and maintenance, further contributing to cost savings.

Ammar et al. (2024) focus on applying DTs to manage safety hardware assets, such as guiderails, in transportation systems. They suggest that DTs enhance asset management cost-effectiveness by enabling predictive and preventive maintenance, thus reducing reactive maintenance costs and extending asset lifespan. Similarly, Xiao et al. (2024) emphasize using DTs and multiobjective optimization for sustainable and cost-effective public transport route planning in smart cit ies. They highlight the MBA in conjunction with a DT model to minimize operational costs and emissions. In a case study on Fuzhou City's public transport system, the MBA achieved a total cost of $100,000, 4.8% lower than the next best algorithm (GA[1] at $102,000) and 9.1% lower than the worst-performing algorithm (BA[2] at $110,000). In emissions reduction, the MBA matched the firefly algorithm (FA) with 500 kg while outperforming others by 2%–4% (PSO[3] at 520 kg, GA at 510 kg, and BA at 508 kg). These improvements demonstrate the MBA's superior performance in cost reduction and emissions minimization, significantly enhancing public transport system sustainability and cost-effectiveness.

DTs enhance the adaptability of urban mobility systems through real-time data integration, simulation, and optimization with AI and ML. Several studies emphasize the DT's role in gathering real-time data from various sources within the transportation system, including sensors, infrastructure, and user devices (Feng, Lv, and Lv 2023; Chen, Sun, and Lin 2024; Kumarasamy et al. 2024;

Xiao et al. 2024). These data create a dynamic virtual representation that mirrors actual traffic conditions, infrastructure status, and user behavior, enabling a thorough understanding of the urban mobility system's current state.

The DT facilitates "what-if" analyses by simulating various scenarios and traffic conditions (Feng, Lv, and Lv 2023; Chen, Sun, and Lin 2024; Kumarasamy et al. 2024; Xiao et al. 2024). This capability tests the impact of different variables, such as accidents, road closures, and changes in traffic patterns, on the transportation system's performance.

The crucial role of AI and ML algorithms in enhancing DT capabilities is evident. These algorithms process vast data collected by the DT to predict future traffic conditions, identify potential bottlenecks, and optimize traffic flow in real time. Chen, Sun, and Lin (2024) introduce the bidirectional anisometric gated recursive unit (BDAGRU) algorithm for predicting travel times and optimizing routes within a DT framework. Their study shows significant improvements in prediction accuracy and route optimization, with BDAGRU outperforming methods like deep artificial neural networks (DANN) and long short-term memory (LSTM) across various error metrics. For instance, with multimodal data, BDAGRU achieved a mean absolute error (MAE) of 13.0785, compared to 14.8143 for DANN and 13.9370 for LSTM. In route optimization, the DT powered by BDAGRU consistently selected the shortest travel time route, with the highest mean information transfer value of 84, compared to 79 for another model.

21.3.4 Social Sustainability Implications

There is limited discussion on DTs and their applications in addressing social equity, equitable access to transportation, and public health. However, some studies provide insights into how DTs can support these areas.

The data-driven nature of DTs can enhance inclusive urban planning. By incorporating diverse datasets reflecting social and economic factors, DTs help identify and address inequalities in urban environments. However, Ziehl et al. (2023) caution against overly technology-centric approaches, emphasizing the need to include social aspects and marginalized groups' needs in DT design, and implementation must consider the complex mobility patterns of unpaid care workers, who are often overlooked in traditional planning. Family caregivers and parents with young children may need frequent stops for shopping or other tasks. The authors emphasize the importance of addressing safety, accessibility, and transportation options for these groups. They also highlight the need to recognize the invisibility of unpaid care work in urban systems and integrate the perspectives of civil society organizations representing these marginalized groups in DT development.

The use of DTs in optimizing public transportation systems is crucial. By analyzing real-time data, DTs can adjust routes and schedules to enhance community service, leading to more equitable access and addressing service coverage disparities. For example, DTs can pinpoint areas with limited public transport access and optimize routes to improve service.

DTs can enhance transportation safety by simulating and evaluating vehicle and infrastructure safety, leading to improved safety measures and fewer accidents (Adibfar and Costin 2022; Wang 2022; Saba, Ullah, and Tariq 2024). They also monitor traffic conditions and predict hazards, contributing to safer roads. Furthermore, the benefits of improved safety and transportation efficiency, such as reduced traffic congestion (Kušić, Schumann, and Ivanjko 2023) and better air quality (Kamal et al. 2024) from optimized traffic flow, can positively impact public health.

21.4 CHALLENGES AND LIMITATIONS

21.4.1 Technological and Practical Challenges

Mitigation and adaptation strategies for addressing DT challenges in sustainable urban mobility, especially scalability and trade-offs, have been discussed. Effectively managing large-scale urban environments requires a modular approach (Ziehl et al. 2023), where DTs consist of interconnected modules representing specific aspects of the transportation system, such as traffic flow, public transport, or parking. This modularity supports scalability by allowing modules to be added or expanded without overhauling the entire system.

Efficient data management is essential for DT scalability (Yeon et al. 2023; Xu et al. 2023). As cities grow and collect more data, DTs must handle and analyze it effectively to deliver real-time insights and support decision-making. Cloud computing, edge computing, and distributed processing are proposed solutions to these challenges. Additionally, open-source platforms and standardized data frameworks are critical for ensuring scalability and interoperability (Wang 2022; Yeon et al. 2023).

Open-source platforms promote adoption and collaboration, while standardized data formats ensure seamless data exchange among systems and stakeholders. However, challenges include license compliance, long-term maintenance, integration costs, project stability, community support, security risks, and retaining essential competencies within the organization (Butler et al. 2022).

Continuous evaluation of components and careful attention to scalability and performance are essential. While these challenges apply to open-source adoption generally, they are especially relevant for urban mobility DTs due to the critical nature of transportation infrastructure, the long lifespan of such systems, and the need for real-time data processing and decision-making.

While DTs hold great potential for optimizing urban mobility, trade-offs exist, particularly between model complexity and computational efficiency (Chang et al. 2024). Complex DTs offer higher accuracy but require substantial computational resources, possibly affecting real-time performance. Striking a balance through model simplification or focusing on key areas is recommended to enhance efficiency.

Data privacy and security pose significant trade-offs (Astarita et al. 2024). Since DTs rely on large-scale data collection, ensuring privacy and security is crucial. Researchers have developed solutions for DTs in manufacturing; for instance, Balta et al. (2024) proposed a framework for detecting cyberattacks in cyber-physical manufacturing systems (CPMS). Their cybersecurity DT uses a detector DT to identify anomalies and a consistency DT to differentiate between expected variations and real attacks. Tested on a three-dimensional printer, the framework effectively detected attacks while accounting for normal process fluctuations. Although similar applications in urban mobility DTs are limited, this framework shows how cybersecurity measures can enhance data protection, traffic management integrity, and connected vehicle communications, increasing the resilience of smart city infrastructure.

A key trade-off involves balancing optimization objectives, such as traffic flow and emissions reduction, with the overall impact on the transportation system. For example, optimizing traffic flow might increase speeds, conflicting with safety goals or raising emissions. A holistic approach that accounts for multiple objectives and trade-offs is essential to achieve sustainable urban mobility.

21.4.2 Governance and Policy Challenges

There are limited insights into governance's role in implementing DTs for urban mobility and balancing regulatory approaches with market-driven solutions. Resolving this conflict is crucial to ensure technological advances promote long-term sustainability while aligning economic, social, and environmental goals (Rodrigue 2024).

The importance of data privacy and security in DT implementation highlights the need for robust cybersecurity frameworks, privacy-preserving technologies, and clear governance policies to mitigate risks (Sahal et al. 2021; Papyshev and Yarime 2021; Banerjee et al. 2023; Jafari et al. 2023; Sellitto et al. 2021; Astarita et al. 2024). Governance structures must address these concerns. Additionally, open-source platforms and data standards (e.g., Ziehl et al. 2023; Yeon et al. 2023) emphasize the role of governance in promoting interoperability and collaboration, possibly through regulations or industry initiatives.

In transportation infrastructure management, discussions on state departments of transportation (DOTs) reveal a growing awareness of DTs for asset management (Ammar, Nassereddine, and Dadi 2022; Ammar et al. 2023; Ammar et al. 2024). This indicates that governance will play an emerging role in adopting these technologies in the public sector. However, challenges may arise, including resource constraints, as implementing DTs requires significant investment in technology and training. DOTs might lack in-house technical expertise, necessitating workforce development or new hires. Data management and privacy concerns are critical due to the vast amounts of real-time data needed. Ensuring interoperability with existing systems and developing appropriate regulatory frameworks may pose further hurdles. Additionally, DOTs must navigate change management and address equity concerns to ensure DT benefits are fairly distributed across communities. Overcoming these challenges requires strategic planning, sustained investment, and collaboration among various transportation stakeholders.

21.5 DIGITAL TWIN MATURITY AND THE CAPABILITIES PERIODIC TABLE

To implement and assess the impact of DTs in urban mobility, understanding their development and effectiveness is crucial. The Digital Twin Consortium's Capabilities Periodic Table (CPT) provides a valuable framework for this evaluation. By examining DT maturity levels and essential capabilities, we can better assess their potential for enhancing sustainable urban mobility systems and guide implementation. This framework is especially useful for urban planners and policymakers in benchmarking and strategizing DT deployments. Additionally, analyzing DT maturity through the CPT allows for more accurate assessments of their sustainability implications, helping identify areas where these technologies can significantly impact the environmental, economic, and social aspects of transportation systems. The CPT outlines essential capabilities for DTs and offers insights into their maturity levels (Schalkwyk, Whiteley, and Goldman 2024). This framework is crucial for understanding the evolution and effectiveness of DTs in urban mobility systems.

21.5.1 Maturity Levels and the CPT

- *Basic*: Focus on fundamental data services and integration capabilities (e.g., data acquisition, streaming, system integration).
- *Intermediate*: Incorporates advanced intelligence and user experience capabilities (e.g., AI, extended reality visualization).
- *Advanced*: Leverages sophisticated capabilities across all CPT categories, including advanced management and trustworthiness features.

21.5.2 Implications for Sustainable Urban Mobility

- *Environmental*: Enhanced intelligence capabilities allow for more accurate forecasts of traffic patterns and energy usage, reducing emissions.
- *Economic*: Improved management capabilities enhance system effectiveness and resource distribution.
- *Social*: Better user experience and integration can increase public engagement and service accessibility.

As DT implementation progresses, its ability to enhance sustainable urban mobility increases. Transport planners and decision-makers can leverage this understanding to guide technology investments and strategies, ensuring that DTs effectively support sustainability objectives in urban transportation systems.

21.6 CHALLENGES AND FUTURE DIRECTIONS

Several challenges need to be addressed in future DT research and development, falling into technological and sociotechnical categories. Data, technology, and social aspects in smart cities are interrelated to ensure productivity, livability, and sustainability (Sabri 2021). Figure 21.2 illustrates these challenges and their interconnections within the sustainability context.

21.6.1 Data Privacy and Security Concerns

Data privacy and security challenges accompany DT implementation, especially in urban transport. Managing vast quantities of sensitive data required for accurate DTs poses significant privacy risks (Astarita et al. 2024). Existing city DTs (CDTs) face limitations in data privacy, availability, and predictive simulation applicability (Papyshev and Yarime 2021). Additionally, CDTs' demand for individualized data, such as real-time traffic simulations, raises safety concerns (Papyshev and Yarime 2021). Ensuring secure data exchange and transaction verification remains a challenge (Banerjee et al. 2023). Balancing model fidelity with computational efficiency further complicates DT adoption (Chang et al. 2024). Security concerns also extend to autonomous vehicles and protecting vehicular ad hoc networks (VANETs) from attacks in electromobility (Ali, Roccotelli, and Fanti 2022). To address these issues, studies recommend robust cybersecurity frameworks, privacy-preserving technologies, and blockchain-based secure data sharing (Banerjee et al. 2023; Astarita et al. 2024).

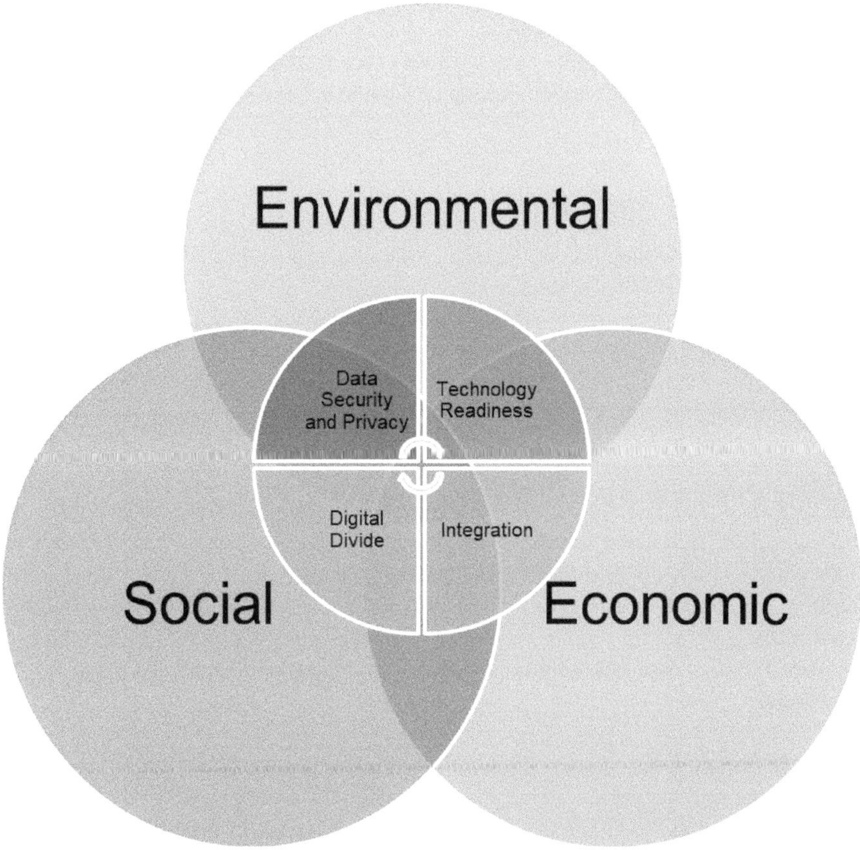

FIGURE 21.2 Challenges and future research areas for sustainable urban digital twins.

These data privacy and security challenges significantly impact the sustainability of DTs in urban transport. Public trust is essential for widespread adoption and long-term success. Inadequate handling of these concerns could result in data breaches, privacy violations, and loss of trust, undermining the sustainability of these technologies. For DTs to support efficient, secure, and sustainable transportation systems, it is crucial to establish strong data governance frameworks, comply with privacy regulations, and implement robust security measures (Astarita et al. 2024).

21.6.2 Technological Barriers and Digital Divide Issues

Technological barriers and the digital divide also hinder the adoption of DTs for urban transport. Although DTs offer substantial benefits, their implementation has lagged, with only 25 countries (13%) utilizing them (Teng et al. 2021). The advantages, such as improved traffic flow, reduced congestion, enhanced energy efficiency, and better urban planning, vary widely based on context. However, quantitative data on these improvements remain limited or context specific.

This limited adoption suggests technological barriers are obstructing broader implementation. A "lab-business social gap" in DT technology highlights the need to move beyond technical benefits, emphasizing engineering, industrialization, and practical problem-solving (Liu et al. 2022). Bridging this gap requires several key steps:

1. *Investment in robust IT infrastructure*: Existing IT infrastructure often falls short for DT data analysis (Jafari et al. 2023). Significant upgrades in computing systems, storage, and networks are essential.
2. *Development of standardized protocols*: A lack of open standards and system interoperability limits progress (Faliagka et al. 2024). Unified standards are needed to enable seamless integration and data sharing.
3. *Capacity building and skills development*: Training professionals in DT technologies, data analysis, and integration is critical to bridge the gap between research and real-world applications.
4. *Addressing data challenges*: Managing vast datasets requires strong information and communication technology infrastructure, which can be costly. Limited resources may create a digital divide, hindering implementation in underfunded regions.
5. *Improving sensor technology*: Reliable sensors, especially in harsh conditions, are essential (Ywet et al. 2024). However, investing in high-quality sensors and robust data processing can raise costs.

Addressing these challenges will help cities overcome technological barriers and digital divides, advancing DT adoption in urban transport systems.

These factors significantly influence the equitable deployment of DT technology. Rapid technological advancements and a lack of interoperability risk making systems obsolete and complicate the integration of new technologies, threatening long-term sustainability. Overcoming these barriers requires investments in robust IT infrastructure and the development of user-friendly, accessible DT platforms to ensure inclusivity and benefit all societal segments.

A digital divide in smart mobility can lead to unequal access to technologies, disproportionately affecting certain communities. For instance, residents unable to afford Internet access may face exclusion from city traffic management systems that rely on DTs and real-time data, worsening social inequalities. It's crucial to assess access levels and the benefits DTs provide to different socioeconomic groups. Bridging this divide through digital literacy initiatives and ensuring equitable technology access are essential for sustainable, inclusive urban transport. Without addressing these issues, DT benefits may be unevenly distributed, further marginalizing disadvantaged communities.

The challenges above directly impact the financial sustainability of DT initiatives in urban transport. Cities with limited resources may struggle to invest continuously in costly technologies and skilled personnel required to address these barriers.

21.6.3 Integration with Other Smart City Initiatives

Integrating DTs for urban transport with other smart city initiatives poses a significant challenge. Mobility DTs often work alongside other transportation modes, such as buses and subways (Yeon et al. 2023). Integrating these existing systems into the DT is essential for realistic simulations and analyses of urban mobility. Without this integration, the DT's ability to provide a comprehensive understanding of the transportation network and optimize mobility may be limited.

There is a critical need for interoperability in developing DTs, especially in smart cities and IoT-based applications (Adibfar and Costin 2022). One example is coordinating electric vehicle charging stations with real-time traffic data. An integrated DT system could predict traffic surges and adjust the availability of charging stations, reducing grid strain during peak hours while ensuring efficient energy use. Similarly, DTs could synchronize with smart waste management systems to adjust collection routes based on traffic patterns, reducing emissions and improving operational efficiency.

The challenge is ensuring that different systems—such as software applications, workflows, and data ontologies—can seamlessly exchange information without significant effort from the end user. Integrating DTs with other smart city initiatives requires interoperability to enable data and insight sharing across various domains.

Finally, considering the broader context of smart city development is essential when implementing DTs. While DTs can optimize individual urban systems like transportation, their true value lies in contributing to the holistic development of smart, sustainable cities. This requires integrating DTs with other smart city initiatives—such as energy and waste management—to create a cohesive, interconnected urban ecosystem.

The lack of seamless integration with other smart city initiatives poses significant challenges to the sustainability of DTs for urban transport. It can lead to redundant investments in data collection and system development, hindering financial sustainability, while well-integrated systems enable cost savings and improved resource utilization. Additionally, DTs must be adaptable and scalable to fit evolving smart city ecosystems; addressing integration challenges is vital to prevent obsolescence and ensure long-term viability. Effective integration is crucial for DTs to realize their full potential in urban mobility, maximizing positive impacts on traffic flow, environmental sustainability, and citizen safety.

21.7 CONCLUSION

This chapter examined the sustainability implications of urban DTs for urban mobility. Key findings indicate that urban DTs significantly enhance environmental sustainability by reducing emissions and optimizing traffic flow. Economically, they promise improved infrastructure management and operational efficiency, while social sustainability benefits include enhanced safety and accessibility in transportation systems.

For urban planners and policymakers, we recommend:

- Prioritizing data privacy and security in DT implementations
- Addressing digital divide issues for equitable access to benefits
- Developing integrated approaches aligning DTs with broader smart city initiatives

Future research should focus on:

- Long-term impacts of DTs on urban mobility patterns
- Strategies for overcoming technological barriers in diverse urban contexts
- Methods for effective integration of DTs with existing urban systems

As cities grapple with mobility challenges, DTs offer a promising tool for sustainable transportation systems. However, successful implementation will require careful consideration of opportunities and challenges, along with ongoing research and innovation in the field.

NOTES

1 Genetic algorithm
2 Original bat algorithm
3 Particle swarm optimization

REFERENCES

Adibfar, Alireza, and Aaron M. Costin. 2022. "Creation of a Mock-up Bridge Digital Twin by Fusing Intelligent Transportation Systems (ITS) Data into Bridge Information Model (BrIM)." *Journal of Construction Engineering and Management* 148 (9). doi: 10.1061/(asce)co.1943-7862.0002332.

Alexandridis, Kostas, Sonya Zhang, Mehrdad Koohikamali, Soheil Sabri, and Erkan Ozkaya. 2024. "Designing and Implementing a Robust, Modular and Interoperable Digital Twin Smart City Framework for Critical Water Spatial Infrastructure."

Ali, Wasim A, Michèle Roccotelli, and Maria Pia Fanti. 2022. "Digital Twin in Intelligent Transportation Systems: A Review." 2022 8th International Conference on Control, Decision and Information Technologies (CoDIT).

American Public Health Association. 2010. The Hidden Health Costs of Transportation.

American Society of Civil Engineers. 2021. Roads. A Comprehensive Assessment of America's Infrastructure.

Ammar, A., H. Nassereddine, and G. Dadi. 2022. "State Departments of Transportation's Vision Toward Digital Twins: Investigation of Roadside Asset Data Management Current Practices and Future Requirements." ISPRS Annals of the Photogrammetry, Remote Sensing and Spatial Information Sciences V-4-2022:319–327. doi: 10.5194/isprs-annals-V-4-2022-319-2022.

Ammar, Ashtarout, Francesca Maier, Rachel Catchings, Hala Nassereddine, and Gabriel Dadi. 2023. "Departments of Transportation Efforts to Digitize Ancillary Transportation Asset Data: A Step Toward Digital Twins." *Transportation Research Record: Journal of the Transportation Research Board* 2677 (11):428–445. doi: 10.1177/03611981231166000.

Ammar, Ashtarout, Francesca Maier, William S. Pratt, Elaine Richard, and Gabriel Dadi. 2024. "Practical Application of Digital Twins for Transportation Asset Data Management: Case Example of a Safety Hardware Asset." *Transportation Research Record: Journal of the Transportation Research Board* 2678 (10). doi: 10.1177/03611981241231804.

Astarita, Vittorio, Giuseppe Guido, Sina Shaffiee Haghshenas, and Sami Shaffiee Haghshenas. 2024. "Risk Reduction in Transportation Systems: The Role of Digital Twins According to a Bibliometric-Based Literature Review." *Sustainability* 16 (8). doi: 10.3390/su16083212.

Balta, Efe C., Michael Pease, James Moyne, Kira Barton, and Dawn M. Tilbury. 2024. "Digital Twin-Based Cyber-Attack Detection Framework for Cyber-Physical Manufacturing Systems." *IEEE Transactions on Automation Science and Engineering* 21 (2):1695–1712. doi: 10.1109/TASE.2023.3243147.

Banerjee, Sourav, Debashis Das, Pushpita Chatterjee, and Uttam Ghosh. 2023. "Blockchain-enabled Digital Twin Technology for Next-Generation Transportation Systems." 2023 IEEE 26th International Symposium on Real-Time Distributed Computing (ISORC).

Brønn, Carl, and Peggy Simcic Brønn. 2018. "Sustainability: A Wicked Problem Needing New Perspectives." In *Business Strategies for Sustainability*, 1–18. Routledge.

Butler, Simon, Jonas Gamalielsson, Björn Lundell, Christoffer Brax, Anders Mattsson, Tomas Gustavsson, Jonas Feist, Bengt Kvarnström, and Erik Lönroth. 2022. "Considerations and Challenges for the Adoption of Open Source Components in Software-Intensive Businesses." *Journal of Systems and Software* 186:111152. doi: 10.1016/j.jss.2021.111152.

Caldarelli, Guido, Elsa Arcaute, Marc Barthelemy, Michael Batty, Carlos Gershenson, Dirk Helbing, Stefano Mancuso, Yamir Moreno, José J Ramasco, and Céline Rozenblat. 2023. "The Role of Complexity for Digital Twins of Cities." *Nature Computational Science* 3 (5):374–381.

Carruthers, R. 2003. The Impacts of Motorization and What can Be Done to Mitigate Them. The World Bank.

Chang, Xiangyu, Rui Zhang, Jianxiao Mao, and Yuguang Fu. 2024. "Digital Twins in Transportation Infrastructure: An Investigation of the Key Enabling Technologies, Applications, and Challenges." *IEEE Transactions on Intelligent Transportation Systems* 25 (7):6449–6471. doi: 10.1109/tits.2024.3401716.

Chen, Claire Y. T., Edward W. Sun, and Yi-Bing Lin. 2024. "Reconciling Spatiotemporal Conjunction With Digital Twin for Sequential Travel Time Prediction and Intelligent Routing." *Annals of Operations Research*. doi: 10.1007/s10479-024-05990-x.

Dembski, Fabian, Uwe Wössner, Mike Letzgus, Michael Ruddat, and Claudia Yamu. 2020. "Urban Digital Twins for Smart Cities and Citizens: The Case Study of Herrenberg, Germany." *Sustainability* 12 (6):2307.

Faiad, Azza, Sarah M. Abdel-Ghany, Mohammed Ayachi, and Shehab Ahmed. 2023. "City Scale Digital Twins for Mobility Emissions Evaluation." 2023 International Wireless Communications and Mobile Computing (IWCMC).

Faliagka, Evanthia, Eleni Christopoulou, Dimitrios Ringas, Tanya Politi, Nikos Kostis, Dimitris Leonardos, Christos Tranoris, Christos P. Antonopoulos, Spyros Denazis, and Nikolaos Voros. 2024. "Trends in Digital Twin Framework Architectures for Smart Cities: A Case Study in Smart Mobility." *Sensors* 24 (5):1665.

Feng, Hailin, Haibin Lv, and Zhihan Lv. 2023. "Resilience Towarded Digital Twins to Improve the Adaptability of Transportation Systems." *Transportation Research Part A: Policy and Practice* 173. doi: 10.1016/j.tra.2023.103686.

Global Infrastructure Outlook. 2024. Forecasting Infrastructure Investment needs and Gaps.

Jafari, Mina, Abdollah Kavousi-Fard, Tao Chen, and Mazaher Karimi. 2023. "A Review on Digital Twin Technology in Smart Grid, Transportation System and Smart City: Challenges and Future." *IEEE Access* 11:17471–17484. doi: 10.1109/access.2023.3241588.

Kajba, Milena, Borut Jereb, and Tina Cvahte Ojsteršek. 2023. "Exploring Digital Twins in the Transport and Energy Fields: A Bibliometrics and Literature Review Approach." *Energies* 16 (9). doi: 10.3390/en16093922.

Kamal, Hani, Wendy Yánez, Sara Hassan, and Dalia Sobhy. 2024. "Digital-Twin-Based Deep Reinforcement Learning Approach for Adaptive Traffic Signal Control." *IEEE Internet of Things Journal* 21946–21953.

Kumarasamy, Vijayalakshmi K., Abhilasha Jairam Saroj, Yu Liang, Dalei Wu, Michael P. Hunter, Angshuman Guin, and Mina Sartipi. 2024. "Integration of Decentralized Graph-Based Multi-Agent Reinforcement Learning With Digital Twin for Traffic Signal Optimization." *Symmetry* 16 (4). doi: 10.3390/sym16040448.

Kušić, Krešimir, René Schumann, and Edouard Ivanjko. 2023. "A Digital Twin in Transportation: Real-Time Synergy of Traffic Data Streams and Simulation for Virtualizing Motorway Dynamics." *Advanced Engineering Informatics* 55:101858.

Liao, Xishun, Ziran Wang, Xuanpeng Zhao, Kyungtae Han, Prashant Tiwari, Matthew J. Barth, and Guoyuan Wu. 2022. "Cooperative Ramp Merging Design and Field Implementation: A Digital Twin Approach Based on Vehicle-to-Cloud Communication." *IEEE Transactions on Intelligent Transportation Systems* 23 (5):4490–4500. doi: 10.1109/tits.2020.3045123.

Litman, Todd. 2025. *Parking Management: Comprehensive Implementation Guide.* Victoria Transport Policy Institute.

Liu, Qian, Xiaoqing Qi, Shanshan Liu, Xiuju Cheng, Xin Ke, and Feng Wang. 2022. "Application of Lightweight Digital Twin System in Intelligent Transportation." *IEEE Journal of Radio Frequency Identification* 6:729–732. doi: 10.1109/jrfid.2022.3212169.

Nieuwenhuijsen, Mark J., and Haneen Khreis. 2016. "Car Free Cities: Pathway to Healthy Urban Living." *Environment International* 94:251–262. doi: 10.1016/j.envint.2016.05.032.

Obeid, Hassan, Michael L. Anderson, Mohamed Amine Bouzaghrane, and Joan Walker. 2024. "Does Telecommuting Reduce Trip-Making? Evidence from a U.S. Panel During the COVID-19 Pandemic." *Transportation Research Part A: Policy and Practice* 180:103972. doi: 10.1016/j.tra.2024.103972.

Papyshev, Gleb, and Masaru Yarime. 2021. "Exploring city Digital Twins as Policy Tools: A Task-Based Approach to Generating Synthetic Data on Urban Mobility." *Data & Policy* 3. doi: 10.1017/dap.2021.17.

Reisi, Marzieh, Soheil Sabri, Muyiwa Agunbiade, Abbas Rajabifard, Yiqun Chen, Mohsen Kalantari, Azadeh Keshtiarast, and Yan Li. 2020. "Transport Sustainability Indicators for an Enhanced Urban Analytics Data Infrastructure." *Sustainable Cities and Society* 59:102095. doi: 10.1016/j.scs.2020.102095.

Rodrigue, Jean-Paul. 2024. "Transport, Energy and Environment." In *The Geography of Transport Systems.* Routledge.

Saba, Irum, Mukhtar Ullah, and Muhammad Tariq. 2024. "Advancing Electric Vehicle Battery Analysis With Digital Twins in Intelligent Transportation Systems." *IEEE Transactions on Intelligent Transportation Systems*:1–10. doi: 10.1109/tits.2024.3361807.

Sabri, S., S. Winter, and A. Rajabifard. 2023. "Creating digital twins to save our cities." https://pursuit.unimelb.edu.au/articles/creating-digital-twins-to-save-our-cities.

Sabri, Soheil. 2021. "Chapter 1 - Introduction: Being Smarter for Productivity, Livability, and Sustainability." In *Smart Cities for Technological and Social Innovation*, edited by Hyung Min Kim, Soheil Sabri and Anthony Kent, 1–8. Academic Press.

Sabri, Soheil, and Abbas Rajabifard. 2019. "Spatial Enablement to Facilitate the New Urban Agenda Commitments for Sustainable Development." In *Sustainable Development Goals Connectivity Dilemma*, 199–211. CRC Press.

Sabri, Soheil, Kostas Alexandridis, Mehrdad Koohikamali, Sonya Zhang, and H. Erkan Ozkaya. 2023. "Designing a Spatially-Explicit Urban Digital Twin Framework for Smart Water Infrastructure and Flood Management." 2023 IEEE 3rd International Conference on Digital Twins and Parallel Intelligence (DTPI).

Sabri, Soheil, Kostas Alexandridis, and Newton Lee. 2024. *Digital Twin: Fundamentals and Applications*, edited by Kostas Alexandridis Soheil Sabri, Newton Lee. 1st ed. Springer Cham.

Sahal, Radhya, Saeed H. Alsamhi, Kenneth N. Brown, Donna O'Shea, Conor McCarthy, and Mohsen Guizani. 2021. "Blockchain-Empowered Digital Twins Collaboration: Smart Transportation Use Case." *Machines* 9 (9). doi: 10.3390/machines9090193.

Schalkwyk, Pieter van, Sean Whiteley, and Marc Goldman. 2024. Digital Twin Capabilities Periodic Table User Guide. Digital Twin Consortium.

Sellitto, Giovanni Paolo, Massimiliano Masi, Tanja Pavleska, and Helder Aranha. 2021. "A Cyber Security Digital Twin for Critical Infrastructure Protection: The Intelligent Transport System Use Case." In *IFIP Working Conference on The Practice of Enterprise Modeling*, pp. 230–244. Springer International Publishing.

Sun, Yuchao, Doina Olaru, Callan Bright, Gary McCarney, Tristan W. Reed, Soheil Sabri, Yiqun Chen, Sam Amirebrahimi, Sharon Biermann, and Abbas Rajabifard. 2023. "Making Accessibility Accessible: A Flexible Planning Tool for Enhanced Urban Analytics." *Research in Transportation Business & Management* 51:101042. doi: 10.1016/j.rtbm.2023.101042.

Teng, Sin Yong, Michal Touš, Wei Dong Leong, Bing Shen How, Hon Loong Lam, and Vítězslav Máša. 2021. "Recent Advances on Industrial Data-Driven Energy Savings: Digital Twins and Infrastructures." *Renewable and Sustainable Energy Reviews* 135:110208. doi: 10.1016/j.rser.2020.110208.

Tu, Z., L. Qiao, R. Nowak, H. Lv, and Z Lv. 2022. "Digital Twins-Based Automated Pilot for Energy-Efficiency Assessment of Intelligent Transportation Infrastructure." *IEEE Transactions on Intelligent Transportation Systems* 23 (11):22320–22330. doi: 10.1109/TITS.2022.3166585.

Tzachor, Asaf, Soheil Sabri, Catherine E Richards, Abbas Rajabifard, and Michele Acuto. 2022. "Potential and Limitations of Digital Twins to Achieve the Sustainable Development Goals." *Nature Sustainability* 5 (10):822–829.

Wang, Ziran. 2022. "Mobility Digital Twin: Concept, Architecture, Case Study, and Future Challenges." *IEEE Internet of Things Journal* 9 (18):17452–17467. doi: 10.36227/techrxiv.16828759.

World Health Organization. 2023. Road traffic injuries. https://www.who.int/news-room/fact-sheets/detail/road-traffic-injuries

Xiao, Ming, Lihua Chen, Haoxiong Feng, Zhigao Peng, and Qiong Long. 2024. "Sustainable and Robust Route Planning Scheme for Smart City Public Transport Based on Multi-Objective Optimization: Digital Twin Model." *Sustainable Energy Technologies and Assessments* 65:103787. doi: 10.1016/j.seta.2024.103787.

Xu, Haowen, Andy Berres, Srikanth B. Yoginath, Harry Sorensen, Phil J. Nugent, Joseph Severino, Sarah A. Tennille, Alex Moore, Wesley Jones, and Jibonananda Sanyal. 2023. "Smart Mobility in the Cloud: Enabling Real-Time Situational Awareness and Cyber-Physical Control Through a Digital Twin for Traffic." *IEEE Transactions on Intelligent Transportation Systems* 24 (3):3145–3156. doi: 10.1109/tits.2022.3226746.

Xu, Z., T. Jiang, and N Zheng. 2022. "Developing and Analyzing Eco-Driving Strategies for on-Road Emission Reduction in Urban Transport Systems - A VR-Enabled Digital-Twin Approach." *Chemosphere* 305:135372. doi: 10.1016/j.chemosphere.2022.135372.

Ye, Xinyue, Jiaxin Du, Yu Han, Galen Newman, David Retchless, Lei Zou, Youngjib Ham, and Zhenhang Cai. 2023. "Developing Human-Centered Urban Digital Twins for Community Infrastructure Resilience: A Research Agenda." *Journal of Planning Literature* 38 (2):187–199.

Yeon, H., T. Eom, K. Jang, and J Yeo. 2023. "DTUMOS, Digital Twin for Large-Scale Urban Mobility Operating System." *Scientific Reports* 13 (1):5154. doi: 10.1038/s41598-023-32326-9.

Ywet, Nan Lao, Aye Aye Maw, Tuan Anh Nguyen, and Jae-Woo Lee. 2024. "YOLOTransfer-DT: An Operational Digital Twin Framework With Deep and Transfer Learning for Collision Detection and Situation Awareness in Urban Aerial Mobility." *Aerospace* 11 (3). doi: 10.3390/aerospace11030179.

Ziehl, Michael, Rico Herzog, Till Degkwitz, Martin Heinrich Niggemann, Gesa Ziemer, and Rosa Thoneick. 2023. "Transformative Research in Digital Twins for Integrated Urban Development." *International Journal of E-Planning Research* 12 (1):1–18. doi: 10.4018/ijepr.333851.

22 From Human Mobility to Social Segregation

What Insights Can Social Media Data Provide?

Meiliu Wu and Qunying Huang

22.1 INTRODUCTION

Urban mobility and segregation are critical topics in geographic information science due to their relevance across a wide spectrum of applications, such as urban planning, transportation management, public health, and social equity and inclusion (Chen and Kwan 2015; Wu and Huang 2022; Wu et al. 2023; Wu, Huang, and Gao 2023). In particular, the characteristics of urban mobility patterns can reflect individual-level segregated experience and influence urban formation, evolvement, and development (Waddell 2002; Gao, Janowicz, and Couclelis 2017). Previous studies have demonstrated that urban movement and segregation patterns are strongly impacted by people's socioeconomic and demographic background (SDB; e.g., employment status, household income, gender, and race-ethnicity) (Bagrow and Lin 2012; Huang and Wong 2016). They are also highly constrained by transport accessibility, commuting facilities, and urban spatial structure (Black and Conroy 1977). The reason is that different SDB groups implicate different adaptabilities to these constraints, resulting in diversified movement patterns and segregated experiences (Dong et al. 2006; Blumenberg and Shiki 2007; Crane 2007).

Traditionally, the trajectory data for urban movement and segregation research based on different SDB groups are collected from surveys, travel diary, or carry-on global positioning system (GPS) devices to represent general patterns of local or national population displacement (Kwan 2008; Antipova, Wang, and Wilmot 2011; Chen et al. 2011). Nowadays, social media has become a popular data source to support the analysis of human mobility due to its exclusive advantages, such as its accessibility to the public, a large amount of participants, high spatio-temporal resolutions, and global coverage (Huang and Wong 2016; Luo et al. 2016; Wu and Huang 2022). Moreover, various statistical and machine learning models (e.g., Liu et al. 2022) have been developed to predict the semantic information of trajectories collected from social media that indicate travel purposes and SDB of travelers, making these datasets valuable for the purpose of human mobility studies. For example, these datasets have more user-generated information (e.g., surnames), which can be used to infer an individual's race-ethnicity, including those of minority groups (Chang et al. 2010; Mislove et al. 2011). In addition, previous studies have shown that geotagged social media data can be used to infer individual economic status by predicting the individual's home location via trajectory mining and attaching the economic statistics by census aggregated units to the predicted home location of each individual (Huang and Wong 2016; Wu and Huang 2022; Wu, Huang, and Gao 2023). These research results can be applied to many areas, including the assessment of social inequality and environmental injustice. For instance, using massive trajectories with inferred race-ethnicity and economic status at the individual level, we can disentangle the effects of race-ethnicity and economic status on individual experienced segregation.

DOI: 10.1201/9781003503262-26

In fact, research on urban segregation has largely been facilitated from this mobility-based perspective based on social media data. As these datasets are generated by users' behaviors with detailed spatiotemporal information (e.g., pings with timestamps and logs, checked-in coordinates, GPS records, or location tags), they enable researchers to expand the spatial extent where segregation could occur from a single sociogeographical space (i.e., residential areas only) to people's activity space, which can be defined as "the subset of all locations within which an individual has direct contact as a result of his or her day-to-day activities" (Golledge and Stimson 1997), such as working, eating in restaurants, buying groceries, exercising, or relaxing in entertainment and recreational places.

However, previous studies in human mobility using social media have mostly focused on individual behavior analysis, e.g., detecting daily frequent trajectory (Simini et al. 2012). Only a few reveal collective travel characteristics of different SDB groups (Candia et al. 2008; Longley, Adnan, and Lansley 2015; Huang and Wong 2016; Wang et al. 2018; Wu and Huang 2022). Although race-ethnicity has been recognized as a relatively sensitive field for research, it is important to dig in deeper based on these findings, especially when few studies have explored the extent of how social media data can help investigate human travel patterns and further explore the travel patterns of different racial-ethnic groups. In addition, previous work using social media data has not revealed the interaction between economic and racial-ethnic factors and which factors can better explain the diverse mobility patterns of different groups, and insufficient efforts have been devoted to revealing the experienced segregation levels of racial-ethnic minorities (e.g., Asian, Hispanic, Native, and multiracial) by developing segregation indices to quantify individual experienced, mobility-based segregation. To fill these research gaps, this study first reveals, analyzes, and compares the movement patterns among different racial-ethnic and economic groups using social media data from the top 50 populated cities in the United States. Next, this study also conducts an empirical case study in Los Angeles–Long Beach–Anaheim to reveal racial-ethnic segregation based on individual-level mobility patterns extracted from social media data.

22.2 SOCIAL MEDIA DATA COLLECTION AND PREPROCESSING

In the first case study, the top 50 populated cities across the United States were chosen as the study areas. To collect the data, we first used Twitter's (now X) streaming application program interface (API) to harvest geotagged tweets globally. Overall, more than 344 million geotagged tweets were collected for 18 months (from December 2013 to May 2015), among which over 110 million are located within the bounding box of the U.S. continent. Because we would infer race-ethnicity for each Twitter user using surnames, we filtered the twitter users based on their usernames and the number of tweets and deleted the invalid ones that (1) were identified as organizations (e.g., companies and advertisements), (2) provided insufficient information to be considered as having valid names (i.e., only one word in the username field, only one letter in the first name or last name, or containing non-English words), and (3) posted more than 10,000 tweets during the time span. Finally, we identified approximately one million unique twitter users with valid first names and last names, who posted more than 37 million tweets within the U.S. top 50 populated cities. In the second study case, we used the subset of these geotagged tweets from the urbanized area of Los Angeles–Long Beach–Anaheim, with 819,618 tweets from 14,510 unique Twitter users after the process of data filtering, in Section 22.3.1.

22.3 FRAMEWORK OF INFERRING INDIVIDUAL MOBILITY PATTERNS, USER PROFILES, AND SEGREGATION

As mentioned in Section 22.1, we aimed to examine individual-level mobility patterns and experienced racial-ethnic segregation based on sociodemographics inferred from social media data. As shown in Figure 22.1, a unified framework is proposed to illustrate the workflow of processing Twitter data to infer users' mobility patterns and profiles.

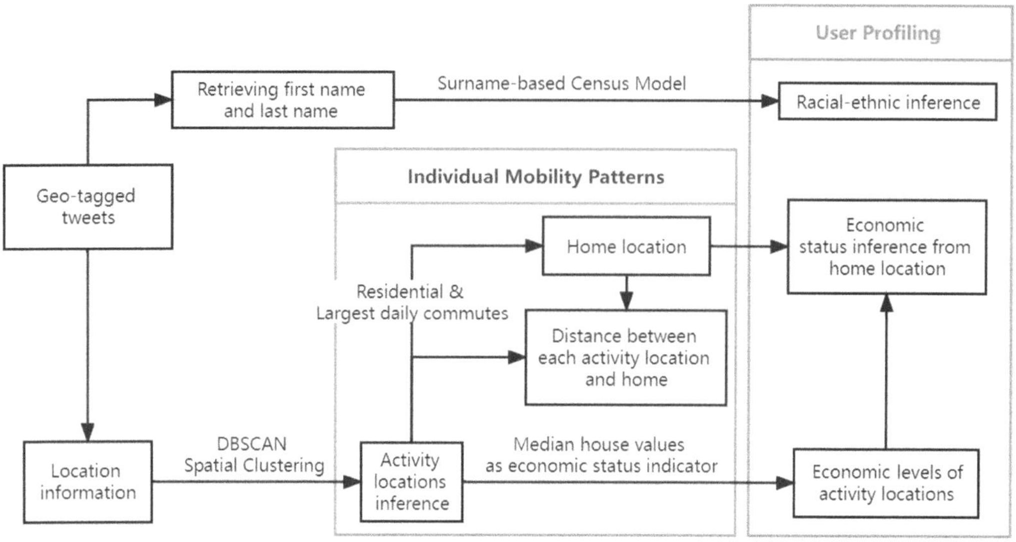

FIGURE 22.1 The framework of processing Twitter data to infer individuals' mobility patterns and user profiles and to estimate experienced racial-ethnic segregation. *Abbreviation:* DBSCAN, density-based spatial clustering of applications with noise.

22.3.1 Mobility Pattern Analysis

22.3.1.1 Activity Zone Identification

First, to differentiate various movement patterns at an individual level, the locations that a Twitter user visits four or more times are identified as activity zones (Huang and Wong 2016; Wu and Huang 2022). To detect these locations, the geotagged tweets from each individual are grouped as different spatial clusters (Huang, Cao, and Wang 2014). While some tweet points were included to form the clusters, dispersed points associated with rare events or activities that deviated from the regular patterns were excluded. Like previous works (Wang et al. 2018; Wu and Huang 2022; Wu et al. 2023), we also applied the density-based spatial clustering of applications with noise (DBSCAN) algorithm to create the spatial clusters for each Twitter user. To obtain a satisfying result of spatial clustering, we referred to previous studies (Borah and Bhattacharyya 2004; Birant and Kut 2007; He et al. 2014) and conducted exploratory experiments using self-adjusted DBSCAN. The optimal result is obtained when the *eps* value is set as 50 meters and the *minpts* is set as 4.

22.3.1.2 Economic Status Inference for Activity Zones

With spatial clusters (i.e., activity zones) detected from the geotagged tweets, the movement pattern can be explored at an individual level. In particular, each activity zone is represented by the median centroid of the zone, the point with the minimum total Euclidean distance to all other points in this zone. The economic status of each activity zone is inferred by the economic status of the census tract that the median centroid of this zone is spatially within. Specifically, the economic status includes lower class, middle class, and upper class, classified according to the median house values of all census tracts within a city: lower-class tracts have a median house value less than the average median house value of the city minus one standard deviation, upper-class tracts have a median house value more than the average median house value of the city plus one standard deviation, and middle-class tracts are in between. Both census tract and urban area data were downloaded from the American Community Survey (ACS; 5-year) 2010–2014. For further exploration of the movement patterns among different racial-ethnic and economic groups, the number of tweets in each activity zone and the number of activity zones of each individual were recorded.

22.3.1.3 Home Location Inference

We assume that an individual leaves and returns home more frequently than any other activity zones in a daily manner. Therefore, the home location of each Twitter user is considered as the activity zone whose median centroid is spatially within residential areas referring to the National Land Cover Database (NLCD) as well as having the largest total number of daily interzone travels (i.e., the daily travels "into" and "out of" this activity zone) among all activity zones (Huang and Wong 2016). Additionally, the cluster size (i.e., the number of tweets in an activity zone) is utilized to break the tie. To obtain the total number of daily in–out travels for each cluster, tweets are first sorted based on the local posted time for each day. Because each tweet has been labeled with the cluster that it belongs to, the total in times and out times for each day are calculated for each cluster. Finally, the total number of daily in–out travels for each cluster is the sum of all days within the whole time span.

22.3.1.4 Collective Trajectory Mining

Individual trajectories are aggregated based on the predicted group, i.e., one of the predefined race-ethnicity and economic status groups. Specifically, this study includes six race-ethnicity classes: (1) non-Hispanic White ("White" for the remainder of the study), (2) non-Hispanic Black or African American ("Black"), (3) non-Hispanic Asian and Native Hawaiian and Other Pacific Islander ("Asian"), (4) non-Hispanic American Indian and Alaska Native ("Native"), (5) non-Hispanic two or more races ("multiracial"), and (6) Hispanic or Latino origin ("Hispanic"), determined by referring to the classification from the U.S. Census Bureau's Frequently Occurring Surname list in 2010, with each surname corresponding to its race-ethnicity probability distribution (Comenetz 2016). The economic status includes lower class, middle class, and upper class, which corresponds to the economic status of the Twitter users. With the cross-classification of economic status and race-ethnicity, we examine 18 (six races-ethnicities × three economic statuses) different groups in total for this study.

For a statistical exploration and comparison of the movement patterns, we observe the demographic differences in (1) number of activity zones, (2) travel distances, and (3) urban mobility spread in different economic destinations. Then, we examine each demographic group in Los Angeles in depth to provide more insight into the local movement patterns of race-ethnicity minorities.

22.3.2 User Profiling Inference

The economic status of each Twitter user is inferred based on the economic status of the predicted home location, which is determined by the methods above. As for individual race-ethnicity inference, there has been a rich history of studies using associations between surnames and races-ethnicities in curated sources such as census data, which is also known as the surname-based cen sus model (Lauderdale and Kestenbaum 2000; Tucker 2005; Ambekar et al. 2009). Therefore, this model would be used to infer races-ethnicities based on the self-provided surnames from the user profiles. Specifically, for each surname with over 100 individuals in the United States, "Frequently Occurring Surnames from the Census 2000"[1] reports the proportions of races-ethnicities for that surname.

22.3.3 Experienced Segregation Indices

A trending direction of segregation studies is to capture the (lack of) physical interaction between individuals, focusing on the question of "to what extent two persons/groups interact physically in their activity spaces," which represents the "exposure" dimension originally defined by Massey and

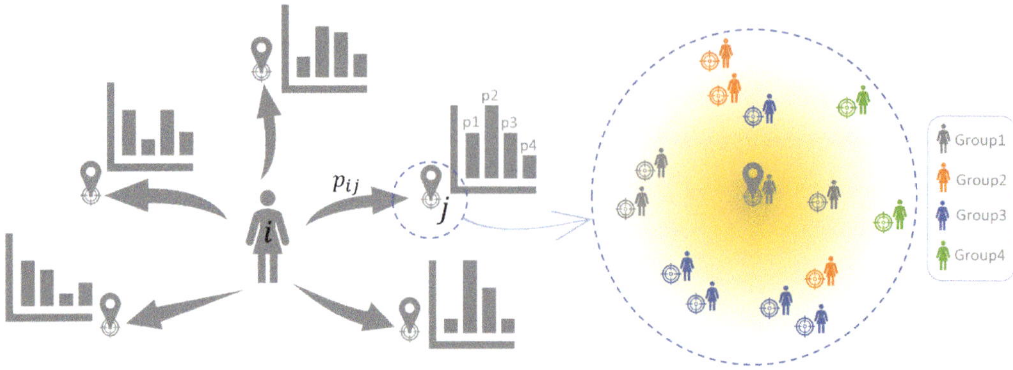

FIGURE 22.2 An illustration of individual experienced segregation. For each individual $\boldsymbol{i}$, multiple activity locations can be captured along with the tweeting frequency proportion of individual $\boldsymbol{i}$ at each activity location $\boldsymbol{j}$, i.e., $\boldsymbol{p_{ij}}$. For each activity location $\boldsymbol{j}$, the proportions of different groups visiting this location will be estimated based on distance-decay functions.

Denton (1988). Although a few person-based segregation indices have been developed by previous studies, they were specifically tailored for the available datasets, most of which were individual-level travel diary surveys (Schnell and Yoav 2001; Park and Kwan 2018). Thus, to capture the interaction between individuals with social media data, this study proposed two experienced segregation indices, i.e., *individual experienced exposure index (EEI) to each race-ethnicity* and *individual experienced diversity index (EDI)*, based on individual-level mobility patterns (i.e., the travel distance weighted by the tweeting frequency proportion and the number of activity locations) and user profiles (i.e., race-ethnicity and economic status) inferred from Twitter data.

Specifically, when measuring the individual-level segregation, existing studies often lack the consideration of spatial proximity of the activity locations (Athey et al. 2021; Moro et al. 2021), largely narrowing down the spatial extent where interaction could exist. Thus, we argue that if segregation measurement focuses more on the degree of potential interaction among different groups (i.e., the experienced segregation level), then the spatial impacts among activity locations should be considered, and that spatial modeling based on distance-decay functions can be employed to address these limitations (Figure 22.2). With the integration of spatial proximity, our proposed indices are demonstrated as below. Note that this approach is conceptually consistent with previous studies (Petrovi, Van Ham, and Manley 2018; Petrović, van Ham, and Manley 2022) applying multiscale measurement and bespoke neighborhoods for residential segregation but not for activity space–based segregation.

22.3.3.1 Individual Experienced Exposure Index to Each Race-Ethnicity

$\breve{L}_j$ is the local population intensity at activity location j of individual i:

$$\breve{L}_j = \sum_{q=1}^{Q} k(p_q) \tag{22.1}$$

$$k(p_q) = p_q * \frac{1}{e^r} \tag{22.2}$$

where Q is the total number of all individuals' activity locations in the local area (i.e., defined as within the 4-km distance threshold centered on the given activity location j), p_q equals the tweeting frequency proportion of an individual at activity location q, and k is the distance-decay kernel estimator that reflects the impacts of the neighboring activity locations on activity location j, where r represents the rank of the distance from q to j (i.e., 1-ranked within 1-km buffer zone, 2-ranked within 2-km buffer zone, etc.) and e is the Euler's number, i.e., the base of the natural logarithms.

$\breve{L}_{jm}$ is the local population intensity of group m visiting activity location j, which could be calculated by replacing the tweeting frequency proportion of an individual at activity location q (p_q) with the total tweeting frequency proportion of group m in activity location q (p_{qm}) in Equation 22.1 above:

$$\breve{L}_{jm} = \sum_{q=1}^{Q} k(p_{qm}) \tag{22.3}$$

Then, the individual EEI of individual i to group m is defined as:

$$\breve{E}_{(i,m)} = \sum_{j=1}^{J_i} p_{ij} \frac{\breve{L}_{jm}}{\breve{L}_j} \tag{22.4}$$

where p_{ij} is the tweeting frequency proportion of individual i at activity location j; J_i is the total number of activity locations of individual i; and when individual i belongs to group n and $m = n$, i.e., $\breve{e}_{(i,n)}$, it measures the individual experienced isolation index, representing the intragroup interaction level. A larger value of $\breve{E}_{(i,m)}$ indicates a higher exposure level of individual i to group m (with a range as [0, 1]).

22.3.3.2 Individual Experienced Diversity Index

In addition to representing the exposure level of an individual to each group by EEI, we can also reflect the overall exposure diversity level of an individual to all different groups using the individual EDI. This index refers to the true diversity index ^tD (Chao, Chiu, and Jost 2016), i.e., the Hill numbers of order t or effective number of species (e.g., different racial-ethnic groups):

$$^tD = \sum_{i=1}^{M} (p_i^t)^{1/(1-t)} \tag{22.5}$$

where p_i is the proportional abundance (i.e., the exposure level represented by the tweeting proportion) of the racial-ethnic group i, and M is richness (i.e., the total number of racial-ethnic groups). When $t = 1$, the weighted geometric mean of the p_i values is used, in which each racial-ethnic group is exactly weighted by its proportional abundance. Specifically, ^1D can be represented as the exponential of the Shannon entropy based on the mathematical limit, as t approaches 1:

$$^1D = \frac{1}{\prod_{i=1}^{M} p_i^{p_i}} = exp\left\{-\sum_{i=1}^{M} p_i \ln(p_i)\right\} \tag{22.6}$$

and the individual EDI of individual i could be measured as:

$$\breve{D}_i = exp\left\{-\sum_{m=1}^{M} \breve{E}_{(i,m)} \ln\left(\breve{E}_{(i,m)}\right)\right\} \tag{22.7}$$

by replacing p_i with $\breve{E}_{(i,m)}$ and M represents the total number of racial-ethnic groups. A larger value of $\breve{D}_i$ indicates a higher exposure diversity level of individual i in terms of interacting with all different groups (with a range as [1, M]).

22.4 CASE STUDIES AND RESULT ANALYSIS

22.4.1 Urban Movement Patterns of Racial-Ethnic and Economic Groups in U.S. Top 50 Populated Cities

Table 22.1 shows the population of sampled valid users of each studied group after the data processing steps from Section 22.3. The sampled distribution of lower-class, middle-class, and upper-class status is 5.6%, 79.6%, and 14.8%, compared with the country level 14.8%, 64.2%, and 21.0%, respectively, reported by the 2014 U.S. Census Bureau's report *Income and Poverty in the United States*. Although the comparison shows that more middle-class Twitter users have been captured while the lower class and upper class are less represented, we are still able to obtain a significant amount of population from the lower-class and upper-class groups. As for race-ethnicity, the sampled distribution of White, Black, Asian, Native, multiracial, and Hispanic is 63.5%, 11.0%, 5.1%, 0.8%, 1.3%, and 18.3%, which is similar to the nationwide distribution of 62.2%, 12.4%, 5.4%, 0.7%, 2.0%, and 17.4%, respectively, from the U.S. Census ACS 2010–2014.

22.4.1.1 Activity Zones

Figure 22.3 indicates the demographic differences in the number of activity zones. In general, lower-class groups have more activity zones than upper-class groups. The average number of activity zones for lower-class groups is 3.42, whereas it is 3.25 for middle-class groups and 3.23 for upper-class groups. This could be due to the fact that upper-class groups post less tweets (see Section 22.4.3), making their activity zones less likely to be captured. However, for the racial-ethnic factor, the average numbers of activity zones range from 3.20 for the multiracial and Hispanic groups to 3.36 for the Black group, indicating that the Black group has 5% more activity zones than the multiracial and Hispanic groups.

Meanwhile, the upper-class Black and lower-class Black groups have the highest values in average number of activity zones among all groups. Within both upper-class and lower-class groups, race-ethnicity has more impact on the variations of the average number of activity zones.

22.4.1.2 Travel Distances

Figure 22.4 indicates more detailed findings of the median travel distances among different racial-ethnic and economic groups. Generally speaking, lower-class groups travel shorter distances to their activity zones than upper-class and middle-class groups, especially the groups that are race-ethnicity minorities such as Asians and Hispanics. Specifically, the median travel distance of lower-class groups is 6,240 m, whereas it is 8,034 m for middle-class groups and 8,872 m for upper-class groups.

TABLE 22.1
Sampled Valid Users of Different Groups in U.S. Top 50 Populated Cities

	Economic Status			
Race-ethnicity	**Lower Class**	**Middle Class**	**Upper Class**	**Subtotal**
White	2,088	33,236	7,104	42,428 (63.5%)
Black	423	5,839	1,118	7,380 (11.0%)
Hispanic	226	2,655	503	3,384 (5.1%)
Asian	37	427	63	527 (0.8%)
Native	62	684	128	874 (1.3%)
Multiracial	900	10,333	989	12,222 (18.3%)
Subtotal	3,736 (5.6%)	53,174 (79.6%)	9,905 (14.8%)	66,815

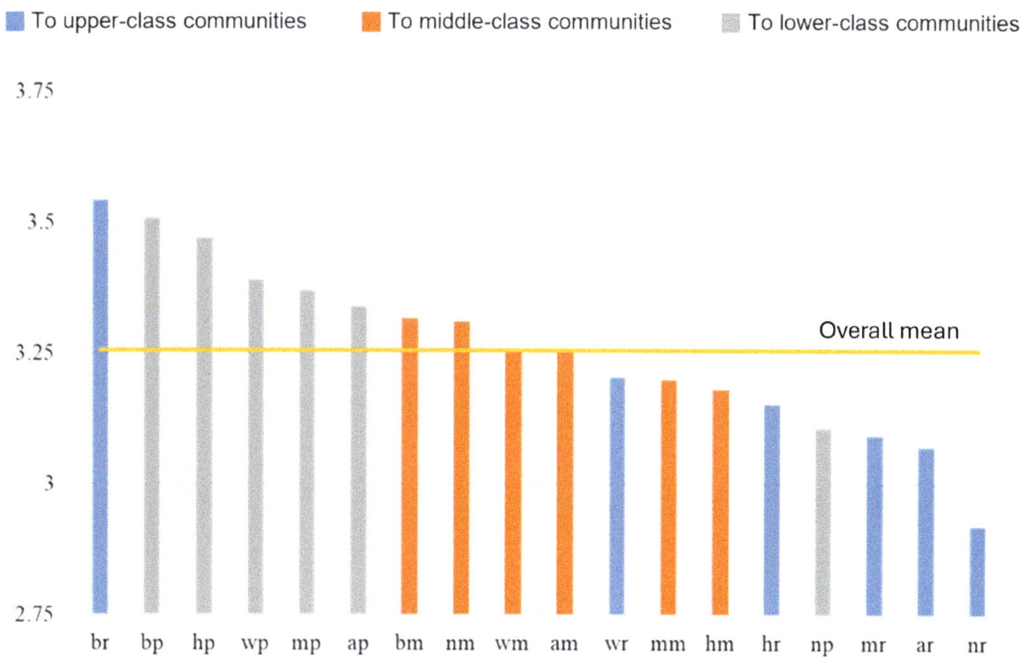

FIGURE 22.3 Average number of activity zones for different groups. Each group is represented by two characters, with the first character indicating race-ethnicity and the second one indicating economic status. Specifically, the corresponding relationships are as follows: w, White; b, Black; a, Asian; n, Native; m, multiracial; h, Hispanic; r, upper class; m, middle class; and p, lower class. For example, wr represents the White and upper-class group.

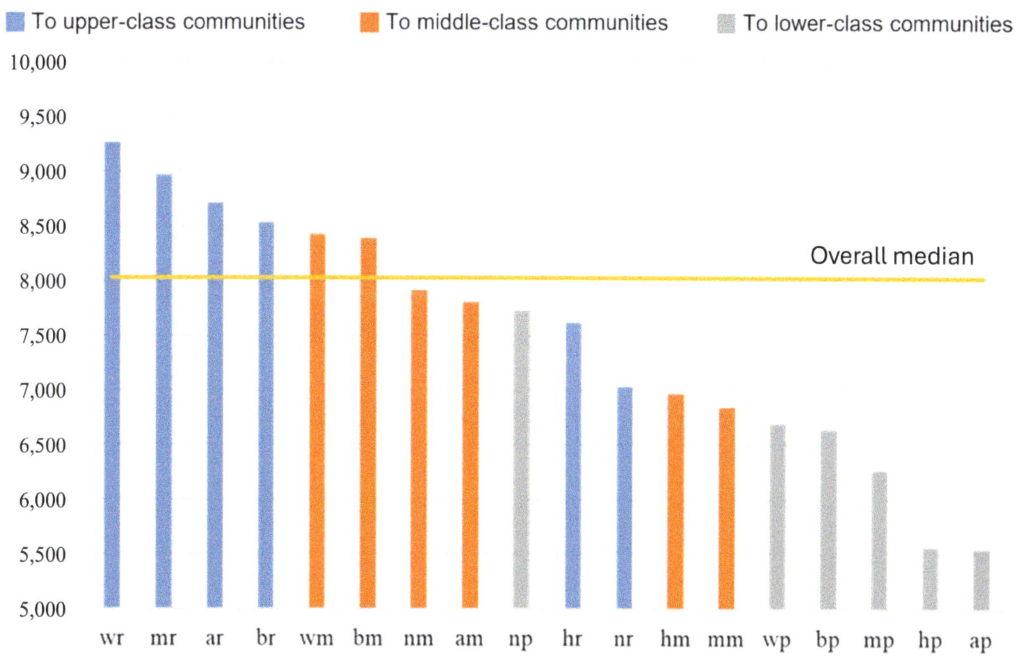

FIGURE 22.4 Median travel distances for different groups.

That is, lower-class groups are 42% shorter in median travel distance than upper-class groups. This finding supports the theories from previous urban and transportation studies: economically disadvantaged populations have less disposable income and have to rely on public transportation more than affluent people; therefore, their mobility is restricted (Murakami and Young 1997; Giuliano 2005; Paulley et al. 2006). Although the race-ethnicity factor can differ in these theories, such as the upper-class Native group having shorter median travel distances than the lower-class Native group, the abnormality might be due to the insufficient amount of social media users identified in this minor race-ethnicity group.

Among different racial-ethnic groups, the median travel distances are 8,450, 8,304, 7,760, 7,682, 7,153, and 6,868 m for the White, Black, Native, Asian, multiracial, and Hispanic groups, respectively. Therefore, White and Black racial-ethnic groups travel longer median distances than other racial-ethnic minorities, which might indicate that racial-ethnic minorities have lower social status in relation to their daily use of space and limited integration into society. In particular, the median travel distance of the White group is 23% longer than that of the Hispanic group, 18% longer than that of the multiracial group, 10% longer than that of the Asian group, and 9% longer than that of the Native group.

22.4.1.3 Urban Mobility Spread in Different Economic Destinations

Figure 22.5a displays an economically segregated movement pattern among different economic groups in the U.S. top 50 populated cities: upper-class neighborhoods are mostly visited by the

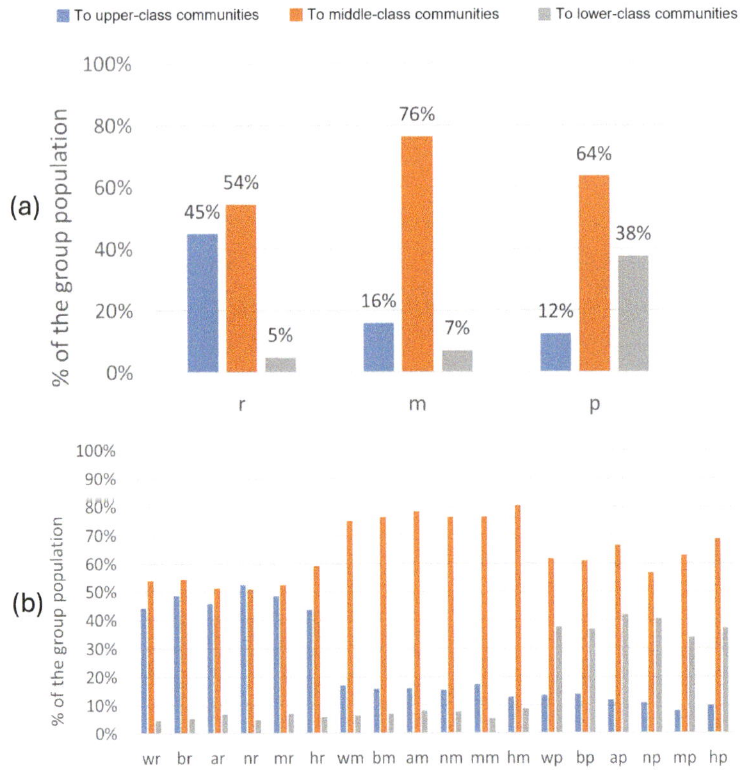

FIGURE 22.5 Percentage of different economic groups traveling to upper-class, middle-class, and lower-class communities in the U.S. top 50 populated cities (a) and percentage of different racial-ethnic and economic groups traveling to upper-class, middle-class, and lower-class communities in the U.S. top 50 populated cities (b). See legend for Figure 22.3 for definitions.

upper class, whereas lower-class neighborhoods are mainly accessed by the lower class. Specifically, 45% of the upper class have access to upper-class neighborhoods as their activity zones, compared to 16% of the middle class and 12% of the lower class; 38% of the lower class travel to lower-class neighborhoods as their activity zones, while only 5% of the upper class and 7% of the middle class visit lower-class neighborhoods. With race-ethnicity as a consideration in Figure 22.5b, there is not much difference in the national scale.

22.4.2 Racial-Ethnic Segregation with Individual Mobility-Based Experience in Los Angeles–Long Beach–Anaheim

Table 22.2 shows the sociodemographics of sampled valid users after the inference of user profiling and mobility patterns in this case study. The total number of sampled people is 7,669, and approximately 21,000 activity locations are captured, with a density of 12.07 activity locations per square mile in this studied area.

22.4.2.1 Spatial Patterns of Experienced Racial-Ethnic Segregation

Figures 22.6a to 22.6f display the spatial patterns of EEIs to different racial-ethnic groups based on the home locations of the sampled valid Twitter users, and Figure 22.6g shows the pattern of EDI. Apparently, the maps in Figures 22.6a–22.6f demonstrate that residential clusters are easily perceptible for most groups, indicating that people living in proximate neighborhoods tend to have similar EEIs to different racial-ethnic groups, which is consistent with the well-known residential segregation phenomena (i.e., people living in proximate neighborhoods tend to be the same race-ethnicity), reported by the Roots of Structural Racism Project in 2019 from the University of California at Berkeley.[2] Although the population in the study area has become more diverse over years according to the 2019 census data,[3] residential communities are not necessarily more diverse, similar to the experienced segregation levels based on people's mobility patterns beyond the residential space (i.e., across different sociogeographical spaces). However, there is also a spatial clustering pattern of the EDI shown in Figure 22.6g, where the points with a lighter color (less exposure diversity) tend to be clustered, the same as the points with a darker color (more exposure diversity), suggesting that people who have more (or less) intergroup interactions tend to live in proximate neighborhoods, too.

Additionally, we discovered some correlations between EEIs and different racial-ethnic groups. For example, Figures 22.6a and 22.6b show that most hot-spot areas of EEIs to White are the cold-spot areas of EEIs to Hispanic, and the same is seen in Figures 22.6b and 22.6c (Hispanic vs. Black). Based on these discoveries, we provide a more in-depth quantitative analysis of their potential correlations in Section 22.4.3.

TABLE 22.2
Sampled Valid Users of Different Groups

	Economic Status			
Race-Ethnicity*	**Lower Class**	**Middle Class**	**Upper Class**	**Subtotal**
White	173	2,549	860	3,582 (46.71%)
Hispanic	227	2,357	180	2,764 (36.04%)
Black	26	472	122	620 (8.08%)
Asian	38	419	84	541 (7.05%)
Multiracial	8	75	24	107 (1.40%)
Native	5	41	9	55 (0.72%)
Subtotal	477 (6.22%)	5,913 (77.10%)	1,279 (16.68%)	7,669

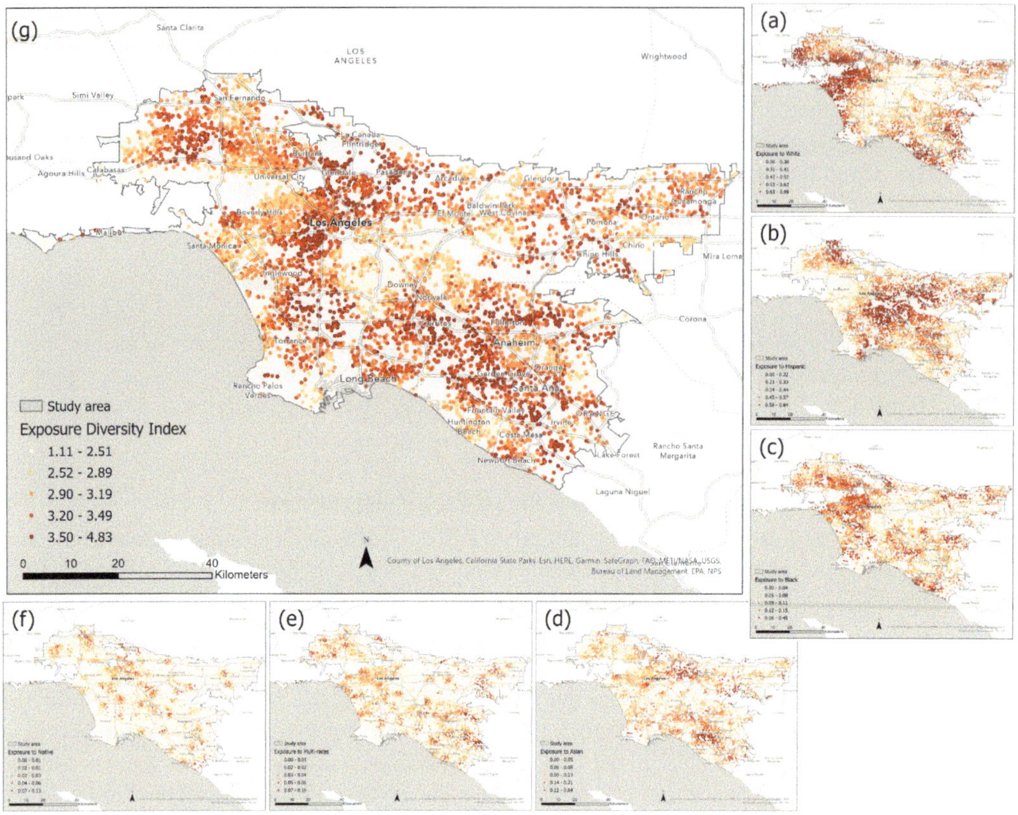

FIGURE 22.6 Spatial clustering patterns of experienced exposure indices (EEIs) to different race-ethnicities based on Twitter users' home locations: (a) the individual EEI to White; (b) the individual EEI to Hispanic; (c) the individual EEI to Black; (d) the individual EEI to Asian; (e) the individual EEI to multiracial; (f) the individual EEI to Native; and (g) the individual experienced diversity index.

22.4.2.2 Racial-Ethnic Differences in Experienced Segregation

Next, the "directed" interactions between any pair of two groups and their differences are probed and plotted in Figure 22.7. While all groups would have the highest mean exposure level to their own group (i.e., experienced isolation), which is consistent with the results of a previous study based on data from the Los Angeles Family and Neighborhood Survey (L.A.FANS) (Jones and Pebley 2014), there is less difference in EEIs to minorities, such as Black, Asian, Native, and multiracial, among different racial-ethnic groups (Figures 22.7b, 22.7c, 22.7d, and 22.7e, respectively). Yet, differences in EEIs to White and Hispanic are clearly perceivable. For instance, Figure 22.7a displays that the groups of White (average 0.52), Black (average 0.51), and multiracial (average 0.48) are more likely to be exposed to the White group (overall 0.47), but the Hispanic group has the most significant lowest exposure to the White group (average 0.40). Similarly, Figure 22.7f shows that the Hispanic group has the most significant highest exposure to itself, i.e., the experienced isolation level (average 0.43), whereas other groups, especially White (average 0.30) and Black (average 0.31), are less likely to be exposed to the Hispanic group (overall 0.36). In Section 22.4.4, we further examine the impact of individual economic status on such disparities in experienced segregation levels.

In addition, the EDIs of different racial-ethnic groups are also plotted in Figure 22.8. The overall average of EDI values is 3.09, meaning that the population in the study area generally would interact with approximately three different groups in their daily activities (including their own racial-ethnic group). With a range of [3.01, 3.20] in EDIs, the Asian group is most diverse in interacting with

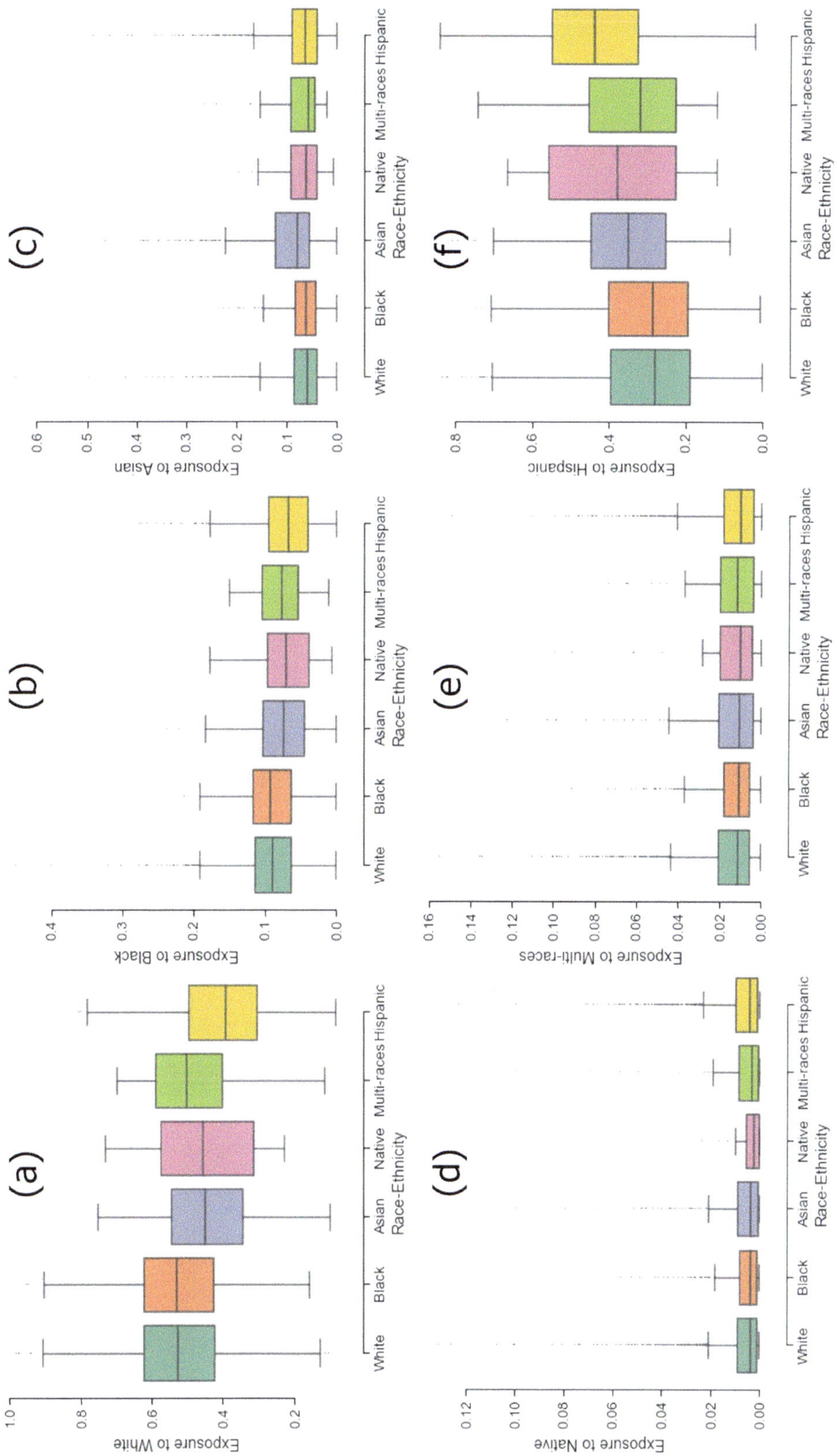

FIGURE 22.7 Box plots of experienced exposure indices, aggregated by different racial-ethnic groups.

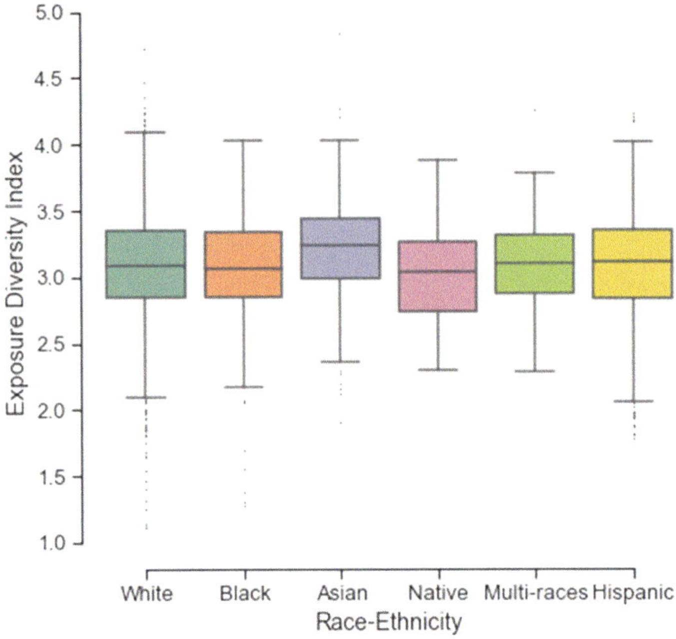

FIGURE 22.8 Box plots of individual experienced diversity index, aggregated by different racial-ethnic groups.

other groups, whereas the Native group is most segregated, although there are just minor differences across the different groups.

22.4.2.3 Disparities in Experienced Segregation Due to Economic Status

For a further analysis, we explored the impact of individual economic status on the disparities in the experienced segregation levels (Figure 22.9). One important finding is that EEIs to White (Figure 22.9a), Black (Figure 22.9b), and Hispanic groups (Figure 22.9f) are markedly influenced by individual economic status. Specifically, both EEIs to the White and Black groups distinctively increase if the individual economic status is higher, whereas EEIs to the Hispanic group are the opposite, with much lower values as the economic status is higher. These patterns are consistent across different racial-ethnic groups.

However, an intuitive finding is that all groups show more intragroup interactions, meaning that the likelihood of exposing the same group is generally high (Figure 22.9). Moreover, Figures 22.9c, 22.9d, and 22.9e, respectively, reveal that the minority groups of Asian, Native, and Multiracial are relatively more likely to have exclusive intragroup interactions, i.e., higher experienced isolation levels, across different economic statuses.

As for the EDIs of different groups (Figure 22.10), the diversity level of interacting with different races-ethnicities decreases as the individual economic status is higher for most racial-ethnic groups, except for the Hispanic group, in which the lowest diversity level is observed in the lower-class category. This finding aligns with a previous one that, on average, increasing socioeconomic status diminishes the likelihood of sharing activity locations with any socioeconomic status group (lower diversity levels) (Browning et al. 2017), which was based on data from the L.A.FANS that identify the locations where residents engage in routine activities (Sastry et al. 2006). In addition, the Asian group generally has the highest diversity level in the intergroup interactions across different economic statuses.

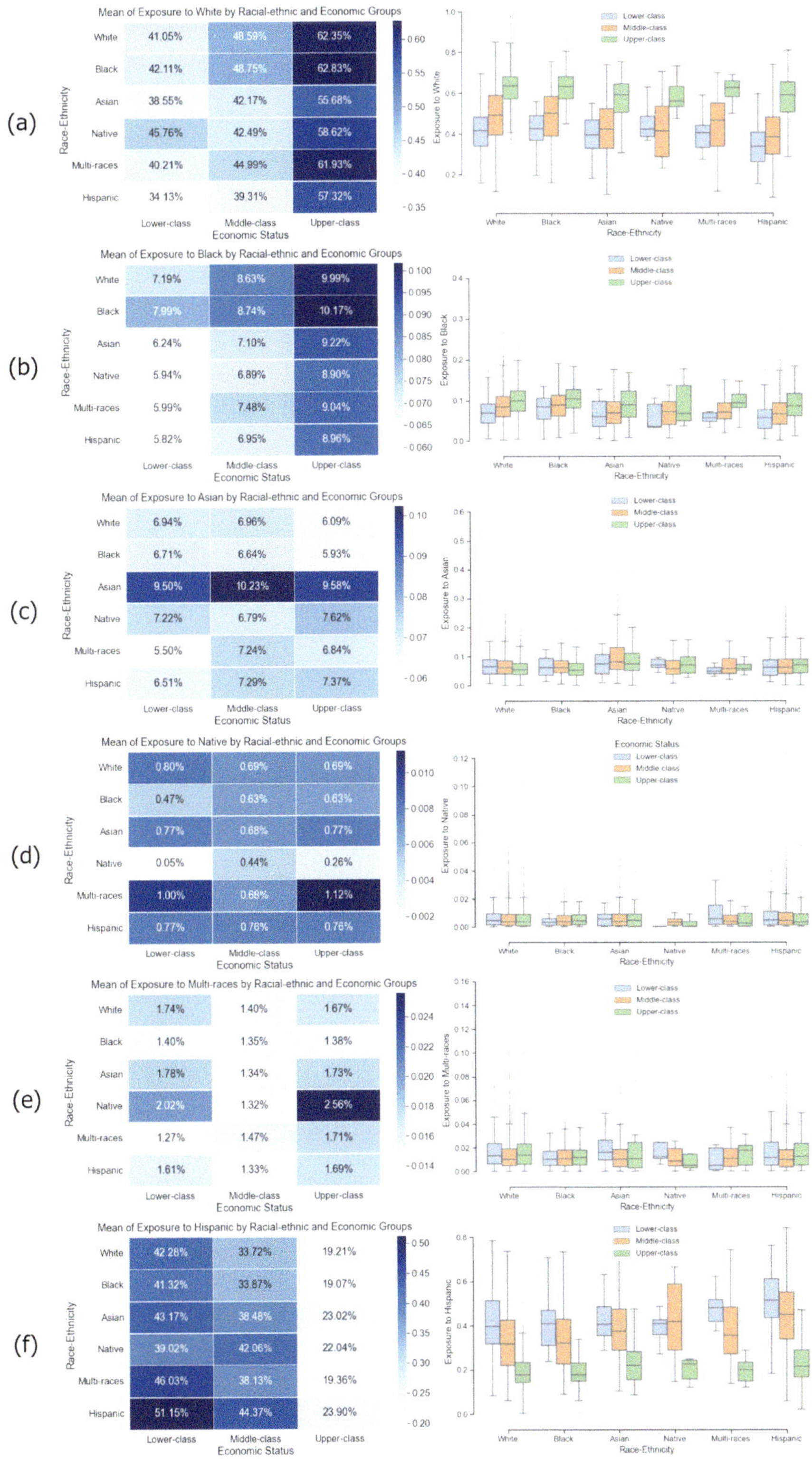

FIGURE 22.9 Disparities in experienced exposure indices of different racial-ethnic and economic groups.

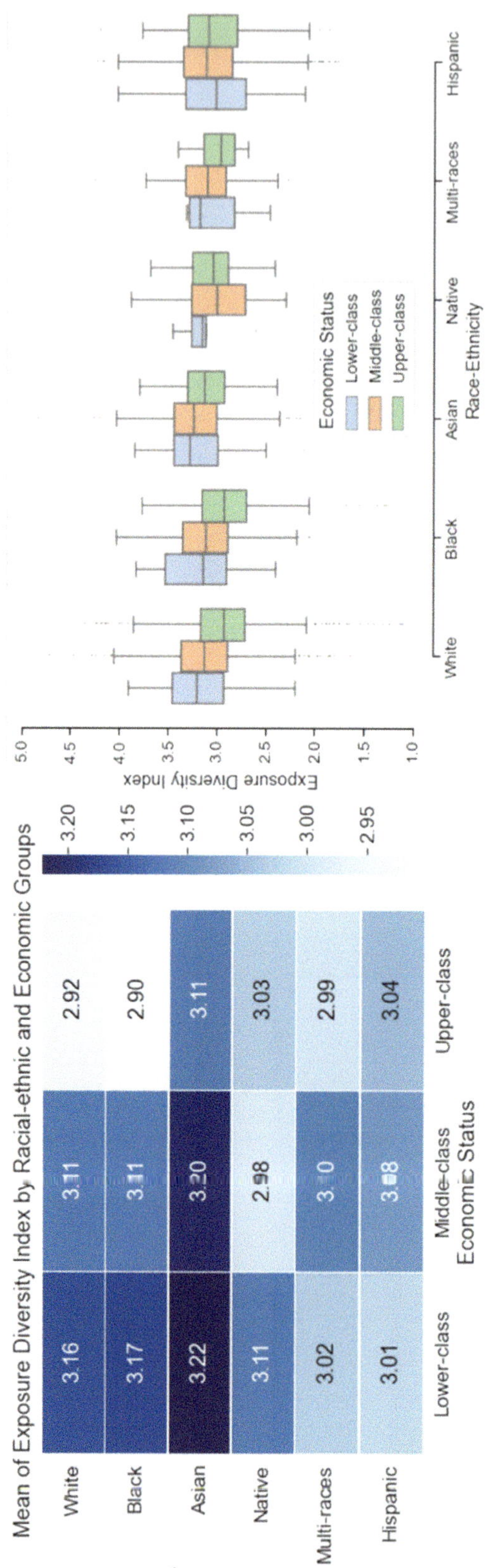

FIGURE 22.10 Disparities in experienced diversity index of different racial-ethnic and economic groups.

22.5 CONCLUSION

22.5.1 Summary of the Study

This study explores the relationship between urban mobility patterns and social segregation using social media data, specifically geotagged tweets from the top 50 populated cities in the United States for the mobility study case and Los Angeles–Long Beach–Anaheim for the segregation study case. The research reveals significant insights into how different racial-ethnic and economic groups move within urban spaces. By inferring user profiles (e.g., race-ethnicity and economic status) from social media data, this study identifies distinct mobility patterns and degrees of segregation experienced by various demographic groups. Through detailed analysis, this study highlights how both socioeconomic status and race-ethnicity influence urban mobility, revealing patterns of spatial restriction and limited interactions between groups, particularly among economically disadvantaged and racial-ethnic minorities.

22.5.2 Advancements and Limitations

This study advances the field by employing a novel approach that uses social media data to infer individual-level mobility patterns and segregation indices, providing a more dynamic understanding of urban segregation beyond traditional residential-based measures. This method allows for the analysis of large-scale data across diverse geographical locations, offering a comprehensive perspective on segregation experienced in daily activities. However, this study also faces several limitations, such as potential biases in social media data due to unequal access and representation across different demographic groups. Additionally, the methods of user information inference may introduce inaccuracies, and the findings might not be generalizable beyond the sampled cities.

22.5.3 Future Directions

Future research in this field could address these limitations by integrating multiple data sources, such as combining social media data with mobile phone data or administrative records, to enhance the accuracy and representativeness of findings. Additionally, developing and comparing more refined methods for inferring user demographics and improving data collection techniques could provide deeper insights into the intersections of mobility, race-ethnicity, and economic status. Expanding studies to include more cities globally and considering longitudinal data could further advance understanding of how urban mobility and social segregation evolve over time, contributing to more inclusive urban planning and policymaking.

NOTES

1 https://www.census.gov/topics/population/genealogy/data/2000_surnames.html
2 https://belonging.berkeley.edu/roots-structural-racism
3 https://censusreporter.org/profiles/05000US06037-los-angeles-county-ca/

REFERENCES

Ambekar, Anurag, Charles Ward, Jahangir Mohammed, Swapna Male, and Steven Skiena. 2009. "Name-Ethnicity Classification from Open Sources." *Proceedings of the ACM SIGKDD International Conference on Knowledge Discovery and Data Mining*, 49–57. https://doi.org/10.1145/1557019.1557032.

Antipova, Anzhelika, Fahui Wang, and Chester Wilmot. 2011. "Urban Land Uses, Socio-Demographic Attributes and Commuting: A Multilevel Modeling Approach." *Applied Geography* 31 (3): 1010–18. https://doi.org/10.1016/j.apgeog.2011.02.001.

Athey, Susan, Billy Ferguson, Matthew Gentzkow, and Tobias Schmidt. 2021. "Estimating Experienced Racial Segregation in US Cities Using Large-Scale GPS Data." *Proceedings of the National Academy of Sciences* 118 (46): e2026160118.

Bagrow, James P., and Yu Ru Lin. 2012. "Mesoscopic Structure and Social Aspects of Human Mobility." *PLoS One* 7 (5): 3–8. https://doi.org/10.1371/journal.pone.0037676.

Birant, Derya, and Alp Kut. 2007. "ST-DBSCAN: An Algorithm for Clustering Spatial-Temporal Data." *Data and Knowledge Engineering* 60 (1): 208–21. https://doi.org/10.1016/j.datak.2006.01.013.

Black, J., and M. Conroy. 1977. "Accessibility Measures and the Social Evaluation of Urban Structure." *Environment and Planning A: Economy and Space* 9 (9): 1013–31. https://doi.org/10.1068/a091013.

Blumenberg, Evelyn, and Kimiko Shiki. 2007. "Transportation Assimilation: Immigrants, Race and Ethnicity, and Mode Choice." *Transportation Research Board 86th Annual Meeting.*

Borah, B., and D. K. Bhattacharyya. 2004. "An Improved Sampling-Based DBSCAN for Large Spatial Databases." *Proceedings of International Conference on Intelligent Sensing and Information Processing, ICISIP 2004*, 92–96. https://doi.org/10.1109/icisip.2004.1287631.

Browning, Christopher R., Catherine A. Calder, Lauren J. Krivo, Anna L. Smith, and Bethany Boettner. 2017. "Socioeconomic Segregation of Activity Spaces in Urban Neighborhoods: Does Shared Residence Mean Shared Routines?" *The Russell Sage Foundation Journal of the Social Sciences* 3 (2): 210–31. https://doi.org/10.7758/rsf.2017.3.2.09.

Candia, Julián, Marta C. González, Pu Wang, Timothy Schoenharl, Greg Madey, and Albert László Barabási. 2008. "Uncovering Individual and Collective Human Dynamics from Mobile Phone Records." *Journal of Physics A: Mathematical and Theoretical* 41 (22): 1–16. https://doi.org/10.1088/1751-8113/41/22/224015.

Chang, Jonathan, Itamar Rosenn, Lars Backstrom, and Cameron Marlow. 2010. "EPluribus : Ethnicity on Social Networks." *ICWSM 2010 - Proceedings of the 4th International AAAI Conference on Weblogs and Social Media*, 18–25.

Chao, Anne, Chun-Huo Chiu, and Lou Jost. 2016. "Phylogenetic Diversity Measures and Their Decomposition: A Framework Based on Hill Numbers." *Biodiversity Conservation and Phylogenetic Systematics* (vol. 14, pp. 141–172). Springer.

Chen, Jie, Shih Lung Shaw, Hongbo Yu, Feng Lu, Yanwei Chai, and Qinglei Jia. 2011. "Exploratory Data Analysis of Activity Diary Data: A Space-Time GIS Approach." *Journal of Transport Geography* 19 (3): 394–404. https://doi.org/10.1016/j.jtrangeo.2010.11.002.

Chen, Xiang, and Mei Po Kwan. 2015. "Contextual Uncertainties, Human Mobility, and Perceived Food Environment: The Uncertain Geographic Context Problem in Food Access Research." *American Journal of Public Health* 105 (9): 1734–37. https://doi.org/10.2105/AJPH.2015.302792.

Comenetz, Joshua. 2016. "Frequently Occurring Surnames in the 2010 Census." United States Census Bureau, no. October, 1–8.

Crane, Randall. 2007. "Is There a Quiet Revolution in Women's Travel? Revisiting the Gender Gap in Commuting." *Journal of the American Planning Association* 73 (3): 298–316. https://doi.org/10.1080/01944360708977979.

Dong, Xiaojing, Moshe E. Ben-Akiva, John L. Bowman, and Joan L. Walker. 2006. "Moving from Trip-Based to Activity-Based Measures of Accessibility." *Transportation Research Part A: Policy and Practice* 40 (2): 163–80. https://doi.org/10.1016/j.tra.2005.05.002.

Gao, Song, Krzysztof Janowicz, and Helen Couclelis. 2017. "Extracting Urban Functional Regions from Points of Interest and Human Activities on Location-Based Social Networks." *Transactions in GIS* 21 (3): 446–67. https://doi.org/10.1111/tgis.12289.

Giuliano, Genevieve. 2005. "Low Income, Public Transit, and Mobility." *Transportation Research Record* 1927 (1): 63–70.

Golledge, R. G., and R. J. Stimson. 1997. *Spatial Behaviour: A Geographic Perspective. Spatial Behaviour: A Geographic Perspective.* Guilford Press. https://doi.org/10.2307/144350.

He, Yaobin, Haoyu Tan, Wuman Luo, Shengzhong Feng, and Jianping Fan. 2014. "MR-DBSCAN: A Scalable MapReduce-Based DBSCAN Algorithm for Heavily Skewed Data." *Frontiers of Computer Science* 8 (1): 83–99.

Huang, Qunying, Guofeng Cao, and Caixia Wang. 2014. "From Where Do Tweets Originate? - A GIS Approach for User Location Inference." *Proceedings of the 7th ACM SIGSPATIAL International Workshop on Location-Based Social Networks, LBSN 2014 - Held in Conjunction with the 22nd ACM SIGSPATIAL GIS 2014*, no. c, 1–8. https://doi.org/10.1145/2755492.2755494.

Huang, Qunying, and David W S Wong. 2016. "Activity Patterns, Socioeconomic Status and Urban Spatial Structure: What Can Social Media Data Tell Us?" *International Journal of Geographical Information Science* 30 (9): 1873–98. https://doi.org/10.1080/13658816.2016.1145225.

Jones, Malia, and Anne R. Pebley. 2014. "Redefining Neighborhoods Using Common Destinations: Social Characteristics of Activity Spaces and Home Census Tracts Compared." *Demography* 51 (3): 727–52. https://doi.org/10.1007/s13524-014-0283-z.

Kwan, Mei Po. 2008. "From Oral Histories to Visual Narratives: Re-Presenting the Post-September 11 Experiences of the Muslim Women in the USA." *Social and Cultural Geography* 9 (6): 653–69. https://doi.org/10.1080/14649360802292462.

Lauderdale, Diane S., and B. Kestenbaum. 2000. "Asian American Ethnic Identification by Surname." *Population Research and Policy Review* 19 (3): 283–300. https://doi.org/10.1023/A:1026582308352.

Liu, Xinyi, Meiliu Wu, Bo Peng, and Qunying Huang. 2022. "Graph-Based Representation for Identifying Individual Travel Activities with Spatiotemporal Trajectories and POI Data." *Scientific Reports* 12 (1). https://doi.org/10.1038/s41598-022-19441-9.

Longley, Paul A, Muhammad Adnan, and Guy Lansley. 2015. "The Geotemporal Demographics of Twitter Usage." *Environment and Planning A* 47 (2): 465–84.

Luo, Feixiong, Guofeng Cao, Kevin Mulligan, and Xiang Li. 2016. "Explore Spatiotemporal and Demographic Characteristics of Human Mobility via Twitter: A Case Study of Chicago." *Applied Geography* 70:11–25. https://doi.org/10.1016/j.apgeog.2016.03.001.

Massey, Douglas S., and Nancy A. Denton. 1988. "The Dimensions of Residential Segregation." *Social Forces* 67 (2): 281–315. https://doi.org/10.1093/sf/67.2.281.

Mislove, Alan, Sune Lehmann, Yong-Yeol Ahn, Jukka-Pekka Onnela, and James Rosenquist. 2011. "Understanding the Demographics of Twitter Users." *Proceedings of the International AAAI Conference on Web and Social Media*, 5:554–57.

Moro, Esteban, Dan Calacci, Xiaowen Dong, and Alex Pentland. 2021. "Mobility Patterns Are Associated with Experienced Income Segregation in Large US Cities." *Nature Communications* 12 (1). https://doi.org/10.1038/s41467-021-24899-8.

Murakami, Elaine, and Jennifer Young. 1997. "Daily Travel by Persons with Low Income." In *Nationwide Personal Transportation Survey Symposium*.

Park, Yoo Min, and Mei Po Kwan. 2018. "Beyond Residential Segregation: A Spatiotemporal Approach to Examining Multi-Contextual Segregation." *Computers, Environment and Urban Systems* 71: 98–108. https://doi.org/10.1016/j.compenvurbsys.2018.05.001.

Paulley, Neil, Richard Balcombe, Roger Mackett, Helena Titheridge, John Preston, Mark Wardman, Jeremy Shires, and Peter White. 2006. "The Demand for Public Transport: The Effects of Fares, Quality of Service, Income and Car Ownership." *Transport Policy* 13 (4): 295–306.

Petrovi, Ana, Maarten Van Ham, and David Manley. 2018. "Multiscale Measures of Population: Within- and between-City Variation in Exposure to the Sociospatial Context." *Annals of the American Association of Geographers* 108 (4): 1057–74. https://doi.org/10.1080/24694452.2017.1411245.

Petrović, Ana, Maarten van Ham, and David Manley. 2022. "Where Do Neighborhood Effects End? Moving to Multiscale Spatial Contextual Effects." *Annals of the American Association of Geographers* 112 (2): 581–601. https://doi.org/10.1080/24694452.2021.1923455.

Sastry, Narayan, Bonnie Ghosh-Dastidar, John Adams, and Anne R Pebley. 2006. "The Design of a Multilevel Survey of Children, Families, and Communities: The Los Angeles Family and Neighborhood Survey." *Social Science Research* 35 (4): 1000–24.

Schnell, Izhak, and Benjamini Yoav. 2001. "The Sociospatial Isolation of Agents in Everyday Life Spaces as an Aspect of Segregation." *Annals of the Association of American Geographers* 91 (4): 622–36. https://doi.org/10.1111/0004-5608.00262.

Simini, Filippo, Marta C. González, Amos Maritan, and Albert László Barabási. 2012. "A Universal Model for Mobility and Migration Patterns." *Nature* 484 (7392): 96–100. https://doi.org/10.1038/nature10856.

Tucker, D. K. 2005. "The Cultural-Ethnic-Language Group Technique as Used in the Dictionary of American Family Names (DAFN)." *Onomastica Canadiana* 87 (2): 71–84.

Waddell, Paul. 2002. "UrbanSim: Modeling Urban Development for Land Use, Transportation, and Environmental Planning." *Journal of the American Planning Association* 68 (3): 297–314. https://doi.org/10.1080/01944360208976274.

Wang, Qi, Nolan Edward Phillips, Mario L. Small, and Robert J. Sampson. 2018. "Urban Mobility and Neighborhood Isolation in America's 50 Largest Cities." *Proceedings of the National Academy of Sciences of the United States of America* 115 (30): 7735–40. https://doi.org/10.1073/pnas.1802537115.

Wu, Meiliu, and Qunying Huang. 2022. "Human Movement Patterns of Different Racial-Ethnic and Economic Groups in U.S. Top 50 Populated Cities: What Can Social Media Tell Us about Isolation?" *Annals of GIS* 28 (5): 1–23. https://doi.org/10.1080/19475683.2022.2026471.

Wu, Meiliu, Qunying Huang, and Song Gao. 2023-July. "Measuring Access Inequality in A Hybrid Physical-Virtual World: : A Case Study of Racial Disparity of Healthcare Access During CoVID-19." *International Conference on Geoinformatics* 1–10. https://doi.org/10.1109/Geoinformatics60313.2023.10247690.

Wu, Meiliu, Xinyi Liu, Yuehan Qin, and Qunying Huang. 2023. "Revealing Racial-Ethnic Segregation with Individual Experienced Segregation Indices Based on Social Media Data: A Case Study in Los Angeles-Long Beach-Anaheim." *Computers, Environment and Urban Systems* 104 (46): 102008. https://doi.org/10.1016/j.compenvurbsys.2023.102008.

23 Measuring Public Transit Resilience with High-Resolution Real-Time Geospatial Data

Luyu Liu and Harvey J. Miller

23.1 INTRODUCTION

Compared with the long history of public transit systems, the time during which we can confidently claim to be able to fully monitor their real-time operation and performance has been relatively short. In other words, despite being a man-made system, we have long lacked a high-fidelity understanding of transit systems' performance until the recent adoption of global positioning system (GPS) system and real-time data (Miller, 2020). The flourishing open data and urban computing techniques have brought a series of new opportunities to address this obstacle. The new data offer high-resolution, real-time, standardized insights into the on-time performance of every bus and trip (Liu & Miller, 2022; Park et al., 2020); this also paves the way for a holistic, customized analysis of mobility and accessibility for each individual transit user. This is game-changing for researchers and planners, who used to rely on small-sample surveys and administrative data to understand the on-time performance of the system.

Among different perspectives of transit performance, resilience, i.e., the ability to keep and recover their functions facing a disruption (Azolin et al., 2020; Holling, 1973), is attracting more attention due to rising risks of system degradation caused by disruptions like pandemics and climate change. System resilience has two cores: robustness, the ability to maintain service during a disruptive event, and recoverability, the ability to land on a new equilibrium. Real-time data are especially advantageous in capturing the dynamic nature of transit systems' robustness, as disruptions typically experience a sharp decline but a slower recovery, i.e., the resilience triangle (Bevilacqua et al., 2018; Zobel, 2010). By providing up-to-date information on system operations during the sharp decline, real-time data offer a new perspective for a more nuanced understanding of how transit networks respond to different disruptions.

A prime example of such disruptions is the COVID-19 pandemic, one of the most critical public health events in this century (Zoumpourlis et al., 2020). COVID-19 had negative impacts on almost every perspectives of public transit systems' operation and useability in a very short amount of time in 2020, which has been extensively reported by prior studies including ridership (Parker et al., 2021; Wilbur et al., 2020), accessibility (Kar et al., 2022), mobility equity (He et al., 2021), relationship with other transportation modes (Meredith-Karam et al., 2021), and user satisfaction (Dong et al., 2021). From an economic perspective, we can generally summarize these different metrics into two categories: *supply side*, which primarily measures the performance of transit systems, and *demand side*, which primarily measures the experience of transit users.

In this chapter, we present a comprehensive system performance analysis focusing on the robustness of transit resilience for four key performance benchmarks, which include two supply-side

DOI: 10.1201/9781003503262-27

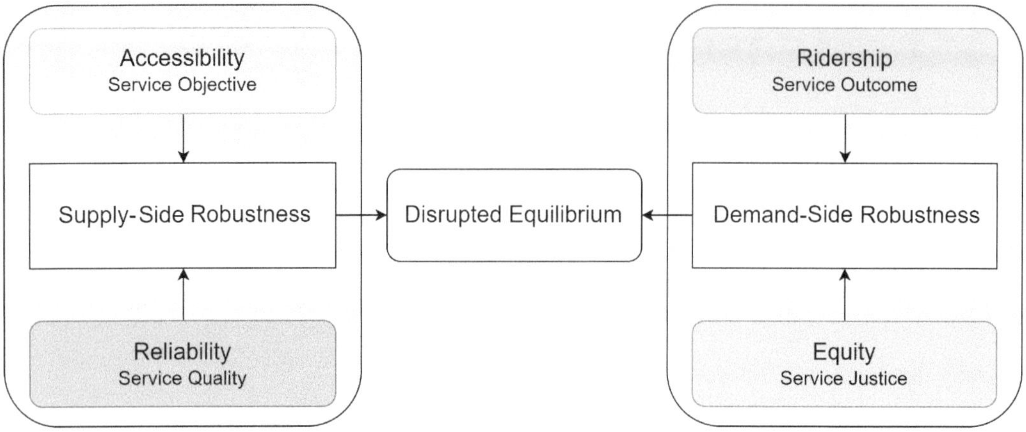

FIGURE 23.1 Structure of the analysis.

metrics, i.e., accessibility and reliability, and two demand-side metrics, i.e., ridership and equity. We have two research objectives as shown in Figure 23.1:

- First, from the perspective of the resilience theory, we focus on the robustness of each benchmark, i.e., how far and how fast each benchmark changed shortly after the outbreak of COVID-19 (Azolin et al., 2020; Gu et al., 2020; Wan et al., 2018). Note that we only offer a short discussion on the recoverability of each benchmark as of the end of 2021, as COVID-19 was a developing disruption, and real-time data are more effective in capturing the evolving nature of such disruptions over short periods.
- Second, from the perspective of econometric analysis, we discuss the disrupted equilibrium during the first phase of the pandemic and the practical insights for transit systems and users. Analyzing the status of equilibrium between supply and demand provides valuable insights into how well transit systems meet their intended functions under pressure and how they react to external shocks.

23.2 SUPPLY-SIDE ROBUSTNESS

Supply-side resilience measures the operation and infrastructure of the transit system, such as schedule, buses, and on time performance. This perspective gauges the mobility that the system can provide, reflecting its ability to maintain service levels, adhere to schedules, and meet demand even under challenging conditions. Factors such as the reliability of the vehicle fleet, the robustness of the scheduling system, and the effectiveness of contingency plans in response to disruptions are all critical components of this aspect of resilience.

Our case study site is the city of Columbus, Franklin County, Ohio, with the Central Ohio Transit Authority (COTA) bus system. Columbus is notably car dependent, with only 2.08% of its 667,000 workers (aged 16 and older) using public transit for commuting. Despite its low usage rate, the COTA system still serves a population of 1.2 million residents and generated 19 million trips in 2019. The COVID-19 pandemic caused significant disruption to the COTA system, including manpower shortage and economic difficulties. As a result, COTA enacted two policies, i.e., major service cuts and a free fare policy, in May 2020 in response to COVID (Ferenchik, 2020; Van Niel, 2021). To measure the change of performance, we have been collecting the general transit feed specification (GTFS) real-time data for COTA, which have recorded the operation status of every bus for every route at every stop at an interval of 1 minute since 2018.

23.2.1 Robustness of Accessibility

Accessibility is the objective of a transit system (Levine, 2020; Saif et al., 2019), i.e., how many opportunities the system can take the passengers to. Due to its fundamental importance, it is also a key issue for sustainable and equitable transportation planning and research (Banister, 2008). There has been abundant literature discussing the measurement of public transit accessibility (Fayyaz et al., 2017; Malekzadeh & Chung, 2020; Saif et al., 2019), including space-time prism, a well-established time geography method to measure the reachable space of a user starting from an origin at a time. The wide adoption of digitalized scheduled data, such as GTFS data, for most of transit systems in the United States has led to the development of new, high-resolution measures of accessibility (Barbeau & Antrim, 2013; Wessel et al., 2017).

However, schedule-based accessibility has a fundamental drawback: schedule data do not record the on-time performance of the system and the delays of buses. They do not represent an accurate, high-fidelity, and detailed accessibility experience of a user. Therefore, Wessel and Farber (2019) introduced a new framework—*retrospective real-time accessibility*—to use the historical actual times derived from real-time data to calculate transit accessibility. Retrospective accessibility essentially substitutes the scheduled times with the corresponding actual times with the same models.

Nevertheless, even with actual arrival times, the retrospective real-time accessibility approach can overestimate accessibility. The primary reason is that it assumes that users can access the real-time information instantly and react instantly to delays and other departures from scheduled service. To address this issue, Liu et al. (2022) developed measures of *realizable accessibility*—a more conservative and realistic real-time accessibility model—to approach the users' accessibility. It follows a two-stage procedure simulating users' decision-making process: first, the routing engine will plan the trips with the scheduled times, which are available before taking the buses. Second, following the planned trips, the routing engine will validate the trips with actual arrival time times and calculate a more accurate travel time with consideration of possibilities of missing buses. Figure 23.2 shows an example of scheduled versus realizable accessibility for one bus stop in the center of the city. In practice, we measure how many bus stops a user can actually access if they follow the schedule.

With both measures of scheduled and realizable real-time accessibility, we conduct a longitudinal analysis from January 2019 to December 2021, as shown in Figure 23.3. Both experienced significant declined during the COVID-19 lockdown period (March–June 2020): realizable and scheduled accessibility dropped by 29% and 18% by the end of April 2020, respectively. The decrease in scheduled accessibility measures reflects the deterioration of transit service caused by service cuts, whereas the decline in realizable accessibility also shows that the actual experienced accessibility also went down with the service cuts. It is important to note that the rapid decline in accessibility did not align perfectly with the onset of the pandemic but occurred around May 2020 following the major service adjustment. Meanwhile, for almost a whole year after the major travel restrictions ended in 2021, the performance of the system stayed down: the average numbers of accessible stops dropped by 17% and 20% for realizable and scheduled accessibility, respectively.

Figure 23.4 visualizes the spatial patterns of the changing rate of realizable accessibility during COVID-19 compared to pre-COVID level. The downtown area, which accounts for most ridership in the system and experienced the fewest service cuts, has less accessibility loss, while the urban outskirts, especially the northeast area of the city, experienced a substantial decline in accessibility.

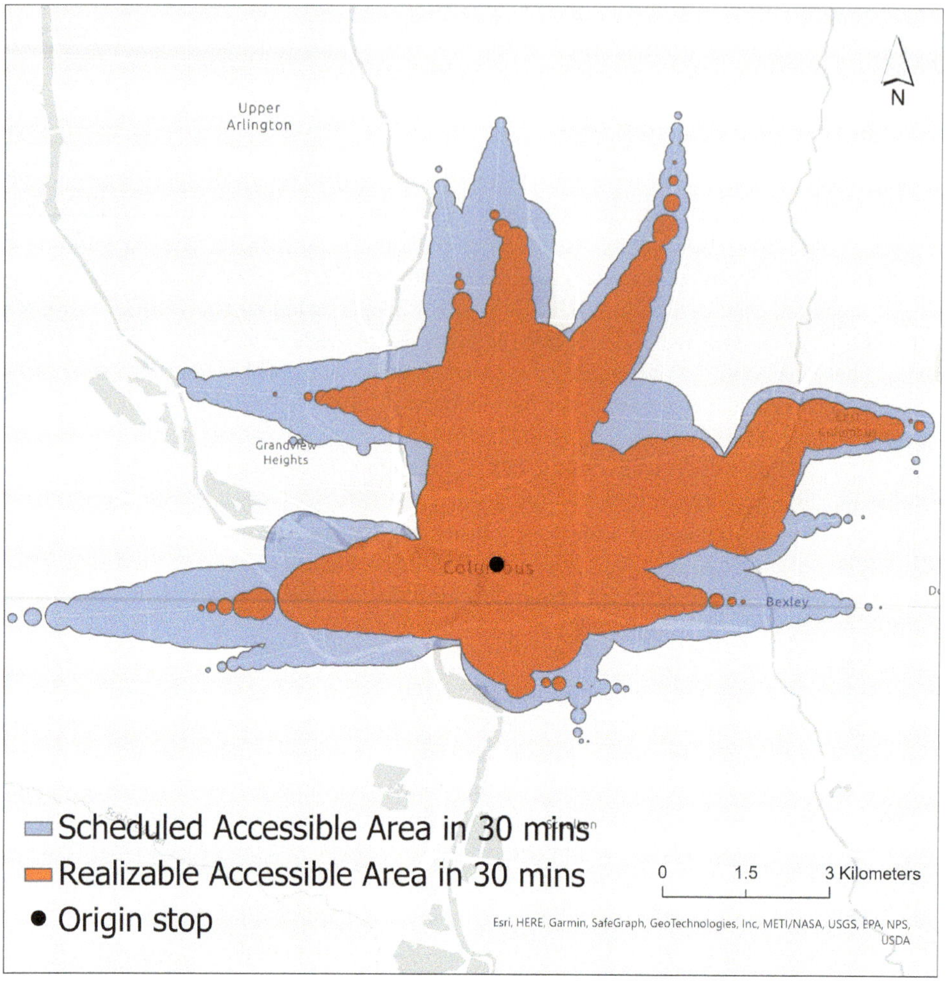

FIGURE 23.2 An example of scheduled and realizable accessibility starting from the center of Columbus, Ohio.

FIGURE 23.3 Scheduled and realizable real-time accessibility from January 2019 to December 2021.

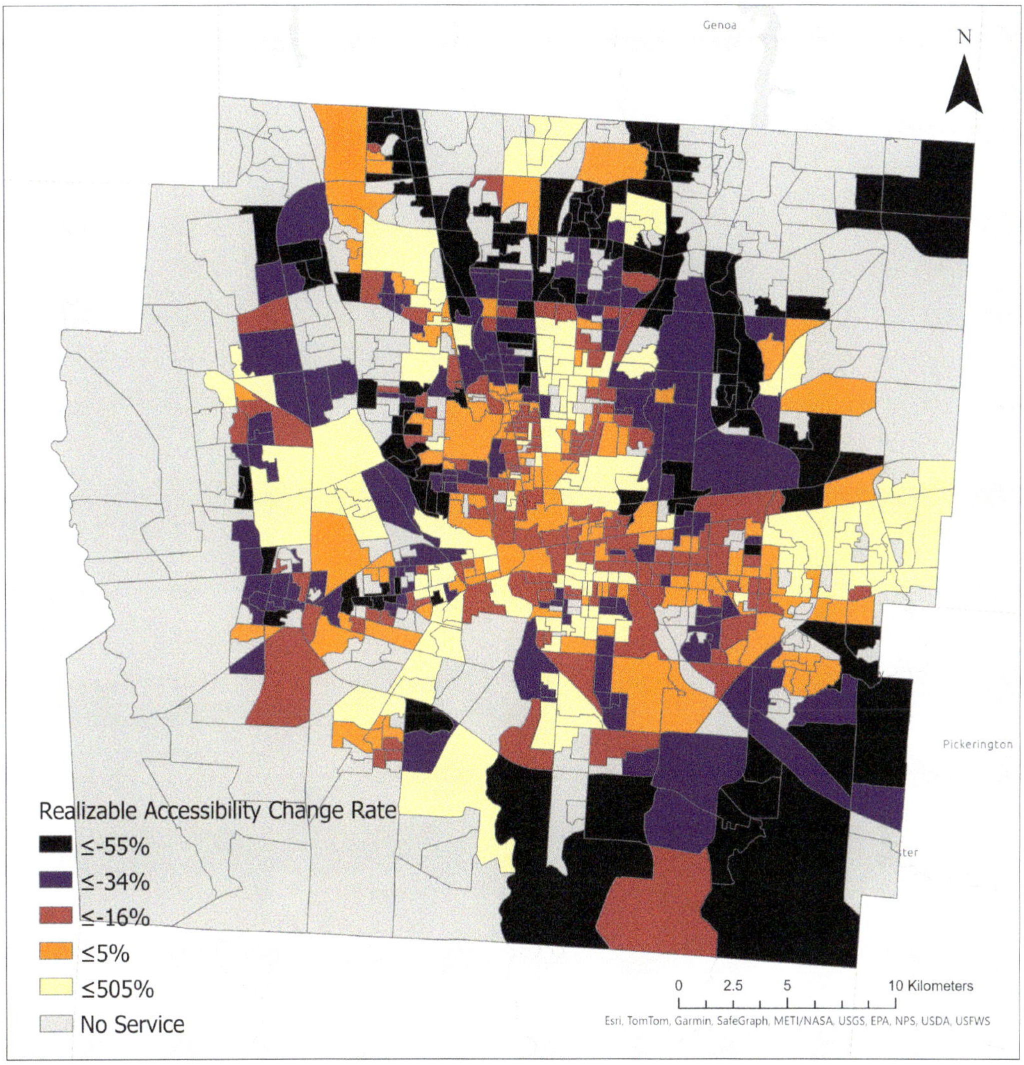

FIGURE 23.4 The change rate of realizable accessibility for each census block in Franklin County, Ohio.

23.2.2 Robustness of Reliability

Note that in Figure 23.2, the realizable accessible area is always smaller than the scheduled accessible area. In the practical sense for an ordinary user, scheduled accessibility represents the promise from the transit service, i.e., how far the user would have been able to access, and realizable accessibility represents the reality of the transit service, i.e., how far the user can actually access. The difference between the two areas, which is always positive, represents the effect of delays and missed buses on the on-time performance of the transit service. Therefore, extending from Liu et al. (2022), we further introduces the concept of reliability as the ratio of realizable accessibility to scheduled accessibility. In a practical sense, it represents the ratio of realized service compared to the promised extent by the transit service.

Figure 23.5 visualizes the change in reliability from 2019 to 2021, and Figure 23.6 visualizes the spatial heterogeneity of the reliability change. First, a major trend is that the reliability of the COTA system did not really experience much change on average. Despite large differences in accessibility,

FIGURE 23.5 Change in reliability from January 2019 to December 2021.

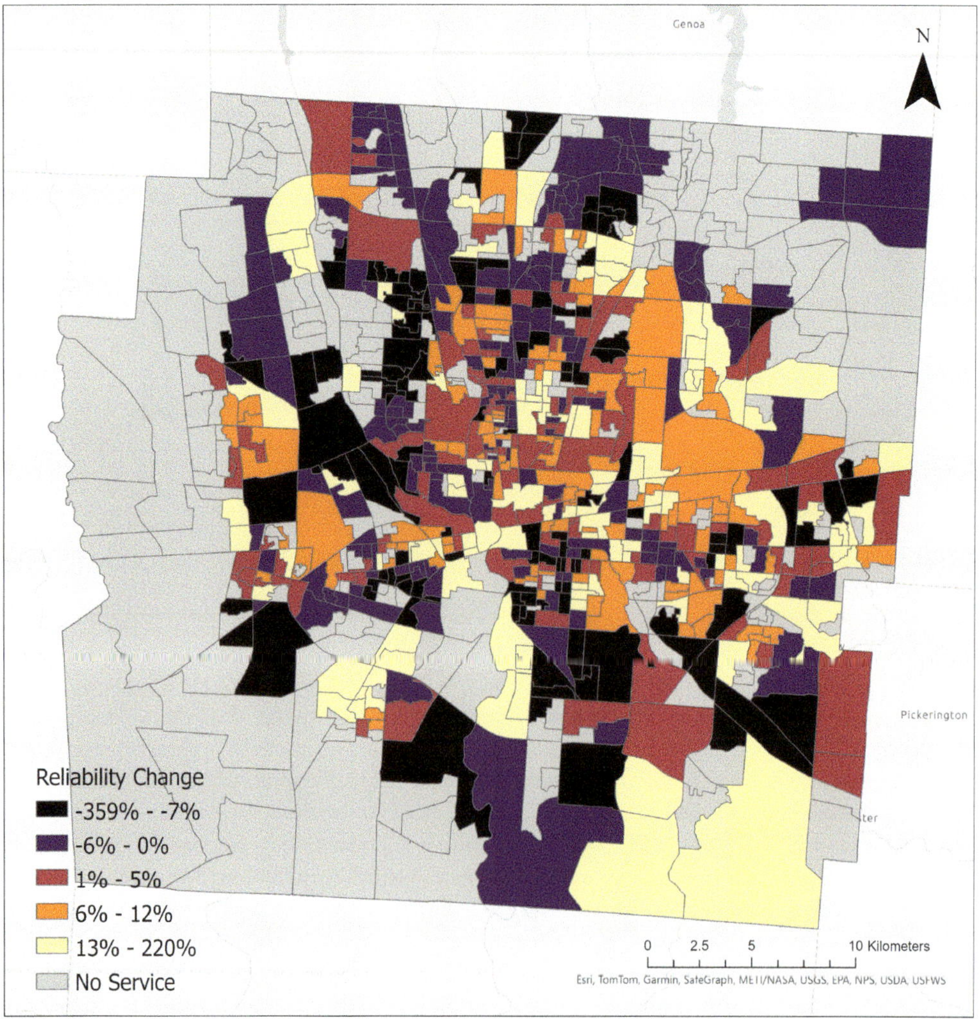

FIGURE 23.6 The difference in reliability between before and after the COVID-19 outbreak. Note that reliability is already a rate, so the values are not normalized again, as in Figure 23.4.

reliability stayed almost the same: the average reliability in 2019 is 59%, while the average reliability in 2020 is 64%. This shows that the reliability was already quite poor before the pandemic, which may trump the impacts of COVID-19 during the process.

Meanwhile, despite little differences on average, the reliability of the system actually increased due to the better traffic situation during the first several weeks of the lockdown era (Lee et al., 2020; Stiles et al., 2023). However, as soon as the service cut in May 2020 was in place, reliability quickly plunged. This shows that COVID-19 had major implications for specific time periods.

As we unravel the supply change of mobility, we can conclude that both the service extent and quality experienced sharp major declines following COVID-19. Despite some recovery in 2021, the average useability of the transit service is still significantly lower than the pre-COVID level. Aside from the empirical findings, the analyses above further exemplify the power of real-time data in the domain of disaster surveillance and response. This is the first time we have had the ability to obtain continuous observation of a transit system's operation status during such a long time in both high spatial and temporal resolution.

23.3 DEMAND-SIDE ROBUSTNESS

The demand-side robustness is all about people and activities. Demand-side resilience measures the degree of riders and served communities to continue to use the service despite the disruptions. This includes a wide spectrum of metrics, including ridership, equity, travel behavior, and satisfaction of the users. In the case of COVID-19, we choose two of the most critical aspects: ridership (how many people were still using the service) and mobility equity (who the system primarily benefited or penalized).

23.3.1 Robustness of Ridership

Ridership is the realized demand for a transit system. In the case of COVID-19, transit ridership plunged during the first few weeks all over the United States due to the travel restriction orders and health concerns. However, declines in transit demand and ridership were also uneven across different social groups and geographic areas. For example, many professionals in fields like technology and management could work remotely, while those in jobs requiring physical presence still needed to commute. Meanwhile, the changes in demand during the first few weeks were highly dynamic and unpredictable, and there was a lack of a scalable, trustworthy, accessible dataset to reflect this change.

To address this gap, we used mobile app usage data collected from Transit App, one of the most popular transit trip-planning apps with millions of users. Because people use the app to obtain real-time data before using the transit service, the app will record and summarize how many devices requested data from the server. This recorded statistic is a measure of transit demand and an approximate proxy for the general ridership. The data cover 113 public transit systems across the United States with usage statistics for each day and each system.

With this high-resolution real-time data, we conducted logistic modeling on the demand profile for each city and obtained the base value of transit usage after the sharp decline when the demand stabilized; Figure 23.7 provides an example. This value represents the number of people who still chose to use the transit service even after the travel restrictions and public health emergencies.

Figure 23.8 shows the spatial distribution of the base value for cities in the United States. One major observation is that despite better service level and more available funding, transit services in big cities did not attract more people to stay in the system, while cities in the Midwest and the South were maintaining higher usage levels during the lockdown. As Kar et al. (2022) pointed out, cities with a lower urban sprawl index, which generally have more accessible and reliable transit service, tended to have less loss in accessibility. From the perspective of a supply-demand relationship, it shows that systems with better services were having lower usage, while systems with more access loss actually kept more people. This is clearly a sign of disruption and disrupted equilibrium, which we further discuss in the next section.

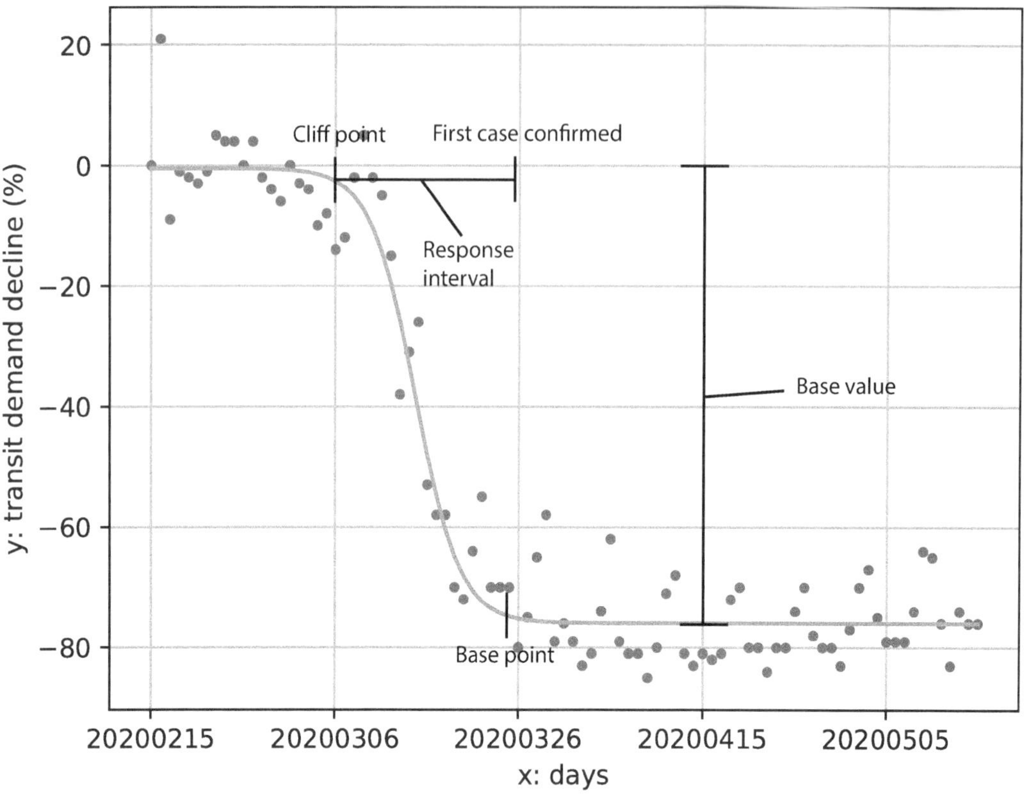

FIGURE 23.7 An example of app usage data in Capital Metro in Austin, Texas (Liu et al., 2020).

23.3.2 Robustness of Mobility Equity

COVID-19 was the direct cause of the disrupted performance, but the root of the disrupted equilibrium can be traced to systemic social equity issues. In Liu, Miller, and Scheff (2020), with the real-time app demand data and an ordinary least squares regression model, we found that the ratios of essential workers, Black people, women, and people over 40 years old are positively correlated with the number of people using transit during the lockdown. This indicates that those who rely on public transit in each city are predominantly from vulnerable communities and essential workers. This conclusion is also supported by multiple survey-based studies. For example, He et al. (2021) used a national survey (n = 500) and found that COVID-19 exacerbated existing transportation burdens for those who have limited mobility freedom including Hispanic riders, females, and nonbinary people. An in-app survey conducted by Transit App (n = 25,000) also revealed that the gender ratio between female and male changed from almost 1:1 to 56:40, with more women relying on transit service during the lockdown.

Among these communities, people with mobility difficulties are some of the most vulnerable. Despite their critical importance to mobility justice, they are often overlooked in most studies and datasets. In Liu et al. (2023), we studied their transit usage during COVID-19, including both fixed-route and paratransit service. We used automatic passenger counter (APC) data from 2018 to 2021, which contain the actual arrival times, stop-level ridership, and Americans with Disabilities Act (ADA) ramp activation data at each stop. Based on how many times the ramp was activated, we can infer the fixed-route transit service usage by people with mobility disabilities. We also used the paratransit ridership data requested from COTA. Figure 23.9 shows how the general ridership

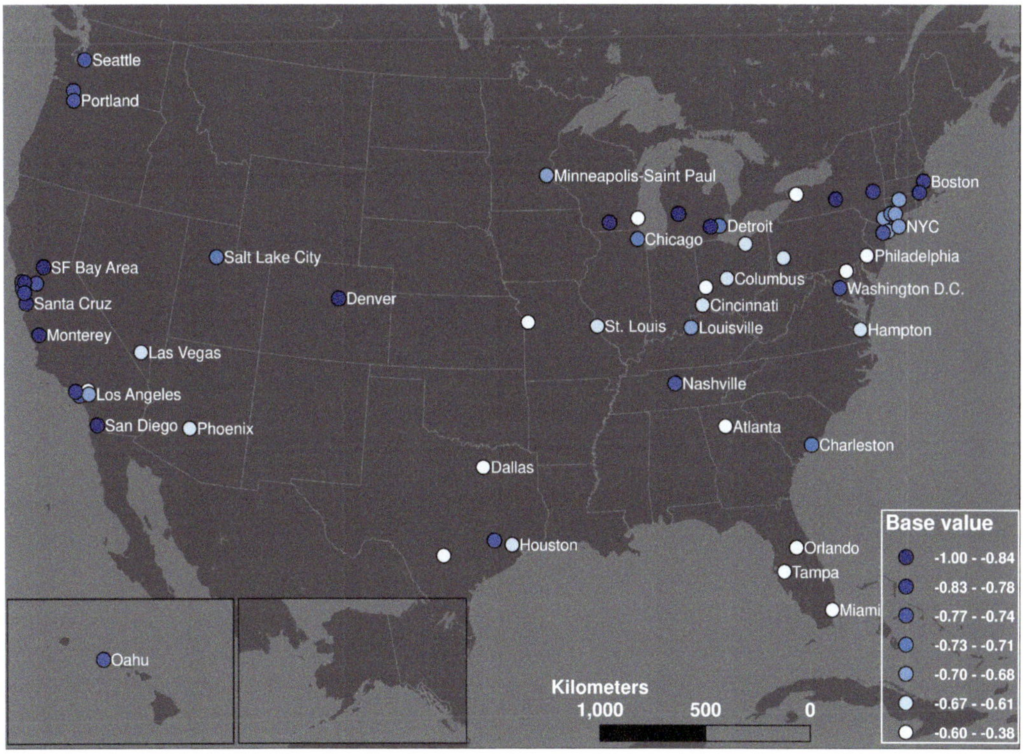

FIGURE 23.8 The spatial distribution of base values across the United States (quantile classification). (From Liu et al., 2020.)

changed from 2018 to 2021, which was mostly stable from 2018 to 2019 with some seasonal variations, such as holiday seasons in both years. In contrast, Figure 23.10 shows the fixed-route ridership for people with mobility disabilities (PwMDs) and paratransit: the fixed-route ridership for PwMDs was highly sensitive to seasonal change and precipitation, while paratransit service was relatively stable due to its door-to-door nature.

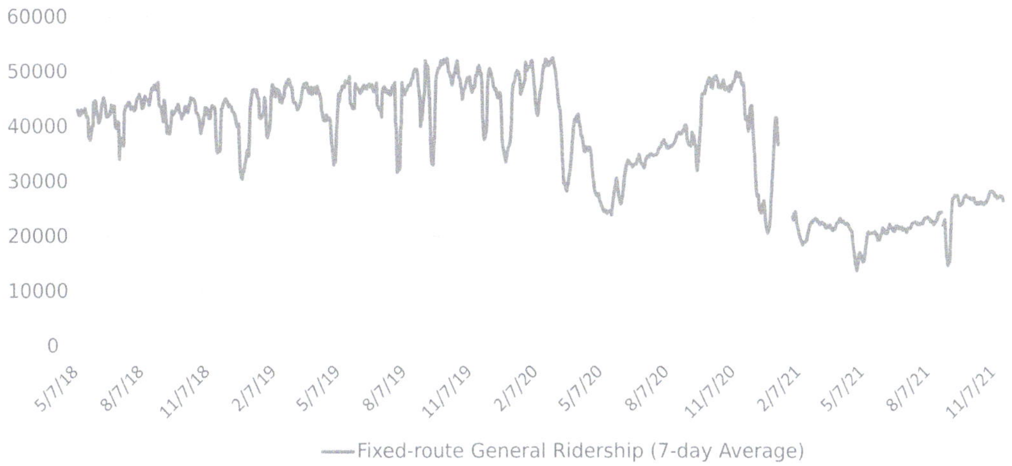

FIGURE 23.9 Trend of general ridership for fixed-route service.

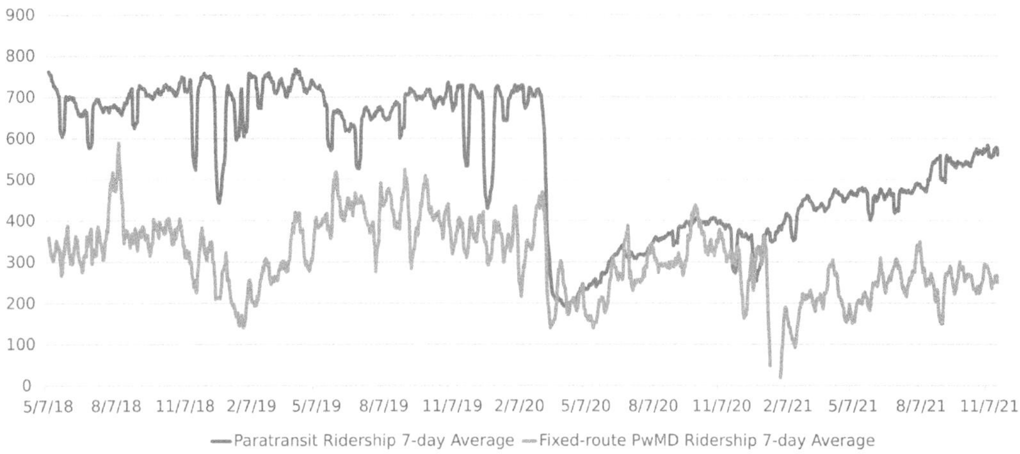

FIGURE 23.10 Trend of paratransit ridership (black lines) and people with mobility disabilities on fixed-route service (grey lines).

For the case of COVID-19, both PwMDs and the general ridership experienced a sharp decline starting in mid-March, but the decrease was more pronounced for wheelchair users, indicating that PwMDs were more cautious about using fixed-route public transit than nondisabled individuals during the early stages of the pandemic. Interestingly, paratransit services saw an even steeper decline than fixed-route services, suggesting a higher dependence on fixed-route transit among PwMD. After hitting the lowest point in May 2020, ridership gradually recovered, nearly reaching prepandemic levels by November 2020, possibly due to the implementation of free fare policies (Ferenchik, 2020). However, general ridership plummeted again to its lowest point at the start of 2021. This sharp decline coincided with the resumption of fare collection on January 11th, leading us to hypothesize that the fare reinstatement may have contributed to the drop (Warren, 2020).

With the rollout of COVID-19 vaccines and boosters in 2021, fixed-route general ridership began to steadily increase, though it remained below the lowest levels seen during the first wave of the pandemic as of November 2021. In contrast, the decline in PwMD fixed-route ridership during the same period was relatively small, and the proportion of PwMD riders actually increased. This may be due to the fact that most PwMDs could continue to ride fixed-route services free of charge under COTA's ADA policy (COTA, 2022), underscoring the significance of economic factors in mobility decisions. This trend also highlights the greater dependence on transit services among PwMDs. Meanwhile, the steady increase in paratransit ridership mirrored the recovery patterns observed in other transit systems, likely because paratransit fares remained unchanged throughout the pandemic.

In conclusion, on the demand side, PwMDs, together with other vulnerable groups with less mobility alternatives, are disproportionately reliant on public transit systems. However, on the supply side, as we showed earlier, the systems often cannot maintain the same level of services as before COVID. Beyond the persistent systemic socioeconomic issues, this imbalance between lower supply and persistent demand is a new source of injustice and a direct consequence of the pandemic. Yet, this new imbalance is likely to persist and will take a long time to recover from.

23.4 SUPPLY-DEMAND EQUILIBRIUM

With the supply-side and demand-side robustness separately discussed, here we offer a combined discussion on their interactions. In other words, we wonder: How did the plunged demand reshape the supply of transit services, and how would the supply guide the recovery of the demand? We continue our discussion in the case of Columbus, Ohio.

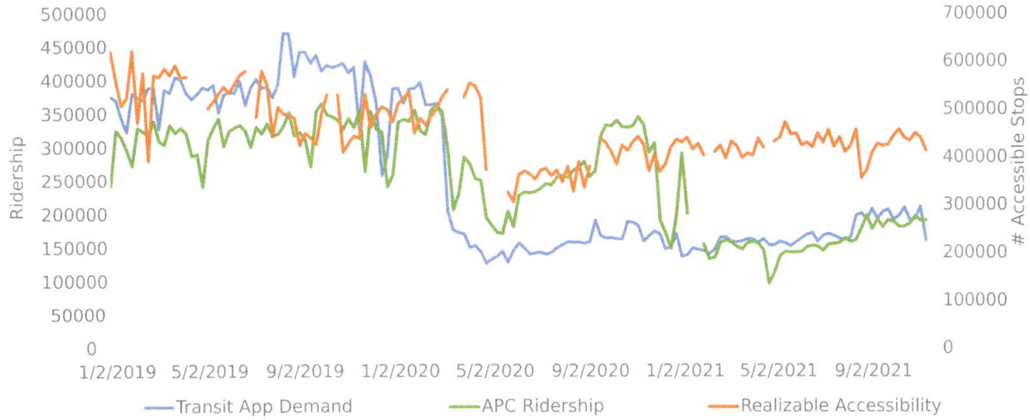

FIGURE 23.11 Trend of average realizable accessibility, app-based transit demand, and official ridership from January 2019 to December 2021 in the Central Ohio Transit Authority system, Columbus, Ohio.

Figure 23.11 visualizes the change of a supply measure, i.e., realizable accessibility (the number of actual accessible stops in 30 minutes from all centroids of census tracts), and two demand measures, i.e., app-based transit demand and official ridership from January 2019 to December 2021 in Columbus, Ohio. Both demand-side metrics showed a sudden decline near the start of the pandemic, which is earlier than the subsequent decline of transit accessibility. As the pandemic disrupted the demand side, it took about 5–6 weeks for the transit authority to adjust the new schedule and respond to the sudden change. Our real-time analyses captured the whole process of this shifted balance.

Meanwhile, note the nuanced differences between the Transit App demand data and the official ridership in late 2020. The two datasets are very close in 2019 and 2021, but official data reported much higher usage than the expected app-based demand, especially after June 2020, after the time period of Liu, Miller, and Scheff (2020). As we discussed earlier, this could be an effect of the free fare policy. This shows that despite decreased accessibility in 2020 and the public health concerns, people still extensively used and relied on transit service. In fact, according to the official ridership data, November 2020 witnessed near pre-COVID level usage. These usages are not reflected in the smartphone app metrics, suggesting high spontaneous demand or demand from people without a smartphone or Internet.

For 2021, compared with the pre-COVID scenario without considering the spatiotemporal heterogeneity, we can say the supply of transit services exceeded the demand. However, the relationship between supply and demand may not be linear, and some areas may face a shortage of supply, especially the areas with recent service cuts. By the end of 2021, both datasets witnessed some increase in usage, showing a slow recovery process. As of the end of 2021, we can see a closing gap between supply and demand. This, again, exemplifies the power of large-scale real-time spatial data analysis and emphasizes the dynamic interplay between transit supply and demand during a major crisis. As transit researchers and authorities continue to monitor these trends, the insights gained from real-time data will be invaluable for optimizing transit services to meet the evolving needs of the population.

23.5 CONCLUSION

In conclusion, this chapter provides a comprehensive analysis of the interplay between transit supply and demand during the COVID-19 pandemic, with a specific focus on Columbus, Ohio. By leveraging multiple real-time data sources, we were able to capture the dynamic shifts in both demand-side metrics and supply-side adjustments. We find that despite reduced system performance and usability, many vulnerable communities continued to rely on public transit due to limited mobility options.

As we move forward, these insights will be crucial for informing transit policies that not only address the immediate impacts of the pandemic but also strengthen the resilience and equity of transit systems in the long term. Future research should continue to explore the intricacies of the differences between Transit App–based demand and official APC ridership. Although COVID-19 as a disruption is no longer an immediate concern, public transit systems are still nowhere near the full recovery or a new balance. The long-term recovery of transit systems will continue to require the integration of real-time analysis and real-time data fusion.

REFERENCES

Azolin, L. G., da Silva, A. N. R., & Pinto, N. (2020). Incorporating public transport in a methodology for assessing resilience in urban mobility. *Transportation Research Part D: Transport and Environment, 85*, 102386.

Banister, D. (2008). The sustainable mobility paradigm. *Transport Policy, 15*(2), 73–80.

Barbeau, S. J., & Antrim, A. (2013). The Many Uses of GTFS Data – Opening the Door to Transit and Multimodal Applications. *ITS America 2013*. http://prezi.com/-69luw8sfabp/the-many-uses-of-gtfs-data-its-america-april-2013/

Bevilacqua, M., Ciarapica, F., & Marcucci, G. (2018). A modular analysis for the Supply Chain Resilience Triangle. *IFAC-PapersOnLine, 51*(11), 1528–1535.

COTA. (2022). *COTA Mainstream/Paratransit*. https://www.cota.com/services/cota-mainstream/

Dong, H., Ma, S., Jia, N., & Tian, J. (2021). Understanding public transport satisfaction in post COVID-19 pandemic. *Transport Policy, 101*, 81–88.

Fayyaz S, S. K., Liu, X. C., & Zhang, G. (2017). An efficient General Transit Feed Specification (GTFS) enabled algorithm for dynamic transit accessibility analysis. *PloS One, 12*(10), e0185333.

Ferenchik, M. (2020). COTA to continue free bus rides amid pandemic. In *The Columbus Dispatch*. https://www.dispatch.com/story/news/local/2020/07/26/cota-to-continue-free-bus-rides-amid-pandemic/42089163/

Gu, Y., Fu, X., Liu, Z., Xu, X., & Chen, A. (2020). Performance of transportation network under perturbations: Reliability, vulnerability, and resilience. *Transportation Research Part E: Logistics and Transportation Review, 133*, 101809.

He, Q., Rowangould, D., Karner, A., Palm, M., & LaRue, S. (2021). Covid-19 pandemic impacts on essential transit riders: Findings from a U.S. survey. *Transportation Research Part D: Transport and Environment, 105*, 103117.

Holling, C. S. (1973). Resilience and stability of ecological systems. *Annual Review of Ecology and Systematics, 4*, 1–23.

Kar, A., Carrel, A. L., Miller, H. J., & Le, H. T. (2022). Public transit cuts during COVID-19 compound social vulnerability in 22 US cities. *Transportation Research Part D: Transport and Environment, 110*, 103435.

Lee, J., Porr, A., & Miller, H. J. (2020). *Evidence of Increased Vehicle Speeding in Ohio's Major Cities during the COVID-19 Pandemic. Findings*, June. https://doi.org/10.32866/001c.12988.

Levine, J. (2020). A century of evolution of the accessibility concept. *Transportation Research Part D: Transport and Environment, 83*, 102309.

Liu, L., Kar, A., Tokey, A. I., Le, H. T., & Miller, H. J. (2023). Disparities in public transit accessibility and usage by people with mobility disabilities: An evaluation using high-resolution transit data. *Journal of Transport Geography, 109*, 103589.

Liu, L., & Miller, H. J. (2022). Measuring the impacts of dockless micro-mobility services on public transit accessibility. *Computers, Environment and Urban Systems, 98*, 101885.

Liu, L., Miller, H. J., & Scheff, J. (2020). The impacts of COVID-19 pandemic on public transit demand in the United States. *Plos One, 15*(11), e0242476.

Liu, L., Porr, A., & Miller, H. J. (2022). Realizable accessibility: Evaluating the reliability of public transit accessibility using high-resolution real-time data. *Journal of Geographical Systems, 25*, 1–23.

Malekzadeh, A., & Chung, E. (2020). A review of transit accessibility models: Challenges in developing transit accessibility models. *International Journal of Sustainable Transportation, 14*(10), 733–748.

Meredith-Karam, P., Kong, H., Wang, S., & Zhao, J. (2021). The relationship between ridehailing and public transit in Chicago: A comparison before and after COVID-19. *Journal of Transport Geography, 97*, 103219.

Miller, H. J. (2020). Geographic information science III: GIScience, fast and slow – Why faster geographic information is not always smarter. *Progress in Human Geography*, *44*(1), 129–138. https://doi.org/10.1177/0309132518799596

Park, Y., Mount, J., Liu, L., Xiao, N., & Miller, H. J. (2020). Assessing public transit performance using real-time data: Spatiotemporal patterns of bus operation delays in Columbus, Ohio, USA. *International Journal of Geographical Information Science*, *34*(2), 367–392. https://doi.org/10.1080/13658816.2019.1608997

Parker, M. E., Li, M., Bouzaghrane, M. A., Obeid, H., Hayes, D., Frick, K. T., Rodríguez, D. A., Sengupta, R., Walker, J., & Chatman, D. G. (2021). Public transit use in the United States in the era of COVID-19: Transit riders' travel behavior in the COVID-19 impact and recovery period. *Transport Policy*, *111*, 53–62.

Saif, M. A., Zefreh, M. M., & Torok, A. (2019). Public transport accessibility: A literature review. *Periodica Polytechnica Transportation Engineering*, *47*(1), 36–43.

Stiles, J., Kar, A., Lee, J., & Miller, H. J. (2023). Lower Volumes, Higher Speeds: Changes to Crash Type, Timing, and Severity on Urban Roads from COVID-19 Stay-at-Home Policies. *Transportation Research Record: Journal of the Transportation Research Board*, *2677*(4), 15–27. https://doi.org/10.1177/03611981211044454

Van Niel, L. (2021, September 29). *COTA faces driver shortages, adjusts services*. The Lantern. https://www.thelantern.com/2021/09/cota-faces-driver-shortages-adjusts-services/

Wan, C., Yang, Z., Zhang, D., Yan, X., & Fan, S. (2018). Resilience in transportation systems: A systematic review and future directions. *Transport Reviews*, *38*(4), 479–498.

Warren, B. (2020, December 8). *COTA Sets Date for Return to Fare Collection*. Columbus Underground. https://columbusunderground.com/cota-sets-date-for-return-to-fare-collection-bw1/

Wessel, N., Allen, J., & Farber, S. (2017). Constructing a routable retrospective transit timetable from a real-time vehicle location feed and GTFS. *Journal of Transport Geography*, *62*, 92–97.

Wessel, N., & Farber, S. (2019). On the accuracy of schedule-based GTFS for measuring accessibility. *Journal of Transport and Land Use*, *12*(1), 475–500.

Wilbur, M., Ayman, A., Ouyang, A., Poon, V., Kabir, R., Vadali, A., Pugliese, P., Freudberg, D., Laszka, A., & Dubey, A. (2020). Impact of COVID-19 on public transit accessibility and ridership. *arXiv Preprint arXiv:2008.02413*.

Zobel, C. W. (2010). Comparative visualization of predicted disaster resilience. Proceedings of the 7th International ISCRAM Conference, Seattle, USA, May 2010.

Zoumpourlis, V., Goulielmaki, M., Rizos, E., Baliou, S., & Spandidos, D. (2020). [Comment] The COVID-19 pandemic as a scientific and social challenge in the 21st century. *Molecular Medicine Reports*, 22(4), 3035–3048. https://doi.org/10.3892/mmr.2020.11393

24 Individual Environmental Exposure Dynamics Through the Lens of Neighborhood Effect Averaging

Jiannan Cai and Mei-Po Kwan

24.1 INTRODUCTION

In the course of daily life, individuals are constantly exposed to a variety of environmental factors, interacting with them in a complex and dynamic manner. Certain positive environmental exposures, such as green (e.g., parks and forests) and blue (e.g., lakes and rivers) spaces, can bestow substantial benefits on the physical and mental well-being of individuals (Chen et al. 2022). In contrast, adverse environmental exposures, such as air and noise pollution, not only precipitate conditions like anxiety and stress but also, in their more severe manifestations, give rise to respiratory diseases, compromised immune systems, and even mortality (Hammer et al. 2014; Kampa and Castanas 2008). Notably, a comprehensive assessment report from the World Health Organization has underscored the alarming fact that a staggering 23% of global deaths can be attributed to these adverse environmental factors, which, importantly, are modifiable (Prüss-Üstün et al. 2016). To develop more effective targeted intervention strategies for promoting public health, it is essential to accurately assess the environmental exposures individuals experience in their daily lives.

In environmental health research, assessing the neighborhood effect is crucial for understanding how neighborhood environments influence health outcomes. Past studies largely assumed that individuals were confined to their residential neighborhoods and treated these neighborhoods as static geographic contexts to derive measures of environmental exposures and assess neighborhood effects. This residence-based approach is effective to some extent for evaluating neighborhood effects on health outcomes that are predominantly shaped by environmental exposures within residential neighborhoods. However, individuals are not static; they typically move between neighborhoods beyond their residential areas to engage in daily activities, thereby being exposed to a wider range of dynamic environmental conditions that could exert additional, nonnegligible influences on their health outcomes. For health outcomes that are also influenced by environmental exposures in nonresidential neighborhoods (e.g., air pollution exposure), this approach may introduce nonnegligible statistical bias in assessing neighborhood effects, leading to inconsistent or even erroneous findings. This methodological issue is termed the Uncertain Geographic Context Problem (UGCoP) (Kwan 2012). It refers to the issue that the assessment of area-based attributes (e.g., air pollution levels) and their effects on individual behaviors or outcomes may be influenced by the geographical boundaries used to define the contextual units or neighborhoods from which environmental variables are derived. All studies that use fixed-area-based environmental variables to estimate individual exposure face this issue, as fixed areas may not accurately represent the true causally relevant geographic context in which environmental factors actually operate and influence individual behaviors or outcomes. To mitigate this issue, it is essential to link contextual information to human mobility in individual environmental exposure assessments, thereby enhancing our understanding of the dynamic nature of individual environmental exposures.

 DOI: 10.1201/9781003503262-28

Kwan (2018) further identified a specific manifestation of individual environmental exposure dynamics: the neighborhood effect averaging. This refers to the tendency of individuals' environmental exposures to converge toward the average exposure level of the study area when considering their daily mobility, compared to the exposure levels associated with their residential neighborhoods. This suggests that the traditional residence-based approach, which overlooks human mobility, is more likely to yield unreliable and often relatively extreme estimates of environmental exposures. Consequently, it may overestimate the neighborhood effects on health outcomes, potentially leading to misguided intervention strategies and policy decisions. This methodological issue has been termed the Neighborhood Effect Averaging Problem (NEAP) (Kwan, 2018). Since the discovery of the NEAP, scholars have conducted a series of empirical studies in various areas and for different types of environmental exposures, providing evidence to support the existence of the NEAP. The latest research by Cai and Kwan (2024) has uncovered the underlying statistical principle, revealing that the NEAP is actually a specific case of the well-known statistical phenomenon of regression to the mean (RTM) (Stigler 1997). Although they share similar manifestations, with both indicating a tendency for extreme values to regress toward the mean, the conditions under which they occur differ significantly. RTM happens in the context of random variation without intervention, whereas the NEAP is constrained by the spatial autocorrelation of environmental factors, making it not a purely random process. This finding offers an important theoretical and statistical foundation for understanding the emergence and universality of the NEAP.

This chapter explains the dynamics of individual environmental exposures from the perspective of neighborhood effect averaging. The remainder of the chapter is organized as follows. Section 24.2 introduces the basic concept of the NEAP. Section 24.3 reviews relevant studies providing evidence for the existence of the NEAP across different types of environmental exposures. Section 24.4 reveals the statistical mechanisms and conditions under which the NEAP occurs. Section 24.5 discusses the social and spatial disparities in the impact of the NEAP on individual environmental exposure estimates. Section 24.6 outlines the policy implications of the NEAP for environmental and health interventions.

24.2 THE NEIGHBORHOOD EFFECT AVERAGING PROBLEM

The NEAP was first discovered and conceptualized by Kwan (2018) as a fundamental methodological issue in environmental health research. The NEAP indicates the tendency for individual environmental exposures to move toward the average exposure level of the population or participants in the study area when considering people's daily mobility. Its initial discovery stemmed from an insightful analysis of existing studies. Existing research found that the probability distribution of individual environmental exposures at their place of residence (e.g., air pollution exposure) typically follows an approximately bell-shaped distribution. This indicates that most individuals' residential environmental exposure levels are near the population average exposure level in the study area, while only a small proportion of individuals at the distribution's tails have residential exposures that deviate significantly from the population mean, either much higher or much lower. For individuals with extreme exposure levels, their daily movements to nonresidential neighborhoods can lead to changes in their environmental exposure levels, with a higher likelihood of these levels shifting closer to the population average in the study area. For example, in the case of air pollution exposure, individuals residing in neighborhoods with lower air pollution levels are more likely to experience higher pollution levels in nonresidential neighborhoods they visit in their daily life. Conversely, individuals residing in neighborhoods with higher air pollution levels are more likely to encounter lower levels of air pollution in nonresidential neighborhoods they visit in their daily life.

The NEAP suggests that neglecting individuals' daily mobility and their nonresidential exposures can lead to inaccurate assessments of environmental exposure and potentially biased associations with health outcomes. If individuals' daily mobility is not taken into account and their residential neighborhoods are used as fixed geographic contexts to assess environmental exposure and its impact on health outcomes, it tends to overestimate the statistical significance and effect size of the neighborhood effect (Kwan 2018). This is because the residence-based exposure estimates ignore the dynamic and complex interactions between human and environmental factors during people's daily mobility. Ignoring the impact of the NEAP on individual environmental exposure estimates may further provide unreliable evidence for environmental management and health intervention policymaking, thereby undermining the effectiveness of corresponding measures. All studies that use people's residences to estimate their exposure to mobility-dependent environmental factors will face this problem. Thus, the NEAP is a fundamental methodological issue in individual environmental exposures.

24.3 THE UNIVERSALITY OF THE NEAP

After the identification of the NEAP, several recent studies have provided further evidence of its presence in specific types of environmental exposures, including air pollution, greenspace, traffic congestion, ethnic segregation, and built-environment quality. For example, Kim and Kwan (2021a) evaluated individual exposures to ground-level ozone using travel diary data collected in the Los Angeles metropolitan area. Their exploratory analysis and spatial regression models revealed a significant trend where individual ozone exposure levels tend to converge toward the average when individuals' daily mobility was considered. Ma et al. (2020) assessed real-time air pollution exposure using data from smart portable sensors and residential-based exposure in a high-pollution community in Beijing. They found that most participants mitigated high residential pollution exposure through daily mobility, bringing their exposures closer to the city-wide average. Similarly, Cai and Kwan (2023) compared perceived air pollution exposure in residential and visited neighborhoods using individual data from two communities in Hong Kong. They discovered that social inequalities in perceived air pollution exposure were less pronounced in nonresidential neighborhoods visited during daily life compared to those in residential neighborhoods. This result indicates the role of daily mobility in reducing extreme perceived exposures. Wang et al. (2024) analyzed residence-based and mobility-based green space exposure using large-scale anonymous mobile phone data in Beijing. The study found that mobility-based green space exposure had a smaller variance and tended to converge toward the mean. Similarly, Xu et al. (2023) investigated vegetation color exposure at both community and individual levels in Shanghai. Their findings showed that individual-level vegetation color exposure also exhibited a tendency to converge toward the mean. Kim and Kwan (2019) used crowdsourced real-time traffic congestion data and individual activity-travel data to evaluate individual traffic congestion exposure in Los Angeles. They compared exposure levels with and without considering individuals' activity-travel patterns. The results indicated that accounting for daily mobility skewed the probability distribution of individual congestion exposure, with levels tending toward the average. Cai and Kwan (2025) studied personal transit availability in the Chicago metropolitan area using transit availability data and individual travel survey data. They found that mobility-based transit availability was more evenly distributed compared to residential transit availability, as it tended toward the mean. Tan et al. (2020) explored ethnic segregation exposure and its relationship with daily activities using census and activity diary data in Xining, China. They reported that participants living in highly mixed neighborhoods were more likely to conduct daily activities in areas with less mixing. Conversely, those from highly segregated neighborhoods were more likely to engage in activities in less segregated areas. Huang and Kwan (2022) assessed individual exposure to case-based and venue-based COVID-19 risk environments using public COVID-19 data and activity records of infected individuals. They also identified the NEAP in individual COVID-19 risk exposure. Residents of high-risk areas often conducted activities in low-risk locations, while those from low-risk areas tended to visit high-risk locations. Wang et al. (2025) used street-view image data and mobile phone signaling data to measure people's exposure to built-environment quality. The study observed the NEAP as a derived phenomenon associated with income groups.

Although these empirical studies have provided evidence for the existence of the NEAP in specific types of individual environmental exposures, it remains unclear whether the NEAP is universally present across other types of environmental exposures. This is because different environmental factors could exhibit diverse spatial and temporal dynamics, and individuals interact with these factors to varying degrees throughout the day. Consequently, individual environmental exposures may follow different distribution patterns, which are not necessarily the bell-shaped distributions commonly observed in most studies. This uncertainty raises questions about whether individual environmental exposures with varying types and distributions consistently align with the patterns suggested by the NEAP.

To further substantiate the universality of NEAP, it is necessary to go beyond empirical studies and delve into the underlying statistical principles. For this purpose, Cai and Kwan (2024) first recognized the striking similarity between the NEAP and the well-known statistical phenomenon of RTM. The RTM indicates that extreme observations in a dataset are likely to shift closer to the average in subsequent measurements, primarily due to random variation or measurement error. The RTM has been widely observed across many practical fields. For example, in genetics, when parents' heights are extremely tall or short, their children's heights tend to be closer to the average height of the population; in medicine, patients with abnormally high blood pressure at their initial measurement tend to see their values return to the normal average level in subsequent measurements, even without medical intervention. Similar to the RTM, the NEAP suggests that extreme environmental exposures experienced by individuals in their residential neighborhoods could be moderated through daily mobility, with exposures tending to shift closer to the average level. To test whether the NEAP shares the same universality as the RTM, Cai and Kwan (2024) examined individuals' exposure to a range of natural, built, and social environments, including green spaces, air pollution, healthy food environments, transit accessibility, and crime rates, using a large human mobility dataset comprising more than 27,000 individuals in the Chicago metropolitan area. The results revealed that, across these different types of individual environmental exposures, individuals' exposures during daily mobility consistently regressed toward the average exposure level, particularly for those with extreme residential exposures. Even though these environmental exposures do not necessarily follow the bell-shaped distributions commonly observed in previous studies (e.g., individual transit accessibility exhibited an M-shaped bimodal distribution), the trend of individual exposure regressing toward the mean remained significant. This provides robust evidence for the universality of the NEAP in mobility-dependent environmental exposures.

It is worth noting that, although the NEAP has been commonly observed at the global level, Cai and Kwan (2024) further discovered that the significant trend of regression toward the mean does not necessarily exist locally for different sections of the value range of individual environmental exposures. For example, in the same dataset mentioned above, for individuals whose residential exposures to green spaces fell within the first and second quintiles, their exposures during daily mobility did not significantly exceed their residential exposures due to regression toward the mean. This suggests that individual exposure data within localized value ranges have limited explanatory power for the NEAP, as they may only reflect the exposure patterns of individuals residing in specific communities with particular environmental conditions, thereby lacking broader statistical generalizability. This also explains why significant NEAP may not be observed in the daily environmental exposures of individuals in small-scale studies focused on specific communities.

24.4 THE MECHANISMS BEHIND THE NEAP

Although empirical studies can provide evidence supporting the existence of the NEAP, the mechanisms underlying its occurrence remain unclear. To better inform interventions targeting environmental factors and human behavior, it is essential to understand the processes behind the NEAP. However, this is challenging when relying solely on observed samples because observed samples may fail to reflect diverse scenarios of environmental and human mobility interactions. Moreover, the interactions between environmental and mobility factors may influence one another, making it difficult to isolate and control them independently. To address this, Cai and Kwan (2024) developed three geographic simulation methods to model environmental and human mobility scenarios under different conditions. These methods employ controllable parameters to independently analyze how specific characteristics of the environment and human mobility contribute to the occurrence of the NEAP in individual environmental exposures, thereby uncovering the conditions under which the NEAP arises.

The above research revealed that the spatial autocorrelation of environmental factors suppresses the occurrence of the NEAP. As the spatial autocorrelation of environmental factors strengthens, individuals experience increasingly similar environmental exposures within neighboring areas. This reduces the differences between mobility-based and residence-based individual exposures, leading to fewer individuals experiencing the NEAP. In addition, the extent to which the NEAP affects individual environmental exposure estimates also decreases. Specifically, by comparing the performance of the NEAP in environmental scenarios with completely spatially random and spatially autocorrelated distributions of environmental factors, it was found that the occurrence of the NEAP is significantly suppressed under spatial autocorrelation. This indicates that, although the NEAP and RTM appear similar in manifestation, their underlying occurrence conditions are fundamentally different. The RTM arises from the random variation of observations without any external intervention, whereas the NEAP does not occur under random conditions. This is because individual mobility is continuous over space and the attributes of environmental factors within continuous spaces exhibit spatial autocorrelation. As a result, the variation in environmental exposure during an individual's daily mobility is intervened by the spatial autocorrelation of environmental factors within continuous spaces, rather than being a completely random process without any external intervention. Based on this finding, Cai and Kwan (2024) were the first to conceptualize the NEAP as a special case of the RTM under the intervention of spatial autocorrelation of environmental factors. This provides a foundational theoretical and statistical explanation for the occurrence of the NEAP, while also clarifying why the NEAP, like the RTM, is universally present in individual environmental exposures, although the degree and frequency of regression to the mean may vary.

Cai and Kwan (2024) further simulated different human mobility scenarios to analyze the impact of travel distance and out-of-home duration on the occurrence of the NEAP. The results revealed that longer travel distance can amplify and promote the effect of the NEAP. The longer the travel distance, the more diverse are the environmental conditions to which individuals are exposed, leading to the increased extent and frequency of the NEAP in individual environmental exposures. Similarly, with the stay location unchanged, an increase in out-of-home duration can further amplify the impact of the NEAP on individual exposures. This is because as individuals spend more time outside their homes, the deviation between their daily exposure and residential exposure becomes greater, thereby intensifying the NEAP effect. However, an increase in out-of-home duration does not necessarily promote the occurrence of the NEAP in more individuals. This promotion effect is primarily significant for relatively dynamic environmental factors (e.g., air pollution), while its impact on relatively static environmental factors (e.g., green space) is not significant. For relatively dynamic environmental factors, even if the visit location remains unchanged, an increase in duration of stay could lead to a broader range of environmental exposures due to the temporal dynamics of these factors. As a result, the environmental exposure estimates of more individuals may be influenced by the NEAP. For relatively static environmental factors, when the visit location remains unchanged, the deviation between estimations of an individual's daily exposure and residential exposure is generally fixed in direction. An increase in out-of-home duration can only amplify the weight of out-of-home exposure in the estimation of overall daily exposure, thereby increasing the deviation from residential exposure, yet it cannot effectively alter the direction of the deviation. As a result, an increase in out-of-home duration does not necessarily lead to more individuals experiencing the NEAP.

24.5 SOCIAL AND SPATIAL DISPARITIES IN THE NEAP's IMPACT

Although existing studies have generally demonstrated the widespread presence of the NEAP in mobility-dependent environmental exposures, its impact on individual exposure estimates may vary considerably across different individuals and spatial contexts. Different social groups tend to exhibit distinct patterns of daily mobility (Jiang et al. 2012). Some groups are confined to limited spatial ranges for their daily activities due to constraints or lack of long-distance travel needs. In contrast, others, benefiting from greater mobility and the need for long-distance travel, tend to have larger activity spaces. These differences can further affect the likelihood and extent to which various social groups can modify their residential environmental exposures through daily mobility. As a result, this leads to social disparities in the impact of the NEAP. For example, in Los Angeles, Kim and Kwan (2021b) identified significant variations in how neighborhood effect averaging manifests in ground-level ozone exposure across different social groups. Their findings revealed that nonworkers (e.g., the unemployed, homemakers, retirees, and students) did not experience downward averaging due to their limited daily mobility, which restricted their ability to alleviate high air pollution exposure in their residential neighborhoods. Similarly, Ma et al. (2020) observed in a high-pollution community in Beijing that three vulnerable groups faced challenges in avoiding high residential air pollution levels because of limited daily mobility. These groups included low-income individuals, blue-collar workers spending extended hours in low-end workplaces, and elderly individuals constrained by household constraints. Likewise, in Beijing, Wang et al. (2024) found that younger individuals and the employed population, who exhibited larger activity spaces and more diverse daily visitation, were more likely to experience a pronounced impact of the NEAP on their greenspace exposure compared to other social groups. In the Chicago metropolitan area, Cai and Kwan (2024) also found significant differences in the extent to which environmental exposure estimates are influenced by the NEAP across different social groups, due to variations in their daily mobility levels. For example, compared to unemployed individuals, employed individuals experience a more pronounced impact of the NEAP on their green space exposure estimates. This is because employed individuals tend to have longer daily commutes and spend more time out of home, which increases the complexity of their environmental exposure during daily mobility.

In addition, existing studies have also found significant spatial heterogeneity in the manifestation and degree of the NEAP in individual environmental exposure due to differences in residential environmental conditions and travel demands across regions. For example, Wang et al. (2024) observed significant spatial heterogeneity in the manifestation of the NEAP in individual green space exposure estimates in Beijing. They found that residents in distant suburban areas, such as Changping and the Beijing Olympic Park, have higher residential green space exposure, while their green space exposure during daily mobility could undergo downward averaging, resulting in lower levels. In contrast, individuals residing in urban areas and near major urban infrastructures tend to experience upward averaging in their green space exposure during daily mobility, resulting in higher exposure levels that exceed their residential exposure. Cai and Kwan (2024) further revealed spatial heterogeneity in the degree of NEAP's impact on environmental exposure estimates. Specifically, they found an inverted U-shaped relationship between the distance from residential locations to the city center and individual exposure bias caused by the NEAP. Taking individual green space exposure in the Chicago metropolitan area as an example, as the distance between an individual's residence and downtown Chicago increases, the bias between mobility- and residence-based exposure estimates also grows. Individuals living farther from the city center typically need longer commutes and greater travel distances to access daily necessities such as employment, entertainment, and healthcare in downtown areas. This increases the likelihood of exposure to more diverse environmental conditions during daily mobility, creating favorable conditions for the occurrence of the NEAP. However, beyond a certain distance, the impact of the NEAP on individual exposure estimates begins to diminish. This is because the attraction of the city center has a distance-based upper bound (Dong et al. 2022; Fujita et al. 2001). If individuals live too far from the city center, they tend to opt for closer subcenters to meet their daily needs, thereby reducing their activity space and minimizing the bias in individual environmental exposure estimates.

24.6 POLICY IMPLICATIONS OF THE NEAP

These findings on the NEAP can not only enhance our understanding of the dynamics of individual environmental exposures but also provide reliable guidance for developing more effective and targeted environmental management policies, urban planning strategies, and recommendations for optimizing individual behaviors. The NEAP suggests that individuals can experience diverse environmental exposures through daily mobility, thereby mitigating extreme exposures encountered in their residential neighborhoods that may negatively affect their health. However, existing studies have found that not all social groups benefit equally from neighborhood effect averaging. For example, Kim and Kwan (2021b) and Ma et al. (2020) found that some vulnerable groups (e.g., nonemployed, low-income, and elderly individuals) living in high-air-pollution communities could be less likely to experience neighborhood effect averaging in their daily exposures due to limited mobility. As a result, their exposures to air pollution throughout the day may not be alleviated. These groups, which face both poor residence-based and mobility-based environmental exposures, are referred to as doubly disadvantaged groups (Kim and Kwan 2021b). To improve the living conditions of such populations, targeted measures to promote mobility should be developed, such as the provision of more age-friendly transportation infrastructure and travel subsidies tailored to specific groups.

In addition, the findings on the spatially heterogeneous effects of the NEAP can provide valuable insights for spatial adjustments of urban resources. Cai and Kwan (2024) found that the NEAP most strongly influences daily greenspace exposure for residents living at the urban periphery, as they have higher mobility needs to travel to the city center for daily activities. In contrast, for city center residents, due to the relative concentration of living resources, there is no need for long-distance travel to meet their daily needs. However, the city center typically has lower greenspace coverage, making it difficult for residents to experience improved greenspace exposure within their limited activity space. A similar situation is observed for residents living in subcenters beyond the main urban boundary. To improve the greenspace exposure for city-center residents, on the one hand, it is necessary to develop more greenspace in specific areas. On the other hand, urban resources (e.g., employment opportunities) should be appropriately decentralized to encourage the expansion of their activity space, thereby promoting greenspace exposure improvements through the neighborhood effect averaging.

REFERENCES

Cai, Jiannan, and Mei-Po Kwan. "Understanding social inequality in individual perceived exposures to air pollution in residential and visited neighborhoods: A study using association rule mining." *Annals of the American Association of Geographers* 113, no. 10 (2023): 2392–2416.

Cai, Jiannan, and Mei-Po Kwan. "The universal neighborhood effect averaging in mobility-dependent environmental exposures." *Environmental Science & Technology* 58, no. 45 (2024): 20030–20039.

Cai, Jiannan, and Mei-Po Kwan. "Revealing the complex dynamics of social disparities in personal transit availability considering human mobility and neighborhood effect averaging." *Annals of the American Association of Geographers*. 115, no. 3 (2025), 620–639.

Chen, Bin, Shengbiao Wu, Yimeng Song, Chris Webster, Bing Xu, and Peng Gong. "Contrasting inequality in human exposure to greenspace between cities of Global North and Global South." *Nature Communications* 13, no. 1 (2022): 4636.

Dong, Lei, Paolo Santi, Yu Liu, Siqi Zheng, and Carlo Ratti. "The universality in urban commuting across and within cities." *arXiv preprint arXiv*:2204.12865 (2022).

Fujita, Masahisa, Paul R. Krugman, and Anthony Venables. *The Spatial Economy: Cities, Regions, and International Trade*. MIT press, 2001.

Hammer, Monica S., Tracy K. Swinburn, and Richard L. Neitzel. "Environmental noise pollution in the United States: Developing an effective public health response." *Environmental Health Perspectives* 122, no. 2 (2014): 115–119.

Huang, Jianwei, and Mei-Po Kwan. "Uncertainties in the assessment of COVID-19 risk: A study of people's exposure to high-risk environments using individual-level activity data." *Annals of the American Association of Geographers* 112, no. 4 (2022): 968–987.

Jiang, Shan, Joseph Ferreira, and Marta C. González. "Clustering daily patterns of human activities in the city." *Data Mining and Knowledge Discovery* 25 (2012): 478–510.

Kampa, Marilena, and Elias Castanas. "Human health effects of air pollution." *Environmental Pollution* 151, no. 2 (2008): 362–367.

Kim, Junghwan, and Mei-Po Kwan. "Beyond commuting: Ignoring individuals' activity-travel patterns may lead to inaccurate assessments of their exposure to traffic congestion." *International Journal of Environmental Research and Public Heal*th 16, no. 1 (2019): 89.

Kim, Junghwan, and Mei-Po Kwan. "How neighborhood effect averaging might affect assessment of individual exposures to air pollution: A study of ozone exposures in Los Angeles." *Annals of the American Association of Geographers* 111, no. 1 (2021a): 121–140.

Kim, Junghwan, and Mei-Po Kwan. "Assessment of sociodemographic disparities in environmental exposure might be erroneous due to neighborhood effect averaging: Implications for environmental inequality research." *Environmental Research* 195 (2021b): 110519.

Kwan, Mei-Po. "The uncertain geographic context problem." *Annals of the Association of American Geographers* 102, no. 5 (2012): 958–968.

Kwan, Mei-Po. "The neighborhood effect averaging problem (NEAP): An elusive confounder of the neighborhood effect." *International Journal of Environmental Research and Public Health* 15, no. 9 (2018): 1841.

Ma, Xinlin, Xijing Li, Mei-Po Kwan, and Yanwei Chai. "Who could not avoid exposure to high levels of residence-based pollution by daily mobility? Evidence of air pollution exposure from the perspective of the neighborhood effect averaging problem (NEAP)." *International Journal of Environmental Research and Public Health* 17, no. 4 (2020): 1223.

Prüss-Üstün, Annette, Jennyfer Wolf, Carlos Corvalán, Robert Bos, and Maria Neira. *Preventing Disease Through Healthy Environments: A Global Assessment of the Burden of Disease from Environmental Risks.* World Health Organization, 2016.

Stigler, Stephen M. "Regression towards the mean, historically considered." *Statistical Methods in Medical Research* 6, no. 2 (1997): 103–114.

Tan, Yiming, Mei-Po Kwan, and Zifeng Chen. "Examining ethnic exposure through the perspective of the neighborhood effect averaging problem: A case study of Xining, China." *International Journal of Environmental Research and Public Health* 17, no. 8 (2020): 2872.

Wang, Jianying, Mei-Po Kwan, Gezhi Xiu, Xia Peng, and Yu Liu. "Investigating the neighborhood effect averaging problem (NEAP) in greenspace exposure: A study in Beijing." *Landscape and Urban Planning* 243 (2024): 104970.

Wang, Linsen, Suhong Zhou, Zhong Zheng, Jiangyu Song, Junwen Lu, and Mei-Po Kwan. "Unveiling the neighborhood effect averaging problem: The role of daily mobility in shaping built-environment quality exposure." *Annals of the American Association of Geographers*115, no. 2 (2025), 264–281.

Xu, Tiantian, Shiyi Wang, Qing Liu, Junghwan Kim, Jingyi Zhang, Yiwen Ren, Na Ta, Xiaoliang Wang, and Jiayu Wu. "Vegetation color exposure differences at the community and individual levels: An explanatory framework based on the neighborhood effect averaging problem." *Urban Forestry & Urban Greening* 86 (2023): 128001.

25 Bike-Sharing

Toward Equitable Urban Transportation

Yunhe Cui

25.1 INTRODUCTION

The growth of cities is shaped by technical advances in urban transportation. Since human settlement and activities are inexorably linked to mobility, the evolution of transportation and mobility services allows urban residents to relocate, thereby expanding urban boundaries. Throughout the urbanization history of the United States, there has been a synergy between urban sprawl and transportation development. The emergence of private automobiles fueled this process (Fairfield 1995), and exurban settlement further expedited it (Clark et al. 2009). Concurrently, public transit ridership experienced a decline over 70 years (Taylor et al. 2009). Due to budget constraints and limited land, expansion of the public transit system could not be unfettered. One notable development gap was the "last-mile problem," which refers to the travel between the trip origin/destination and the public transit terminal (Shaheen and Chan 2016). As a result, emerging mobility services, such as bike-sharing, have provided flexible and environmentally friendly solutions for accommodating short-distance travel.

Cycling has long been seen as an affordable form of nonmotorized transportation and as one of the effective solutions to the challenges presented by car dependency, such as traffic congestion in metropolitan areas (Li and Kamargianni 2018). A growing interest in bicycle use led to the launch of bike-sharing programs in European cities in the 1960s (Fishman 2016). Contemporary bike-sharing programs involve the provision of bikes for short-term rental, either between docking stations or within certain service areas.

The bike-sharing system has undergone four generations of evolution (DeMaio 2009). The prototype bike-sharing system, known as the White Bikes, debuted in Amsterdam on July 28, 1965. Along with many avant-gardes (e.g., Yellow Bikes in Austin, Texas), this first-generation system adopted an unregulated business model, where bikes were left unlocked and restricted within a bounded area (Shaheen, Guzman, and Zhang 2010). In the early 1990s, a second generation of bike-sharing was developed, integrating a coin deposit station, where the station served as a hub for payment, pick-up, return, and storage of bikes. However, these first two generations suffered from rampant bike theft and vandalism, primarily due to the lack of administration and unrestricted bike usage.

As the coin deposit system began to wane, a third generation, the automated dock-based bike-sharing system (hereafter, "dock-based bike-sharing system") emerged (DeMaio 2009). The dock-based bike-sharing system monitored users' identities and travel histories via magnetic stripe cards or cell phones to mitigate the risk of bike misuse. With its initial success, the dock-based bike-sharing system expanded to about 125 world cities as of 2010 (Shaheen, Guzman, and Zhang 2010) and remains active in world megacities today. Studies found that the largest barrier to users' rental experience was the difficulty of accessing the docking stations (Fuller et al. 2011; Fishman et al. 2014). Therefore, docking stations were strategically placed near transport hubs to facilitate mode switching. Additionally, location-allocation

DOI: 10.1201/9781003503262-29

models were employed to optimize the placement of docking stations based on socioeconomic data and user demands (Martinez et al. 2012; García-Palomares, Gutiérrez, and Latorre 2012; Lin, Yang, and Chang 2013).

The emergence and prevalence of wireless technology brought a renaissance to the bike-sharing industry. Equipped with global positioning system (GPS) tracking devices, smart locks, and electronic payment systems, the bike-sharing system overcame the disadvantages of earlier generations. The fourth generation, the dockless bike-sharing system, saw dramatic expansions in the 2010s, serving over 200 Chinese cities with more than 23 million bikes (Gu, Kim, and Currie 2019). Although this new generation revolutionized the industry by eliminating the need for docking stations for bike retrieval, return, and payment (Shaheen, Zhang, Martin, and Guzman 2011), only a few U.S. cities, such as Seattle and Washington, have adopted dockless bike-sharing systems, compared to their rapid expansion in East Asia. The explosive and unregulated growth brought challenges: the profit generated from the bike rental was barely sufficient to offset the cost of fleet maintenance, and other issues like bike theft and destruction exacerbated the financial plight (Tu et al. 2019). More critically, as the market faded in some cities, the oversupply of bikes led to them being abandoned, piled up, and dismantled, generating nondisposable "bike litter" (Li and Kamargianni 2018; Chen 2019).

The rapid adoption of the bike-sharing system posed not only emerging opportunities for sustainable transportation development but also operational challenges for urban planning. Stakeholders have continuously formulated policies to guide the market's healthy growth and minimize conflicts with urban functionality (Wang and Zhou 2017; Shi et al. 2018). While bike-sharing is viewed as a sustainable mode of transportation, the operation of these systems has raised environmental concerns. As the supply and demand for shared bikes are often unmatched, excessive carbon footprints could be generated from bike rebalancing using automobiles (Luo 2019).

Several studies investigating bike-sharing ridership have tried to find concerning factors hypothesized to influence ridership. A city's geographical context, such as topography, weather, daylight hours, and climate, significantly impacts bicycle usage, frequency of bike-sharing usage, and bike-sharing system design (Sears et al. 2012). Additionally, it is worth pointing out that land use and the built environment influence travel behavior and mode choice. In a review paper published in 2001, researchers stated that population density, employment density, and mixed land use patterns can lead to an increase in trip frequency (Ewing and Cervero 2001). Specifically, compared to suburban neighborhoods, transit, bike, and walk mode shares tend to increase in mixed land use and high-population or high-employment-density zones, while the car dependency of those areas tends to decrease.

Many studies further indicated that station location, the surrounding built environment, and bike infrastructure are the key factors affecting bike-sharing usage (Antonakos 1993; Dill and Carr 2003; Dill 2009; Tilahun, Levinson, and Krizek 2007; El-Assi, Salah Mahmoud, and Nurul Habib 2017; Shen, Zhang, and Zhao 2018; Scott and Ciuro 2019). First, the location of bike-sharing stations within a city determines its spatiotemporal usage pattern and the user structures, referring to who uses the bike-share and where (Faghih-Imani and Eluru 2015). Bike-sharing usage is expected to increase at stations near city centers, recreation areas, and transit, while docking stations and dockless bike fleets distributed in higher population density areas are more likely to have more bike users than those in less populated areas. The station's proximity to residential units also plays an important role (El-Assi, Mahmoud, and Habib 2017; Shen, Zhang, and Zhao 2018). Moreover, in many large North American cities such as New York, Toronto, and Chicago, researchers found that a considerable portion of bike-sharing usage is contributed by tourists rather than local commuters, and their usage is sensitive to the bike-sharing station locations and the surrounding points of interest (Scott and Ciuro 2019). Second, according to Dill and Carr's (2003) analysis of bike use patterns in 43 U.S. cities, a higher level of bike infrastructure led to more bike usage. Roads with unmarked and mixed-use lanes shared by bikes and motor vehicles are the least preferred routes, whereas separate and protected bike lanes are the most preferred and have a positive relationship with an increase in bike usage (Antonakos 1993; Tilahun, Levinson, and Krizek 2007; Dill 2009).

However, the equity issues related to bike-sharing remain underexplored in existing literature. Public bike-sharing systems have rapidly expanded across the United States in recent years, yet there is evidence that a significant portion of the population, including people of color, lower-income individuals, females, senior citizens, and those with lower levels of education, are underrepresented among bike-share users (Buck et al., 2013). While the lack of physical access to shared bikes in neighborhoods with underserved groups is one factor, it does not fully explain the disparities in usage. Financial issues, information barriers, and cultural factors may also play important roles in preventing socioeconomically disadvantaged populations from using bike-sharing services (Golub et al., 2016). With the growing awareness of climate change in recent years, bike-sharing has emerged as a low-carbon-emission alternative for transit, eagerly promoted to improve transportation system efficiency, enhance the travel experience, and reduce the overall environmental impact of transit (Slowik and Kamakaté 2017). Thus, exploring the equity issue associated with uneven bike-sharing accessibility and usage is becoming increasingly important.

This chapter aims to enhance the understanding of bike-sharing system usage and equity with the Citi Bike case study. The subsequent sections explore the interplay between the dock-based bike-sharing system and the public transit system and how this interaction affects the performance of the bike-sharing system.

25.2 CITI BIKE IN NEW YORK CITY

Two years after the first bike-sharing system debuted in North America in Washington D.C., New York City (NYC) launched Citi Bike in May 2013. The NYC bike-sharing system exemplifies a successful private-public partnership. The NYC Department of Transportation (DOT) played a crucial role in system planning and public outreach, while Lyft Inc. took the responsibility for daily operations and maintenance. After a decade of operation, Citi Bike has become an integral part of NYC's landscape, standing as the nation's largest bike-sharing program. It has grown rapidly, becoming the city's fastest-expanding transportation network with over 200 million rides since its inception.

In 2013, following several delays including equipment damage caused by Superstorm Sandy and software problems (Kaufman, Gordon-Koven, Levenson, and Moss 2015), Citi Bike built 332 stations and brought 6,000 bikes to lower Manhattan and the neighborhoods near downtown Brooklyn (New York City Department of Transportation 2023). This new system drew significant public attention. By 5 p.m. on the launch day, 16,000 people had signed up as annual members, and over 6,000 trips were made.

The period from 2015 to 2017 marked the second phase of Citi Bike expansion. The system grew to 750 stations and 12,000 bikes, extending its service area into Queens, midtown Manhattan, and deeper into Brooklyn. In 2018, pedal-assist e-bikes, with a top speed of 20 miles per hour and mechanisms engaged only during pedaling for safety, were introduced. However, due to the oversensitivity of the front brake and resulting injuries, Lyft recalled over 1,000 e-bikes from Citi Bike in late April 2019. Similar precautions were subsequently taken for its bike-share programs in Washington, D.C., and the San Francisco Bay Area (Pager 2019). Then, in July 2019, battery fires in San Francisco's Lyft e-bikes further delayed the relaunch of e-bikes in NYC (Yobbi 2019). Ten months after the recall, Lyft unveiled the redesigned Citi Bike e-bike, featuring a longer battery life and a smoother hydraulic rear brake. Soon after the e-bike relaunch, e-bike rides increased from less than 14% of rides in 2020 to over 46% in 2023, despite e-bike only making up about one-fifth of the fleet (City of New York 2023).

NYC mayor Bill de Blasio announced the details of the program's growth, known as phase 3, in July 2019. Unlike the previous phases, phase 3 was planned with two distinct but interrelated efforts: expansion and infill. The system growth includes doubling the service area by 35 square miles and increasing its fleet from 6,000 bikes at launch in May 2013 to 40,000 bikes (New York City Department of Transportation 2024). At the completion of phase 3 in 2024, more than 2,000 stations now serve New Yorkers, focusing on both reaching more NYC neighborhoods and meeting the increasing demand in the system's core area (particularly south of 60th Street, Manhattan) (New York City Department of Transportation 2024).

25.3 DOCK-BASED BIKE-SHARING SYSTEM AND PUBLIC TRANSPORTATION[1]

One emerging trend in modern transportation is the sharing economy. While traditional modes of mass transit (e.g., buses, subways) enable the sharing of mobility services by lowering individual travel costs, they are unable to fulfill the need for short-distance travel between the transit station and the traveler's origin/destination, which is generally known as "the last mile problem" (Liu, Jia, and Cheng 2012; Shaheen and Chan 2016; Moorthy et al. 2017). Recent urban and transportation planning initiatives have taken strides to promote shared mobility services, such as bike-sharing. Bike-sharing accommodates individual travel by enabling more flexibility in mode-switching while alleviating environmental impacts from automobiles, such as traffic congestion and air pollution. It has gained considerable popularity in world megacities, including Beijing, Paris, and NYC, where the public transit infrastructure has reached its full capacity for expansion (DeMaio 2009; Liu, Jia, and Cheng 2012).

Location-based bike usage and tracking data derived from dock-based and dockless bike-sharing systems enable the further exploration of bike usage. These spatiotemporal analyses are generally split into three groups of studies. The first group explored how the physical and built environments impact bike usage. These environmental variables included weather and seasonality (Nankervis 1999; Miranda-Moreno and Nosal 2011; Sears et al. 2012; Martinez 2017; Kim 2018), land use and infrastructure (Pucher, Dill, and Handy 2010; Kim et al. 2012; Habib et al. 2014; Schoner and Levinson 2014; Lu, Scott, and Dalumpines 2018), and public transportation (Fishman et al. 2014; Guo et al. 2017; Zhang and Zhang 2018). The second group explored the improvement of system efficiency. As the supply and demand of the bikes did not align spatially, optimization models were employed to improve the placement of the docking stations (García-Palomares, Gutiérrez, and Latorre 2012; Martinez et al. 2012; Lin, Yang, and Chang 2013) or dynamically rebalancing the bikes (Ghosh and Varakantham 2017; Chiariotti et al. 2018; Legros 2019). The third group explored the environmental impacts of bike-sharing, such as reducing carbon footprints and air pollution. It was found that bike-sharing as a sustainable transportation mode can have potential environmental benefits by reducing energy use, carbon dioxide emissions, and harmful air pollutants (e.g., nitrogen oxide), under the premise that bike-sharing ridership was sourced from private automobile or public transit (Zhang and Mi 2018; Li et al. 2021; Chen et al. 2022). However, the operation of the system raised environmental concerns due to the carbon footprint generated by bike rebalancing (Luo 2019; Luo et al. 2020; D'Almeida, Rye, and Pomponi 2021). More severely, with the bike-share market diminishing in some cities, oversupplied bikes were abandoned, piled up, and dismantled, generating nondisposable "bike litter" and thus substantial environmental problems (Chen 2019).

One underexplored area in the literature is how bike-sharing intertwines with the public transit system to solve "the last mile problem." This relationship can be built on the theory of spatial interaction in transportation geography: transportation services can either compete with or complement each other depending on the spatial interaction between supply and demand, further driven by distance decay (Ullman and Boyce 1980). These notions, known as complementarity and competition, can be extended to the discussion of bike-sharing. For example, it was unclear if bike-sharing was diverting urban commuters from public transit or supplementing the unfulfilled travel demand. Although numerous studies attempted to demystify the synergy between bike-sharing and public transit, the findings were rather inconsistent. For example, a case study in Washington, D.C., revealed that a 10% increase in bike trips generated a 2.8% increase in public transit ridership (Ma, Liu, and Erdoğan 2015). Singleton and Clifton (2014) noted that bike-sharing and public transit had both competition and integration relationships. Another study in four U.S. cities showed that the bike-sharing system was largely integrated with the public transit system (Kong, Jin, and Sui 2020). However, a separate study across major U.S. cities highlighted that bike-sharing can negatively impact bus ridership (Graehler, Mucci, and Erhardt 2019). On a regional scale and relevant to this article, the relationship between bike-sharing and public transit in NYC has not been well elucidated. For example, Campbell and Brakewood (2017) found a significant decrease in bus ridership as a result of expanding the bike-sharing system in NYC. However, Ashraf et al. (2021) found a 2.3% increase in subway ridership when there was a 10% increase in bike-sharing trips in NYC.

While it is relatively challenging to rationalize these inconsistencies, one path forward is to explore the long-term spatiotemporal bike usage patterns. Existing studies employed either primary trip data for a limited period (Saberi et al. 2018; Kong, Jin, and Sui 2020) or secondary bike usage or summary data (Shaheen, Guzman, and Zhang 2010; Cheng et al. 2018). Also, the majority of these studies were focused on the statistical analysis of the trip data while overlooking the spatiotemporal dynamics. To fill the void, we solicit 1-year trip data from Citi Bike and then reveal the three different roles (i.e., competition, integration, and complementation) that dock-based bike-sharing plays in relation to public transit. These revelations will provide new insights into the roles of bike-sharing in urban transportation.

First, 1-year bike-sharing data containing about 17 million trips in 2018 within the five NYC boroughs were collected from Citi Bike's system data website (Citi Bike 2018). Each trip entry included the start time, end time, start docking station ID, end docking station ID, and trip duration. Citi Bike does not release the path information related to an individual trip. However, this data restriction can be overcome by analyzing the bike usage pattern at docking stations, which is a method employed by other dock-based bike-sharing studies (Rixey 2013; Faghih-Imani et al. 2014; Li et al. 2015; O'Mahony and Shmoys 2015; Faghih-Imani et al. 2017).

Then, we employed K-means clustering, a well-known unsupervised machine learning algorithm to further delve into the bike usage patterns at docking stations. Six docking station types were derived with unique bike usage patterns. The spatial distribution of the clustered stations is shown in Figure 25.1, while the temporal patterns are shown in Figure 25.2 with normalized temporal indices. For comparison, Figure 25.2 also includes the hourly vehicle traffic volume (NYC OpenData 2020). Table 25.1 summarizes the spatiotemporal patterns of bike usage for each station type.

Based on the clustered spatiotemporal patterns, the relationships between the dock-based system and the public transit system were further examined based on three criteria: bike trip duration, proximity to a public transit station, and bike trip length. The first criterion compares the duration of a bike trip to its alternative public transit trip. The second criterion identifies if the start or end of a bike trip is near a public transit station. Specifically, 100 meters is considered to be a competitive distance for mode-switching. The third criterion is based on the bike trip length, which can largely influence travelers' mode choices (Scheiner 2010). It was found that most bike-sharing trips are under 2 miles (Taylor and Mahmassani 1996; Hochmair 2015; Lee, Choi, and Leem 2016; Zuo, Wei, and Rohne 2018). Thus, 2 miles, which corresponds to 14.45 minutes based on Citi Bike users' average riding speed (i.e., 8.3 miles per hour) (Schneider 2016) is considered to be the threshold for the preference of bike usage. Based on these three criteria, three different relationships are proposed between bike-sharing and public transit, as shown below and in Figure 25.3.

- *Competition*: A bike trip is competing with public transit if the trip can be completed by its public transit alternative in a shorter time. The competition trip signifies that travelers may prefer bike-sharing even if it is more time costly and energy intensive than public transit.

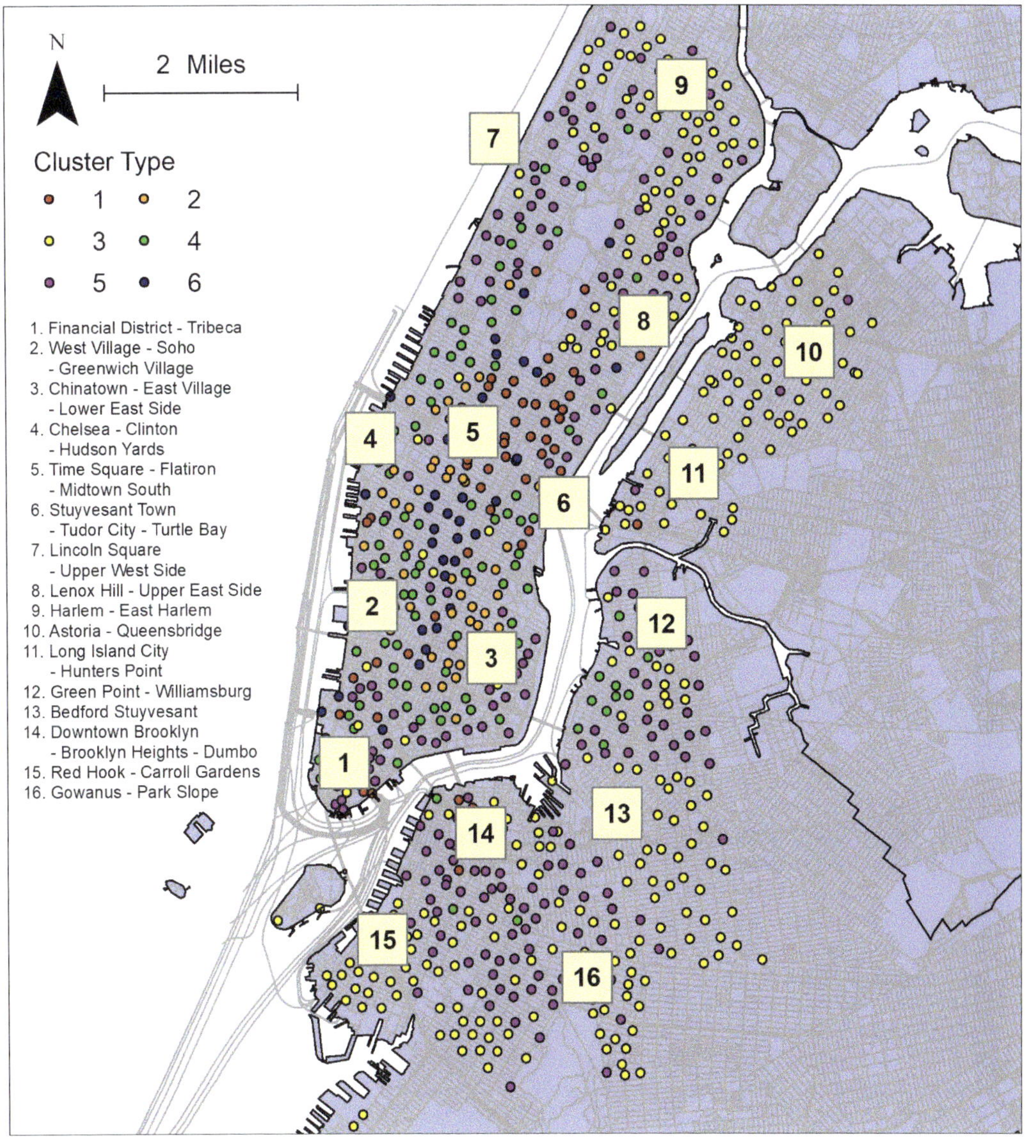

FIGURE 25.1 Spatial distribution of six types of docking stations.

- *Integration*: A bike trip is considered integrated with the public transit system if (1) the trip is less time costly than its public transit alternative, (2) either the origin or destination of the trip is within the 100-m buffer of a public transit station, and (3) the trip is under 2 miles. The integration trip is a short-distance trip that can facilitate mode-switching from and to the public transit system.
- *Complementation*: A bike trip is considered complementary to the public transit system if it is beyond the abovementioned scenarios. The complementation trip is less time consuming than its public transit alternative. However, such a trip is either too far away from a public transit station or a long-distance trip, making its trip purpose unlikely to be mode-switching.

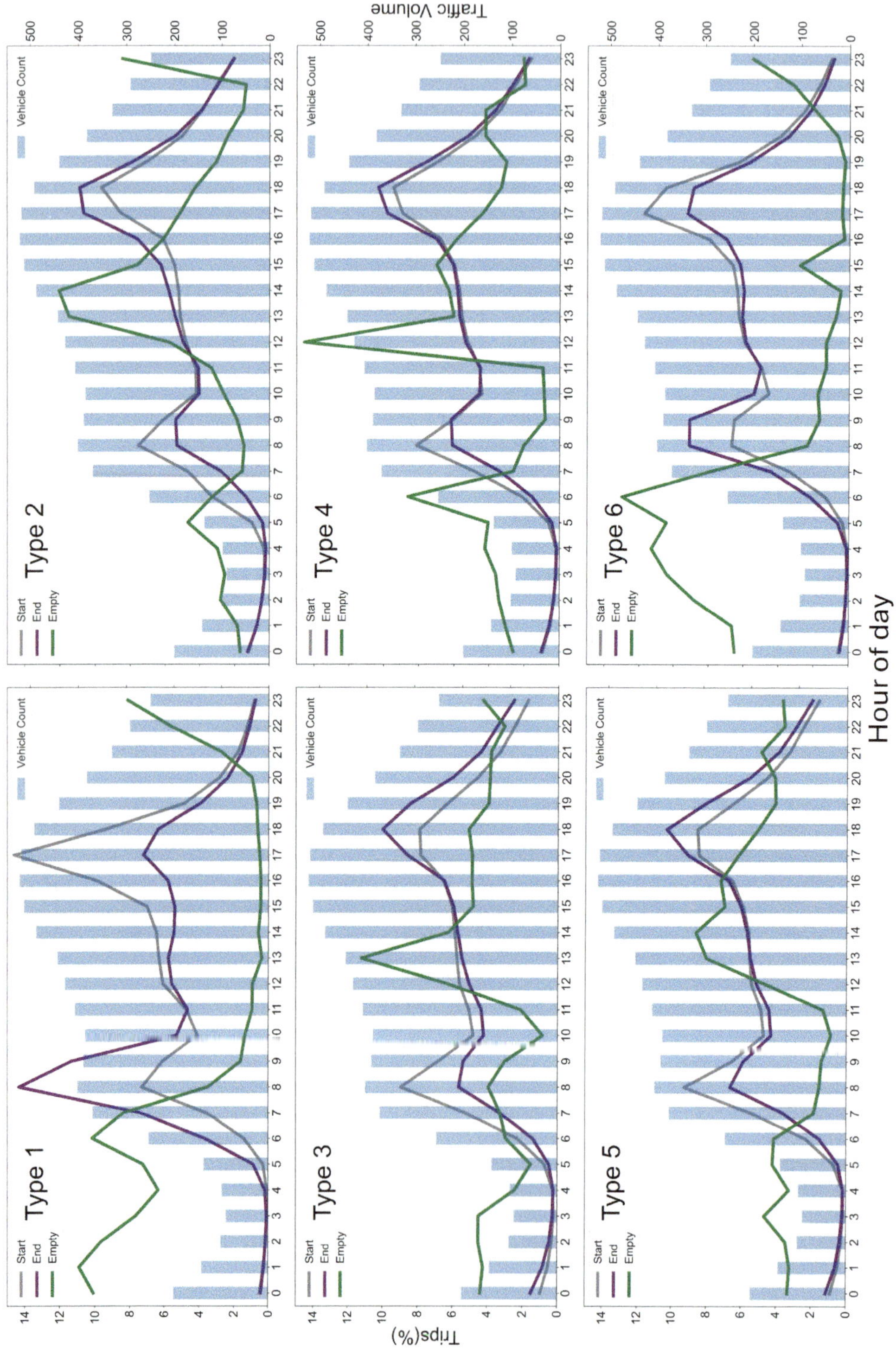

FIGURE 25.2 Daily bike usage pattern by station type (curves) and the hourly vehicle traffic volume (bars).

TABLE 25.1
Spatiotemporal Patterns of Bike Usage by Station Type

Type	Bike Usage Pattern	Major Neighborhoods*	Major Land Use	Station Count (%)	Cumulative Daily *Start***	Cumulative Daily *End***
1	Morning: *Start* << *End* Evening: *Start* >> *End* *Empty* spikes in the midnight and morning rush hours	5, 6	Commercial	56 (7.3%)	96	97
2	Morning: *Start* > *End* Evening: *Start* < *End* *Empty* spikes at noon and before the midnight	2, 3, 4	Mixed residential	53 (6.9%)	158	158
3	Morning: *Start* > *End* Evening: *Start* < *End* *Empty* spikes at noon	8, 9, 10, 11, 13, 15, 16	Residential (low to medium density)	310 (40.4%)	17	17
4	Morning: *Start* > *End* Evening: *Start* < *End* *Empty* spikes in the morning rush hours and at noon	1, 2, 3, 4, 7	Commercial (near transit)	95 (12.4%)	101	102
5	Morning: *Start* > *End* Evening: *Start* < *End* *Empty* is slightly high at noon and relatively stable throughout the day	1, 3, 7, 8, 12, 14, 16	Residential (high density); mixed residential and commercial	216 (28.1%)	48	47
6	Morning: *Start* < *End* Evening: *Start* > *End* *Empty* spikes in the midnight and morning rush hours	2, 5	Commercial	38 (4.9%)	195	197

* Refer to Figure 25.1 for the neighborhood.
** The index is the cumulative count over 24 hours, averaged among the stations of the same type.

To derive the alternative public transit time for each bike-sharing trip, the Google Maps' Directions Application Programming Interface (Directions API) was used. Then, the type of each bike-sharing trip was classified by building spatial queries. Finally, the three types of bike trips were further aggregated at the station level based on their trip origins, as shown in Table 25.2.

Table 25.2 shows that the dominant bike trips (61.47%) are competition trips, followed by complementation (29.03%) and integration (9.50%). The trip composition does not vary substantially by station type. Since competition trips are more time costly than their alternative trips that could be otherwise completed by public transit, the result suggests that bike-sharing can largely compete with public transit and attract potential public transit users. One piece of evidence refers to the trips starting at type 3 stations, where the degree of competition slightly intensifies (63.28%).

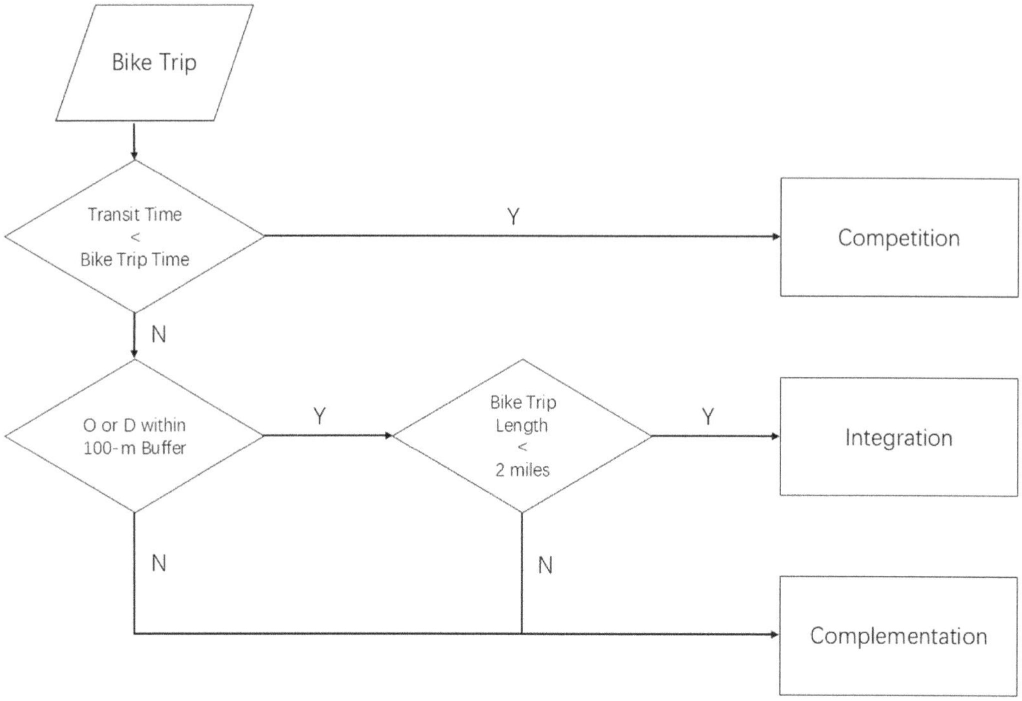

FIGURE 25.3 Flowchart of trip classification. *Abbreviations:* D, destination; O, origin.

Type 3 stations are primarily distributed in low- to medium-density residential areas in Queens and Brooklyn, and these areas have adequate coverage of bus stations but limited coverage of metro stations. Since the bus schedule is considerably less frequent, travelers in these areas may have a stronger preference for bike-sharing due to its more flexible nature, eventually making bike-sharing more competitive.

The study reveals the unique mobility roles of bike-sharing in NYC. While past studies argued that bike-sharing aims to solve "the last mile problem" (Liu, Jia, and Cheng 2012; Shaheen and Chan 2016; Moorthy et al. 2017), this is not the case in NYC. According to the results, the integration trips, which are short-distance trips starting or ending near a public transit station, are considerably

TABLE 25.2
Trip Composition by Station Type

	Competition	Integration	Complementation
Station Type	**Trip Composition for the Same Station Type (%)**		
1	63.06	9.92	27.02
2	60.60	8.03	31.37
3	63.28	10.93	25.79
4	60.66	9.75	29.59
5	61.73	10.15	28.11
6	60.69	8.59	30.72
All types	61.47	9.50	29.03

limited (9.5%). The result signifies that most users do not adopt bike-sharing for connecting to public transit or for the convenience of mode-switching. This result can be further explained by Citi Bike's site planning strategy. Specifically, Citi Bike employed a community-based participatory model in the initial planning phase—New Yorkers were asked to propose candidate docking station sites by placing pins on a map, and the sites receiving high preference were further refined by professionals to ensure an equitable spatial distribution (e.g., one within each 1,000-square-feet grid) (New York City Department of Transportation 2013). This participatory model, which emphasized community members' travel demands instead of proximity to public transit, is a possible cause of the low integration.

Revealing the mixed relationships between bike-sharing and public transit is important, as it helps to demystify the trip purpose of bike-sharing, which is not necessarily to minimize travel time. In addition to flexibility, bike-sharing has a financial competitive edge, especially for frequent riders. For example, a monthly pass for unlimited Citi Bike use is only $15 on average, which is less than one-eighth the cost of a monthly metro pass in NYC (Citi Bike 2022; Metropolitan Transportation Authority 2022). Also, various community programs have been implemented to further reduce bike fares for socioeconomically disadvantaged commuters, such as those living in public housing units or receiving Supplemental Nutrition Assistance Program (SNAP) benefits (Citi Bike 2021). Therefore, the existence of competition could mean a certain degree of transportation equity in the city (Lee, Sener, and Jones 2017), as bike-sharing provides a low-cost commuting option for those experiencing financial hardships.

25.4 CONCLUSION

The bike-sharing system plays a critical role in the urban system by enhancing intracity travel connectivity and accessibility. This chapter enhances the understanding of bike-sharing system usage and the related research, analyzes the spatiotemporal patterns of Citi Bike, and proposes three different relationships between bike-sharing and public transit: competition, integration, and complementation. The result reveals a primary competitive role that bike-sharing plays in relation to public transit. The existence of such competition can benefit socioeconomically disadvantaged commuters, eventually facilitating a certain degree of transportation equity.

NOTE

1 This section is based on Cui, Y., Chen, X., Chen, X., & Zhang, C. (2023). Competition, integration, or complementation? exploring dock-based bike-sharing in New York City. The Professional Geographer, 75(1), 65–75.

REFERENCES

Antonakos, Cathy L. 1993. *Environmental and Travel Preferences of Cyclists*: University of Michigan.

Ashraf, Md Tanvir, Md Amdad Hossen, Kakan Dey, Sarah El-Dabaja, Moathe Aljeri, and Bhaven Naik. 2021. "Impacts of bike sharing program on subway ridership in New York City." *Transportation Research Record* 2675 (9):924–934.

Buck, Darren, Ralph Buehler, Patricia Happ, Bradley Rawls, Payton Chung, and Natalie Borecki. 2013. "Are bikeshare users different from regular cyclists? A first look at short-term users, annual members, and area cyclists in the Washington, DC, region." *Transportation Research Record* 2387 (1):112–119.

Campbell, Kayleigh B, and Candace Brakewood. 2017. "Sharing riders: How bikesharing impacts bus ridership in New York City." *Transportation Research Part A: Policy and Practice* 100:264–282.

Chen, Ruoying. 2019. ""Bike litter" and obligations of the platform operators: Lessons from China's dockless sharing bikes." *Computer Law & Security Review* 35 (5):105317.

Chen, Yan, Yongping Zhang, D'Maris Coffman, and Zhifu Mi. 2022. "An environmental benefit analysis of bike sharing in New York City." *Cities* 121:103475.

Cheng, Peng, Ji Hu, Zidong Yang, Yuanchao Shu, and Jiming Chen. 2018. "Utilization-aware trip advisor in bike-sharing systems based on user behavior analysis." *IEEE Transactions on Knowledge and Data Engineering* 31 (9):1822–1835.

Chiariotti, Federico, Chiara Pielli, Andrea Zanella, and Michele Zorzi. 2018. "A dynamic approach to rebalancing bike-sharing systems." *Sensors* 18 (2):512.

Citi Bike. 2018. "City Bike system data." https://www.citibikenyc.com/system-data.

Citi Bike. 2021. "Community programs." https://citibikenyc.com/community-programs/programs.

Citi Bike. 2022. "Pricing." https://citibikenyc.com/pricing.

City of New York. 2023. "Mayor Adams, DOT Commissioner Rodriguez, Lyft Announce Expansion, Improvements to CITI Bike System as Ridership Reaches Record Highs."

Clark, Jill K, Ronald McChesney, Darla K Munroe, and Elena G Irwin. 2009. "Spatial characteristics of exurban settlement pattern in the United States." *Landscape and Urban Planning* 90 (3-4):178–188.

D'Almeida, Léa, Tom Rye, and Francesco Pomponi. 2021. "Emissions assessment of bike sharing schemes: The case of Just Eat Cycles in Edinburgh, UK." *Sustainable Cities and Society* 71:103012.

DeMaio, Paul. 2009. "Bike-sharing: History, impacts, models of provision, and future." *Journal of Public Transportation* 12 (4):3.

Dill, Jennifer. 2009. "Bicycling for transportation and health: The role of infrastructure." *Journal of Public Health Policy* 30:S95–S110.

Dill, Jennifer, and Theresa Carr. 2003. "Bicycle commuting and facilities in major US cities: If you build them, commuters will use them." *Transportation Research Record* 1828 (1):116–123.

El-Assi, Wafic, Mohamed Salah Mahmoud, and Khandker Nurul Habib. 2017. "Effects of built environment and weather on bike sharing demand: A station level analysis of commercial bike sharing in Toronto." *Transportation* 44:589–613.

Ewing, Reid, and Robert Cervero. 2001. "Travel and the built environment: A synthesis." *Transportation Research Record* 1780 (1):87–114.

Faghih-Imani, Ahmadreza, and Naveen Eluru. 2015. Analyzing destination choice preferences in bicycle sharing systems: An investigation of Chicago's Divvy System. *Transportation Research Board 94th Annual Meeting* No. 15-2959.

Faghih-Imani, Ahmadreza, Naveen Eluru, Ahmed M El-Geneidy, Michael Rabbat, and Usama Haq. 2014. "How land-use and urban form impact bicycle flows: Evidence from the bicycle-sharing system (BIXI) in Montreal." *Journal of Transport Geography* 41:306–314.

Faghih-Imani, Ahmadreza, Robert Hampshire, Lavanya Marla, and Naveen Eluru. 2017. "An empirical analysis of bike sharing usage and rebalancing: Evidence from Barcelona and Seville." *Transportation Research Part A: Policy and Practice* 97:177–191.

Fairfield, John D. 1995. "Rapid transit: Automobility and settlement in urban America." *Reviews in American History* 23 (1):80–85.

Fishman, Elliot. 2016. "Bikeshare: A review of recent literature." *Transport Reviews* 36 (1):92–113.

Fishman, Elliot, Simon Washington, Narelle Haworth, and Armando Mazzei. 2014. "Barriers to bikesharing: An analysis from Melbourne and Brisbane." *Journal of Transport Geography* 41:325–337.

Fuller, Daniel, Lise Gauvin, Yan Kestens, Mark Daniel, Michel Fournier, Patrick Morency, and Louis Drouin. 2011. "Use of a new public bicycle share program in Montreal, Canada." *American Journal of Preventive Medicine* 41 (1):80–83.

García-Palomares, Juan Carlos, Javier Gutiérrez, and Marta Latorre. 2012. "Optimizing the location of stations in bike-sharing programs: A GIS approach." *Applied Geography* 35 (1–2):235–246.

Ghosh, Supriyo, and Pradeep Varakantham. 2017. "Incentivizing the use of bike trailers for dynamic repositioning in bike sharing systems." Proceedings of the International Conference on Automated Planning and Scheduling.

Golub, Aaron, M. Hoffmann, Adonia Lugo, and G. Sandoval. 2016. *Bicycle Justice and Urban Transformation*: Routledge.

Graehler, Michael, Richard Alexander Mucci, and Gregory D Erhardt. 2019. "Understanding the recent transit ridership decline in major US cities: Service cuts or emerging modes." 98th Annual Meeting of the Transportation Research Board, Washington, DC.

Gu, Tianqi, Inhi Kim, and Graham Currie. 2019. "To be or not to be dockless: Empirical analysis of dockless bikeshare development in China." *Transportation Research Part A: Policy and Practice* 119:122–147.

Guo, Yanyong, Jibiao Zhou, Yao Wu, and Zhibin Li. 2017. "Identifying the factors affecting bike-sharing usage and degree of satisfaction in Ningbo, China." *PloS one* 12 (9):e0185100.

Habib, Khandker Nurul, Jenessa Mann, Mohamed Mahmoud, and Adam Weiss. 2014. "Synopsis of bicycle demand in the City of Toronto: Investigating the effects of perception, consciousness and comfortability on the purpose of biking and bike ownership." *Transportation Research Part A: Policy and Practice* 70:67–80.

Hochmair, Hartwig H. 2015. "Assessment of bicycle service areas around transit stations." *International Journal of Sustainable Transportation* 9 (1):15–29.

Kaufman, Sarah M, Lily Gordon-Koven, Nolan Levenson, and Mitchell L Moss. 2015. "Citi Bike: The first two years." *The Rudin Center for Transportation Policy and Management.*

Kim, DongJun, H Shin, Hyeongjun Im, and Junsik Park. 2012. "Factors influencing travel behaviors in bike-sharing." *Transportation Research Board 91st Annual Meeting.*

Kim, Kyoungok. 2018. "Investigation on the effects of weather and calendar events on bike-sharing according to the trip patterns of bike rentals of stations." *Journal of Transport Geography* 66:309–320.

Kong, Hui, Scarlett T Jin, and Daniel Z Sui. 2020. "Deciphering the relationship between bikesharing and public transit: Modal substitution, integration, and complementation." *Transportation Research Part D: Transport and Environment* 85:102392.

Lee, Jaeyeong, Keechoo Choi, and Yountaik Leem. 2016. "Bicycle-based transit-oriented development as an alternative to overcome the criticisms of the conventional transit-oriented development." *International Journal of Sustainable Transportation* 10 (10):975–984.

Lee, Richard J, Ipek N Sener, and S Nathan Jones. 2017. "Understanding the role of equity in active transportation planning in the United States." *Transport Reviews* 37 (2):211–226.

Legros, Benjamin. 2019. "Dynamic repositioning strategy in a bike-sharing system; how to prioritize and how to rebalance a bike station." *European Journal of Operational Research* 272 (2):740–753.

Li, Aoyong, Kun Gao, Pengxiang Zhao, Xiaobo Qu, and Kay W Axhausen. 2021. "High-resolution assessment of environmental benefits of dockless bike-sharing systems based on transaction data." *Journal of Cleaner Production* 296:126423.

Li, Weibo, and Maria Kamargianni. 2018. "Providing quantified evidence to policy makers for promoting bike-sharing in heavily air-polluted cities: A mode choice model and policy simulation for Taiyuan-China." *Transportation Research Part A: Policy and Practice* 111:277–291.

Li, Yexin, Yu Zheng, Huichu Zhang, and Lei Chen. 2015. "Traffic prediction in a bike-sharing system." Proceedings of the 23rd SIGSPATIAL International Conference on Advances in Geographic Information Systems.

Lin, Jenn-Rong, Ta-Hui Yang, and Yu-Chung Chang. 2013. "A hub location inventory model for bicycle sharing system design: Formulation and solution." *Computers & Industrial Engineering* 65 (1):77–86.

Liu, Zhili, Xudong Jia, and Wen Cheng. 2012. "Solving the last mile problem: Ensure the success of public bicycle system in Beijing." *Procedia-Social and Behavioral Sciences* 43:73–78.

Lu, Wei, Darren M Scott, and Ron Dalumpines. 2018. "Understanding bike share cyclist route choice using GPS data: Comparing dominant routes and shortest paths." *Journal of Transport Geography* 71:172–181.

Luo, Hao. 2019. "*Environmental Impact Assessment and Improved Design of Bike Sharing System from the Life Cycle Perspective*." Master's Thesis, Purdue University Graduate School, West Lafayette, IN, USA.

Luo, Hao, Fu Zhao, Wei-Qiang Chen, and Hua Cai. 2020. "Optimizing bike sharing systems from the life cycle greenhouse gas emissions perspective." *Transportation Research Part C: Emerging Technologies* 117:102705.

Ma, Ting, Chao Liu, and Sevgi Erdoğan. 2015. "Bicycle sharing and public transit: Does capital bikeshare affect metrorail ridership in Washington, DC?" *Transportation Research Record* 2534 (1):1–9.

Martinez, Luis M, Luís Caetano, Tomás Eiró, and Francisco Cruz. 2012. "An optimisation algorithm to establish the location of stations of a mixed fleet biking system: An application to the city of Lisbon." *Procedia-Social and Behavioral Sciences* 54:513–524.

Martinez, Mark. 2017. "The impact weather has on NYC Citi Bike share company activity." *Journal of Environmental and Resource Economics at Colby* 4 (1):12.

Metropolitan Transportation Authority. 2022. "Fares." https://www.mta.info/fares-tolls.

Miranda-Moreno, Luis F, and Thomas Nosal. 2011. "Weather or not to cycle: Temporal trends and impact of weather on cycling in an urban environment." *Transportation Research Record* 2247 (1):42–52.

Moorthy, Aditi, Robert De Kleine, Gregory Keoleian, Jeremy Good, and Geoff Lewis. 2017. "Shared autonomous vehicles as a sustainable solution to the last mile problem: A case study of Ann Arbor-Detroit area." *SAE International Journal of Passenger Cars-Electronic and Electrical Systems* 10 (2017-01-1276):328–336.

Nankervis, Max. 1999. "The effect of weather and climate on bicycle commuting." *Transportation Research Part A: Policy and Practice* 33 (6):417–431.

New York City Department of Transportation. 2013. NYC Bike Share: Designed by New Yorkers.: NYCDOT.

New York City Department of Transportation. 2023. "Citi Bike Expansion Updated Plan." https://nycdotbikeshare.info/sites/default/files/2023-06/3O_Updated%20Plan%20Presentaiton%20_20230613.pdf.

New York City Department of Transportation. 2024. "Citi Bike Phase 3 Overview." accessed January, 13. https://nycdotprojects.info/project/931/overview.

NYC OpenData. 2020. "Traffic Volume Counts (2014-2019)." https://data.cityofnewyork.us/Transportation/Traffic-Volume-Counts-2014-2019-/ertz-hr4r.

O'Mahony, Eoin, and David Shmoys. 2015. "Data analysis and optimization for (citi) bike sharing." Proceedings of the AAAI Conference on Artificial Intelligence.

Pager, Tyler. 2019. Citi Bike Pulls New Electric Bikes Off Streets, Citing Safety Concerns. *The New York Times*.

Pucher, John, Jennifer Dill, and Susan Handy. 2010. "Infrastructure, programs, and policies to increase bicycling: An international review." *Preventive Medicine* 50:S106–S125.

Rixey, R Alexander. 2013. "Station-level forecasting of bikesharing ridership: Station network effects in three US systems." *Transportation Research Record* 2387 (1):46–55.

Saberi, Meead, Mehrnaz Ghamami, Yi Gu, Mohammad Hossein Sam Shojaei, and Elliot Fishman. 2018. "Understanding the impacts of a public transit disruption on bicycle sharing mobility patterns: A case of Tube strike in London." *Journal of Transport Geography* 66:154–166.

Scheiner, Joachim. 2010. "Interrelations between travel mode choice and trip distance: Trends in Germany 1976–2002." *Journal of Transport Geography* 18 (1):75–84.

Schneider, Todd W. 2016. "A Tale of Twenty-Two Million Citi Bike Rides: Analyzing the NYC Bike Share System." https://toddwschneider.com/posts/a-tale-of-twenty-two-million-citi-bikes-analyzing-the-nyc-bike-share-system/.

Schoner, Jessica E, and David M Levinson. 2014. "The missing link: Bicycle infrastructure networks and ridership in 74 US cities." *Transportation* 41 (6):1187–1204.

Scott, Darren M, and Celenna Ciuro. 2019. "What factors influence bike share ridership? An investigation of Hamilton, Ontario's bike share hubs." *Travel Behaviour and Society* 16:50–58.

Sears, Justine, Brian S Flynn, Lisa Aultman-Hall, and Greg S Dana. 2012. "To bike or not to bike: Seasonal factors for bicycle commuting." *Transportation Research Record* 2314 (1):105–111.

Shaheen, Susan, and Nelson Chan. 2016. "Mobility and the sharing economy: Potential to facilitate the first- and last-mile public transit connections." *Built Environment* 42 (4):573–588.

Shaheen, Susan A, Stacey Guzman, and Hua Zhang. 2010. "Bikesharing in Europe, the Americas, and Asia: Past, present, and future." *Transportation Research Record* 2143 (1):159–167.

Shaheen, Susan A, Hua Zhang, Elliot Martin, and Stacey Guzman. 2011. "China's Hangzhou public bicycle: Understanding early adoption and behavioral response to bikesharing." *Transportation Research Record* 2247 (1):33–41.

Shen, Yu, Xiaohu Zhang, and Jinhua Zhao. 2018. "Understanding the usage of dockless bike sharing in Singapore." *International Journal of Sustainable Transportation* 12 (9):686–700.

Shi, Jian-gang, Hongyun Si, Guangdong Wu, Yangyue Su, and Jing Lan. 2018. "Critical factors to achieve dockless bike-sharing sustainability in China: A stakeholder-oriented network perspective." *Sustainability* 10 (6):2090.

Singleton, Patrick A, and Kelly J Clifton. 2014. "Exploring synergy in bicycle and transit use: Empirical evidence at two scales." *Transportation Research Record* 2417 (1):92–102.

Slowik, Peter, and Fanta Kamakaté. 2017. "New mobility: Today's technology and policy landscape." *White Paper. International Council on Clean Transportation, Washington, DC.*

Taylor, Brian D, Douglas Miller, Hiroyuki Iseki, and Camille Fink. 2009. "Nature and/or nurture? Analyzing the determinants of transit ridership across US urbanized areas." *Transportation Research Part A: Policy and Practice* 43 (1):60–77.

Taylor, Dean, and Hani Mahmassani. 1996. "Analysis of stated preferences for intermodal bicycle-transit interfaces." *Transportation Research Record* 1556 (1):86–95.

Tilahun, Nebiyou Y, David M Levinson, and Kevin J Krizek. 2007. "Trails, lanes, or traffic: Valuing bicycle facilities with an adaptive stated preference survey." *Transportation Research Part A: Policy and Practice* 41 (4):287–301.

Tu, Yuanjie, Peng Chen, Xu Gao, Jiawen Yang, and Xiaohong Chen. 2019. "How to make dockless bikeshare good for cities: Curbing oversupplied bikes." *Transportation Research Record* 2673 (6):618–627.

Ullman, Edward Louis, and Ronald R Boyce. 1980. *Geography as Spatial Interaction*: University of Washington Press.

Wang, Mingshu, and Xiaolu Zhou. 2017. "Bike-sharing systems and congestion: Evidence from US cities." *Journal of Transport Geography* 65:147–154.

Yobbi, Dean. 2019. Lyft e-bikes out of service again after 2 bikes catch fire. *Bicycle Retailer and Industy News.*

Zhang, Yongping, and Zhifu Mi. 2018. "Environmental benefits of bike sharing: A big data-based analysis." *Applied Energy* 220:296–301.

Zhang, Yuanyuan, and Yuming Zhang. 2018. "Associations between public transit usage and bikesharing behaviors in the United States." *Sustainability* 10 (6):1868.

Zuo, Ting, Heng Wei, and Andrew Rohne. 2018. "Determining transit service coverage by non-motorized accessibility to transit: Case study of applying GPS data in Cincinnati metropolitan area." *Journal of Transport Geography* 67:1–11.

26 Ridesourcing

A Game-Changer for Urban Mobility

Sicheng Wang

26.1 INTRODUCTION

The advancement of information and communication technologies (ICT) has enabled and driven the rise of the sharing economy (or gig economy), where consumption occurs through the sharing or renting of goods via consumer-to-consumer transactions on Internet-based platforms (Puschmann and Alt 2016). Various new business models in urban transportation, known as shared mobilities, have emerged as part of the sharing economy (Shaheen, Cohen, and Zohdy 2016). These include docked and dockless bike-sharing, e-scooter sharing, car-sharing, truck-sharing, food delivery, and ridesourcing, which provide mobility services to transport passengers and goods across different transportation modes (Machado et al. 2018). As one of the most influential shared mobilities, ridesourcing has revolutionized urban transportation systems worldwide (S. T. Jin et al. 2018).

Ridesourcing is also known as ride-hailing, ride-sharing, e-hailing, transportation network company (TNC) service, or mobility-as-a-service (MaaS) in different contexts. It refers to on-demand mobility services where private vehicle drivers are matched with passengers through mobile applications (e.g., Uber, Lyft, and Didi) based on their dynamic locations and times. Unlike traditional taxis or chauffeur services, which use commercial vehicles and are subject to stringent entry requirements and strict industry regulations, ridesourcing operates with more flexibility (Wang and Yang 2019). Empowered by innovative features such as real-time driver-passenger matching, advanced route-finding algorithms, cashless payments, and dynamic pricing, ridesourcing has swiftly filled a critical gap in meeting people's mobility needs and has experienced rapid global growth since its inception. As of May 2024, Uber, the oldest and largest TNC, has operated in more than 10,000 cities across 70 countries, with 30,400 employees, 7.1 million drivers, and 149 million users served. (Dean 2024). Lyft, the second-largest TNC in the United States, reported 1.4 million drivers serving 23 million users in the United States and Canada in 2023 (Campbell 2023). Didi, the largest TNC in China, reported 15 million annual active drivers and 600 million users globally in 2021 (Njanja and Kene-Okafor 2021). Bolt, an Estonia-based TNC that primarily operates in European countries, served about 75 million users in 2021 (Smith 2024). Other international ridesourcing providers, including Ola (from India), Grab (from Singapore), BlaBlaCar (from France), Gett (from the United Kingdom), and Cabify (from Spain), have divided up the remaining major ridesourcing markets globally.

Ridesourcing transforms urban mobility in various dimensions. First, it has altered many people's transportation mode choices, leading urban travelers to switch from private vehicles, public transportation, and traditional taxis to ridesourcing services (Rayle et al. 2016; Wang, Chen, and Chen 2019; Tarabay and Abou-Zeid 2020). Second, its scalability and availability enable it to expand the services to areas that lack access to private vehicles, public transportation, and taxis, reshaping people's accessibility to essential destinations (e.g., jobs, grocery stores, healthcare services, restaurants) (Wang 2021; Brown 2018; Hughes and MacKenzie 2016) Third, it has significantly contributed to increasingly aggravated urban traffic (Wang, Du, and Lee 2024; Li et al. 2022). Finally,

DOI: 10.1201/9781003503262-30

ridesourcing can be considered a precursor to automated taxis (e.g., robotaxis), which may play an essential role in the future of urban transportation systems and smart cities (Cervero 2017).

Considering its profound impact on urban mobility, ridesourcing has garnered significant interest from policymakers, planners, and researchers across various domains, including but not limited to traffic modeling, travel demand forecasting and management, pricing and surcharges, social equity in usage and accessibility of ridesourcing, sustainable transportation solutions, smart and connected cities, competition with public transportation, workforce impacts, and data security.

In this chapter, we discuss seven topics related to ridesourcing, including history of ridesourcing, ridesourcing data and analytics, impacts on the existing urban mobility systems, policy and regulatory challenges, and directions of ridesourcing in smart cities.

26.2 THE RISE AND EVOLUTION OF RIDESOURCING

The history of ridesourcing began at the end of the 2010s. Although car-sharing was not a new concept, the rise of ridesourcing was enabled by advancements in Internet technology and smartphones. The first ridesourcing company, Uber, was founded in San Francisco in 2009, with its beta launch following in May 2010 (Blystone 2024). Uber introduced a new business model by offering a smartphone app that served as a platform for both drivers and passengers, allowing people to order a ride, match with a driver, monitor the trip, make payments, and rate the service. It rapidly expanded to other major U.S. cities, such as New York, Seattle, Boston, Chicago, and Washington, D.C. By the end of 2018, Uber had entered more than 280 metropolitan areas in the United States, covering almost all urban areas (Wang and Smart 2020). Its major competitor, Lyft, launched in 2012 and expanded to 656 cities across the United States and Canada as of August 2023 (Greiner et al. 2019; Dean 2023). The Uber platform reached about 150 million customers and 7.1 million drivers, with $68.9 billion in total for ridesourcing bookings in 2023 (Dean 2024). Didi was also founded in 2012 in China and has become one of the world's largest ridesourcing companies, now serving about 500 million customers across Asia-Pacific, Africa, South America, Central Asia, and Russia (Ciaccia 2022).

The emergence and growth of ridesourcing disrupted the traditional on-demand passenger mobility market, leading to job displacement and income decline for taxi drivers and limousine chauffeurs (Wang and Smart 2020). Workers in these traditional transportation sectors have gone on strike and protested against ridesourcing globally (McKane and Hess 2023). Nevertheless, a trend of merging between the two industries is emerging. Many traditional taxi services have adopted Uber-like apps, allowing users to order rides via their phones. For example, people can use the app Curb to request and pay for taxis in New York. Beginning in September 2022, users have also been able to order taxis through Uber in New York (Chapman 2022). In China, there are multiple apps (e.g., Didi) available to hail taxis.

Ridesourcing companies gradually developed different types of services to target various populations and demands. Uber launched UberPool (later renamed to UberX Share), and Lyft launched Lyft Line in 2014. Both services offer passengers the option to share a vehicle with others during their rides. These services, known as ride-pooling or ride-splitting, match passengers based on the proximity of their origins and destinations and the similarity of their routes, offering lower fares to customers. Ride-pooling trips, if successfully matched, typically have detours and longer wait and travel times (Wang and Noland 2021b). Ride-pooling services are only available in a few large cities (e.g., New York, Los Angeles, Chicago, San Francisco, Phoenix, San Diego, Portland, Indianapolis, and Pittsburgh). Uber and Lyft also provide wheelchair-accessible vehicles (WAVs) for people in need, but WAVs are undersupplied compared to regular ridesourcing vehicles (Hassanpour, Bigazzi, and MacKenzie 2021).

The COVID-19 pandemic, which began in early 2020, significantly disrupted the ridesourcing market worldwide. Overall mobility and travel demand declined due to lockdowns, quarantine policies, and the prevalence of remote work (Huang et al. 2022). People shifted from public transit and ridesourcing to private car driving and active transportation (e.g., walking and bicycling) due to concerns about disease transmission (Loa et al. 2022). Ridesourcing companies paused ride-pooling services due to the social distancing requirements. Many drivers also left the industry (Fielbaum et al. 2023). The impact of the pandemic on ridesourcing usage persisted even after the concern about the disease diminished. In most urban areas, demand for ridesourcing services has not returned to pre-COVID levels quickly (Jin et al. 2023; Wang, Huang, and Shen 2024). The Singapore-based international ridesourcing company Grab reported that its gross merchandise value (GMV) recovered to 85% of pre-COVID levels as of August 2023 (Chiang 2023).

Ridesourcing drivers have reported a sharp decline in earnings despite the increase in demand in recent years (Zinkula and Sheidlower 2024). There are several potential reasons for this. First, companies are taking a greater share of the customers' fares from drivers (Zinkula and Sheidlower 2024). Second, in some regions such as China, the number of ridesourcing drivers is severely over-saturated, leading to increasingly fierce competition (Sina News 2024). Finally, the expenses for gas and maintenance have increased dramatically since the pandemic (Zinkula and Sheidlower 2024). Recently, ridesourcing drivers have started to protest and resist the threat of autonomous taxis (also known as robotaxis) to their job opportunities and incomes, much like taxi drivers once protested against ridesourcing services several years ago (Mao 2024).

26.3 RIDESOURCING DATA AND ANALYTICS

Ridesourcing heavily relies on positioning, navigation, and timing (PNT), high-speed Internet, and mobile applications, all of which generate massive spatiotemporal data such as geotagged trip trajectories. In addition to real-time timestamps and geotagged locations recorded during each trip, ridesourcing trips produce transaction data, including pricing and payment information (e.g., fares, tips, surcharges), types of services, the number of passengers, and customer reviews. Finally, the platforms retrieve personal information on both drivers and customers. This information includes name, gender, date of birth, home address, phone number, email address, credit card information, and bank account details, all of which are highly sensitive. Didi also collects audio recordings during rides for safety purposes.

Most ridesourcing service data are owned by private ridesourcing companies, and there is a lack of effective regulation governing data collection and usage within these companies. As a result, there is widespread concern about data security and privacy protection related to ridesourcing data. In July 2022, the Cyberspace Administration of China imposed administrative penalties on Didi, fining the company more than 8 billion yuan for a series of violations concerning data security and privacy protection. These violations included the illegal and excessive collection of users' mobile photos, facial recognition data, age information, occupational data, family relationships, and educational background.

Furthermore, there is a lack of transparency regarding how these companies use the data. Data analytics teams within these companies utilize the data to develop and improve their predictive supply management systems and matching algorithms, which predict and determine demand, dispatches, traffic, routes, pickup and drop-off times, and prices. However, these algorithms are black boxes for both drivers and passengers.

Data ownership also creates barriers to integrating ridesourcing into regional and local public transportation systems. Private companies like Uber and Lyft typically treat their operational data as proprietary, limiting the ability of public agencies to access the operational data needed for multimodal transportation coordination (McCoy et al. 2018). Some regions and cities have enacted laws requiring ridesourcing companies to share desensitized data for analytical and planning purposes. For example, the city of Chicago has released Transportation Network Providers (TNP) trip data since November 2018. The New York Taxi and Limousine Commission (TLC) has published trip records for app-based ridesourcing vehicles since 2016, first included into the category of for-hire vehicles (FHVs) and later defined as high-volume for-hire vehicles (HVFHVs) if a company completes 10,000 or more trips since 2019. These ongoing data collection programs have aggregated information on locations and times of trips for privacy protection purposes. For example, Chicago's TNP data aggregate the origins and destinations of trips to the centroids of census tracts and community areas and rounds the timestamps of pickups and drop-offs to the nearest 15 minutes. The New York data, on the other hand, aggregate the pickup and drop-off locations to the IDs of taxi zones, which are a type of traffic analysis zone designated based on the New York City Department of City Planning's neighborhood tabulation areas (NTAs).

The HVFHV trip records in New York and the TNP trip records in Chicago are the most well-known ridesourcing datasets covering all ridesourcing companies in these two major cities, and they are constantly updated. Consequently, most empirical studies on ridesourcing are based in these cities. However, some researchers have obtained operational data from specific companies through restricted collaborations or agreements.

Additionally, some companies have also made limited desensitized data publicly available, primarily for analytical use. RideAustin, a local nonprofit provider in Austin, Texas, released all its trip record data from its founding in June 2016 until its closure in June 2020. Didi initiated the GAIA Open Data Program and published 1 month's trip records from Chengdu, China, in 2019, which are now unavailable. In 2017, Uber developed an application programming interface (API) integrated with a mapping dashboard known as Uber Movement, which provided historical data on average travel times for rides between any origin and destination within selected cities across the world; this service is now unavailable. For example, the most recent two U.S. National Household Travel Survey (NHTS) datasets (2017 and 2022) reported data about the use of ridesourcing services (defined as rideshare in the datasets). However, these travel survey data usually lack in-depth and insightful information on ridesourcing, which limits the utility and value of the data. In conclusion, the research community increasingly faces difficulties in accessing ridesourcing data, which are more scarce than public transportation usage and ridership data.

Geotagged trip records have been widely used in analyzing the spatiotemporal patterns of ridesourcing use, the factors associated with the behavioral patterns, and the social, economic, and environmental outcomes of ridesourcing. Relevant studies include trip generation and travel demand (Yan, Liu, and Zhao 2020; Wang and Noland 2021a; Yu and Peng 2020; Marquet 2020; Wang and Noland 2021b; Wang, Huang, and Shen 2024), wait times (Hughes and MacKenzie 2016; Yang, Liang, and Yang 2021), congestion ride-pooling (Wang and Noland 2021b; Feng et al. 2024), pricing (Yan et al. 2020), and relationships with public transportation (Diao, Kong, and Zhao 2021; Jin, Kong, and Sui 2019). In addition to trip record datasets, recent travel surveys conducted by transportation agencies have included ridesourcing into their questionnaires. In the next section, we will conclude the primary impacts of ridesourcing on urban mobility based on prior empirical analyses.

26.4 IMPACTS ON URBAN MOBILITY SYSTEMS

Ridesourcing has transformed urban mobility systems in four primary ways: (1) its interaction with the existing public transportation systems, (2) its impact on road traffic and congestion, (3) its impact on vehicle usage and vehicle miles traveled (VMT), and (4) its impact on destination accessibility.

26.4.1 Interaction with Public Transportation

There has been a long-standing concern about the competition and substitution effects of ridesourcing on urban public transportation since the service emerged. As ridesourcing platforms have become more popular, they have increasingly attracted passengers away from traditional public transportation means (Rayle et al. 2016). However, researchers have also found that ridesourcing can complement public transportation in areas and periods where facilities are less accessible (Rayle et al. 2016; Jin, Kong, and Sui 2019). Moreover, ridesourcing is used to connect people to public transportation stations or stops, enhancing public transport access and solving first-/last-mile problems (Rayle et al. 2016; Qin, Ke, and Yang 2022). This is referred to as the complementary effect of ridesourcing (McKane and Hess 2023).

A large body of literature has explored the competitive/substitution effects versus the complementary effects of ridesourcing. Researchers typically assess the proximity of public transport facilities to ridesourcing pickup and drop-off locations to determine whether ridesourcing trips have potentially substituted for public transport options. Researchers also measure the change in public transport ridership after ridesourcing services enter an area. While the results are complicated and depend on the contexts, competition between ridesourcing and public transport becomes more evident (Cats et al. 2022; Jin, Kong, and Sui 2019; Diao, Kong, and Zhao 2021; Babar and Burtch 2020). Most existing studies have shown that a significant portion of ridesourcing trips has replaced what would have otherwise been trips on public transport. This shift has led to a decline in public transportation ridership, potentially resulting in reduced revenue for local and regional public transportation agencies. This, in turn, can affect the funding and resources available for maintaining and improving public transport infrastructure and services.

26.4.2 Impact on Urban Traffic and Congestion

The impact of ridesourcing on urban congestion has been another concern. Researchers compare the traffic speeds and delays before and after ridesourcing enters the area to examine how ridesourcing contributes to congestion. Evidence indicates that the growing prevalence of ridesourcing has exacerbated traffic congestion regarding the intensity and duration (Diao, Kong, and Zhao 2021; Erhardt et al. 2019; Nie 2017). Although ridesourcing provides an alternative mode of transportation, it also introduces more vehicles onto already crowded urban roads, especially during peak hours and in high-demand urban areas. According to the New York City TLC, the average number of ridesourcing vehicles (HVFHVs) running on roads per day in New York in May 2024 was over 60,000, which increased by 738% compared to January 2015 (New York City Taxi & Limousine Commission, n.d.). The congestion effect of ridesourcing may have led to increased travel times and delays for all road users, including public transport vehicles, further diminishing the appeal of public transportation.

26.4.3 Impact on Vehicle Usage and Vehicle Miles

Relatedly, ridesourcing has profound impacts on vehicle usage and vehicle miles traveled (VMT). A large number of ridesourcing trips have generated substantial deadheading and cruising mileages (Henao and Marshall 2019; Clewlow and Mishra 2017). It was estimated that ridesourcing increased VMT by 3.5% in New York as of 2017 (Schaller Consulting 2017). Given the growing VMT, there are concerns regarding the environmental impact of ridesourcing caused by greenhouse gas (GHG) emissions. However, observations show that an increasing number of ridesourcing vehicles are electric. In China, it is estimated that new-energy vehicles, including electric vehicles, accounted for 87% of all ridesourcing vehicles as of 2023, and that number was 12% in 2016 (Qichetoutiao 2024). This trend may mitigate the negative environmental impact of ridesourcing to some extent. Additionally, the popularity of ridesourcing may lead to a reduction in private vehicle ownership, which is also beneficial for the environment. A study suggests that the entry of ridesourcing is associated with an average decline of 3% in vehicle registrations in the United States (Ward et al. 2022). Finally, the ride-pooling option can potentially reduce emissions from ridesourcing vehicles (Zhu and Mo 2022).

26.4.4 Impact on Destination Accessibility

The popularity of ridesourcing has enhanced auto-mobility and reshaped destination accessibility, especially for people who lack access to private vehicles and in areas that are underserved by traditional taxis and public transportation facilities (Brown 2018; Wang, Huang, and Shen 2024;

Rayle et al. 2016; Jin, Kong, and Sui 2019). A case study based in Rio de Janeiro suggested that ridesourcing can enhance job accessibility in commercial centers with a high number of jobs (de Oliveira Souza et al. 2021). Furthermore, research demonstrated that integrating ridesourcing into the existing public transport system could potentially improve job accessibility (Hu and Sun 2023). Additionally, a Chicago-based study highlighted that ridesourcing provides significantly greater accessibility to healthcare facilities, grocery stores, and restaurants compared to public transportation in high-income, privileged neighborhoods in urban center areas. However, this advantage does not extend to low-income, minority communities, where traditional public transportation offers more equitable access to destinations (Wang 2021). Ridesourcing has also been found to improve accessibility to both work and nonwork opportunities for people with disabilities (Ong, Loa, and Nurul Habib 2024). Overall, destination/opportunity accessibility provided by ridesourcing is relevant to the costs and affordability.

However, it should be noted that ridesourcing may create new forms of social disparities by marginalizing certain groups. In particular, older individuals may face significantly greater difficulty using ridesourcing, as they are less likely to be fluent in using smartphones and e-payment technologies (Agrawal et al. 2020; Young and Farber 2022). This issue extends beyond ridesourcing and affects other transportation modes, including car driving and public transportation, as they increasingly rely on digital technologies.

26.5 GOVERNING RIDESOURCING

The rapid growth of ridesourcing has introduced complexity to the mobility pattern, challenging the policy and regulatory frameworks of urban transportation systems. To address these challenges, governments at different levels and various administrative entities have implemented policies regarding licensing, operation, the workforce of ridesourcing, and integration of ridesourcing into public transport.

Unlike most taxis, which had stringent licensing and registration requirements, ridesourcing did not require special licenses in the beginning, enabling the fast growth in the number of drivers. Eventually, local governments imposed new regulations on ridesourcing providers. While varying across area, the regulations and requirements may include completing driver background checks, obtaining commercial/business permits or licenses, obtaining special vehicle license plates, and displaying the trade dress (Beer et al. 2017). For example, in August 2018, New York City enacted a law requiring all ridesourcing providers to obtain an HVFHV license from the TLC. In China, a driver applying for a ridesourcing permit must meet the following requirements: (1) possess a valid driver's license and have 3 or more years of driving experience; (2) have no record of traffic accidents involving criminal liability, dangerous driving offenses, drug use, driving under the influence of alcohol, or accumulating 12 points in the last three consecutive penalty cycles; (3) have no record of violent crimes; (4) comply with other local laws and policies; and (5) in some cities, locally register both themselves and the vehicle (The Central People's Government of China 2022).

There are various regulations aimed at addressing the negative impacts of ridesourcing on urban transportation systems, particularly regarding congestion. The approaches include capping the number of ridesourcing vehicles, imposing congestion pricing and surcharges, designating drop-off/pickup zones in high-demand areas (e.g., commercial centers, railway stations, airports), and providing incentives for ride-pooling.

There are also policies aimed at regulating the ridesourcing workforce. Some cities and regions have implemented minimum wages for drivers. In February 2019, New York City began requiring a minimum earning of $17.22 after expenses, becoming the first city to impose a minimum wage for ridesourcing drivers. In most areas, ridesourcing drivers are classified as self-employed contractors, while some local governments (e.g., London) have designated them as employees (or paid workers) to ensure they receive job benefits.

Government agencies have begun integrating ridesourcing into public transportation systems. The U.S. Department of Transportation announced the Livability, Sustainability, and Smart Cities initiative, aimed at developing Management and Operations (M&O) strategies to promote livability and sustainability, while fostering smart, connected communities through collaboration with Transportation Systems Management and Operations (TSMO) (Federal Highway Administration, n.d.). The U.S. Federal Transit Administration (FTA) supports this initiative through various programs, including Integrated Mobility Innovation (IMI), Accelerating Innovative Mobility (AIM), Enhancing Mobility Innovation (EMI), and Mobility on Demand (MOD) (Federal Transit Administration, n.d.). These programs are designed to shift public transportation agencies into new roles as "mobility managers" within the "diverse mobility ecosystem," which includes ridesourcing services. Nevertheless, efforts to integrate ridesourcing into public transportation are still in the conceptual phase and far from full implementation.

26.6 DIRECTIONS OF RIDESOURCING IN SMART CITIES

There are four directions for ridesourcing: (1) hybrid multiplatform collaboration, (2) partnership with other transportation modes, (3) evolution toward self-driving ridesourcing, and (4) connection to smart infrastructures.

26.6.1 Hybrid Multiplatform Collaboration

Ridesourcing drivers often work for multiple platforms (e.g., Uber and Lyft). Additionally, many drivers also work for Internet-based delivery services for food and goods (e.g., Amazon, DoorDash, Uber Eats, Instacart, Walmart, and Grubhub) to diversify their income sources. These multiplatform drivers, often referred to as gig drivers, are blurring the boundaries of ridesourcing and transforming both human mobility and the urban supply chain. Previously, drivers working for multiple platforms had to switch apps from one to another manually based on their experience. Currently, there are third-party apps available that assist with integrating, tracking, and managing multiple platforms for drivers. These apps, such as Gridwise, Maxymo, and SherpaShare, are designed to track earnings, taxes, traffic, mileage, and driving times, as well as allow drivers to accept or decline orders and conduct real-time analytics on data to support informed decision-making. This enables drivers to enhance efficiency by reducing deadheading times and distance.

26.6.2 Partnership with Other Transportation Modes

There are various approaches to connecting ridesourcing with other transportation modes. In New York City, Uber has added real-time public transportation information, including subway, buses, New Jersey Transit, Long Island Railroad, Metro-North Railroad, and the New York City Ferry, to its app interface. When users plan a ride, the app displays the costs and estimated arrival times for public transportation, in addition to the costs and arrival times for different types of ridesourcing vehicles (e.g., UberX, UberPool). This information can increase the likelihood of users choosing public transportation after considering the trade-off between travel time and costs. In Atlanta, Uber has been integrated into the public transportation app MARTA On the Go. According to Uber's report, a substantial number of ridesourcing trips have been started or ended at MARTA stations in Atlanta (Cmar, n.d.). Didi developed a smartphone app, Didi Transit, in 2016 to provide real-time public transportation information in multiple Chinese cities and gained more than 10 million users. However, the app was banned by the Chinese government due to Didi's aforementioned privacy breach violation.

Ridesourcing also partners with shared micromobility options, such as bikes and e-scooters. Lyft users are able to use the app for Citi Bike, the bike-sharing system in New York City. Lyft also provides dockless e-scooters in multiple big U.S. cities. Lime, one of the largest e-scooter providers in the United States, acquired Uber's e-bike and e-scooter business in 2020. Didi has developed its own bike-sharing service, Didi Qingju, which is one of the largest bike-sharing providers in China.

Paratransit, also known as intermediate public transport, provides on-demand, non-fixed-route public transportation services, particularly for people in need (e.g., people with disabilities), and has begun partnering with ridesourcing services as well. In 2016, the Massachusetts Bay Transportation Authority announced it would subsidize paratransit trips provided by ridesourcing vehicles (Cmar, n.d.).

WeChat, the most popular social networking super app in China, has embedded and connected diverse mobility and travel planning micro-apps, including ridesourcing, car rental, bike-sharing, and flight and train ticket booking. More importantly, WeChat has its own e-pay system that enables a seamless experience when paying for all these mobility services.

26.6.3 Evolution toward Self-driving Ridesourcing

Another direction for ridesourcing is the adoption of automated vehicles (AVs) and the transition to on-demand shared AV services (SAV, e.g., robotaxis). AV technologies have rapidly advanced and been widely tested in many cities, with developers now deploying shared AV services worldwide. Waymo, a subsidiary of Alphabet (Google's parent company) and a leading pioneer in AV technologies, is providing ridesourcing services through its self-driving vehicles in Phoenix, San Francisco, Los Angeles, and Austin. Another major AV developer, Cruise, acquired by General Motors, has offered its supervised automated driving vehicles for ridesourcing in Phoenix, Houston, and Dallas, following a 5-month pause in self-driving ridesourcing services due to an accident involving one of its vehicles in October 2023. Robotgo, a self-driving ridesourcing service by the Chinese technology giant Baidu, has been available in 11 Chinese cities as of July 2024. Although access to the services is still limited in these cities, the number of users is rapidly growing.

A primary advantage of self-driving ridesourcing is the reduction in costs by decreasing the demand for human labor. As SAVs potentially eliminate the expense of paying drivers, ridesourcing providers can reduce fares and increase profitability, making the service more affordable. Furthermore, self-driving ridesourcing can extend operational hours without breaks, sleep, shift handovers, or the limited work hours required by human drivers. This will optimize vehicle utilization and overall efficiency in the long run. Additionally, the large-scale adoption of self-driving ridesourcing may significantly reduce the demand for parking in urban areas, which could further increase the operational efficiency of the transportation system.

Admittedly, AV technologies are still immature and lack the standards for full commercialization, universal deployment, and effective regulations. For example, the average speed of SAVs is slower than that of human-driven ridesourcing vehicles; the right of way for SAVs is not yet defined; and the liability for accidents involving SAVs remains unclear. However, policymakers should proactively prepare for future urban mobility systems, where self-driving ridesourcing will play an essential role.

26.6.4 Connection to Smart Infrastructures

Ridesourcing services will increasingly rely on smart urban infrastructures, such as street cameras, real-time traffic detectors, road condition sensors, smart signals, smart parking, real-time traffic data, police networks, and advanced communication technologies. The connection between ridesourcing vehicles and infrastructures (vehicle-to-infrastructure, or V2I) can optimize routes, reduce congestion, improve operational efficiency, enhance safety and security, and help synergize the multimodal transportation system.

However, there are still several crucial barriers preventing seamless V2I connectivity. First, the development of smart infrastructures is not evenly distributed across cities, causing inconsistencies in the V2I connectivity. Second, there is a lack of regulations and standards for data sharing, especially for the data exchange between public sectors (e.g., transportation agencies, police departments) and private sectors (e.g., ridesourcing companies). Moreover, the real-time data exchange between ridesourcing vehicles and infrastructures raises concerns over data security and the protection of user privacy. Third, the implementation and maintenance of smart infrastructures and V2I technologies, such as urban sensors and advanced communication systems, require significant financial investments from both government agencies and service providers.

26.7 CONCLUSION

Ridesourcing has been in the market for almost 15 years, significantly transforming people's travel behavior and urban mobility systems worldwide. This chapter presents an overview of the history, impact, empirical data analysis, governance, and directions and challenges of ridesourcing. It plays an important role in the future smart cities by connecting and reshaping diverse transportation modes, ranging from walking and bicycling to car-driving and railways. However, many aspects of ridesourcing are still underexplored. In particular, more research is needed to understand, test, and evaluate regulations for ridesourcing to ensure it provides equitable, effective, and sustainable mobility services for everyone in society.

REFERENCES

Agrawal, Asha Weinstein, Manish Shirgaokar, Aditi Misra, Martin Wachs, and Bonnie M. Dobbs. 2020. "Will Ride-Hailing Enhance Mobility for Older Adults? A California Survey." 20–30. *Mineta Consortium for Transportation Mobility*. https://doi.org/10.31979/mti.2020.1815.

Babar, Yash, and Gordon Burtch. 2020. "Examining the Heterogeneous Impact of Ride-Hailing Services on Public Transit Use." *Information Systems Research* 31 (3): 820–34. https://doi.org/10.1287/isre.2019.0917.

Beer, Rachel, Candace Brakewood, Subrina Rahman, and Jennifer Viscardi. 2017. "Qualitative Analysis of Ride-Hailing Regulations in Major American Cities." *Transportation Research Record* 2650 (1): 84–91. https://doi.org/10.3141/2650-10.

Blystone, Dan. 2024. "The History of Uber." Investopedia. https://www.investopedia.com/articles/personal-finance/111015/story-uber.asp.

Brown, Anne Elizabeth. 2018. "Ridehail Revolution: Ridehail Travel and Equity in Los Angeles." UCLA. https://escholarship.org/uc/item/4r22m57k.

Campbell, Harry. 2023. "List of Ridesharing Companies in 2024." The Rideshare Guy. 2023. https://therideshareguy.com/ridesharing-companies/.

Cats, Oded, Rafal Kucharski, Santosh Rao Danda, and Menno Yap. 2022. "Beyond the Dichotomy: How Ride-Hailing Competes with and Complements Public Transport." *PLOS ONE* 17 (1): e0262496. https://doi.org/10.1371/journal.pone.0262496.

Cervero, Robert. 2017. "Mobility Niches: Jitneys to Robo-Taxis." *Journal of the American Planning Association* 83 (4): 404–12. https://doi.org/10.1080/01944363.2017.1353133.

Chapman, Michelle. 2022. "You Can Now Order an NYC Taxi on the Uber App." *NBC New York* (blog). September 9, 2022. https://www.nbcnewyork.com/news/local/you-can-now-order-an-nyc-taxi-on-the-uber-app/3858428/.

Chiang, Sheila. 2023. "Singapore's Grab Says Its Ride-Hailing Unit Is on Track to Hit Pre-Covid Levels by the End of the Year." CNBC. August 24, 2023. https://www.cnbc.com/2023/08/24/grab-ride-hailing-unit-on-track-to-hit-pre-covid-levels-by-end-2023.html.

Ciaccia, Chris. 2022. "DiDi Chuxing: The Chinese Ride-Sharing Giant." Investopedia. https://www.investopedia.com/articles/small-business/012517/didi-chuxing.asp.

Clewlow, Regina R., and Gouri S. Mishra. 2017. "Disruptive Transportation: The Adoption, Utilization, and Impacts of Ride-Hailing in the United States," October. https://escholarship.org/uc/item/82w2z91j.

Cmar, Wyatt. n.d. "How Cities Are Integrating Rideshare and Public Transportation." GovTech. Accessed September 13, 2024. https://www.govtech.com/dc/articles/How-Cities-are-Integrating-Rideshare-and-Public-Transportation.html.

Dean, Brian. 2023. "Lyft 2023 User and Revenue Stats." Backlinko. March 27, 2023. https://backlinko.com/lyft-users.

Dean, Brian. 2024. "Uber Statistics 2024: How Many People Ride With Uber?" Backlinko. https://backlinko.com/uber-users.

Diao, Mi, Hui Kong, and Jinhua Zhao. 2021. "Impacts of Transportation Network Companies on Urban Mobility." *Nature Sustainability* 4 (6): 494–500. https://doi.org/10.1038/s41893-020-00678-z.

Erhardt, Gregory D., Sneha Roy, Drew Cooper, Bhargava Sana, Mei Chen, and Joe Castiglione. 2019. "Do Transportation Network Companies Decrease or Increase Congestion?" *Science Advances* 5 (5): eaau2670. https://doi.org/10.1126/sciadv.aau2670.

Federal Highway Administration. n.d. "Livability, Sustainability and Smart Cities | Organizing and Planning for Operations - FHWA Office of Operations." Accessed September 11, 2024. https://ops.fhwa.dot.gov/plan4ops/livability_sus.htm.

Federal Transit Administration. n.d. "Integrated Mobility Innovation." Accessed September 11, 2024. https://www.transit.dot.gov/IMI.

Feng, Xuan, Qinping Lin, Ning Jia, and Junfang Tian. 2024. "The Actual Impact of Ride-Splitting: An Empirical Study Based on Large-Scale GPS Data." *Transport Policy* 147 (March): 94–112. https://doi.org/10.1016/j.tranpol.2023.12.008.

Fielbaum, Andrés, Felipe Ruiz, Giorgio Boccardo, Daniela Rubio, Alejandro Tirachini, and Jorge Rosales-Salas. 2023. "The Job of Public Transport, Ride-Hailing and Delivery Drivers: Conditions during the COVID-19 Pandemic and Implications for a Post-Pandemic Future." *Travel Behaviour and Society* 31 (April): 63–77. https://doi.org/10.1016/j.tbs.2022.11.004.

Greiner, Andrew, Matt McFarland, Ivory Sherman, and Jen Tse. 2019. "A History of Lyft, from Fuzzy Pink Mustaches to Global Ride Share Giant." 2019. https://www.cnn.com/interactive/2019/03/business/lyft-history/index.html.

Hassanpour, Amir, Alexander Bigazzi, and Don MacKenzie. 2021. "Equity of Access to Uber's Wheelchair Accessible Service." *Computers, Environment and Urban Systems* 89 (September): 101688. https://doi.org/10.1016/j.compenvurbsys.2021.101688.

Henao, Alejandro, and Wesley E. Marshall. 2019. "The Impact of Ride-Hailing on Vehicle Miles Traveled." *Transportation* 46 (6): 2173–94. https://doi.org/10.1007/s11116-018-9923-2.

Hu, Lingqian, and Sai Sun. 2023. "Integrating Transit and TNC Services to Improve Job Accessibility: Scenario Analysis with an Equity Lens." *Journal of Transport and Land Use* 16 (1): 43–65.

Huang, Xiao, Junyu Lu, Song Gao, Sicheng Wang, Zhewei Liu, and Hanxue Wei. 2022. "Staying at Home Is a Privilege: Evidence from Fine-Grained Mobile Phone Location Data in the United States during the COVID-19 Pandemic." *Annals of the American Association of Geographers* 112 (1): 286–305. https://doi.org/10.1080/24694452.2021.1904819.

Hughes, Ryan, and Don MacKenzie. 2016. "Transportation Network Company Wait Times in Greater Seattle, and Relationship to Socioeconomic Indicators." *Journal of Transport Geography* 56 (October): 36–44. https://doi.org/10.1016/j.jtrangeo.2016.08.014.

Jin, Scarlett T., Hui Kong, and Daniel Z. Sui. 2019. "Uber, Public Transit, and Urban Transportation Equity: A Case Study in New York City." *The Professional Geographer* 71 (2): 315–30. https://doi.org/10.1080/00330124.2018.1531038.

Jin, Scarlett T., Hui Kong, Rachel Wu, and Daniel Z. Sui. 2018. "Ridesourcing, the Sharing Economy, and the Future of Cities." *Cities* 76 (June): 96–104. https://doi.org/10.1016/j.cities.2018.01.012.

Jin, Tanhua, Long Cheng, Sicheng Wang, Jun Cao, Tim Schwanen, and Frank Witlox. 2023. "Do Built Environment Factors Have Different Effects on Ridesourcing Usage before and after the COVID-19 Pandemic?" *Cities* 141: 104520.

Li, Ziru, Chen Liang, Yili Hong, and Zhongju Zhang. 2022. "How Do On-demand Ridesharing Services Affect Traffic Congestion? The Moderating Role of Urban Compactness." *Production and Operations Management* 31 (1): 239–58. https://doi.org/10.1111/poms.13530.

Loa, Patrick, Sanjana Hossain, Yicong Liu, and Khandker Nurul Habib. 2022. "How Has the COVID-19 Pandemic Affected the Use of Ride-Sourcing Services? An Empirical Evidence-Based Investigation for the Greater Toronto Area." *Transportation Research Part A: Policy and Practice* 155 (January): 46–62. https://doi.org/10.1016/j.tra.2021.11.013.

Machado, Cláudia A. Soares, Nicolas Patrick Marie De Salles Hue, Fernando Tobal Berssaneti, and José Alberto Quintanilha. 2018. "An Overview of Shared Mobility." *Sustainability* 10 (12): 4342. https://doi.org/10.3390/su10124342.

Mao. 2024. "China's Robotaxi Replacing Taxi Drivers: The Huge Impact on People at the Grassroots Level." July 14, 2024. https://www.registrationchina.com/articles/chinas-robotaxi-taxi-drivers/.

Marquet, Oriol. 2020. "Spatial Distribution of Ride-Hailing Trip Demand and Its Association with Walkability and Neighborhood Characteristics." *Cities* 106 (November): 102926. https://doi.org/10.1016/j.cities.2020.102926.

McCoy, Kevin, James Andrew, Russell Glynn, William Lyons, and John A. Volpe National Transportation Systems Center (U.S.). 2018. "Integrating Shared Mobility into Multimodal Transportation Planning: Improving Regional Performance to Meet Public Goals." DOT-VNTSC-FHWA-18-13. https://rosap.ntl.bts.gov/view/dot/34689.

McKane, Rachel G., and David Hess. 2023. "The Impact of Ridesourcing on Equity and Sustainability in North American Cities: A Systematic Review of the Literature." *Cities* 133 (February): 104122. https://doi.org/10.1016/j.cities.2022.104122.

New York City Taxi & Limousine Commission. n.d. "Aggregated Reports - TLC." Accessed September 9, 2024. https://www.nyc.gov/site/tlc/about/aggregated-reports.page.

Nie, Yu (Marco). 2017. "How Can the Taxi Industry Survive the Tide of Ridesourcing? Evidence from Shenzhen, China." *Transportation Research Part C: Emerging Technologies* 79 (June): 242–56. https://doi.org/10.1016/j.trc.2017.03.017.

Njanja, Annie, and Tage Kene-Okafor. 2021. "Didi Expands, inDriver Monetizes to Rival Uber, Bolt in Africa." *TechCrunch*. October 28, 2021. https://techcrunch.com/2021/10/28/didi-expands-indriver-monetizes-to-rival-uber-bolt-in-africa/.

Oliveira Souza, Camilla de, Gabriella Vitorino Guimarães, Luiz Emerson da Cruz Saldanha, Filipe Almeida Corrêa do Nascimento, Tálita Floriano dos Santos, and Marcelino Aurélio Vieira da Silva. 2021. "Analysis of Job Accessibility Promoted by Ride Hailing Services: A Proposed Method." *Journal of Transport Geography* 93 (May): 103048. https://doi.org/10.1016/j.jtrangeo.2021.103048.

Ong, Felita, Patrick Loa, and Khandker Nurul Habib. 2024. "Ride-Sourcing Demand in Metro Vancouver: Looking through the Lens of Disability." *Transportation Research Part A: Policy and Practice* 181 (March): 103984. https://doi.org/10.1016/j.tra.2024.103984.

Puschmann, Thomas, and Rainer Alt. 2016. "Sharing Economy." *Business & Information Systems Engineering* 58 (1): 93–99. https://doi.org/10.1007/s12599-015-0420-2.

Qichetoutiao. 2024. "2023 年网约车市场, 埃安和比亚迪的占据' 半壁江山.'" 2024. https://m.qctt.cn/news/1620388.

Qin, Xiaoran, Jintao Ke, and Hai Yang. 2022. "Government Regulations for Ride-Sourcing Services as Substitute or Complement to Public Transit." SSRN Scholarly Paper. Rochester, NY. https://doi.org/10.2139/ssrn.4129034.

Rayle, Lisa, Danielle Dai, Nelson Chan, Robert Cervero, and Susan Shaheen. 2016. "Just a Better Taxi? A Survey-Based Comparison of Taxis, Transit, and Ridesourcing Services in San Francisco." *Transport Policy* 45 (January): 168–78. https://doi.org/10.1016/j.tranpol.2015.10.004.

Schaller Consulting. 2017. "Unsustainable? The Growth of App-Based Ride Services and Traffic, Travel and the Future of New York City." 2017. http://www.schallerconsult.com/rideservices/unsustainable.htm.

Shaheen, Susan, Adam Cohen, and Ismail Zohdy. 2016. "Shared Mobility: Current Practices and Guiding Principles." United States. Federal Highway Administration. https://rosap.ntl.bts.gov/view/dot/42193.

Sina News. 2024. "司机严重饱和！想开网约车, 难了_手机新浪网." 2024. https://news.sina.cn/gn/2024-05-22/detail-inawaarw6012242.d.html.

Smith, Craig. 2024. "Bolt." DMR. https://expandedramblings.com/index.php/taxify-facts-and-statistics/.

Tarabay, Rana, and Maya Abou-Zeid. 2020. "Modeling the Choice to Switch from Traditional Modes to Ridesourcing Services for Social/Recreational Trips in Lebanon." *Transportation* 47 (4): 1733–63. https://doi.org/10.1007/s11116-019-09973-x.

The Central People's Government of China. 2022. "网络预约出租汽车经营服务管理暂行办法_交通运输部_中国政府网." 2022. https://www.gov.cn/zhengce/2022-12/06/content_5730384.htm.

Wang, Hai, and Hai Yang. 2019. "Ridesourcing Systems: A Framework and Review." *Transportation Research Part B: Methodological* 129 (November): 122–55. https://doi.org/10.1016/j.trb.2019.07.009.

Wang, Sicheng. 2021. "*Travel Behavior of Ridesourcing: A Data-Driven Analysis*." PhD Thesis, Rutgers The State University of New Jersey, School of Graduate Studies. https://search.proquest.com/openview/85407ea0a205feff15fc7a509d3c1490/1?pq-origsite=gscholar&cbl=18750&diss=y&casa_token=1XtHehbs5BsAAAAA:xlMTOqEeDuiD11gDqVpnut8p3V3L7chveKF5degz9CbZ_Vp0XcwkqU1eD4lzAwIS0QrMMj01nUw.

Wang, Sicheng, Rui Du, and Annie S. Lee. 2024. "Ridesourcing Regulation and Traffic Speeds: A New York Case." *Journal of Transport Geography* 116 (April): 103846. https://doi.org/10.1016/j.jtrangeo.2024.103846.

Wang, Sicheng, Xiao Huang, and Qing Shen. 2024. "Disparities in Resilience and Recovery of Ridesourcing Usage during COVID-19." *Journal of Transport Geography* 114 (January): 103745. https://doi.org/10.1016/j.jtrangeo.2023.103745.

Wang, Sicheng, and Robert B. Noland. 2021a. "Variation in Ride-Hailing Trips in Chengdu, China." *Transportation Research Part D: Transport and Environment* 90: 102596.

Wang, Sicheng, and Robert B. Noland. 2021b. "What Is the Elasticity of Sharing a Ridesourcing Trip?" *Transportation Research Part A: Policy and Practice* 153: 284–305.

Wang, Sicheng, and Michael Smart. 2020. "The Disruptive Effect of Ridesourcing Services on For-Hire Vehicle Drivers' Income and Employment." *Transport Policy* 89: 13–23.

Wang, Ze, Xiaowei Chen, and Xiqun (Michael) Chen. 2019. "Ridesplitting Is Shaping Young People's Travel Behavior: Evidence from Comparative Survey via Ride-Sourcing Platform." *Transportation Research Part D: Transport and Environment* 75 (October): 57–71. https://doi.org/10.1016/j.trd.2019.08.017.

Ward, Kanesha, Sanjyot Vagholkar, Fareeya Sakur, Neha Nafees Khatri, and Annie Y. S. Lau. 2022. "Visit Types in Primary Care with Telehealth Use During the COVID-19 Pandemic: Systematic Review." *JMIR Medical Informatics* 10 (11): e40469. https://doi.org/10.2196/40469.

Yan, Chiwei, Helin Zhu, Nikita Korolko, and Dawn Woodard. 2020. "Dynamic Pricing and Matching in Ride-Hailing Platforms." *Naval Research Logistics (NRL)* 67 (8): 705–24. https://doi.org/10.1002/nav.21872.

Yan, Xiang, Xinyu Liu, and Xilei Zhao. 2020. "Using Machine Learning for Direct Demand Modeling of Ridesourcing Services in Chicago." *Journal of Transport Geography* 83 (February): 102661. https://doi.org/10.1016/j.jtrangeo.2020.102661.

Yang, Hongtai, Yuan Liang, and Linchuan Yang. 2021. "Equitable? Exploring Ridesourcing Waiting Time and Its Determinants." *Transportation Research Part D: Transport and Environment* 93 (April): 102774. https://doi.org/10.1016/j.trd.2021.102774.

Young, Mischa, and Steven Farber. 2022. "Ride-Hailing Platforms Are Shaping the Future of Mobility, but for Whom." *The Platform Economy and the Smart City: Technology and the Transformation of Urban Policy*, 84–107. Montreal: McGill-Queen's University Press. https://doi.org/10.1515/9780228007944-006

Yu, Haitao, and Zhong-Ren Peng. 2020. "The Impacts of Built Environment on Ridesourcing Demand: A Neighbourhood Level Analysis in Austin, Texas." *Urban Studies* 57 (1): 152–75. https://doi.org/10.1177/0042098019828180.

Zhu, Pengyu, and Haoyu Mo. 2022. "The Potential of Ride-Pooling in VKT Reduction and Its Environmental Implications." *Transportation Research Part D: Transport and Environment* 103 (February): 103155. https://doi.org/10.1016/j.trd.2021.103155.

Zinkula, Jacob, and Noah Sheidlower. 2024. "The 3 Worst Things about Being an Uber or Lyft Driver." *Business Insider*. https://www.businessinsider.com/top-uber-lyft-driver-concerns-low-earnings-declining-pay-safety-2024-8.

27 A Comparative Analysis of Governance of Urban Public Transport in Asian Cities

Madan B. Regmi

27.1 INTRODUCTION

Asia is urbanizing with more than 2.2 billion urban dwellers living in Asian cities. Asia's urban population will surge by 50%, adding an additional 1.2 billion people by 2050 (UNHABITAT, n.d.). This growth presents both opportunities and challenges. One of the most urgent challenges that Asian cities face is the rapid rise in demand for urban public transport, with the number of vehicles doubling every 6 years (ADB, n. d.). The passenger demand is anticipated to soar by 79%, while freight demand will nearly double by 2050 (ITF, 2023).

Various form of mass transit systems such as metro, light rail transits, urban rails, bus rapid transits, buses, and minibuses are catering to the mobility needs of urban dwellers in Asian cities. In addition, various form of private transport modes using cars, two- and three-wheelers, and informal transport are being used in Asian cities. However, Asian cities are faced with the challenges of insufficient public transit, growing congestion, increasing road fatalities, growing emissions, and air pollution.

Urban mobility is essential for city residents, as it facilitates access to employment, education, healthcare, and other vital services. Many Asian cities are planning and implementing policies, programs, and projects to improve urban public transport systems. Urban public transport is an important aspect of sustainable urban development, influencing economic growth, environmental health, and social equity. Effective governance of urban public transport systems is essential to ensure that they are sustainable, inclusive, and accessible.

Sustainability in urban transport ensures that these systems do not harm the environment and remain viable for future generations. This involves the use of public transit and eco-friendly vehicles, reducing emissions, and promoting nonmotorized transport options like cycling and walking.

Inclusiveness guarantees that transport systems and services are available to all segments of society, including marginalized groups such as low-income individuals, the elderly, and people with disabilities. This means designing transport systems that are physically accessible, affordable, and equitable. Ensuring inclusiveness might involve implementing subsidized fare structures, enhancing the accessibility of public transit vehicles and infrastructure, and providing comprehensive coverage that reaches underserved areas.

Accessibility ensures that transport services are convenient and affordable for everyone. This involves strategic planning to reduce travel times, optimize routes, and maintain affordability to prevent cost from being a barrier to mobility. It also means integrating various modes of transport to create a seamless journey for users, making it easy to switch from buses to trains to bicycles, for instance. Additionally, accessibility encompasses the digital aspect, such as providing real-time information and easy-to-use apps for route planning and ticketing, which can enhance the user experience and make the transport system more user-friendly.

Effective urban transport governance can significantly enhance the quality of life of urban residents by ensuring that transport systems are efficient, reliable, and safe. Given the increasing

DOI: 10.1201/9781003503262-31

frequency and intensity of natural and climate events in Asia, there is also a need to strengthen the resiliency of transport infrastructure and adapt to these events and make transition to low-carbon mobility.

By focusing on these key aspects of sustainability, inclusiveness, and accessibility, urban transport governance can create systems that not only meet the current mobility needs of city residents but also contribute to the long-term health, equity, and efficiency of urban environments.

Each Asian city is in different stages of urbanization and development, providing a diverse perspective on urban transport governance. Local contexts play important role in planning and deciding for sustainable transport solutions (World Bank and ADB, 2012). Governance of urban mobility in most of Asian cities is fragmented with competing interest among various institutions and limited coordination among related sectors such as land use, transport planning, and energy (Regmi, 2024). In some cases, existing city, state, and federal authorities have overlapping responsibilities without clear-cut demarcation. Some cities have created new urban transport institutions with a view to improving coordination.

This chapter focuses on the governance of urban public transport in Asian cities. It discusses theoretical aspects of urban governance, and by taking a case study approach, it explores how Asian cities are aspiring to provide seamless mobility solutions to their residents. It will look at key attributes of sustainable transport governance as well as funding mechanisms. The chapter investigates strategies and challenges to create more sustainable, accessible, affordable, and inclusive public transport systems and provides insights for future policy and practice in urban transport planning, governance, and implementation. By doing so, it reveals some of the good practices and challenges Asian cities face. After identifying the main issues and challenges, some policy suggestions will be offered to improve urban mobility and governance in Asian cities.

27.2 LITERATURE REVIEW

27.2.1 State of Public Transport in Asian Cities

Asian cities are being rapidly urbanized and facing urgent challenges to improve urban mobility. Rapid economic growth and population growth are leading to a significant increase in private vehicle use. We are witnessing severe congestion on streets, increased travel time and costs, worsening road safety, poor air quality, and growing greenhouse gas emissions.

As cities grow, introducing rapid transit modes becomes necessary to make public transport competitive. Strong urban institutions and governance are essential to tackle the problem of managing and operating multiple public transport modes and providing seamless and efficient services to commuters.

Many forms of mass transit systems (metro, buses, bus rapid transit, light rail transit, and subways) and private vehicles are meeting the mobility needs of the urban dwellers. Given the growth of the urban population and growing number of cities, in order to make Asian transport systems sustainable, the urban public transport system has to be sustainable. Figure 27.1 shows the share of public transport and active mobility in selected Asian cities. It shows a low share of passenger transport and a high share of active mobility. One of the challenges for countries and cities has been to increase the share of public transport despite considerable investment in public transport systems. The high share of active mobility also indicates the need for good infrastructure for walking and cycling.

There is wider agreement on how we can make passenger transport more sustainable. The narrative is to improve accessibility (Boisjoly and El-Geneidy, 2017) by developing mass transit systems, increase the share of public transport, reduce use of private cars, promote active mobility and cycling (Buehler and Pucher, 2021), and promote stakeholders' engagement and advocacy (Foltýnová et al., 2020). But cities are taking different paths, processes (Beukers et al., 2012), and approaches to move toward sustainability.

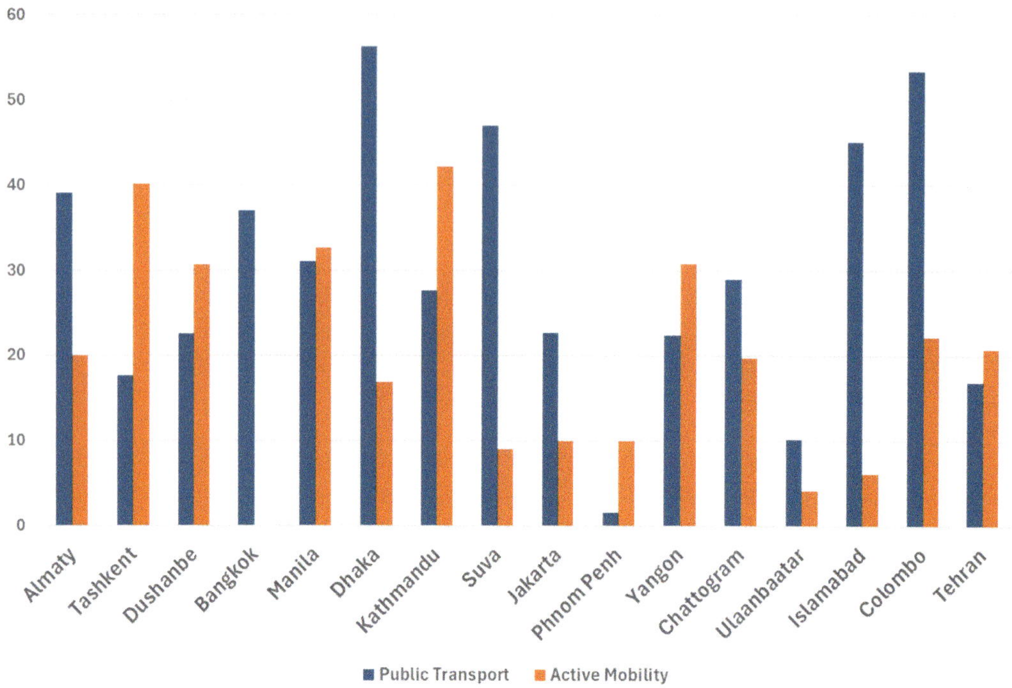

FIGURE 27.1 Mode share of public transport and active mobility. (From ESCAP SUTI Assessment reports.)

27.2.2 Urban Transport Governance

Urban transport governance involves the frameworks and processes through which public transport systems are planned, implemented, and managed. Effective governance is crucial for ensuring that urban transport systems are sustainable, efficient, and accessible to all. Several governance theories emphasize the roles of institutions, stakeholders, and policy mechanisms in achieving these goals. Governance is about the processes by which public policy decisions are made and implemented (UNDP, 2009). The formulation and implementation of public transport policies is difficult and challenging in Asian cities. Although it is well recognized that local governments are effective in developing and implementing sustainable mobility agendas (World Bank and ADB, 2012), federal governments and related departments often remain prominently involved in decision-making and implementation. Governance of urban mobility in most of Asian cities is fragmented, with competing interest among city, state, and federal authorities and limited coordination between land use and transport planning (Regmi, 2024). For example, in Kathmandu, all three levels of government are involved in urban transport with overlapping functions and without a designated lead authority.

Authority over the various facets of transport plans is commonly dispersed among numerous agencies that do not operate in a coordinated fashion. Even within metropolitan areas, power over transit planning is widely dispersed. Other deep-rooted issues about how governments operate include: limited planning capacity, lack of public participation, lack of national guidance, inertia, and aversion to making any decisions.

Regmi (2024) looked at the governance of urban mobility and used an evaluation framework to analyze the cases of four Asian cities (Dhaka, Greater Jakarta, Kathmandu, and Surat). For cities to perform rational planning and decision-making, they should be encouraged to make evidence-based decision-making by providing data and analysis on the existing state of public transport with key indicators to inform the policy decisions. Additionally, robust data collection and analysis should be undertaken to inform policy decisions, by identifying emerging trends, pressures, and challenges and then setting objectives and clearly defined measures for the impact of interventions.

Many argue that governmental organizations involved in transport and land use planning, provision of transport services, and operations should be consolidated. Given the need to utilize renewable energy in transport, many researchers have called for multisectoral collaboration among transport, energy, environment, and social dimensions. This would help to develop a long-term sustainable and inclusive mobility vision. Where consolidation is not possible, various models to enhance coordination in planning and management of urban mobility are being used. The Unified Metropolitan Transport Authority in Indian cities, Dhaka Transport Coordination Authority, and Greater Jakarta Transport Authority are some authorities aiming to enhance coordination.

Another important aspect of urban governance is stakeholder consultation. This would ensure that mobility planning is inclusive and caters to the needs of the whole urban population including all sectors of society.

One of the key models is the Sustainable Urban Mobility Plan (SUMP) framework, which focuses on integrated and sustainable transport planning. SUMP emphasizes the importance of long-term planning that considers the environmental, social, and economic impacts of transport systems. It advocates for the coordination of various transport modes and the inclusion of all relevant stakeholders, including local governments, the private sector, and the public, in the planning process (Rupprecht Consult, 2019).

Another important approach is Public Value Management (PVM), which emphasizes the creation of public value through effective governance. PVM focuses on the outcomes of governance, particularly the extent to which public transport systems meet the needs of the population and contribute to the overall well-being of society. This approach stresses the importance of transparency, accountability, and citizen participation in governance processes, aiming to maximize the public benefits derived from urban transport systems (Moore, 1995; Stoker, 2006).

In Asian cities, urban transport governance faces challenges due to rapid urbanization, high population density, congestion, and significant environmental impacts. Earlier studies demonstrate the roles of different governance models in addressing urban challenges and highlight the need for a coordinated approach that integrates various transport modes, promotes public transport over private vehicle use, and ensures that governance frameworks are adaptable to the fast-changing urban environments typical of many Asian cities (ADB, 2012; World Bank, 2017).

Global studies on urban public transport governance have consistently highlighted the importance of integrated planning, stakeholder engagement, and robust funding mechanisms. The United Nations Economic and Social Commission for Asia and the Pacific (ESCAP) has recently published a guidebook for policymakers on integration of public transport systems (ESCAP, 2024a) and a compendium of cases from the Asia-Pacific region; examples of institutions and governance from Singapore, Bengaluru, and Surat are also presented (ESCAP, 2024b).

Cities with strong governance frameworks tend to have more efficient, sustainable, and resilient transport systems. Effective governance involves not just the technical aspects of transport planning but also the institutional arrangements, policy mechanisms, and stakeholder relationships that support the implementation and management of these systems (UITP, 2019).

In the context of Asian cities, Singapore, Seoul, Tokyo, and Hong Kong, offer good practices of public transport planning, services, and governance. These cities have efficiently integrated different modes of public transport systems and services and excelled in urban governance. However, they each have a different form of governance. Singapore offers a unique example of integrated land use, transport planning, and transport governance, where planning, infrastructure development, implementation, and operation are handled by the Land Transport Authority (LTA, n.d.)

Singapore provides a unique example of urban governance and has undertaken comprehensive initiatives to ensure that its public transport system is highly accessible to all users, including those with disabilities, the elderly, and families with young children. These measures include the installation of barrier-free access points, which remove physical obstacles that might impede movement

(LTA, 2020). For example, bus stops and train stations are designed with features such as ramps and lifts that allow individuals in wheelchairs or with mobility challenges to navigate the transport system with ease. Barrier-free access encompasses various elements aimed at providing seamless travel. Ramps are strategically placed to facilitate easy boarding and alighting from buses and trains. Lifts are installed at all train stations, ensuring that individuals who cannot use stairs or escalators can still access the platforms. Additionally, tactile paving guides are used to assist visually impaired commuters, enabling them to move around stations and platforms independently and safely. By integrating these features, the LTA has significantly enhanced the inclusivity and usability of public transport in Singapore. This comprehensive approach ensures that all segments of the population can benefit from the public transport system, making it one of the most accessible transit networks in the Asia-Pacific region. The LTA's commitment to accessibility demonstrates a broader dedication to social equity and the well-being of its citizens, aligning with global standards and setting a benchmark for other cities to follow.

Similarly, Tokyo's public transport system is highly accessible due to advanced features like tactile paving, audio-visual aids, and platform screen doors that ensure safe boarding and alighting for visually impaired passengers (Kawakami, 2017). Tactile paving helps passengers, especially passengers who have difficulties with vision, to navigate stations by providing textured ground surfaces that they can feel. Audio-visual aids deliver important travel information through both sound and sight, helping people with hearing or visual disabilities. Platform screen doors prevent accidents by aligning with train doors, ensuring safety while getting in and off the train. These features enhance safety, independence, and efficiency for all passengers, making Tokyo's transit system a model of inclusivity and accessibility.

Smart technologies are significantly enhancing accessibility in urban public transport, particularly benefiting visually impaired individuals. In Hong Kong, mobile applications provide real-time updates and navigation assistance tailored for these users, offering audio cues and step-by-step directions (Yau, 2019). This innovation allows visually impaired passengers to navigate transit systems independently and safely, reducing their reliance on others. Additionally, these technologies improve overall safety and convenience for all passengers by offering accessible navigation and timely information, making the public transport system more user-friendly and inclusive. By bridging the accessibility gap, smart technologies ensure greater independence and ease of travel for individuals with disabilities, fostering social inclusion and equality.

Seoul has an extensive network of public transport that includes rail-based and bus-based systems. The Seoul Metropolitan Government has introduced structural changes to upgrade and integrate Seoul's public transport infrastructure and operations (SMG, 2020). The Seoul Metropolitan Subway consists of multiple systems that form a larger network, including mass rapid transit, light metro, commuter rail, and automated people movers. Seoul's bus services are classified into trunk, feeder, interregional, and circular lines.

In Jakarta, the TransJakarta bus rapid transit system has introduced various strategies to ensure that public transport remains affordable for all residents, especially those who are economically disadvantaged. One such strategy is offering subsidized fares for specific groups such as students, seniors, and low-income individuals. By providing discounted fares to these groups, TransJakarta aims to make public transport more accessible and financially viable for those who may struggle to afford regular fares (Bappenas, 2019). This initiative not only promotes social equity but also ensures that everyone, regardless of their financial situation, can access essential services and opportunities. It also encourages greater use of public transport, thereby reducing congestion and environmental impact. Overall, these efforts contribute to a more inclusive and sustainable urban transport system in Jakarta.

Likewise, in Seoul, innovative fundings and revenue models are very important for keeping urban transit projects affordable. Public-private partnerships have been particularly effective in financing these projects, reducing the financial burden on users (Lee and Hwang, 2014). These partnerships involve collaboration between government agencies and private entities to fund, build, and operate transit infrastructure. By sharing the costs and risks, public-private partnerships enable the development of transit systems without depending fully on government funds or increasing fares for passengers.

Multimodal integration in Melbourne's public transport system, which seamlessly connects buses, trams, and trains, has significantly improved accessibility for people living in suburban areas, making the transit system more inclusive (Currie, 2016). This integration makes travel easier by reducing wait times and simplifying transfers, benefitting everyone, including those with disabilities, the elderly, and families with young children.

27.2.3 Reflection on Literature Review

The challenge for Asian urban governments and institutions is to strengthen not only policymaking capacities but also their ability to deliver, enforce, and implement sustainable and inclusive urban mobility policies. This requires more effective coordination and information-sharing mechanisms across sectors (urban, transport, energy, finance, etc.), as well as more stable financing models to pay for infrastructure investments.

Government agencies need to focus on establishing cross-sectoral, cross-jurisdictional teams to facilitate strong horizontal integration in the process of policy formulation for the creation of integrated urban mobility masterplans and project implementation. This should be accompanied by strong vertical integration mechanisms such as feedback channels for public participation and engagement in core aspects of the planning process.

Although the well-known examples from Singapore, Seoul, Tokyo, and Jakarta provide some insights on urban governance and provision of public transport services, additional cases of governance of urban and public transport planning would provide new insights and challenges regarding urban mobility. In addition to studying capital cities, it would be helpful to explore cases from smaller and secondary Asian cities.

27.3 METHODOLOGY

27.3.1 Data Collection Methods

The aim was to explore cases of urban governance from other Asian cities of different sizes and cultural contexts. The cities of Pokhara, Bangkok, Kuala Lumpur, and Thimphu have been selected as case studies because they represent different geographic regions and have varying levels of urban public transport development. To understand how urban transport is managed in these cities, data were collected from two main sources: primary and secondary.

Primary data were gathered through interactions with key public transport stakeholders in each city. The interactions were conducted during visits to observe public transport systems in these cities as well as interactions and discussions during urban transport events. The stakeholders included transport planners, policymakers, public transport operators, and drivers. These interactions provided firsthand insights into the current state of urban transport and governance in each city. Similarly, secondary data were collected from existing reports, research articles, and policy documents published by government bodies and academic institutions. This type of data helped provide context and background information, allowing for a deeper understanding of the policies and strategies in place.

27.3.2 Comparative Approach

To evaluate and compare the urban transport systems in these four cities, a comparative method was used. This approach looked at several key factors, including how policies are structured, how institutions work together, how different stakeholders (like the public and private sectors) are involved, and how the cities fund their transport projects.

By comparing these elements across the four cities, it will be possible to identify which strategies work best and what challenges are commonly faced. This comparison also highlights unique practices that could be applied in other contexts to improve urban transport governance. The focus was on how the form of governance has influenced overall sustainability, inclusiveness, and accessibility of urban public transport systems.

For the qualitative analysis, the focus is on understanding the deeper aspects of urban transport governance. This involved reviewing policy documents, analyzing interaction logs, and interpreting the structures and processes that guide public transport planning and implementation in each city. This analysis provided a nuanced understanding of how governance works in practice and what factors contribute to its success or limitations.

27.4 CASES OF GOVERNANCE OF URBAN PUBLIC TRANSPORT IN ASIAN CITIES

27.4.1 Bangkok

Bangkok operates many forms of public urban transit, including elevated sky trains (BTS), subways (MRT), airport links, commuter trains, and buses. However, the city faces significant challenges in integrating services and fares across these various forms of public transit. The safety record is very poor in Bangkok despite many policy efforts. The poor safety in Bangkok and Thailand is largely due to poor driver behavior during two long holidays. The city also hosts a variety of informal transport options and boats and ferries along the Chao Phraya River and canals. The use of technology, including on-demand taxis and Grab bikes, has improved accessibility. Despite many plans, fare integration among various public transport is overdue in Bangkok. Bangkok could learn lessons from Brisbane on integration of public transport, non-motorized transport, and operation of river ferry boats. Convenience and safety are what commuters are looking for while choosing public transport. Despite substantial investment in public transport, commuters are forced to pay more while combining different modes for a single trip in Bangkok.

Bangkok's urban transport governance is characterized by a highly centralized system under the Bangkok Metropolitan Administration (BMA). The BMA has played a critical role in coordinating and implementing large-scale transport projects, including the expansion of the city's mass transit systems, such as the BTS Skytrain and MRT subway. Feedback from the commuters were used to improve accessibility to the BTS Skytrain by adding elevators and ramps, improved lighting and security for safety, and reserving seats for women and elderly passengers (Sung et al., 2015). The centralized governance model has enabled Bangkok to pursue ambitious infrastructure projects, supported by substantial government funding and public-private partnerships (OTP, 2017; BMA, 2020).

However, Bangkok's transport governance also faces significant challenges. The rapid pace of urbanization has led to severe traffic congestion and pollution, issues that the existing governance structures have struggled to fully address. Research suggests that while the centralization of governance has allowed for swift decision-making and implementation, it has also led to coordination challenges between various agencies and stakeholders involved in urban transport. Additionally, the focus on large-scale infrastructure projects has sometimes overshadowed the need for more integrated and sustainable transport planning, as advocated by frameworks like SUMP (Rupprecht Consult, 2019).

Bangkok's governance model is centralized under BMA, which oversees urban transport planning and execution. The centralized model allows for efficient decision-making and large-scale infrastructure projects, such as the expansion of the BTS Skytrain and MRT subway systems. However, this model faces challenges in integrating various transport modes and engaging local stakeholders effectively (BMA, 2020).

Bangkok has long struggled with severe traffic congestion and high pollution levels, largely due to rapid urbanization and an increasing number of private vehicles on the road. The city's burgeoning population and economic activities have exacerbated the strain on its road networks, leading to chronic traffic jams and deteriorating air quality. To address these issues, the city has focused on expanding its mass transit systems, particularly the BTS Skytrain and MRT. These projects have been supported by public-private partnerships, which have played a crucial role in funding and implementing the necessary infrastructure (BMA, 2020). The expansion of these transit systems has helped alleviate some of the congestion by providing residents with alternative means of transport. However, despite these improvements, the city still faces significant challenges in integrating these transit systems with other modes of transport. The lack of seamless connectivity between different transport options continues to hinder the overall effectiveness of the city's transport network, highlighting the need for a more holistic approach to transport planning that considers all aspects of urban mobility (OTP, 2017).

Urban transport development in various cities showcases a range of successes and challenges, deeply influenced by each city's governance model, resource availability, and planning priorities. Bangkok has achieved notable success in implementing large-scale infrastructure projects, such as the expansion of its mass transit systems. The city's centralized planning model, led by the Office of Transport and Traffic Policy and Planning, enables rapid project completion and efficient allocation of resources. However, this approach often struggles with the integration of these systems, leading to a transport network that is not as seamless as intended. Furthermore, the top-down model may sometimes overlook the specific needs of marginalized communities, who may not benefit equally from these developments (OTP, 2017).

27.4.2 Thimphu

Thimphu's governance of urban transport is still in the initial stages of development, guided by Bhutan's broader philosophy of gross national happiness (GNH). This unique approach to governance emphasizes sustainability, environmental protection, and the well-being of the population. The Ministry of Information and Communications (MOIC) is responsible for developing and overseeing transport policies, with a strong focus on creating a cohesive and environmentally friendly transport system (MOIC, 2018). Strong leadership and a well-developed governance framework are essential for implementation of mobility plans (Tønnesen et al., 2020).

Given Thimphu's small size and low level of urbanization compared to other cities in the region, its transport governance has had to adapt to limited resources and infrastructure. International aid and donor funding play a crucial role in supporting Thimphu's transport projects, which aim to balance modernization with the preservation of Bhutan's environmental and cultural values. The governance challenges here include the need to build capacity within local institutions and to develop transport systems that are both sustainable and capable of meeting future urbanization pressures (RGB, 2021).

Thimphu, the capital of Bhutan, faces unique challenges in developing a comprehensive transport system due to its limited infrastructure and institutional capacity. The city's small size and mountainous terrain further complicate the implementation of large-scale transport projects. To navigate these challenges, Thimphu has placed a strong emphasis on sustainable, eco-friendly transport options. This approach aligns with Bhutan's GNH philosophy, which prioritizes environmental preservation and the well-being of its citizens (RGB, 2021). Significant reliance on international aid has been necessary to fund these initiatives, given the country's limited financial resources. While the focus on sustainability is commendable, there are concerns about the long-term viability of these efforts, particularly in terms of maintaining and expanding the infrastructure without continued external support, making it imperative for Thimphu to develop more self-sufficient strategies for its transport system (MOIC, 2018).

Thimphu's governance is still developing and is heavily influenced by Bhutan's GNH philosophy, which emphasizes sustainability and environmental protection. MOIC plays a pivotal role in transport governance, with a focus on creating an eco-friendly and cohesive transport system aligned with the country's environmental and cultural values. The city's reliance on international aid highlights the need for capacity building within local institutions to manage and sustain urban transport initiatives effectively (MOIC, 2018).

27.4.3 Pokhara

Pokhara's approach to urban transport governance is characterized by a decentralized model, with significant decision-making power vested in local municipalities. This governance structure is designed to promote local participation and ensure that transport policies reflect the specific needs of the community. Pokhara's transport policies emphasize the integration of nonmotorized transport options, such as cycling and walking, into the city's transport network (PMC, 2017).

However, Pokhara faces several challenges in implementing effective transport governance. The city has limited financial resources and relies heavily on international aid and donor funding to support its transport projects. Additionally, the decentralized governance model, while promoting local autonomy, can sometimes lead to coordination challenges and inefficiencies in project implementation. Studies suggest that for Pokhara to develop a more cohesive and efficient transport system, there is a need for stronger coordination between local governments and greater investment in infrastructure (GoN, 2018).

Pokhara's governance model is characterized by its decentralized structure, where local municipalities hold significant decision-making authority. This model aims to encourage local participation and ensure that transport policies are tailored to community-specific needs. However, the decentralized approach can lead to coordination challenges among different municipalities, particularly in project planning and implementation (PMC, 2017).

Pokhara, a city in Nepal, operates under a decentralized governance model, which has presented both opportunities and challenges in the development of its urban transport system. The city's governance structure allows for greater local autonomy, enabling tailored solutions that address specific local needs. However, this decentralized approach also creates coordination challenges, particularly in aligning transport initiatives across different levels of government. Additionally, resource constraints have limited the city's ability to develop a comprehensive transport network. To overcome these challenges, Pokhara has focused on integrating nonmotorized transport options, such as cycling and walking, into its urban mobility plans. These efforts have been supported by international aid, which has provided crucial funding for infrastructure development (PMC, 2017). While these initiatives have improved local mobility and contributed to a more sustainable urban environment, the city continues to face significant challenges in scaling and sustaining these efforts. The reliance on external funding and the need for better coordination among various stakeholders highlight the ongoing difficulties in achieving a fully integrated and sustainable transport system in Pokhara (GoN, 2018).

Pokhara faces even more pronounced challenges, primarily due to its limited financial resources and technical expertise. The city operates under a decentralized governance model, which aims to involve local communities and tailor solutions to local needs. However, this decentralized approach sometimes leads to inefficiencies in project execution and coordination, particularly when it comes to large-scale transport projects. The lack of sufficient funding and expertise exacerbates these issues, making it difficult for Pokhara to implement and sustain its transport initiatives effectively (GoN, 2018).

27.4.4 Kuala Lumpur

Kuala Lumpur's urban transport governance is based on a multiagency approach, with several governmental bodies and agencies involved in the planning and management of the city's transport systems. This governance model is supported by significant investments from both the government and the private sector, with a strong emphasis on sustainability and inclusiveness. Kuala Lumpur's public transport network includes extensive mass rapid transit (MRT) and light rail train (LRT) systems, which are designed to reduce dependency on private vehicles and promote more sustainable modes of transport (PMB, 2020).

One of the key challenges in Kuala Lumpur's governance of urban transport is the integration of various transport modes into a seamless and efficient system. The city's rapid growth and urban sprawl have made it difficult to ensure that all parts of the city are well-served by public transport. Additionally, while the multiagency governance model allows for the specialization and expertise of different bodies, it also requires effective coordination to avoid overlapping responsibilities and inefficiencies. Research indicates that improving interagency coordination and ensuring that governance frameworks are flexible enough to adapt to the city's changing needs are crucial for the continued success of Kuala Lumpur's transport systems (MOT, 2019).

Kuala Lumpur employs a multi-agency governance model involving various governmental bodies and agencies, each with specific roles in transport planning and management. This model benefits from specialization and significant investments but requires robust coordination to avoid inefficiencies and overlapping responsibilities. The city's emphasis on sustainability and inclusiveness is evident in its extensive MRT and LRT networks (PMB, 2020).

In Kuala Lumpur, the primary challenge lies in integrating the various modes of transport into a unified, efficient system. The city's rapid urbanization has led to a complex transport landscape where different modes of transport, including buses, trains, and private vehicles, often operate in silos. To address this issue, the Klang Valley Integrated Transit System was implemented, aiming to improve connectivity and integration between these different transport modes (PMB, 2020). This system has been a significant step forward in enhancing the city's public transport network, making it easier for commuters to switch between different modes of transport. The initiative, led by Prasarana Malaysia Berhad (PMB), has led to improvements in the overall efficiency and reliability of the city's transit system. However, challenges remain, particularly in areas experiencing rapid urban expansion where the demand for transport services is continually outpacing supply. The ongoing efforts to enhance multimodal integration are essential for accommodating the city's growth and ensuring a sustainable transport system for the future (MOT, 2019).

Kuala Lumpur has also made considerable strides in infrastructure development, driven by substantial investments from both public and private sectors. This funding has supported the expansion and modernization of the city's transport infrastructure, particularly in enhancing public transit options. Despite these advancements, Kuala Lumpur faces ongoing challenges in integrating various modes of transport into a cohesive and seamless system. The complexity of coordinating multiple transport modes remains a significant hurdle in achieving a fully unified transport network (PMB, 2020).

27.5 FINDINGS AND DISCUSSION

Each city has unique approaches to managing urban transport, shaped by its specific socioeconomic, geographic, and political contexts. For instance, Bangkok has focused on expanding its mass transit systems, including the BTS Skytrain and MRT subway, to tackle severe traffic congestion (BMA, 2020). Kuala Lumpur, on the other hand, has implemented comprehensive plans that emphasize sustainability and inclusiveness, with significant investments in public transport infrastructure such as the MRT and LRT networks (PMB, 2020). Thimphu's transport policies are in their initial stage but are guided by Bhutan's GNH philosophy, aiming to develop a cohesive and environmentally friendly transport system (MOIC, 2018). Pokhara's transport policies aim to spread decision-making power, decentralizing management and integrating nonmotorized transport options to enhance mobility (PMC, 2017).

Urban transport planning varies significantly across different cities, reflecting their unique challenges and governance structures. In Bangkok, the focus is on expanding mass transit infrastructure to address severe traffic congestion, with the planning process primarily driven by the BMA. This top-down approach allows for efficient decision-making but often limits direct input from local communities (BMA, 2020). Kuala Lumpur, on the other hand, integrates sustainability into its planning, with policies designed to reduce car dependency and promote public transport. The city's planning process is more inclusive, involving multiple stakeholders, including the private sector, which contributes to a more comprehensive and holistic strategy (MOT, 2019). Thimphu's planning is guided by principles of environmental sustainability and cultural preservation, aiming to minimize the negative impacts of urbanization. However, the process is still evolving, and the city's limited capacity poses challenges to effective implementation (RGB, 2021). In Pokhara, the emphasis is on incorporating local input into the planning process, reflecting its decentralized governance model. While this approach ensures that local needs are considered, the effectiveness of these plans is often hindered by a lack of resources and technical expertise (PMC, 2017).

Improving urban mobility in the four cities—Pokhara, Bangkok, Kuala Lumpur, and Thimphu—presents several challenges, each unique to the city's context. Bangkok faces severe traffic congestion and pollution due to rapid urbanization and overreliance on private vehicles (World Bank, 2018). Kuala Lumpur's challenges include integrating various modes of transport and ensuring accessibility for all segments of society, despite having a well-developed public transport system (ADB, 2019). Thimphu, being a smaller and less developed city, struggles with limited infrastructure and resources to develop a comprehensive public transport system (RGB, 2021). Pokhara similarly faces challenges related to limited funding and the need to expand its transport infrastructure to meet growing demands (PMC, 2017). Identifying these challenges will provide a deeper understanding of the urban mobility issues these cities face and the context-specific factors that influence them.

The governance strategies and funding mechanisms employed by Pokhara, Bangkok, Kuala Lumpur, and Thimphu significantly impact the effectiveness of their urban transport systems. Bangkok's centralized governance model under the BMA has allowed for more coordinated and large-scale transport projects, supported by substantial government funding and public-private partnerships (BMA, 2020). Kuala Lumpur's transport governance involves multiple agencies under a coordinated framework, with significant investments from both government and private sectors, emphasizing sustainability and inclusiveness (PMB, 2020). Thimphu's transport governance is still developing, with a focus on sustainability guided by national policies and reliant on international aid and donor funding (MOIC, 2018). Pokhara's decentralized governance involves local municipalities with limited financial resources, relying on international aid and donor funding to implement transport projects (PMC, 2017). Analyzing these governance strategies and funding mechanisms will highlight the strengths and weaknesses of each approach and provide insights into best practices for effective urban transport governance.

27.5.1 Comparative Insights and Governance Challenges

When comparing the governance of urban public transport across Bangkok, Thimphu, Pokhara, and Kuala Lumpur, several key insights emerge. Each city's governance model reflects its unique socioeconomic and political context, which in turn shapes the effectiveness of its transport policies. Bangkok's centralized model allows for large-scale infrastructure projects but faces challenges in achieving integration and sustainability. Thimphu's governance, guided by the GNH philosophy, prioritizes environmental sustainability but must overcome limitations in capacity and resources. Pokhara's decentralized approach promotes local participation but requires stronger coordination and investment. Kuala Lumpur's multiagency model benefits from substantial investments and a focus on sustainability but needs better integration and coordination among different transport modes.

Across all four cities, common challenges include the need for more integrated planning, effective stakeholder engagement, and sustainable funding mechanisms. As studies have shown, cities with strong governance frameworks that emphasize these elements tend to have more efficient and sustainable transport systems. Therefore, enhancing the governance structures in these cities, through the adoption of frameworks like SUMP and approaches like PVM, could lead to significant improvements in their urban transport systems.

27.5.2 Proposed Solutions for Common Challenges

To improve urban transport systems, cities should focus on strengthening integration mechanisms, enhancing funding strategies, and promoting inclusiveness through policy. A key approach to achieving better integration is the adoption of robust frameworks like SUMP, which ensures that different transport modes are well-coordinated and accessible, thereby facilitating smoother transitions between them (Rupprecht Consult, 2019). Additionally, diversifying funding sources and promoting public-private partnerships are essential for providing sustainable financial support for urban transport projects. For instance, cities like Thimphu and Pokhara could greatly benefit from capacity-building initiatives aimed at enhancing local institutions' abilities to manage and allocate resources more effectively (UITP, 2019). Equally important is the promotion of inclusiveness through targeted policies that address the needs of marginalized groups. Governments should implement measures such as subsidized fares, improved accessibility features, and expanded service coverage to underserved areas. Cities like Bangkok and Kuala Lumpur can further enhance inclusiveness by incorporating more community feedback into their planning processes, ensuring that the voices of all residents are heard and considered in transport development (World Bank, 2018). The capacity of local authorities also needs to be strengthened for the development of policies and strategies and their implementation (Woodcock et al., 2023).

27.6 CONCLUSIONS

The comparative study of urban transport governance in Asian cities reveals that each city's approach is deeply influenced by its unique socioeconomic and political context. The effectiveness of urban transport governance is closely linked to the governance model and the level of stakeholder engagement. While centralized models allow for rapid infrastructure development, they often struggle with integration and inclusiveness. Decentralized and multiagency models, on the other hand, promote local participation and specialization but require strong coordination and sustainable funding mechanisms to be effective.

Centralized models like Bangkok's allow for large-scale infrastructure development but may struggle with inclusiveness and integration. In contrast, decentralized models like Pokhara's promote local participation but require stronger coordination and resource allocation to be effective. Multiagency models, such as Kuala Lumpur's, benefit from specialization but need robust coordination mechanisms to avoid inefficiencies.

All four cities face challenges in integrating different modes of transport into a seamless system. This is particularly evident in Bangkok and Kuala Lumpur, where the rapid expansion of infrastructure has not been matched by equally effective integration strategies (OTP, 2017; PMB, 2020). Bangkok, despite various efforts, has not been able to introduce a combined and integrated fare for different public transport systems. Commuters need to pay multiple fares for using the Skytrain, buses, and subway systems. Bangkok has invested heavily in the development of mass transit systems, but that has not led to an increase in ridership.

Ensuring that transport systems are inclusive and accessible to all segments of society is a major challenge. Bangkok and Kuala Lumpur have made strides in this area but still struggle to fully meet the needs of marginalized groups. Thimphu and Pokhara face more fundamental challenges due to their limited resources and infrastructure (MOT, 2019; BMA, 2020).

Sustainable funding is a common challenge, especially for cities like Thimphu and Pokhara, which rely heavily on international aid. Even in Bangkok and Kuala Lumpur, where funding is more robust, ensuring that resources are allocated efficiently to meet long-term sustainability goals is a challenge (PMC, 2017; MOIC, 2018).

Addressing common challenges such as integration, funding, and inclusiveness requires a comprehensive approach that leverages best practices from each governance model. Cities must adopt integrated planning frameworks like SUMP, diversify their funding strategies, and implement policies that ensure transport systems are accessible and equitable for all. By doing so, these cities can develop urban transport systems that are not only efficient and sustainable but also resilient and inclusive, meeting the needs of current and future generations.

To address these challenges, future research should focus on exploring sustainable technologies such as electric vehicles and renewable energy sources. These innovations can significantly enhance the sustainability and efficiency of public transport systems, reducing reliance on fossil fuels and lowering emissions. Long-term studies on the impact of policy interventions on affordability and inclusiveness will also be crucial. These studies can provide valuable insights into what works and what doesn't, helping policymakers design more effective strategies. Furthermore, cross-cultural studies can help elucidate how different cultural contexts affect the implementation and success of transport policies. By comparing experiences across different regions, researchers and policymakers can identify best practices and develop more culturally sensitive and effective transport strategies.

ACKNOWLEDGEMENT

This chapter uses secondary data and policy documents available on urban public transport and governance. This is part of ongoing research at ESCAP to improve overall sustainability and inclusiveness of the transport sector and urban governance in Asia. The author thanks Sanya Adhikari for providing research support, data collection, and analysis. The views expressed in this paper are those of the author and do not necessarily reflect the views of the United Nations.

REFERENCES

ADB. (n. d.). https://www.adb.org/what-we-do/topics/transport/overview

ADB. (2012). Sustainable Transport Initiative Operational Plan. ADB.

Asian Development Bank (ADB). (2019). Kuala Lumpur Urban Transport Development. ADB.

Bangkok Metropolitan Administration (BMA). (2020). Urban Transport Development and Management. BMA.

Bappenas. (2019). TransJakarta Bus Rapid Transit. Retrieved from https://www.bappenas.go.id

Beukers, E., Bertolini, L., and Te Brömmelstroet, M. (2012). Why Cost Benefit Analysis Is Perceived as a Problematic Tool for Assessment of Transport Plans: A Process Perspective, Transportation Research Part A 46, 68–78.

Bhutan Ministry of Information and Communications (MOIC). (2018). Transport Policy Guidelines under Gross National Happiness. MOIC.

Boisjoly, G.; & El-Geneidy, A. M. (2017). How to Get There? A Critical Assessment of Accessibility Objectives and Indicators in Metropolitan Transportation Plan, Transport Policy 55, 38–50

Buehler, R.; & Pucher, J. (2021). Cycling to a More Sustainable Transport Future, In Cycling for Sustainable Cities, edited by R. Buehler; & J. Pucher, MIT Press.

Currie, G. (2016). Multimodal Transport Systems in Melbourne. Journal of Transport and Land Use, 9(1), 75–91.

ESCAP (2024a). Integration of Public Transport Systems: A Guidebook for Policymakers, ESCAP, Bangkok.

ESCAP (2024b). Integration of Public Transport Systems: A Compendium of Good Practices from Asia and the Pacific, ESCAP, Bangkok.

ESCAP SUTI Assessment reports, available from: https://www.unescap.org/our-work/transport/suti

Foltýnová, H. B., Vejchodská, E., Rybová, E., and Květoň, V. (2020). Sustainable Urban Mobility: One Definition, Different stakeholders' Opinions, Transportation Research Part D, 87, 102465

Government of Nepal (GoN). (2018). Urban Transport Planning and Development in Pokhara. Kathmandu: Government of Nepal.

International Transport Forum (ITF). (2023). ITF Transport Outlook, ITF-OECD

Kawakami, A. (2017). Accessible Public Transportation in Tokyo. Japanese Journal of Public Health, 64(1), 45–58.

Land Transport Authority (LTA). (2020). Accessibility in Singapore's Public Transport. Retrieved from https://www.lta.gov.sg

Lee, J.; & Hwang, K. (2014). Public-Private Partnerships in Seoul's Transportation System. Asian Journal of Public Administration, 36(3), 347–364.

LTA, Singapore (n.d.). Getting Around. Land Transport Authority. Available at https://www.lta.gov.sg/content/ltagov/en/getting_around.html

Ministry of Information and Communications (MOIC), Bhutan. (2018). Bhutan Transport 2040: Integrated Strategic Vision. Thimphu: Royal Government of Bhutan.

Ministry of Transport (MOT), Malaysia. (2019). Integrated Public Transport Planning and Implementation in Kuala Lumpur. MOT Malaysia.

Moore, M. H. (1995). Creating Public Value: Strategic Management in Government. Harvard University Press.

Office of Transport and Traffic Policy and Planning (OTP). (2017). Bangkok's Urban Mobility Plan: Challenges and Solutions. OTP.

Pokhara Metropolitan City (PMC). (2017). Integrated Urban Transport Policy and Planning. Pokhara: PMC.

Prasarana Malaysia Berhad (PMB). (2020). Annual Report 2020: Sustainability in Urban Transit. Kuala Lumpur: Prasarana Malaysia Berhad.

Regmi, M. B. (2024). Governance of Urban Mobility Policies and Plans in Asian Cities, Transport Policy, 149, 108–121

Royal Government of Bhutan (RGB). (2021). Urban Transport Strategy and Capacity Development. Thimphu: RGoB.

Rupprecht Consult. (2019). Guidelines for Developing and Implementing a Sustainable Urban Mobility Plan (SUMP). Rupprecht Consult GmbH.

Seoul Metropolitan Government (SMG) (2020). Seoul Public Transportation. Retrieved from https://www.seoulsolution.kr/sites/default/files/gettoknowus/Seoul%20Public%20Transportation_0.pdf

Stoker, G. (2006). Public Value Management: A New Narrative for Networked Governance? American Review of Public Administration, 36(1), 41–57.

Sung, H., Lee, S., and Choi, K. (2015). Community Engagement in the Bangkok Mass Transit System. Urban Studies, 52(9), 1685–1703.

Tønnesen, A., Hanssen, G. S., Hansen, K. B., and Valencia, S. C. (2020). Integrative Climate Leadership in Multi-Level Policy Packages for Urban Mobility - A Study of Governance Systems in Two Nordic Urban Regions, Transport Policy 128, 309–317

UITP. (2019). Global Report on Urban Transport Governance. International Association of Public Transport.

UNDP. (2009). A Users' Guide to Measuring Local Governance, UNDP Oslo Governance Centre.

UNHABITAT (n.d.). https://unhabitat.org/asia-and-the-pacific-region

Woodcock, A., Saunders, J., Fadden-Hopper, K., and O'Connell, E. (eds.) (2023). Capacity Building in Local Authorities for Sustainable Transport Planning, Springer.

World Bank. (2017). The Evolution of Urban Transport Governance in Asian Cities. World Bank.

World Bank. (2018). Bangkok's Traffic and Pollution Management: Policy Recommendations. World Bank.

World Bank & ADB. (2012). Cities at Crossroads: Unlocking the potential for green urban transport

Yau, W. (2019). Smart Mobility Solutions for the Visually Impaired in Hong Kong. Journal of Urban Technology, 26(2), 95–113.

28 The Future of Urban Human Mobility Studies

Xiao Huang

28.1 INTRODUCTION

Urban human mobility studies have emerged as a critical field for understanding the complex dynamics of modern cities. As urban populations continue to grow, the movement patterns of individuals within these environments become increasingly significant for policymakers, researchers, and urban planners. The advent of advanced tracking technologies and data analytics has enabled the collection and analysis of vast amounts of mobility data, providing granular insights into how people interact with their surroundings. This comprehensive understanding is essential for addressing a multitude of societal challenges that urban areas face today.

This chapter explores the important role of human mobility science in tackling pressing urban issues. It delves into how mobility data can be leveraged to address socioeconomic inequalities by identifying disparities in access to essential services and infrastructure. The discussion extends to enhancing public health research through mobility insights, particularly in tracking the spread of infectious diseases and planning effective interventions. The role of mobility science in improving disaster response and emergency management is also examined, highlighting its value in evacuation planning and resource allocation. Furthermore, the chapter discusses how mobility science promotes environmental sustainability by informing strategies to reduce greenhouse gas emissions and improve urban livability. It also explores the advancement of urban planning and the development of smart cities through detailed mobility data, enabling optimized infrastructure development and resource allocation. Finally, the chapter addresses the challenges and biases associated with mobility data, including data privacy and ethical considerations, sampling bias, data quality issues, and technical limitations. By understanding and mitigating these challenges, the effective utilization of mobility science can significantly enhance the quality of urban life and contribute to the sustainable development of cities.

28.2 LEVERAGING HUMAN MOBILITY SCIENCE FOR SOCIETAL CHALLENGES

Human mobility science has emerged as a transformative field that offers profound insights into the movement patterns of populations within urban environments. By systematically analyzing these patterns, researchers and policymakers can address a multitude of societal challenges that cities face today. The advent of advanced tracking technologies and data analytics has made it possible to collect and interpret vast amounts of mobility data, providing a granular understanding of how people interact with their surroundings (Huang et al., 2024). This knowledge is instrumental in tackling issues such as socioeconomic inequalities, public health crises, disaster response, environmental sustainability, and the development of smart cities. The following subsections delve into specific applications of mobility science, illustrating its pivotal role in shaping the future of urban living and addressing some of the most pressing challenges of our time.

28.2.1 Addressing Socioeconomic Inequalities through Mobility Science

Mobility science plays a critical role in identifying and mitigating socioeconomic inequalities within urban areas. These movement patterns allow researchers to uncover disparities in access to a wide range of essential services, e.g., education, healthcare, employment, and recreational facilities

DOI: 10.1201/9781003503262-32

(Chen et al., 2018; Lu et al., 2024; Ren et al., 2024). Furthermore, mobility patterns highlight areas where infrastructure investments are lacking. If certain neighborhoods exhibit low mobility flows, it may indicate insufficient transportation options or unsafe conditions that deter movement. By pinpointing these areas, policymakers can prioritize infrastructure development, such as adding bus routes and bike lanes or improving pedestrian pathways, to enhance connectivity.

Mobility science also informs targeted social programs. By understanding the travel behaviors of specific demographic groups, interventions can be designed to meet their unique needs (Huang et al., 2022b). Programs like subsidized transit fares for low-income individuals or the introduction of community shuttles can significantly improve accessibility. Moreover, integrating mobility data with socioeconomic indicators enables a more comprehensive analysis of inequality (Chukwu et al., 2024). Researchers can correlate mobility constraints with factors such as income levels, educational attainment, or health outcomes. This holistic approach facilitates the development of multifaceted strategies that address the root causes of inequality rather than just the symptoms.

28.2.2 Enhancing Public Health Research through Mobility Insights

Mobility data offer significant benefits for public health research by providing detailed information on how populations move and interact within urban environments. Understanding these patterns is essential for tracking the spread of infectious diseases, planning effective interventions, and effectively allocating healthcare resources.

During epidemics, mobility insights are invaluable for modeling disease transmission. For example, analyzing mobility flows between different regions allows health authorities to identify potential hot spots and predict where outbreaks are likely to occur next (Kraemer et al., 2020; Xiong et al., 2020), enabling the implementation of targeted measures such as travel advisories, quarantine zones, or focused vaccination campaigns to contain the spread of disease effectively. Beyond infectious diseases, mobility data can shed light on environmental health concerns. For instance, by mapping population movements in areas with high air pollution, researchers can assess exposure levels and their potential impact on respiratory health, informing policies aimed at reducing pollution or protecting vulnerable populations (Park and Kwan, 2017). Mobility insights also contribute to understanding lifestyle-related health issues. Examining access to recreational spaces, grocery stores, or healthcare facilities, public health officials allow researchers to identify "food deserts" or areas lacking in essential services. Interventions can thus be designed to improve access, such as promoting the development of community gardens or incentivizing the establishment of health clinics in underserved areas.

28.2.3 Improving Disaster Response and Emergency Management

In disaster response and emergency management, mobility science serves as a powerful tool for enhancing preparedness, operational efficiency, and resilience. Analyzing both real-time and historical mobility data allows scholars to gain critical insights into evacuation behaviors, identify potential bottlenecks, and optimize emergency routes (Song et al., 2014). These capabilities are invaluable in ensuring a timely and coordinated response to various disasters, empowering emergency planners to develop strategies that minimize traffic jams and ensure safe, efficient evacuations.

In addition to evacuation planning, mobility insights are instrumental in resource allocation. By understanding population concentrations at different times of the day or week, emergency services can strategically position personnel and equipment. In urban areas prone to earthquakes or floods, this proactive deployment of resources can significantly reduce response times and improve incident outcomes. Real-time mobility data also enhance situational awareness during ongoing disasters, allowing authorities to monitor evacuation progress, identify areas where people may be trapped, and adjust their response efforts dynamically. This adaptive approach improves coordination among agencies and ensures that assistance reaches affected populations swiftly.

After a disaster, mobility data are critical for assessing the impact on infrastructure and services. Comparing mobility patterns before and after a disaster allows researchers to pinpoint areas where infrastructure has been damaged or where normal movement has not resumed (Liu et al., 2023). These insights highlight ongoing issues that need urgent attention to restore normalcy. Additionally, mobility science plays a crucial role in guiding recovery efforts. Understanding how mobility patterns have shifted after a disaster informs the restoration of transportation networks and public services, ultimately enhancing community resilience against future disasters.

28.2.4 Promoting Environmental Sustainability through Mobility Science

Mobility science plays an important role in advancing environmental sustainability through the provision of data-driven strategies aimed at reducing greenhouse gas emissions, improving air quality, and enhancing urban livability. The analysis of transportation patterns enables cities to identify areas with high traffic congestion and vehicle emissions, which are significant contributors to environmental degradation.

Insights derived from mobility data enable urban planners to develop targeted interventions, including congestion pricing, traffic signal optimization, and the establishment of car-free zones, to alleviate traffic congestion and reduce emissions. Mobility science informs the expansion and optimization of public transportation networks through the alignment of transit routes and schedules with actual demand patterns, thereby enhancing the convenience and attractiveness of public transit as an alternative to private vehicle use. Additionally, it supports the promotion of active transportation modes, such as walking and cycling, through the identification of areas with high pedestrian or cyclist activity, allowing cities to prioritize infrastructure development including bike lanes, pedestrian pathways, and secure bike parking facilities (Bibri, 2020). Furthermore, mobility data guide land use planning; the data highlight regions where mixed-use developments could reduce the necessity for long commutes, thus facilitating the creation of more compact and sustainable urban forms through the integration of residential, commercial, and recreational spaces.

Moreover, mobility science is instrumental in monitoring the environmental impact of transportation policies. Tracking changes in mobility patterns over time enables cities to assess the effectiveness of interventions and make data-driven adjustments to their sustainability strategies. In essence, the utilization of mobility science allows cities to design and implement policies that promote environmental sustainability, mitigate the impacts of climate change, and enhance the overall quality of life for their residents.

28.2.5 Advancing Urban Planning and Smart Cities

Detailed mobility data provides urban planners with a comprehensive understanding of human interactions with the built environment, thereby informing decisions on infrastructure development, zoning regulations, and resource allocation. Incorporating mobility patterns into planning processes enables cities to optimize the placement of amenities such as schools, hospitals, and commercial centers to ensure maximum accessibility for the populations that require them most. For instance, if mobility data indicate that a significant number of residents travel long distances to access healthcare services, planners might consider establishing new clinics in underserved areas (Sharma and Patil, 2021).

Mobility insights are instrumental in supporting the integration of technology into urban systems. In smart cities, data collected from sensors, connected devices, and Internet of Things (IoT) platforms are utilized to enhance the efficiency of services. Mobility data can be integrated into intelligent transportation systems to manage traffic in real time, adjust public transit schedules according to demand, and provide dynamic routing information to commuters. Public safety initiatives can also benefit from mobility data because they enable the identification of areas with high pedestrian traffic that may require increased surveillance or improved lighting.

Within the context of urban development, mobility data facilitate the creation of mixed-use neighborhoods that reduce the necessity for long commutes and promote community engagement. A comprehensive understanding of individual movement patterns within and between different areas allows planners to foster connectivity and cohesion within the urban fabric. The utilization of mobility data enables cities to enhance operational efficiency, promote sustainable development, and improve the overall quality of urban life. This data-driven approach is essential for constructing the smart cities of the future, where technology and innovation are leveraged to meet the evolving needs of urban populations.

28.3 UNDERSTANDING AND MITIGATING CHALLENGES AND BIASES IN MOBILITY DATA

28.3.1 Data Privacy and Ethical Considerations

The collection and analysis of mobility data raise significant privacy concerns, particularly because personal location information is highly sensitive. Various sources contribute to the wealth of mobility data available today, including global positioning system (GPS) data from smartphones, ride-hailing services, public transit records, social media platforms, smart card transactions, and cellular network data. For instance, smartphones equipped with GPS capabilities track users' real-time locations, and apps often request access to these data for services like navigation or local recommendations. Ride-hailing services such as Uber and Lyft collect detailed trip information, including pick-up and drop-off locations, routes taken, and timestamps (Acheampong et al., 2020). Similarly, smart card systems used in public transit—like London's Oyster card or Hong Kong's Octopus card—record entry and exit stations, times, and sometimes personal identifiers linked to payment methods. Social media platforms may collect location data through check-ins or geotagged posts, while mobile network operators gather information based on cell tower connections, approximating a user's location and movement patterns (Li et al., 2021a; Huang et al., 2022a).

The misuse or unauthorized access to such data can lead to privacy breaches and an erosion of trust. Studies have shown that anonymized mobility data can often be re-identified when combined with other datasets, as demonstrated in cases where supposedly anonymous data were linked back to specific individuals (e.g., Rocher et al., 2019). Ethical considerations must therefore guide the handling of mobility data, ensuring compliance with data protection regulations. In the European Union, the General Data Protection Regulation (GDPR) imposes strict rules on collecting and processing personal data, including location information (Hoofnagle et al., 2019). Organizations must obtain explicit consent from users, provide clear information on data usage, and allow users to access, correct, or delete their data. In the United States, the California Consumer Privacy Act (CCPA) grants similar rights, reflecting a growing trend toward stricter data privacy laws (Goldman, 2020).

Implementing anonymization techniques and obtaining informed consent are essential steps in safeguarding individual privacy while enabling valuable research. Anonymization methods such as data aggregation reduce the risk of identifying individuals, while differential privacy adds statistical noise to datasets to balance data utility with privacy protection. Obtaining informed consent involves transparently communicating to users what data are being collected, how the data will be used, and who the data will be shared with, giving users the option to agree or decline. Furthermore, adopting privacy-by-design principles—embedding privacy considerations into every stage of product development and data processing activities—aligns with regulatory expectations and helps build user trust. In summary, while mobility data holds immense potential for advancing urban studies and addressing societal challenges, it is imperative to handle such data ethically and responsibly, adhering to current policies and ensuring the protection of individual privacy.

28.3.2 Sampling Bias and Representativeness

Sampling bias and representativeness are critical issues in mobility data analysis, as the sources of such data often do not capture the full spectrum of the population. Mobility data are frequently derived from smartphone applications, GPS devices, social media platforms, or transportation services, which may predominantly reflect the behaviors of specific demographic groups. For instance, individuals who own smartphones and actively use apps are more likely to be younger, more affluent, and technologically savvy. Conversely, elderly populations, low-income individuals, or those living in rural areas may be underrepresented due to limited access to technology or digital services. It is important to note, however, that the severity of this issue depends on the region of study. For example, statistics from 2020 indicate significant variations between countries: in high-income countries, there are 122 mobile phone subscriptions per 100 people, while in low-income countries, the number is only 59 (Pappalardo et al., 2023).

One major consequence is the reinforcement of existing socioeconomic inequalities. For instance, if mobility data primarily capture the movements of urban, tech-savvy individuals who use smartphones and mobility apps, it may overlook the mobility patterns of low-income groups, the elderly, or those in rural areas who may not have access to such technology (Hu et al., 2021; Huang et al., 2021). Policies developed based on this skewed data might prioritize infrastructure and services that cater to the needs of the overrepresented groups, neglecting marginalized communities. This can exacerbate disparities in access to transportation, employment opportunities, and essential services. In public health research, failure to address sampling bias can impede effective disease control and prevention efforts (Li et al., 2021b). During an epidemic, if mobility data do not accurately represent the movements of all demographic groups, especially those at higher risk, health authorities may not identify critical transmission pathways. This can result in inadequate allocation of resources, misdirected interventions, and ultimately, a failure to contain the spread of disease. For example, during the COVID-19 pandemic, underrepresentation of certain communities in mobility data could have led to overlooked hot spots and insufficient support for vulnerable populations. In disaster response and emergency management, biased mobility data can compromise the effectiveness of evacuation plans and resource distribution. If evacuation models are based on data that exclude segments of the population, emergency services may not anticipate the needs of all affected individuals, leading to insufficient shelter provisions, delayed assistance, and increased risk of harm. This oversight can have dire consequences during natural disasters when timely and inclusive response strategies are critical for saving lives.

To address these challenges, it is essential to implement strategies that enhance the representativeness of mobility data. Combining multiple data sources can help mitigate sampling bias by capturing a broader range of mobility patterns across different demographic groups. Integrating data from public transportation records, surveys, and census information with traditional mobility datasets can provide a more comprehensive view (Huang et al., 2020). Additionally, applying statistical techniques such as weighting adjustments can correct for known biases by giving appropriate emphasis to underrepresented groups in the analysis.

28.3.3 Data Quality and Reliability Issues

Data quality and reliability are fundamental concerns in urban human mobility studies because they directly impact the validity of research findings and the effectiveness of policy recommendations. Mobility data are often collected from various sources, with each of these sources presenting unique challenges that can compromise data quality, including inaccuracies, inconsistencies, missing data, and temporal or spatial errors.

One major issue is the accuracy of location data. GPS signals, for instance, can be affected by environmental factors such as urban canyons, where tall buildings obstruct satellite signals, leading to inaccurate positioning. This can result in erroneous mapping of individuals' movements, affecting analyses of travel patterns and congestion hot spots. Mobile phone data, while providing broader coverage, often has a lower spatial resolution, as the location is approximated based on cell tower triangulation rather than precise coordinates. This limitation can hinder fine-grained studies that require exact location information. Temporal resolution is another aspect of data quality that poses challenges. Data collected at irregular or infrequent intervals may miss critical movements or fail to capture the dynamics of urban mobility accurately. Consistency in data collection methods and timestamping is crucial for reliable temporal analyses.

Data completeness and coverage significantly impact the reliability of analyses. Data gaps can arise from various factors, such as technical malfunctions, user behavior (e.g., disabling location services), or network connectivity issues. These gaps often introduce biases, particularly when certain time periods, locations, or demographic groups are systematically underrepresented. For example, data reliability may decrease during peak hours due to network congestion or in areas with poor connectivity—typically lower-income neighborhoods—exacerbating inequalities in urban planning decisions. Another critical challenge is the heterogeneity of data sources. Integrating data from multiple platforms, each with distinct formats, resolutions, and quality standards, can be complex and error-prone. Discrepancies in data schemas, units of measurement, or definitions of key variables can lead to inaccuracies in data fusion and analysis. Without rigorous preprocessing and standardization, these inconsistencies may undermine the validity and integrity of the study's findings.

To mitigate data quality and reliability issues, it is imperative to implement robust data validation and cleansing procedures. This includes cross-verifying data with multiple sources when possible, employing statistical techniques to identify and correct anomalies, and thoroughly documenting all data processing steps. Furthermore, advances in technologies such as artificial intelligence and advanced machine learning algorithms can aid in detecting patterns indicative of errors or inconsistencies.

28.3.4 Technical Limitations and Analytical Challenges

The analysis of urban human mobility data presents a range of technical limitations and analytical challenges that can impede the extraction of accurate and actionable insights. Understanding these challenges is crucial for researchers and practitioners aiming to leverage mobility data effectively in urban planning and policymaking.

One of the foremost technical challenges is the sheer volume of data generated by mobility tracking technologies. High-frequency data collection from numerous sources—such as GPS devices, mobile phones, and transportation systems—results in massive datasets that require substantial computational resources for storage, processing, and analysis. Handling such big data necessitates advanced infrastructure, including high-performance computing systems and cloud-based platforms, as well as scalable algorithms capable of efficiently managing and analyzing data at scale.

Another significant analytical challenge lies in modeling the dynamic and stochastic nature of human mobility patterns. Human movements are influenced by a multitude of factors, including individual preferences, social interactions, environmental conditions, and unexpected events like natural disasters or pandemics. Capturing these complex patterns demands advanced analytical methods, such as machine learning algorithms, deep learning models, and spatiotemporal modeling techniques. However, these methods can be computationally intensive and often require large amounts of labeled data for training—data that are frequently scarce or incomplete.

Data sparsity and noise further complicate the analysis. Incomplete or noisy data can result from signal loss, device limitations, inconsistent data collection intervals, or user nonparticipation (e.g., disabling location services). These gaps and inaccuracies introduce biases into the dataset, potentially leading to erroneous conclusions. Additionally, the high dimensionality of mobility data—with numerous variables and temporal-spatial dimensions—poses challenges for traditional analytical methods, which may struggle with scalability and computational efficiency when handling such complex datasets. This complexity requires sophisticated data preprocessing and dimensionality reduction techniques to make the data tractable for analysis.

Mobility data often originate from diverse sources, each with its own format, resolution, and data quality standards. Integrating these heterogeneous datasets is a significant technical hurdle. Discrepancies in data schemas, units of measurement, and definitions of key variables can lead to errors in data fusion and analysis. Without rigorous preprocessing and standardization, inconsistencies may undermine the validity and integrity of study findings. The lack of standardized data formats and protocols complicates collaborative efforts and inhibits data interoperability among researchers and organizations.

To address the technical limitations and analytical challenges in mobility data analysis, researchers are adopting strategies like utilizing high-performance computing and cloud-based platforms to efficiently process and analyze large-scale data, thereby reducing processing times and enhancing capabilities (e.g., Li et al., 2021c). Advanced data mining techniques and machine learning algorithms—such as deep learning, clustering methods, and spatiotemporal models—are employed to capture complex patterns inherent in the nonlinear and stochastic nature of human mobility (Luca et al., 2021). Implementing robust data validation, cleaning, imputation, and normalization procedures mitigates the effects of data sparsity and noise, enhancing data quality and reliability. Developing standardized data formats and protocols facilitates the integration of heterogeneous data sources, while collaborative efforts to establish common frameworks improve data interoperability and promote effective data sharing among researchers and organizations. Despite significant obstacles, ongoing advancements in computational technologies and analytical methodologies are progressively overcoming these hurdles, enabling researchers to unlock the full potential of mobility data. These innovations, coupled with collaborative approaches, foster a deeper understanding of urban dynamics and support informed policy decisions that enhance urban living and address societal challenges.

28.4 CONCLUSION

Urban human mobility studies stand at a critical intersection of technology, data science, and societal needs. By harnessing mobility data, researchers and policymakers can address socioeconomic inequalities, enhance public health initiatives, improve disaster response, promote environmental sustainability, and advance urban planning and smart city development. However, the effective use of mobility data is accompanied by significant challenges. Ethical considerations regarding data privacy demand strict adherence to regulations, necessitating anonymization techniques and the obtaining of informed consent to protect individual privacy. Sampling bias and representativeness issues underscore the importance of strategies that ensure data inclusivity, preventing the marginalization of underrepresented groups. Data quality and reliability concerns highlight the need for robust validation and cleansing procedures to maintain the integrity of research findings. Technical limitations and analytical challenges call for advanced computational resources and sophisticated analytical methods to manage and interpret large, complex datasets effectively. As technological advancements continue, integrating innovative solutions with collaborative efforts and standardized protocols will be essential. By proactively addressing these challenges, the future of urban human mobility studies holds immense potential to create more equitable, efficient, and sustainable urban environments, ultimately enhancing the quality of life for all city inhabitants.

REFERENCES

Acheampong, Ransford A., Alhassan Siiba, Dennis K. Okyere, and Justice P. Tuffour. "Mobility-on-demand: An empirical study of internet-based ride-hailing adoption factors, travel characteristics and mode substitution effects." *Transportation Research Part C: Emerging Technologies* 115 (2020): 102638.

Bibri, Simon Elias. "Compact urbanism and the synergic potential of its integration with data-driven smart urbanism: An extensive interdisciplinary literature review." *Land Use Policy* 97 (2020): 104703.

Chen, Bi Yu, Yafei Wang, Donggen Wang, Qingquan Li, William HK Lam, and Shih-Lung Shaw. "Understanding the impacts of human mobility on accessibility using massive mobile phone tracking data." *Annals of the American Association of Geographers* 108, no. 4 (2018): 1115–1133.

Chukwu, Michaelmary, Xiao Huang, Siqin Wang, Xiao Li, and Hanxue Wei. "Urban park accessibility assessment using human mobility data: A systematic review." *Annals of GIS* 30, no. 2 (2024): 181–198.

Goldman, Eric. "An introduction to the California consumer privacy act (CCPA)." *Santa Clara Univ. Legal Studies Research Paper* (2020).

Hoofnagle, Chris Jay, Bart Van Der Sloot, and Frederik Zuiderveen Borgesius. "The European Union general data protection regulation: What it is and what it means." *Information & Communications Technology Law* 28, no. 1 (2019): 65–98.

Hu, Tao, Siqin Wang, Bing She, Mengxi Zhang, Xiao Huang, Yunhe Cui, Jacob Khuri et al. "Human mobility data in the COVID-19 pandemic: Characteristics, applications, and challenges." *International Journal of Digital Earth* 14, no. 9 (2021): 1126–1147.

Huang, X., Z. Li, Y. Jiang, X. Li, and D. Porter. "Twitter reveals human mobility dynamics during the COVID-19 pandemic." *PLoS One* 15, no. 11 (2020): e0241957.

Huang, Xiao. "Computational urban science needs to go beyond computational." *Computational Urban Science* 4, no. 1 (2024): 1–5.

Huang, Xiao, Zhenlong Li, Yuqin Jiang, Xinyue Ye, Chengbin Deng, Jiajia Zhang, and Xiaoming Li. "The characteristics of multi-source mobility datasets and how they reveal the luxury nature of social distancing in the US during the COVID-19 pandemic." *International Journal of Digital Earth* 14, no. 4 (2021): 424–442.

Huang, Xiao, Yago Martin, Siqin Wang, Mengxi Zhang, Xi Gong, Yue Ge, and Zhenlong Li. "The promise of excess mobility analysis: Measuring episodic-mobility with geotagged social media data." *Cartography and Geographic Information Science* 49, no. 5 (2022a): 464–478.

Huang, Xiao, Yuhui Zhao, Siqin Wang, Xiao Li, Di Yang, Yu Feng, Yang Xu, Liao Zhu, and Biyu Chen. "Unfolding community homophily in US metropolitans via human mobility." *Cities* 129 (2022b): 103929.

Kraemer, Moritz UG, Chia-Hung Yang, Bernardo Gutierrez, Chieh-Hsi Wu, Brennan Klein, David M. Pigott, Open COVID-19 Data Working Group et al. "The effect of human mobility and control measures on the COVID-19 epidemic in China." *Science* 368, no. 6490 (2020): 493–497.

Li, Xiao, Haowen Xu, Xiao Huang, Chenxiao Guo, Yuhao Kang, and Xinyue Ye. "Emerging geo-data sources to reveal human mobility dynamics during COVID-19 pandemic: Opportunities and challenges." *Computational Urban Science* 1 (2021b): 1–9.

Li, Zhenlong, Xiao Huang, Tao Hu, Huan Ning, Xinyue Ye, Binghu Huang, and Xiaoming Li. "ODT FLOW: Extracting, analyzing, and sharing multi-source multi-scale human mobility." *PLoS One* 16, no. 8 (2021c): e0255259.

Li, Zhenlong, Xiao Huang, Xinyue Ye, Yuqin Jiang, Yago Martin, Huan Ning, Michael E. Hodgson, and Xiaoming Li. "Measuring global multi-scale place connectivity using geotagged social media data." *Scientific Reports* 11, no. 1 (2021a): 14694.

Liu, Xiaoyan, Saini Yang, Xiao Huang, Rui An, Qiangqiang Xiong, and Tao Ye. "Quantifying COVID-19 recovery process from a human mobility perspective: An intra-city study in Wuhan." *Cities* 132 (2023): 104104.

Lu, Junyu, Xiao Xiao, Xiao Huang, Xiaowei Chuai, Zhenlong Li, Hanxue Wei, and Sicheng Wang. "Big data insights into urban park use in the pandemic: Changes in visitation patterns and exacerbated social inequalities in the US." *Cities* 152 (2024): 105204.

Luca, Massimiliano, Gianni Barlacchi, Bruno Lepri, and Luca Pappalardo. "A survey on deep learning for human mobility." *ACM Computing Surveys (CSUR)* 55, no. 1 (2021): 1–44.

Pappalardo, Luca, Ed Manley, Vedran Sekara, and Laura Alessandretti. "Future directions in human mobility science." *Nature Computational Science* 3, no. 7 (2023): 588–600.

Park, Yoo Min, and Mei-Po Kwan. "Individual exposure estimates may be erroneous when spatiotemporal variability of air pollution and human mobility are ignored." *Health & Place* 43 (2017): 85–94.

Ren, Mengyao, Yang Xu, Sangwon Park, Xiao Huang, Mengqi Sun, Jizhe Xia, and Sun-Young Koh. "Consumer spending during COVID-19 in a tourism city." *Annals of Tourism Research* 109 (2024): 103830.

Rocher, Luc, Julien M. Hendrickx, and Yves-Alexandre De Montjoye. "Estimating the success of re-identifications in incomplete datasets using generative models." *Nature Communications* 10, no. 1 (2019): 1–9.

Sharma, Gajanand, and Gopal R. Patil. "Public transit accessibility approach to understand the equity for public healthcare services: A case study of Greater Mumbai." *Journal of Transport Geography* 94 (2021): 103123.

Song, Xuan, Quanshi Zhang, Yoshihide Sekimoto, and Ryosuke Shibasaki. "Prediction of human emergency behavior and their mobility following large-scale disaster." In *Proceedings of the 20th ACM SIGKDD international conference on Knowledge discovery and data mining*, pp. 5–14. 2014.

Xiong, Chenfeng, Songhua Hu, Mofeng Yang, Weiyu Luo, and Lei Zhang. "Mobile device data reveal the dynamics in a positive relationship between human mobility and COVID-19 infections." *Proceedings of the National Academy of Sciences* 117, no. 44 (2020): 27087–27089.

Index

C

D

E

Evaluation 14, 21, 23–24, 57–58, 91–92, 144, 158, 169, 185, 198–199, 229, 256–257, 332

S

T

U

V

W

Y

Z